J. G. Heywood K. Masuda
R. Rautmann V. A. Solonnikov (Eds.)

The Navier-Stokes Equations II – Theory and Numerical Methods

Proceedings of a Conference held in Oberwolfach, Germany, August 18-24, 1991

Springer-Verlag
Berlin Heidelberg New York
London Paris Tokyo
Hong Kong Barcelona
Budapest

Editors

John G. Heywood
Department of Mathematics
University of British Columbia
Vancouver B. C. V6T 1Y4, Canada

Kyûya Masuda
Department of Mathematics
Rikkyo University
3-34-1 Nishi-Ikebukuro, Toshimaku
Tokyo, Japan

Reimund Rautmann
Fachbereich Mathematik-Informatik
Universität-Gesamthochschule Paderborn
Warburger Str. 100
W-4790 Paderborn, Germany

Vsevolod A. Solonnikov
St. Petersburg Branch of V. A. Steklov
Mathematical Institute of the
Russian Academy of Sciences
Fontanka 27, St. Petersburg, Russia

Mathematics Subject Classification (1991): 00B25, 35Q30, 35Q35, 35R35, 35S10, 35S15, 35B40, 35B45, 35C15, 35D10, 60M15, 60M30, 65M06, 65M12, 65M25, 65M60, 65M70, 76D05, 76D07, 76U05

ISBN 3-540-56261-3 Springer-Verlag Berlin Heidelberg New York
ISBN 0-387-56261-3 Springer-Verlag New York Berlin Heidelberg

Typesetting: Camera-ready by author/editor
46/3140-543210 - Printed on acid-free paper

This volume is dedicated

to the memory of

Professor Charles J. Amick
(1951, † 1991).*

Preface

The Navier-Stokes equations have long been regarded with scientific fascination for the wide variety of physical phenomena that come within their governance. But the rigorous mathematical treatment of these phenomena remained out of reach until quite recently, for the lack of a sufficiently developed basic theory of the equations. During the first half of this century, new roads were opened to a basic theory by the pioneering works of Oseen, Odqvist, Leray, and Hopf. This research accelerated during the fifties and sixties, and finally, during the last twenty five years, the basic theory of the equations has developed and matured to a point that makes possible its application to the rigorous understanding of many widely diverse phenomena. Researchers are now undertaking the study of flows with free surfaces, flows past obstacles, jets through apertures, heat convection, bifurcation, attractors, turbulence, etc., on the basis of an exact mathematical analysis. At the same time, the advent of high speed computers has made computational fluid dynamics a subject of the greatest practical importance. Hence, the development of computational methods has become another focus of the highest priority for the application of the mathematical theory. It is not surprising, then, that there has been an explosion of activity in recent years, in the diversity of topics being studied, in the number of researchers who are involved, and in the number of countries where they are located.

Inevitably, it has become difficult for researchers in one area to keep up with even basic developments arising in another. The Navier-Stokes theory is beginning to suffer the same branching into separate and isolated streams that has befallen twentieth century mathematics as a whole. The organizers of the Oberwolfach meetings on the Navier-Stokes equations of 1988 and 1991 have endeavored to bring together leading researchers from all parts of the world, and from all areas of research that are intimately connected with the basic mathematical theory of the equations. They also included representatives from the engineering community, who presented experimental and numerical works of high theoretical interest.

These proceedings contain most of the new results presented during the conference and, in addition, some contributions quite recently given from participants to the conference's field.

Vancouver, Tokyo, Paderborn, and St. Petersburg, June 1992.

J. G. Heywood, K. Masuda, R. Rautmann, V. A. Solonnikov.

CONTENTS

Preface

Free boundary problems

Antanovskii, L.K.:
Analyticity of a free boundary in plane quasi-steady flow of a liquid form subject to variable surface tension.......... 1

Socolowsky, J.:
On a free boundary problem for the stationary Navier-Stokes equations with a dynamic contact line.......... 17

Solonnikov, V.A.:, Tani, A.:
Evolution free boundary problem for equations of motion of viscous compressible barotropic liquid.......... 30

Wolff, M.: Heat-conducting fluids with free surface in the case of slip-condition on the walls....... 56

Problems in unbounded domains

Borchers, W., Miyakawa, T.:
On some coercive estimates for the Stokes problem in unbounded domains.......... 71

Chang, H.: The steady Navier-Stokes problem for low Reynolds number viscous jets into a half space. 85

Farwig, R., Sohr, H.:
An approach to resolvent estimates for the Stokes equations in L^q-spaces. 97

Galdi, G.P.: On the Oseen boundary-value problem in exterior domains.......... 111

Salvi, R.: The exterior problem for the stationary Navier-Stokes equations: on the existence and regularity.......... 132

Schonbek, M.E. :
Some results on the asymptotic behaviour of solutions to the Navier-Stokes equations.146

Wiegner, M.: Approximation of weak solutions of the Navier-Stokes equations in unbounded domains.
161

Numerical methods

Rannacher, R.:
On Chorin's projection method for the incompressible Navier-Stokes equations........ 167

Süli, E., Ware, A .:
Analysis of the spectral Lagrange-Galerkin method for the Navier-Stokes equations...184

Varnhorn, W.:
A fractional step method for regularized Navier-Stokes equations..............................196

Wetton, B.T.R.:
Finite difference vorticity methods.. 210

Statistical methods

Fursikov, A.V.:
The closure problem for the chain of the Friedman-Keller moment equations in the case of large Reynolds numbers.. 226

Inoue, A.: A tiny step towards a theory of functional derivative equations - A strong solution of the space-time Hopf equation..246

General qualitative theory

Grubb, G.: Initial value problems for the Navier-Stokes equations with Neumann conditions..........262

Mogilevskii, I.:
Estimates in $C^{21,1}$ for solution of a boundary value problem for the nonstationary Stokes system with a surface tension in boundary condition..284

Schmitt, B.J., Wahl v.,W.:
Decomposition of solenoidal fields into poloidal fields, toroidal fields and the mean flow. Applications to the Boussinesq-equations..291

Walsh, O.: Eddy solutions of the Navier-Stokes equations..306

Xie, W.: On a three-norm inequality for the Stokes operator in nonsmooth domains.................. 310

List of participants **317**

ANALYTICITY OF A FREE BOUNDARY IN PLANE QUASI-STEADY FLOW OF A LIQUID FORM SUBJECT TO VARIABLE SURFACE TENSION

Leonid K. ANTANOVSKII

Lavrentyev Institute of Hydrodynamics
630090 Novosibirsk, Russia†

Abstract - The plane quasi-steady flow of an incompressible viscous fluid completely bounded by a free surface and driven solely by variable surface tension, is analyzed. The mathematical problem is decomposed into an auxiliary elliptic problem for Stokes equations in a fixed region with an imposed dynamic condition, whose solution leads to a Cauchy problem for the free boundary governed by a kinematic condition. Using the bianalytic stress-stream function and time-dependent conformal mapping of the unit disk onto the flow domain sought, the auxiliary problem is reduced to a Fredholm boundary integral equation for the normal velocity of the free boundary. The existence theorem is obtained in a class of analytic curves prescribing the free-boundary position.

Introduction. The quasi-steady approximation can be applied for description of viscous flows in domains of small size in one, or several directions, such as films, capillaries, and droplets, where the surface-tension and viscosity forces are dominant over the inertia and gravity ones. This approach leads to a non-standard problem for the Stokes system, which can be formally reduced to a Cauchy problem for the free boundary, involving the so-called "normal velocity" operator [2-4]. To define this operator, it is necessary to solve an auxiliary problem for Stokes equations, in which the free boundary is prescribed as a known smooth curve, and the kinematic condition is temporarily ignored.

Introducing the time-dependent conformal mapping of the unit

---

†Present address: *Microgravity Advanced Research and Support center, Via Diocleziano 328, 80125 Naples, Italy*

disk of a parametric plane onto the flow domain, a boundary integral equation for solution of the auxiliary problem is constructed. This procedure is based on the preliminary consideration of two Schwartz problems for analytic functions in the unit disk in combination with an appropriate normalization of the Stokes solution to avoid appearance of a rigid-body velocity [4]. As a result, a Fredholm equation is constructed directly for the normal velocity of the free boundary which defines its evolution by virtue of the kinematic condition.

The intention is to obtain the Cauchy problem in the special case of the flow of a liquid form completely bounded by the free surface and driven solely by capillarity. The existence theorem is formulated in the class of analytic curves prescribing the free boundary position. In particular, analyticity of the free boundary takes place regardless to variable (non-smooth) surface tension which is assumed to be a given function of the parametric plane.

For steady-state flows governed by the Navier-Stokes equations with constant surface tension, there was proved infinite smoothness of the free boundary in [18] (2D case), and analyticity in [1] (2D case) and [6] (3D case). The solvability and large-time regularity theorems for an unsteady surface-tension-driven flow were obtained in [21, 22]. The quasi-steady evolution of the shape of a liquid form was analyzed theoretically and numerically in [12, 14].

1. Formulation. Let a fluid of constant viscosity μ and variable surface tension σ occupy a domain Ω bounded by the free surface $\Gamma = \partial\Omega$. The quasi-steady problem implies finding the domain $\Omega(t)$, velocity $\mathbf{v}(\mathbf{z},t)$, and pressure $p(\mathbf{z},t)$, depending on time t and a point $\mathbf{z} \in \Omega(t)$, as the solution of the following system of equations,

$$\nabla p = \mu\Delta\mathbf{v}, \quad \nabla\cdot\mathbf{v} = 0 \quad \text{in } \Omega, \tag{1.1}$$

$$p\mathbf{n} - \mu(\mathbf{n}\cdot\nabla\mathbf{v} + \nabla\mathbf{v}\cdot\mathbf{n}) = \nabla_\Gamma\cdot(\sigma\nabla_\Gamma\mathbf{z}) \quad \text{on } \Gamma, \tag{1.2}$$

$$V_n = \mathbf{v}\cdot\mathbf{n} \quad \text{on } \Gamma, \tag{1.3}$$

$$\Omega = \Omega_* \quad \text{at } t = 0. \tag{1.4}$$

Here V_n is the normal speed of Γ along its inward normal $\mathbf{n}$; Ω_*, the given initial position of Ω; the dot is used to denote the inner product.

Due to the well-known formula $\Delta_\Gamma \mathbf{z} = C_\Gamma \mathbf{n}$ (C_Γ is the curvature of Γ), the dynamic (force balance) condition (1.2) yields the Laplace formula $p = \sigma C_\Gamma$ at equilibrium. In the general case of variable σ, the tangential stress $\nabla_\Gamma \sigma$ induces a flow caused by capillarity. For instance, thermocapillary convection is the result of dependence of σ on temperature [10] for which one is able to formulate a boundary-value problem and solve it in association with the equations (1.1)-(1.4). To concentrate our attention on the hydrodynamic part of the problem, the given function $\sigma = \sigma(\mathbf{z},t)$ will be assumed for definiteness. The case of constant surface tension is also of interest [12, 14].

2. Stress-stream function. For description of plane flows, let us identify each two-dimensional vector with a complex number. In particular, we introduce $z = x + iy$, $v = v_x + iv_y$ instead of $\mathbf{z} = (x,y)$, $\mathbf{v} = (v_x,v_y)$. Defining the differential form

$$F(\mathrm{d}z) = i\left(p(z,t)\mathrm{d}z + 2\mu\frac{\partial v(z,t)}{\partial \bar{z}}\mathrm{d}\bar{z}\right),$$

we rewrite the dynamic condition (1.2) for the solenoidal velocity field as $F(\mathrm{d}z) = \mathrm{d}\left(\sigma(z,t)\frac{\partial z}{\partial s}\right)$ where s is arc length. We have adopted that the inward normal is connected with the tangential vector by the formula $n = i\frac{\partial z}{\partial s}$ which corresponds to the usual choice of the orientation of $\partial\Omega$ (the positive direction of the contour $\partial\Omega$ leaves Ω on the left). Herein and throughout the overbar denotes the complex conjugate.

By the definition, a complex-valued function $w = \varphi + i\psi$ is bianalytic if it satisfies the equation $\partial^2 w/\partial\bar{z}^2 = 0$ [8]. The latter is equivalent to the well-known Goursat representation $w(z,t) = w_0(z,t) + w_1(z,t)\bar{z}$ where $w_0(z,t)$ and $w_1(z,t)$ are analytic within $\Omega(t)$.

Theorem 1. *The solutions of the Stokes equations* (1.1) *have the representations* $v = i\nabla\psi$, $p = -\mu\Delta\varphi$, $F(dz) = -2\mu i d(\nabla\varphi)$, *where* $w = \varphi + i\psi$ *is a bianalytic stress-stream function defined by the velocity* $v(z,t)$ *and pressure* $p(z,t)$ *apart from a linear function* $\mathrm{Re}\left(\overline{a(t)z}\right) + b(t)$ *with* $a(t), b(t) \in \mathbb{C}$.

In terms of the analytic functions $w_0(z,t)$ and $w_1(z,t)$, the representations of Theorem 1 coincide with those of Kolosov [13] and Muskhelishvili [16] in the plane theory of elasticity when Poisson's ratio is equal to 1/2.

Integrating the dynamic condition (1.2) along Γ, we find the equation $2\mu\nabla\varphi = \sigma n$ on Γ. Therefore the boundary conditions (1.2) and (1.3) take the form [2]

$$\varphi = 0, \quad 2\mu\frac{\partial\varphi}{\partial n} = \sigma(z,t) \quad \text{on } \Gamma, \tag{2.1}$$

$$V_n = \frac{\partial\psi}{\partial s} \quad \text{on } \Gamma. \tag{2.2}$$

Since the stress-stream function is defined apart from a linear function, integration constants are dropped.

Lemma 1. *The integral identity*

$$\iint\limits_{\Omega}\left(\left|\frac{\partial^2\varphi}{\partial\bar{z}^2}\right|^2 + \left|\frac{\partial^2\psi}{\partial\bar{z}^2}\right|^2 - \left|\frac{\partial^2 w}{\partial\bar{z}^2}\right|^2\right)dxdy + \mathrm{Re}\int\limits_{\partial\Omega}\frac{\partial\psi}{\partial\bar{z}}d\left(\frac{\partial\varphi}{\partial z}\right) = 0$$

is valid for a smooth function $w = \varphi + i\psi$.

Lemma 2. *A real bianalytic function* $c(z)$ *has the general form* $c(z) = c_0(x^2+y^2) + c_1 x + c_2 y + c_3$ *where* $c_j \in \mathbb{R}$.

Due to Lemmas 1 and 2 proved in [3, 4], the homogeneous auxiliary problem (2.1) has the only solution $w = ic(z)$ which is associated with $v_\infty(z,t) = 2ic_0(t)z - c_2(t) + ic_1(t)$ (rigid-body velocity). It is easy to check that the flow field $v_\infty(z,t)$ is orthogonal to capillary forces, *viz.*,

$$\mathrm{Re}\int\limits_{\Gamma}\overline{v_\infty(z,t)}\,d\left(\sigma(z,t)\frac{\partial z}{\partial s}\right) = 0.$$

This is the necessary condition of solvability of the inhomogeneous auxiliary problem.

Let us demonstrate that, by choosing a relevant moving coordinate system centered on $0 \in \Omega(t)$, it is possible to require the following normalization conditions

$$\frac{\partial w(0,t)}{\partial \bar{z}} = 0, \quad \operatorname{Im}\frac{\partial^2 w(0,t)}{\partial z \partial \bar{z}} = 0, \quad \operatorname{Im} w(0,t) = 0, \tag{2.3}$$

which result in the uniqueness of the auxiliary problem (2.1), and, in particular, eliminate $v_\infty(z,t)$. Indeed, for a given solution $\{\Omega(t),\ w(z,t)\}$, consider the path $z = q(t)$ whose points satisfy the Cauchy problem

$$\frac{dq(t)}{dt} = 2\frac{\partial w(q(t),t)}{\partial \bar{z}}, \quad q(0) = q_*,$$

where q_* is a fixed point of Ω_*. Using the simple identity

$$2\bar{n}\frac{\partial w}{\partial \bar{z}} = \frac{\partial w}{\partial n} - i\frac{\partial w}{\partial s} \quad \text{on } \Gamma, \tag{2.4}$$

which is in accordance with the choice $\frac{\partial z}{\partial n} = n = i\frac{\partial z}{\partial s}$ on Γ, we find the inequality

$$2\operatorname{Re}\left(\bar{n}\frac{\partial w(z,t)}{\partial \bar{z}}\right) = \frac{\sigma(z,t)}{2\mu} + V_n > V_n \quad \text{on } \Gamma(t)$$

owing to the boundary conditions (2.1) and (2.2) (surface tension is positive). Therefore, $q(t)$ must belong $\Omega(t)$ for all t. Let us rewrite the problem in the coordinate system centered on $q(t)$ and revolving by the angle

$$\beta(t) = \frac{1}{2}\int_0^t \Delta\psi(q(t),t)\,dt.$$

Upon changing the variables as is shown below,

$$z = e^{i\beta(t)}z^* + q(t), \quad w(z,t) = w^*(z^*,t) + ic(z^*,t),$$

$$c(z^*,t) = \frac{1}{2}\frac{d\beta(t)}{dt}|z^*|^2 + \mathrm{Im}\left(e^{-i\beta(t)}\frac{dq(t)}{dt}\overline{z}^*\right) + \psi(q(t),t),$$

and omitting the asterisk, we obtain the same problem (2.1), (2.2) with the additional condition (2.3) provided that σ is a given positive function of the new coordinate z and time t [4].

Using the identity (2.4), we rewrite (2.1)-(2.3) in terms of $w_0(z,t)$ and $w_1(z,t)$,

$$\mathrm{Im}\left(\frac{\partial \overline{z}}{\partial s}\frac{dz}{dt}\right) = \frac{\partial \psi(z,t)}{\partial s}, \quad z \in \Gamma(t), \tag{2.5}$$

$$\mathrm{Re}\left(w_0(z,t) + w_1(z,t)\overline{z}\right) = 0, \quad z \in \Gamma(t), \tag{2.6}$$

$$2\mathrm{Im}\left(\frac{\partial \overline{z}}{\partial s}w_1(z,t)\right) = \frac{\partial \psi(z,t)}{\partial s} + \frac{\sigma(z,t)}{2\mu}, \quad z \in \Gamma(t), \tag{2.7}$$

$$w_1(0,t) = 0, \quad \mathrm{Im}w_1'(0,t) = 0, \quad \mathrm{Im}w_0(0,t) = 0, \tag{2.8}$$

where $\psi(z,t) = \mathrm{Im}\left(w_0(z,t) + w_1(z,t)\overline{z}\right)$. Of course, the initial curve $\Gamma_* = \Gamma(0)$ has to enclose the origin of the flow plane.

3. Time-dependent conformal mapping. Let us introduce the conformal mapping $z = z(\zeta,t)$ of the unit disk $G = \{|\zeta| < 1\}$ onto the flow domain Ω with the normalization conditions

$$z(0,t) = 0, \quad z'(0,t) > 0, \tag{3.1}$$

whose differentiating by t provides the following

$$\frac{\partial z(0,t)}{\partial t} = 0, \quad \mathrm{Im}\frac{\partial z'(0,t)}{\partial t} = 0. \tag{3.2}$$

Let us change the variables in (2.5)-(2.8), $z = z(\zeta,t)$, retaining the same notations for the unknown functions, *e.g.*,

$$w(\zeta,t) = \varphi(\zeta,t) + i\psi(\zeta,t) = w_0(\zeta,t) + w_1(\zeta,t)\overline{z(\zeta,t)}.$$

Also we will assume that σ is a given function of ζ and t. This procedure yields the following problem

$$\mathrm{Re}\frac{\partial z(\tau,t)/\partial t}{\tau z'(\tau,t)} + u(\tau,t) = 0, \quad \tau = e^{i\theta}, \tag{3.3}$$

$$\mathrm{Re}\left(w_0(\tau,t) + w_1(\tau,t)\overline{z(\tau,t)}\right) = 0, \quad \tau = e^{i\theta}, \tag{3.4}$$

$$2\mathrm{Re}\frac{w_1(\tau,t)}{\tau z'(\tau,t)} + u(\tau,t) + \mathbf{A}(z|\tau,t) = 0, \quad \tau = e^{i\theta}, \tag{3.5}$$

$$w_1(0,t) = 0, \quad \mathrm{Im}\frac{w_1'(0,t)}{z'(0,t)} = 0, \quad \mathrm{Im}w_0(0,t) = 0, \tag{3.6}$$

where

$$\mathbf{A}(z|\tau,t) = \frac{\sigma(\tau,t)}{2\mu|z'(\tau,t)|}, \quad u(\tau,t) = \frac{\partial\psi(\tau,t)/\partial\theta}{|z'(\tau,t)|^2}.$$

In order to solve explicitly the problem (3.3)-(3.6) for temporarily given $u(\tau,t)$, let us introduce the Schwartz integral and Hilbert transform,

$$\mathbf{S}(u|\zeta) = \frac{1}{2\pi}\int_{\partial G} S(\tau,\zeta)u(\tau)\mathrm{d}\theta, \quad S(\tau,\zeta) = \frac{\tau+\zeta}{\tau-\zeta},$$

$$\mathbf{H}(u|\tau) = \frac{1}{2\pi}\mathrm{V.p.}\int_{\partial G} H(\tau,\lambda)u(\lambda)\mathrm{d}\nu, \quad H(\tau,\lambda) = -iS(\lambda,\tau),$$

where $|\zeta| < 1$, $\tau = e^{i\theta}$, $\lambda = e^{i\nu}$ (due to the Plemelj formulas [8], $\mathbf{S}(u|\zeta) \longrightarrow u(\tau) + i\mathbf{H}(u|\tau)$ when $\zeta \longrightarrow \tau$). The kinematic condition (3.3), having the form of the Schwartz problem, can be uniquely solved in the disk G,

$$\frac{\partial z(\zeta,t)}{\partial t} + \zeta z'(\zeta,t)\mathbf{S}(u(\cdot,t)|\zeta) = 0, \quad z(\zeta,0) = z_*(\zeta). \tag{3.7}$$

Here the identities (3.2) have been taken into consideration, and the initial conformal map $z_*(\zeta)$ of G onto Ω_*, satisfying the normalization condition (3.1), has been introduced. The intention is to construct an integral equation for $u(\tau,t)$ which completely determines the time-dependent conformal mapping due to (3.7).

4. Boundary integral equation. In the following we will suppress explicit dependence on time t in the notation whenever is possible without danger of confusion. Solving the Schwartz problem (3.4) with respect to $w_0(\zeta)$, we get

$$\psi(\tau) = \mathrm{Im}\left(w_0(\tau) + w_1(\tau)\overline{z(\tau)}\right) = -\mathbf{H}\left(\mathrm{Re}(w_1\bar{z})\,|\,\tau\right) +$$

$$+ \mathrm{Im}\left(w_1(\tau)\overline{z(\tau)}\right) = \mathrm{Re}\left(\mathbf{H}(w_1|\tau)\overline{z(\tau)} - \mathbf{H}(w_1\bar{z}|\tau)\right),$$

because $w_1(\tau) \equiv i\mathbf{H}(w_1|\tau)$ due to $w_1(0) = 0$. Therefore, denoting temporarily $f(\tau) = u(\tau) + \mathbf{A}(z|\tau)$, we find [4]

$$\psi(\tau) \equiv \mathrm{Re}\left[\frac{1}{2\pi}\int_{\partial G} H(\lambda,\tau)\left(\overline{z(\tau)} - \overline{z(\lambda)}\right)w_1(\lambda)\,d\nu\right] =$$

$$= \mathrm{Re}\left[\frac{i}{\pi}\iint_G S(\zeta,\tau)\overline{z'(\zeta)}\frac{w_1(\zeta)}{\zeta}d\xi d\eta\right] =$$

$$= \mathrm{Im}\left[\frac{1}{2\pi}\iint_G S(\zeta,\tau)|z'(\zeta)|^2\left(\frac{1}{2\pi}\int_{\partial G} S(\lambda,\zeta)f(\lambda)\,d\nu\right)d\xi d\eta\right], \quad \zeta = \xi + i\eta.$$

Using the identity

$$\mathrm{Im}\left(S(\lambda,\zeta)S(\zeta,\tau)\right) + H(\tau,\lambda)\mathrm{Re}\left(S(\tau,\zeta) - S(\lambda,\zeta)\right) = 0,$$

and introducing the function

$$Z(\tau) = \frac{1}{2\pi}\iint_G \frac{1 - |\zeta|^2}{|\tau - \zeta|^2}|z'(\zeta)|^2 d\xi d\eta, \tag{4.1}$$

we obtain

$$\psi(\tau) = \frac{1}{2\pi}\int_{\partial G} H(\tau,\lambda)\left(Z(\lambda) - Z(\tau)\right)f(\lambda)\,d\nu \equiv [\mathbf{H},Z](f|\tau),$$

where $[\mathbf{H},Z]$ denotes the commutator of the Hilbert transform $\mathbf{H}$ and the operator of multiplication by $Z(\tau)$ [3]. Therefore, differentiating $\psi(\tau)$ by θ, we obtain the following boundary integral equation for $u(\tau)$

$$u(\tau) = \mathbf{K}[z](u + \mathbf{A}(z)|\tau) \tag{4.2}$$

where

$$\mathbf{K}[z](f|\tau) = \frac{1}{|z'(\tau)|^2}\frac{\partial[\mathbf{H},Z](f|\tau)}{\partial\theta}.$$

Due to Kellog's theorem [11], $z'(\tau)$ does not vanish if Γ is a simple Lyapunov curve. The normalized solution of (4.2) is obviously unique, therefore, it exists for an arbitrarily fixed conformal mapping provided that (4.2) is a Fredholm integral equation. Note that the operator $\mathbf{K}[z]$ is the same for the whole class of similar domains, *i.e.*, it does not change when $z(\zeta)$ is replaced by $Az(\zeta) + B$ with $A \neq 0$.

In terms of the Taylor coefficients of $z(\zeta)$, $\{z_k\}$, we can easily get the representation [4]

$$Z(\tau) = Z_0 + 2\mathrm{Re}\sum_{k=1}^{\infty} Z_k\tau^k, \quad Z_k = \frac{1}{2}\sum_{m=1}^{\infty} m\bar{z}_m z_{k+m}. \tag{4.3}$$

Insofar as $Z(\tau)$ is constant when $z(\zeta)$ is a linear function, the exact solution $z(\zeta) = a\zeta$, $u(\tau) = 0$ of the problem (3.7), (4.2) exists for any $\sigma(\tau)$ $(z_*(\zeta) = a\zeta)$. Our intention is to solve the problem (3.7), (4.2) in the neighborhood of this exact solution. For constant σ, the linearized problem (3.7), (4.2) at this solution can be explicitly solved with the use of (4.3). The resulting conformal mapping has the form

$$z(\zeta,t) = e^{At}z_*(e^{-At}\zeta), \quad A = \frac{\sigma}{2\mu a} > 0.$$

Note that $z(\zeta,t)$ admits an analytical continuation onto the expanding disk $\{|\zeta| < e^{At}\}$, so it exponentially tends to the linear function $z_*'(0)\zeta$ when t goes to infinity ($z_*(0) = 0$).

5. Auxiliary estimates. Let us introduce the scale of Banach spaces $B_\rho(\partial G)$, $\rho \in \mathbb{R}$, whose elements are represented by the Fourier series

$$f(\tau) = \sum_{m=-\infty}^{\infty} f_m e^{im\theta}, \quad \tau = e^{i\theta},$$

with the norm [17]

$$\|f(\tau)\|_\rho = \sum_{m=-\infty}^{\infty} |f_m| e^{|m|\rho} < \infty.$$

Given $f(\tau)$, the map $\rho \longmapsto \|f(\tau)\|_\rho$ is a convex, increasing, analytic function in its domain of definition. For $\rho \geq 0$, $B_\rho(\partial G)$ is a Banach algebra, *i.e.*, the inequality

$$\|f(\tau)g(\tau)\|_\rho \leq \|f(\tau)\|_\rho \|g(\tau)\|_\rho$$

is valid. The compact imbedding $B_\rho(\partial G) \subset B_r(\partial G)$ takes place for $\rho > r$, and

$$\|\partial f(\tau)/\partial\theta\|_\rho = \frac{\partial}{\partial\rho}\|f(\tau)\|_\rho, \qquad \|\mathbf{H}(f|\tau)\|_\rho = \|f(\tau) - f_0\|_\rho.$$

Obviously, the space $B_0(\partial G)$ is contained in the set of continuous functions. For $\rho > 0$, the functions of $B_\rho(\partial G)$ admit an analytical continuation from the circumference ∂G onto the annulus $\{e^{-\rho} < |\zeta| < e^\rho\}$.

It is straightforward to obtain the following inequalities for $\rho \geq 0$

$$\frac{\partial^n}{\partial\rho^n}\|Z(\tau) - Z_0\|_\rho \leq \frac{1}{\rho}\Big(\|z(\tau)\|_0 - \|z(\tau)\|_{-\rho}\Big)\frac{\partial^n}{\partial\rho^n}\|z(\tau) - z'(0)\tau\|_\rho, \tag{5.1}$$

$$\frac{\partial^n}{\partial\rho^n}\|[\mathbf{H},Z](f|\tau)\|_\rho \leq \|f(\tau)\|_{-\rho}\,\frac{\partial^n}{\partial\rho^n}\|Z(\tau) - Z_0\|_\rho, \tag{5.2}$$

$$\frac{\partial}{\partial\rho}\|1/|z'(\tau)|^2\|_\rho \le \|1/z'(\tau)\|_\rho^4 \frac{\partial^2}{\partial\rho^2}\left(\|z(\tau)\|_\rho^2\right). \tag{5.3}$$

Using (5.1)-(5.3), one proves

Theorem 2. *The following estimates are valid for* $\rho \ge 0$

$$\|\mathbf{K}[z](f|\tau)\|_\rho \le \varepsilon_0(z,\rho)\|f(\tau)\|_{-\rho}, \tag{5.4}$$

$$\frac{\partial}{\partial\rho}\|\mathbf{K}[z](f|\tau)\|_\rho \le \varepsilon_1(z,\rho)\|f(\tau)\|_{-\rho}, \tag{5.5}$$

where

$$\varepsilon_0(z,\rho) = \|1/z'(\tau)\|_\rho^2 \frac{1}{\rho}\left(\|z(\tau)\|_0 - \|z(\tau)\|_{-\rho}\right)\frac{\partial}{\partial\rho}\|z(\tau)-z'(0)\tau\|_\rho,$$

$$\varepsilon_1(z,\rho) = \|1/z'(\tau)\|_\rho^2 \frac{1}{\rho}\left(\|z(\tau)\|_0 - \|z(\tau)\|_{-\rho}\right)\left[\frac{\partial^2}{\partial\rho^2}\|z(\tau)-z'(0)\tau\|_\rho + \right.$$

$$\left. + \|1/z'(\tau)\|_\rho^2 \frac{\partial^2}{\partial\rho^2}\left(\|z(\tau)\|_\rho^2\right)\frac{\partial}{\partial\rho}\|z(\tau)-z'(0)\tau\|_\rho\right].$$

6. Existence theorem. Due to Theorem 2, the operator $\mathbf{K}[z]$ is a linear completely continuous transformation of $B_\rho(\partial G)$ provided that $z'(\tau) \in B_\rho(\partial G)$, $\rho \ge 0$, and $z'(\tau) \ne 0$. Moreover, the estimate (5.4) allows us to represent the solution of (4.2) through the series

$$u(\tau) = \sum_{m=1}^{\infty} \mathbf{K}[z]^m \circ \mathbf{A}(z|\tau) \tag{6.1}$$

which converges if merely $\varepsilon_0(z,\rho) < 1$. The latter takes place when the norm $\|z'(\tau)-z'(0)\|_\rho$ is sufficiently small. Inserting (6.1) into (3.7), we get the nonlocal Cauchy problem for the time-dependent conformal mapping.

Let us demonstrate that the equation (3.7) generates a Cauchy problem for the real function $h(\tau) = \mathrm{Re}\left(\frac{z(\tau)}{a\tau} - 1\right)$ where a is the average radius of Ω, *viz.*,

$$a^2 = \frac{1}{2\pi i}\int_\Gamma \bar{z}dz. \tag{6.2}$$

Indeed, taking into consideration (3.1), we find

$$z(\zeta) = a\zeta\left(1 + \mathbf{S}(h|\zeta)\right), \tag{6.3}$$

and the condition (6.2) yields

$$h_0(t) = \left(1 - 4\sum_{k=1}^{\infty}(k+1)|h_k(t)|^2\right)^{1/2} - 1. \tag{6.4}$$

Here the normalization condition (3.1) was adopted for extracting an appropriate square root of a polynomial for $h_0(t)$. In particular, using (6.4), we get the estimate

$$\|h(\tau)\|_\rho \le \left(1 + \frac{\partial}{\partial\rho}\|h(\tau)\|_\rho\right)\frac{\partial}{\partial\rho}\|h(\tau)\|_\rho, \quad \rho \ge 0.$$

Inserting (6.3) into (3.7), we obtain

$$\frac{\partial h}{\partial t} + u + \mathbf{B}(h,u) = 0, \quad h|_{t=0} = h_*, \tag{6.5}$$

where $\mathbf{B}(h,u) = \left(\frac{\partial\mathbf{H}(h)}{\partial\theta} + h\right)u + \left(\frac{\partial h}{\partial\theta} - \mathbf{H}(h)\right)\mathbf{H}(u)$; $h_*(\tau)$ is a given continuous function related to $z_*(\tau)$.

Due to (6.1) and (6.3), the function $u(\tau)$ depends on $h(\tau)$ which is assumed sufficiently small. Let us write the problem (6.5), (4.2) in the form

$$\frac{\partial h}{\partial t} + \frac{\partial}{\partial\theta}\left[\mathbf{H}(Ah) - h\mathbf{H}(A)\right] = \mathbf{N}(h), \quad h|_{t=0} = h_*, \tag{6.6}$$

where $A(\tau) = \frac{\sigma(\tau)}{2\mu a}$; $\mathbf{N}(h|\tau)$ denotes the residual nonlinear terms.

Applying (5.2), Theorem 2, and the inequality

$$\frac{\partial^n}{\partial\rho^n}\|Z(\tau)-a^2h(\tau)\|_\rho \le$$

$$\le a^2\left[\|h(\tau)\|_0 + \frac{1}{\rho}\left(\|h(\tau)\|_0-\|h(\tau)\|_{-\rho}\right)\right]\frac{\partial^n}{\partial\rho^n}\|h(\tau)\|_\rho,\ n \ge 1,$$

it is straightforward to obtain the estimate

$$\frac{\partial}{\partial\rho}\|\mathbf{N}(h|\tau))\|_\rho \le \gamma\frac{\partial}{\partial\rho}\|h(\tau)\|_\rho\ \frac{\partial^2}{\partial\rho^2}\|h(\tau)\|_\rho \tag{6.7}$$

where the constant γ is bounded provided that

$$\frac{\partial}{\partial\rho}\|h(\tau)\|_\rho \le q,\quad \rho \ge 0, \tag{6.8}$$

for a sufficiently small positive q (in other words, γ depends on q only). Differentiating (6.6) by θ and implementing standard manipulations with the Fourier coefficients of $h(\tau,t)$, it is straightforward to get the differential inequality

$$\frac{\partial}{\partial t}\left(\frac{\partial}{\partial\rho}\|h(\tau,t)\|_\rho\right) + \alpha\frac{\partial^2}{\partial\rho^2}\|h(\tau,t)\|_\rho \le \frac{\partial}{\partial\rho}\|\mathbf{N}(h|\tau,t)\|_\rho \tag{6.9}$$

where

$$\alpha = \min_{t\ge 0}\left(A_0(t) - \sum_{k\ne 0}|A_k(t)|e^{-r|k|}\right)$$

is assumed positive for a fixed r, $0 \le r \le \rho$. Let β and q satisfy the conditions $0 < \beta < \alpha$, $0 < q \le (\alpha - \beta)/\gamma$. Then the inequalities (6.7) and (6.9) yield the following bound

$$\frac{d}{dt}\|\partial h(\tau,t)/\partial\theta\|_{r+\beta t} \le 0.$$

Therefore, we find the final *a-priori* estimate of the solution of (6.5) and (4.2),

$$\|\partial h(\tau,t)/\partial\theta\|_{r+\beta t} \le \|\partial h_*(\tau)/\partial\theta\|_r \le q$$

which *a-posteriori* gives (6.8) with $\rho = r + \beta t$ for all $t \ge 0$. Note that $\|\partial h(\tau,t)/\partial\theta\|_r \le e^{-\beta t}\|\partial h(\tau,t)/\partial\theta\|_{r+\beta t}$, hence

$$\|h(\tau,t)\|_r \le \left(1 + qe^{-\beta t}\right)qe^{-\beta t} \longrightarrow 0 \quad \text{when } t \longrightarrow \infty.$$

Theorem 3. *Let the Fourier coefficients $\sigma_k(t)$ of the surface tension $\sigma(\tau,t)$ be Lipschitz functions satisfying the condition*

$$\alpha = \frac{1}{2\mu a}\min_{t\ge 0}\left(\sigma_0(t) - \sum_{k\ne 0}|\sigma_k(t)|e^{-r|k|}\right) > 0$$

for a given $r \ge 0$. Then, for a fixed $\beta \in (0,\alpha)$, there exists $q > 0$ such that, for each initial function $h_*(\tau)$ satisfying (6.4) *and the inequality $\|\partial h_*(\tau)/\partial\theta\|_r \le q$, the problem* (6.5), (4.2) *has a unique solution $h(\tau,t) \in B_{r+\beta t}(\partial G)$, $\|\partial h(\tau,t)/\partial\theta\|_{r+\beta t} \le q$, vanishing when t tends to infinity.*

The proof follows from the analysis of the problem (6.6) using the stereotyped technique of solving the abstract nonlinear parabolic equations by reducing them to a fixed point theorem [15]. Of course, the conformal mapping (6.3) is one-sheet for a sufficiently small positive q. It is worth noting that, under the conditions of Theorem 3 with $r > 0$, it is, in principle, possible to prove the existence theorem for $t < 0$ until the inequality $r + \beta t > 0$ holds [17].

Concluding remarks. The representation of the Stokes solutions through two analytic functions has been explored for investigation of viscous free-boundary flows in [1, 5, 7, 9, 12, 19, 20]. In these papers the method of Kolosov [13] and Muskhelishvili [16], extensively used in the plane theory of elasticity, was adapted to creeping viscous flows. Instead of two analytic functions, it is more convenient to use the bianalytic stress-stream function [2, 3, 7].

The integral equation (4.2) is somewhat simpler than those based on the theory of hydrodynamic potentials [5] because (a) the unknown function is real, and (b) the equation is not singular though the homogeneous problem (1.1) and (1.2) admits the rigid-body velocity.

Compactness of the operator $\mathbf{K}[z]$ clarifies analyticity of Γ for arbitrary surface tension. This fact can be also formulated as the following form: *the smoothness of the free boundary in quasi-steady viscous flows does not depend on the smoothness of the rotational component of the velocity field.*

References

1. Antanovskii, L.K. Complex representation of the solutions of the Navier-Stokes equations for a low Reynolds number. *Dinamika Sploshnoi Sredy* **51**, 3-16 (1981)
2. Antanovskii, L.K. Stability of a liquid film equilibrium under the action of thermocapillary forces in quasisteady approximation. *Zh. Prikl. Mekh. Tekh. Fiz.* No.2, 47-53 (1990)
3. Antanovskii, L.K. Boundary integral equations in quasisteady problems of capillary fluid mechanics. Part 2: Application of the stress-stream function. *Meccanica - J. Ital. Assoc. Theoret. Appl. Mech.* **26**(1), 59-65 (1991)
4. Antanovskii, L.K. Bianalytic stress-stream function in plane quasi-steady problems of capillary fluid mechanics. *Sibirsk. Matem. Zh.* **33**(1), 3-15 (1992)
5. Belonosov, S.M. & Chernous, K.A. *Boundary-Value Problems for the Navier-Stokes Equations*. Moscow: Nauka (1985)
6. Bemelmans, J. & Friedman, A. Analyticity for the Navier-Stokes equations governed by surface tension on the free boundary. *J. Diff. Equat.* **55**(1), 135-150 (1984)
7. Clarke, N.S. Two-dimensional flow under gravity in a jet of viscous liquid. *J. Fluid. Mech.* **31**(3), 481-500 (1968)
8. Gakhov, F.D. *Boundary-Value Problems*. Oxford: Pergamon Press (1966)
9. Garabedian, P.R. Free boundary flows of viscous liquid. *Comm. Pure Appl. Math.* **19**(4), 421-434 (1966)
10. Gibbs, J.W. *Collected Works*, **1**. New Haven: Yale University Press (1948)
11. Goluzin, G.M. *Geometric Theory of Functions of Complex Variable*. Providence, Rhode Island: Amer. Math. Soc. (1969)
12. Hopper, R.W. Plane Stokes flow driven by capillarity on a free surface. *J. Fluid Mech.* **213**, 349-375 (1990)

13. Kolosov, G.V. *Application of the Complex Variable to the Theory of Elasticity*. Moscow-Leningrad: ONTI (1935)
14. Kuiken, H.K. Viscous sintering: the surface-tension-driven flow of a liquid form under the influence of curvature gradients at its surface. *J. Fluid Mech.* **214**, 503-515 (1990)
15. Martin, R.H.Jr. *Nonlinear Operators and Differential Equations in Banach Spaces.* Interscience, Wiley (1976)
16. Muskhelishvili, N.I. *Some Basic Problems of Mathematical Theory of Elasticity*. Groningen-Holland: Noordhoff (1953)
17. Ovsiannikov, L.V. A nonlinear Cauchy problem in a scale of Banach spaces. *Dokl. Akad. Nauk SSSR* **200**(4), 789-792 (1971)
18. Pukhnachov, V.V. On smoothness of the steady-state solutions of the Navier-Stokes equations near the free boundary. *Dinamika Sploshnoi Sredy* **15**, 133-144 (1973)
19. Richardson, S. Two-dimensional bubbles in slow viscous flows. *J. Fluid Mech.* **33**, 476-493 (1968)
20. Richardson, S. Two-dimensional bubbles in slow viscous flows. Part 2. *J. Fluid Mech.* **58**, 115-127 (1973)
21. Solonnikov, V.A. Unsteady motion of a finite mass of fluid, bounded by a free surface. *Zap. Naucn. Sem. LOMI* **152**, 137-157 (1986)
22. Solonnikov, V.A. On the evolution of an isolated volume of a viscous incompressible capillary liquid for large values of time. *Vestnik LGU. Ser.1*, No.3, 49-55 (1987)

This paper is in final form and no similar paper has been or is being submitted elsewhere.

ON A FREE BOUNDARY PROBLEM FOR THE STATIONARY NAVIER-STOKES EQUATIONS WITH A DYNAMIC CONTACT LINE

Jürgen Socolowsky
Technische Hochschule Merseburg

1. Introduction and function spaces

In this paper we consider the plane steady-state problem with two free boundaries for the Navier-Stokes equations describing a coating process with a heavy viscous incompressible fluid onto a horizontally moving rigid wall. The problem is considered in the plane $\mathbb{R}^2$ with a fixed Cartesian coordinate system $x=(x_1, x_2)$. The fluid fills the infinite region V bounded by the straight line $\Sigma_1 = \left\{x \in \mathbb{R}^2;\, x_2=0\right\}$ and the half-lines

$$\Sigma_2=\left\{x\in\mathbb{R}^2;\, x_1\leq 0,\ \ x_2=h_1-x_1\tan\,\alpha\right\},\ \Sigma_3=\left\{x\in\mathbb{R}^2;\ x_1\leq 0,\ \ x_2=h_2-x_1\tan\,\alpha,\, h_2>h_1>0\right\},$$

where α is a fixed real number with $0\leq\alpha<\pi/2$ (see Fig. 1). The region V is then the union of the first quadrant of $\mathbb{R}^2$ with the half-strip between Σ_2 and Σ_3.

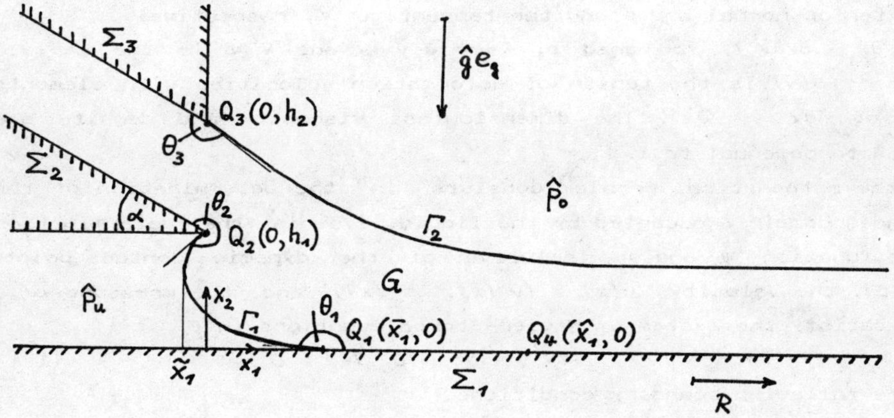

Fig. 1

The force of gravity is directed along the vector $e_g=(0,-1)^T$. We suppose that the flow of the liquid is generated by the stream $F^0(\alpha)$ and by the motion of the lower rigid wall Σ_1 with constant velocity R in x_1-direction. The values R and $F^0(\alpha)$ are assumed to be positive. Under the given assumptions the free surface Γ_1 is compact and the free surface Γ_2 is non-compact. In agreement with many experimental studies we further suppose that the free surfaces separate from the rigid walls at the "sharp" corners Q_2 and Q_3. The free surface Γ_1 "touches" the moving rigid wall Σ_1 at the a priori unknown point $Q_1(\bar{x}_1,0)$. In the literature this point is called dynamic contact point (or line) and the angle θ_1 between the positive x_1-axis and the left tangent to Γ_1 at Q_1 is called dynamic contact angle. We assume that the free boundary Γ_1 is described as the graph of a function ψ_1 with respect to $x_2\in[0,h_1]$. This assumption makes physically sense and it is a key to handling the lower free surface.

The following notations are used:

Γ_m $(m=1,2)$ – free surfaces with the representations $x_1=\psi_1(x_2)$ and $x_2=\varphi_2(x_1)$,

$G=\left\{x\in V;\ x_2<\varphi_2(x_1) \text{ for } x_1>0 \text{ and } x_1>\psi_1(x_2) \text{ for } x_2<h_1\right\}$ – flow – domain of the fluid, $\sigma(t)=\left\{x\in G;\ x_1=t\right\}$; n and τ are unit vectors directed along the exterior normal and along the tangent to ∂G, respectively.

$\nabla=(\partial/\partial x_1,\ \partial/\partial x_2)$, $\nabla p=$ grad p, $\nabla\cdot v=$ div v and $\nabla^2=\Delta$ is the Laplace operator; $S(v)$ is the tensor of deformation velocities with elements $S_{ij}=\partial v_i/\partial x_j+\partial v_j/\partial x_i$. The dimensionless viscosity and density are assumed to be equal to 1.

Thus the mathematical problem consists in the determination of the unbounded domain G occupied by the liquid, i.e. in the determination of the functions ψ_1 and φ_2 inclusive of the dynamic contact point $Q_1(\bar{x}_1,0)$, the velocity $v(x)=(v_1(x),\ v_2(x))^T$ and the pressure $\hat{p}(x)$ which satisfy the system of Navier-Stokes equations in G.

$$(v\cdot\nabla)v-\nabla^2 v+\nabla p=0,\ \nabla\cdot v=0 \tag{1.1}$$

and the following boundary conditions

$$v\big|_{\Sigma_j}=0\ (j=2,3),\quad v\cdot n\big|_{\Gamma_m}=0,\quad \tau\cdot S(v)n\big|_{\Gamma_m}=0\ (m=1,2),$$

$$\tilde{B}\ \tau\cdot S(v)n+\left[v-(R,0)^T\right]\cdot\tau=0,\quad v\cdot n=0\quad\left(x\in\Sigma_1;\bar{x}_1\le x_1\le\hat{x}_1\right),$$

$$v\big|_{\Sigma_1} = (R,0)^T \quad \left(x\in\Sigma_1,\ x_1 \geq \hat{x}_1\right) \tag{1.2}$$

$$\frac{d}{dx_2}\,\frac{\psi_1'(x_2)}{\left[1+(\psi_1'(x_2))^2\right]^{1/2}} + \beta x_2 = W\left[\bar{p} + p - n\cdot S(v)n\right],$$

$$\frac{d}{dx_1}\,\frac{\varphi_2'(x_1)}{\left[1+(\varphi_2'(x_1))^2\right]^{1/2}} - \beta\varphi_2(x_1) = W\left[-p + n\cdot S(v)n\right]. \tag{1.3}$$

At $x_1=0$ and $x_1=\bar{x}_1$ the functions φ_2 and ψ_1 must fulfil the boundary conditions

$$\varphi_2(0) = h_2,\ \psi_1(h_1) = 0,\ \psi_1(0) = \cot\theta_1 = -A. \tag{1.4}$$

The fact that the motion is caused by a non-zero stream $F^0(\alpha)$ is mathematically formulated in the form

$$\int_{\sigma(t)} v_1(t,x_2)\,dx_2 = F^0(\alpha)\ . \quad (t\in\mathbb{R}) \tag{1.5}$$

The restriction

$$|p(x)| \leq \text{const as } x_1 \longrightarrow +\infty \tag{1.6}$$

completes the free *boundary value problem* (=BVP) $\left[(1.1)-(1.6)\right]$. In this problem the notation $a\cdot b = a_1b_1 + a_2b_2$ is used. In order to obtain Eqs. (1.1) and (1.3) in the presented form a transformation $p(x) := \hat{p}(x) + \hat{g}x_2 - \hat{p}_0$ for the original pressure $\hat{p}$ was performed.
In the problem $\left[(1.1)-(1.6)\right]$ which we call the BVP-1 in the sequel, the values $\tilde{B}$, $\hat{g}$, $\hat{p}_u$, $\hat{p}_0$, W, A, β are constant and positive. The value $\bar{p} := \hat{p}_0 - \hat{p}_u$ is also constant. Here $\hat{g}$ denotes the acceleration of gravity; $\hat{p}_u$ and $\hat{p}_0$ denote the athmospheric pressures outside Γ_1 and Γ_2, respectively, and W contains the surface tension. Obviously, $\bar{p}>0$ is equivalent to $\hat{p}_0>\hat{p}_u$. The dynamic contact angle θ_1 (cf. Fig. 1) is also prescribed with $\pi/2 < \theta_1 \leq \pi$. This physically makes sense and it is equivalent to the condition $0<A\leq+\infty$ for A. The condition in the second line of Eqs. (1.2) is called slip-condition. In our problem BVP-1 the positive parameters $\tilde{B}$=const., $\hat{x}_1$ = const. are also prescribed a priori. The slip-condition is necessary in that case where $A< +\infty$ corresponding to the dynamic contact angle is given. More

details about the slip boundary condition or about the correct model for the slip are contained in [1, 2, 7, 8].

In the papers [5, 7, 8,14] a slip boundary condition was incorporated in numerical studies of flow problems with static or dynamic contact lines by FEM. Kröner [3, 4] performed analytical investigations for stationary flow in an infinite channel with a moving wall. In that model problem it was possible to bound the flow domain and to prescribe a priori the dynamic contact point with the help of a simultaneously moving coordinate system. Kröner obtained local statements of existence and uniqueness and also asymptotical expansions for the solution. He used the slip-condition from Eq. (1.2).

In [8] the same problem was solved numerically. Analytical studies of free BVP with a priori unknown dynamic contact points or lines do not exist.

Now we introduce the notations $\Sigma_{11}:=\left\{x\in\Sigma_1;\ \bar{x}_1<x_1<\hat{x}_1\right\}$ and $\Sigma_{12}:=\left\{x\in\Sigma_1; x_1>\hat{x}_1\right\}$. By $Q^*:=Q_1\cup Q_2\cup Q_3\cup Q_4$ we mean the union of all 4 singular points. Then for $x\in G$ we define

$$d_{Q^*}(x) = \text{dist}\,(x, Q^*):= \inf_{y\in Q^*} |x-y|$$

Let B be an arbitrary domain in $\mathbb{R}^2$ and $N\subset\bar{B}$ a manifold of dimension $n<2$.

- By $H^k_\mu(B,N)$, for arbitrary $\mu\in\mathbb{R}$ and integer $k\geq 0$, we understand the set of functions with

$$\|u\|^2_{H^k_\mu(B,N)}:= \sum_{0\leq|\beta|\leq k} \int_B |D^\beta u(x)|^2 d_N(x)^{2\mu-2(k-|\beta|)}\, dx < \infty$$

where $\beta=(\beta_1, \beta_2)$ is a multiindex in this formula with

$$|\beta|=\beta_1+\beta_2 \text{ and } D^\beta u:=\frac{\partial^{|\beta|}u}{\partial x_1^{\beta_1}\partial x_2^{\beta_2}} \quad \left(\beta_i\in \mathbb{N}\cup\{0\}\right).$$

The space H^k_μ is a weighted Sobolev space.

The weighted Hölder spaces $\overset{\circ}{C}{}^l_s(B,N)$ and $C^l_s(B,N)\,(l>s; l,s$ non-integer) are defined with the help of the weight $d_N(x)$ in the usual way (cf. [12,13, 9]).

Finally, we define the weighted Hölder spaces to which the generalized solutions of BVP-1 belong. For this purpose let us use the following notations: $\hat{Q}:=Q_1 \cup Q_2$, $J_1:=\left]0,h_1\right[$, $Y_1:=\left\{0,h_1\right\}$, $G^0:=\left\{x\in G;\ \tilde{x}_1-2<x_1<\hat{x}_1+2\right\}$, $G^+:=\left\{x\in G;\ x_1>\hat{x}_1+1\right\}$, $G^-:=\left\{x\in G; x_1<\tilde{x}_1-1\right\}$, $J_2^0:=\left\{x_1\in\mathbb{R};\ 0<x_1<\hat{x}_1+2\right\}$, $J_2^+:=\left\{x_1\in\mathbb{R}; x_1>\hat{x}_1+1\right\}$. Here $\tilde{x}_1$ denotes the value $\tilde{x}_1:=\min\limits_{x_2\in[0,h_1]}\psi_1(x_2)$.

- For an arbitrary real number $z>0$ we define the space

$$C^l_{s,z}(G):=\left\{u(x);\ u|_{G^0}\in C^l_s(G^0,Q^*),\ u\exp(zx_1)|_{G^+}\in C^l(G^+),\right.$$
$$\left.u\exp(-zx_1)|_{G^-}\in C^l(G^-)\right\} \text{ with the norm}$$
$$\|u\|^{l,z}_{G,s}:=|u|_{C^l_s(G^0,Q^*)}+|e^{zx_1}u|^{(l)}_{G^+}+|e^{-zx_1}u|^{(l)}_{G^-}.$$

- For functions $f(x_1)$ defined in $\Delta:=\mathbb{R}^1_+$ we introduce the space $C^l_{s,z}(\Delta)$ with the norm

$$\|f\|^{l,z}_{\Delta,s}:=\ |f|_{C^l_s(J^0_2,0)}+|f(x)e^{zx_1}|^{(l)}_{J^+_2}.$$

2. Linear and nonlinear auxiliary problems

Define the sector

$$d:=\ d(\theta):=\left\{(r,\varphi)\in\mathbb{R}^2;\quad 0<r<\infty,\quad 0<\varphi<\theta\right\},$$ where $\pi/2<\theta<\pi$ holds and θ is fixed. Consider in d the mixed BVP-2 with fixed boundary for the Stokes equations

$$-\nabla^2 v+\nabla p=f,\ \nabla\cdot v=\hat{r}, \tag{2.1}$$

$$\left.\frac{\partial v_1}{\partial x_2}\right|_{\varphi=0}=v_2\Big|_{\varphi=0}=0,\quad v\cdot n\Big|_{\varphi=\theta}=\tau\cdot S(v)n\Big|_{\varphi=\theta}=0. \tag{2.2}$$

By a well-known procedure (transition to polar coordinates, application of the Mellin transform) the BVP-2 can be transformed to a BVP for a system of ordinary differential equations on the interval $]0,\theta[$ which can be solved analytically (cf. [2, 8,10]).

The corresponding homogeneous problem has nontrivial solutions if $\delta:=-\lambda i\neq 0$ is a root of the following equation

$$H_1(\delta):=\cos 2\theta-\cos 2\theta\delta=0, \tag{2.3}$$

where λ is the parameter from the Mellin-transformation. The following theorem can be proved in the same way as in [13] or [8].

Theorem 2.1. *For arbitrary* $f=(f_1,f_2)^T\in H^k_\mu(d,0)$, $\hat{r}\in H^{k+1}_\mu(d,0)$ *the BVP-2 has a unique solution with* $v\in H^{k+2}_\mu(d,0)$, $p\in H^{k+1}_\mu(d,0)$ *and*

$$\|v\|_{H^{k+2}_\mu}+\|p\|_{H^{k+1}_\mu}\le c_0\left(\|f\|_{H^k_\mu}+\|\hat{r}\|_{H^{k+1}_\mu}\right), \tag{2.4}$$

provided $1+k-\mu\ne \mathrm{Re}\,\delta$, *where* $\delta\ne 0$ *is a root of Eq.(2.3) (*c_0 *does not depend on* f *and* $\hat{r}$*).*

Lemma 2.1. *There exists a positive real number* $s_1<1$ *such that the complex function* $H_1(\delta)$ *from (2.3) does not have roots on the interval* $0<\mathrm{Re}\,\delta<s_1$.

The proof of this lemma is carried out with the help of elementary methods (cf.[8]).Now we are able to formulate the principal result for the BVP-2.

Theorem 2.2. *Let* l *be a positive non-integer and* $s\ne\mathrm{Re}\,\delta$ *where* δ *is a non-zero root of Eq.(2.3). For arbitrary* $f\in\mathring{C}^{l}_{s-2}(d,0)$, $\hat{r}\in\mathring{C}^{l+1}_{s-1}(d,0)$ *the BVP-2 has a unique solution* $v\in\mathring{C}^{l+2}_{s}(d,0)$, $p\in\mathring{C}^{l+1}_{s-1}(d,0)$ *and*

$$|v|_{\mathring{C}^{l+2}_{s}}+|p|_{\mathring{C}^{l+1}_{s-1}}\le c_1\left(|f|_{\mathring{C}^{l}_{s-2}}+|\hat{r}|_{\mathring{C}^{l+1}_{s-1}}\right) \tag{2.5}$$

where c_1 *does not depend on* f *and* $\hat{r}$.

This theorem can be proved in the same way as in [13] (see Theorem 3.3, pp. 360-363).

Recall the domain G described in Sect.1 and suppose that the boundaries Γ_1 and Γ_2 are given and fixed. Let Γ_m have unique tangents at their endpoints Q_1, Q_2 and Q_3. Then by BVP-3 we denote the problem

$$-\nabla^2 v+\nabla p=f,\ \nabla\cdot v=r\quad (x\in G), \tag{2.6}$$

$$v=0\ (x\in\Sigma_{12}\cup\Sigma_2\cup\Sigma_3),\quad v\cdot n=0\ (x\in\Gamma_1\cup\Gamma_2\cup\Sigma_{11}),$$

$$\tau\cdot S(v)n\big|_{\Gamma_m}=0\ (m=1,2),\ \tilde{B}\tau\cdot S(v)n+v\cdot\tau=0\ (x\in\Sigma_{11}). \tag{2.7}$$

A concept of a generalized solution to BVP-3 can be defined with the help of an appropriate bilinear form as in [8,10]. Using Korn's inequality which is valid in the unbounded domain G too and Lemmas 4.2, 4.3 from [10] one can prove the existence and uniqueness of a weak solution (v,p) to BVP-3.

In [13] model problems like BVP-2 where the slip condition was replaced by a no-slip condition $v = 0$ are studied using Kondrat'ev's theory. There the existence of such numbers $s_j (j=2,3,4)$ with $0<s_j<1$ was shown that the expression $H_j(\delta)=\sin 2\delta\theta_j-\delta \sin 2\theta_j$ does not have roots δ_0 with real parts on the interval $]o,s_j[$. Here $\theta_4:=\pi$ was set. Define $s_*:= \min(s_1,s_2,s_3,s_4)$. By $G(\lambda,t)$ we denote the set $G(\lambda,t):= \{x\in G;\ \lambda-t < x_1 < \lambda+1\}$.

Theorem 2.3. *There exists a positive number* z_* *such that for arbitrary* $f\in C^s_{s-2,z}(G)$ *and* $r\in C^{s+1}_{s-1,z}(G)$ *with* $z\in\ ]0,z_*[$ *and* $s\in\ ]o,s_*[$ *the generalized solution of BVP-3 satisfies the inequality*

$$\int_{G(\lambda,1)} |v_x|^2 dx \le c_2 e^{-2z|\lambda|} \left[\left(\|f\|^{s,z}_{G,s-2}\right)^2 + \left(\|r\|^{s+1,z}_{G,s-1}\right)^2\right] \tag{2.8}$$

where $-2+\tilde{x}_1<\lambda<\hat{x}_1+2$ *holds and the constant* c_2 *does not depend on* λ.

The proof of this basic theorem is carried out in the same way as in the papers [8, 9]. Th. 2.3 enables us to formulate the main result of this section.

Theorem 2.4. *Let* $h^*_2>0$ *be a real constant and* $(\varphi_2-h^*_2)\in C^{3+s}_{1+s,z}(\Delta)$, $\psi_1\in C^{3+s}_{1+s}(J_1,Y_1)$. *For arbitrary* $f\in C^s_{s-2,z}(G)$, $s\in\ ]0,s_*[$, $z\in]0,z_*[$ *and* $r\in C^{s+1}_{s-1,z}(G)$ *satisfying the condition*

$$\int_G r(x)dx = 0 \tag{2.9}$$

the BVP-3 has a unique solution (v,p) *such that* $v\in C^{s+2}_{s,z}(G)$, $\nabla p\in C^s_{s-2,z}(G)$. *Moreover,*

$$\|v\|^{s+2,z}_{G,s} + \|\nabla p\|^{s,z}_{G,s-2} \le c_3 \left(\|f\|^{s,z}_{G,s-2} + \|r\|^{s+1,z}_{G,s-1}\right). \tag{2.10}$$

The proof is realized in an analogous way as in [9].

Let the fixed boundaries $\Gamma_m (m=1,2)$ be the same as in Th. 2.4. We regard the BVP-4 consisting of the nonlinear Eq.(1.1) and the conditions (1.2), (1.5), i.e.

$$(v\cdot\nabla)v - \nabla^2 v + \nabla p = 0,\ \operatorname{div} v = 0\ (x\in G), \tag{2.11}$$

$$v\Big|_{\Sigma_j}=0\ (j=2,3),\quad v\cdot n\Big|_{\Gamma_m}=0,\quad \tau\cdot S(v)n\Big|_{\Gamma_m}=0\ (m=1,2),$$

$$\tilde{B}\,\tau\cdot S(v)n+[v-(R,0)^T]\cdot\tau=0,\quad v\cdot n=0\ (x\in\Sigma_{11}),\quad v\Big|_{\Sigma_{12}}=(R,0)^T, \tag{2.12}$$

$$\int_{\delta(t)} v_1(t,x_2)\,dx_2=F^0(\alpha). \tag{2.13}$$

In order to solve BVP-4 we study two auxiliary problems. In the domain G^- (cf. Sect.1) one seeks a solution $\left(u^{(-)},\ p^{(-)}\right)$ to Eqs. (2.11) satisfying the conditions

$$u^{(-)}\Big|_{\Sigma_j}=0\ (j=2,3),\qquad \int_{\delta(t)} u_1^{(-)}(t,x_2)\,dx_2=F^0(\alpha). \tag{2.14}$$

In $\hat{G}^+$ we are looking for a solution $(u^{(+)},\ p^{(+)})$ to Eqs.(2.11) fulfilling the conditions

$$u^{(+)}\Big|_{\Sigma_{12}}=(R,0)^T,\quad u_2^{(+)}\Big|_{x_2=h_2^*}=0,\quad \frac{\partial u_1^{(+)}}{\partial x_2}\Bigg|_{x_2=h_2^*}=0,$$

$$\int_0^{h_2^*} u_1^{(+)}(x_1,x_2)dx_2=F^0(\alpha), \tag{2.15}$$

where $\hat{G}^+$ denotes the domain $\hat{G}^+:=\left\{x\in\mathbb{R}^2;\ x_1>\hat{x}_1+1,\ 0<x_2<h_2^*\right\}$. Problem (2.11), (2.14) permits the following solution

$$u_1^{(-)}=\frac{6F^0(\alpha)}{(h_2-h_1)^3}\,(x_2-h_1+x_1\tan\alpha)(h_2-x_2-x_1\tan\alpha),$$

$$u_2^{(-)}=-\frac{6F^0(\alpha)\sin\alpha}{(h_2-h_1)^3\cos\alpha}\,(x_2-h_1+x_1\tan\alpha)(h_2-x_2-x_1\tan\alpha),$$

$$p^{(-)}=-\frac{12F^0(\alpha)}{(h_2-h_1)^3\cos^2\alpha}\left[x_1-(x_2-h_1)\tan\alpha\right]. \tag{2.16}$$

In (2.16) an additive constant for the pressure was neglected. Problem (2.11), (2.15) has a solution of the form

$$u_1^{(+)} = R + \frac{3}{(h_2^*)^2}\left(R - \frac{F^0(\alpha)}{h_2^*}\right) x_2\left(\frac{x_2}{2} - h_2^*\right), \; u_2^{(+)} \equiv 0,$$

$$p^{(+)} = \frac{3}{(h_2^*)^2}\left(R - \frac{F^0(\alpha)}{h_2^*}\right) x_1. \tag{2.17}$$

Now by formula (2.17) the solution $\left(u^{(+)},\ p^{(+)}\right)$ hitherto only defined on $\hat{G}^+$ can be extended or restricted to $G^+ \neq \hat{G}^+$. In this case the notation is preserved. Formulae (2.16), (2.17) are verified by direct calculations. Notice that $M_R^F := \max\left[F^0(\alpha),\ R\right]$. Further let g be a smooth real function with $g(t)=0$ as $t \le \hat{x}_1+1$ and $g(t)=1$ as $t \ge \hat{x}_1+2$. Now we can formulate the main result for the nonlinear auxiliary problem.

Theorem 2.5. *Let z and s be the same as above. For a sufficiently small number M_R^F the BVP-4 has a unique solution (v,p) permitting the representation*

$$v = g(-x_1)u^{(-)} + g(x_1)u^{(+)} + w, \quad p = g(-x_1)p^{(-)} + g(x_1)p^{(+)} + q, \tag{2.18}$$

where $u^{(+)}$, $u^{(-)}$, $p^{(+)}$, $p^{(-)}$ are given by (2.16), (2.17) and $w \in C^{s+2}_{s,z}(G)$, $\nabla q \in C^{s}_{s-2,z}(G)$. Furthermore, the inequalities

$$\|w\|^{s+2,z}_{G,s} + \|\nabla q\|^{s,z}_{G,s-2} \le c_4(F^0(\alpha),\ R), \; |v|_{C^{s+2}_s(G^0, Q^*)} + |v|_{C^{s+2}(G^+ \cup G^-)} +$$

$$+ \; |\nabla p|_{C^s_{s-2}(G^0,\ Q^*)} + |\nabla p|_{C^s(G^+ \cup G^-)} \le c_5(F^0(\alpha), R) \tag{2.19}$$

hold, where $c_4, c_5 \longrightarrow 0$ as $M_R^F \longrightarrow 0$.

This theorem can be proved in a standard way as in [9,10].

3. The determination of the free surfaces

In this section we want to explain a sketch of the solution to the complete problem BVP-1. For this purpose it is necessary to find solutions for the free boundaries. As in [6] or [9] we make use of a statement about the dependence of the solution to the nonlinear auxiliary BVP-4 on small variations of the "fixed" boundaries Γ_1 and Γ_2. Next, initial solutions Γ^0_1 and Γ^0_2 for both free surfaces Γ_1 and Γ_2 have to be determined. For this purpose we temporarily assume that

$F^0(\alpha)=R=0$ hold. In this case $v\equiv 0$, $p=p_c=$const. is a solution to BVP-4 (i.e. Eqs. (2.11)-(2.13)) and from Eqs.(1.3) we receive the following equations corresponding to Γ_1^0 and Γ_2^0

$$\frac{d}{dx_2}\frac{\bar{\psi}_1'(x_2)}{\left[1+(\bar{\psi}_1'(x_2))^2\right]^{1/2}}+\beta x_2 = W(p_c+\bar{p}) \quad (x_2\in\,]0,h_1[\,), \tag{3.1}$$

$$\frac{d}{dx_1}\frac{\bar{\varphi}_2'(x_1)}{\left[1+(\bar{\varphi}_2'(x_1))^2\right]^{1/2}}-\beta\bar{\varphi}_2(x_1) = -Wp_c \quad (x_1\in\Delta) \tag{3.2}$$

The corresponding boundary conditions have the form

$$\bar{\psi}_1(h_1)=0,\ \bar{\psi}_1'(0) = -A = \cot\theta_1 \tag{3.3}$$

$$\bar{\varphi}_2(0) = h_2,\ \bar{\varphi}_2(x_1) \longrightarrow h_2^* \text{ as } x_1\longrightarrow +\infty, \tag{3.4}$$

We define $\bar{c}_0 := Wp_c/\beta$, $\bar{c} := W\bar{p}/\beta$ and $\bar{c}_1 := \bar{c}_0+\bar{c}$. The BVP (3.2), (3.4) is solvable iff we set $\bar{c}_0=h_2^*$. This statement immediately follows from the second condition (3.4). In [8, 9] the following lemma was proved.

Lemma 3.1. *For $\bar{c}_0=h_2^*$ the BVP (3.2),(3.4) possesses an infinitely differentiable solution $\bar{\varphi}_2(x_1)$ having the properties*

1) $\bar{\varphi}_2$ is a strongly $\begin{cases}\text{decreasing}\\ \text{increasing}\end{cases}$ function as $\begin{cases}h_2>h_2^*\\ h_2<h_2^*\end{cases}$

2) $|\bar{\varphi}_2(x_1)-h_2^*|$ tends to zero like $e^{-\sqrt{\beta}x_1}$ as $x_1\longrightarrow +\infty$,

provided that the condition

$$|h_2-h_2^*| < \sqrt{2/\beta} \tag{3.5}$$

is fulfilled.

Let $\tilde{A}$ denote the value $\tilde{A} := A(1+A^2)^{-1/2}$

Lemma 3.2. *For any h_1 and $\bar{c}_1 := h_2^*+W(\hat{p}_0-\hat{p}_u)/\beta$ satisfying the conditions*

$$0 < h_1 < \sqrt{\frac{2(\tilde{A}+1)}{\beta}},\ h_1 \le \bar{c}_1 < \frac{h_1}{2}+\frac{\tilde{A}+1}{\beta h_1} \tag{3.6}$$

the BVP (3.1), (3.3) possesses a unique infinitely differentiable solution $\bar{\psi}_1$.

A complete discussion of BVP (3.1), (3.3) is given by the author in a separate paper [11].

Let $\bar{\psi}_1$ be the solution to BVP (3.1), (3.3) in the case $\bar{c}_0 = h_2^* = F^0(\alpha)/R$. In this case we receive $\bar{c}_1 = \bar{c}_0 + \bar{c} = F^0(\alpha)/R + W\bar{p}/\beta$. Further we denote by ψ the solution which has to be found and by $\omega(x_2) = \psi_1(x_2) - \bar{\psi}_1(x_2)$ the difference of both functions. Subtracting Eq.(3.1) (for $\bar{c}_1$ given above) from the first Eq.(1.3) and taking into account the corresponding boundary conditions we obtain the following BVP with respect to ω

$$\frac{d}{dx_2}\, \omega'(x_2)\tilde{Q}'(\bar{\psi}_1') = \tilde{g}(x_2), \tag{3.7}$$

$$\omega(h_1) = 0\,, \quad \omega'(0) = 0\,. \tag{3.8}$$

This BVP is shortly denoted by BVP-5. It contains the expressions $Q(t) := t(1+t^2)^{-1/2}$ and

$$\tilde{g}(x_2) := W(-p_0 + n\cdot S(v)n)\Big|_{x_2 = \bar{\psi}_1(x_2) + \omega(x_2)} -$$

$$- \frac{d}{dx_2}\left[(\omega'(x_2))^2 \int_0^1 (1-\lambda)\, Q''(\bar{\psi}_1' + \lambda\omega')\, d\lambda \right].$$

Here p_0 denotes that pressure from auxiliary BVP-4 which is normalized by the condition $p_0 \longrightarrow 0$ as $x_1 \longrightarrow +\infty$.

Theorem 3.1. *For an arbitrary function* $\tilde{g} \in C^{s+1}_{s-1}(J_1, Y_1)$ *the BVP-5 has a unique solution* ω *and the inequality*

$$|\,\omega\,|_{C^{3+s}_{1+s}(J_1,Y_1)} \le c_2\, |\,\tilde{g}\,|_{C^{s+1}_{s-1}(J_1,Y_1)} \tag{3.9}$$

holds.

The proof of Th. 3.1 is carried out using methods from [10,12].
The following main result of this paper can be proved with the help of Banach's fixed point theorem as in [9] or [10].

Theorem 3.2. *There exist positive real numbers* $s_0 < s_*$, $z_0 \le \min(z_*, \sqrt{\beta})$ *and* M_0 *such that for arbitrary* $s \in]0, s_0[$, $z \in]0, z_0[$, $M_R^F < M_0$ *and for positive* $h_1, h_2, F^0(\alpha), R$ *satisfying the conditions*

$$\left| h_2 - \frac{F^0(\alpha)}{R} \right| < \sqrt{\frac{2}{\beta}}\,, \quad 0 < h_1 < \sqrt{\frac{2(\tilde{A}+1)}{\beta}}\,,$$

$$h_1 \le \frac{F^0(\alpha)}{R} + \frac{(\hat{p}_0-\hat{p}_u)}{\hat{g}} < \frac{h_1}{2} + \frac{\tilde{A}+1}{\beta h_1}\,, \tag{3.10}$$

the BVP-1 (i.e. Eqs. (1.1)-(1.6)) is uniquely solvable. The solution $\{v, p, \psi_1, \varphi_2\}$ *can be represented by*

$v = g(-x_1)u^{(-)} + g(x_1)(R,0)^T + w, p(x) = g(-x_1)p^{(-)} + p_0(x) + \beta F^{\circ}(\alpha)/(WR)$,

$\varphi_2(x_1) = \overline{\varphi}_2(x_1) + \omega_2(x_1)\ ,\ \psi_1(x_2) = \overline{\psi}_1(x_2) + \omega_1(x_2)\ ,$

where $g, \{u^{(-)}, p^{(-)}\}$ *are given in Th.2.5. The functions* $\overline{\varphi}_2$ [$\overline{\psi}_1$, *respectively*] *are the solutions to BVP (3.2),(3.4)* [*BVP (3.1),(3.3), respectively*] *for* $\overline{c}_0 = h_2^* = = F^0(\alpha)/R$. *Furthermore, the relations*

$w \in C^{s+2}_{s,z}(G),\ p_0 \in C^{s+1}_{s-1,z}(G^0 \cup G^+),\ \nabla p_0 \in C^{s}_{s-2,z}(G),\ \omega_2 \in C^{3+s}_{1+s,z}(\Delta),$

and $\omega_1 \in C^{3+s}_{1+s}(J_1, Y_1)$ *hold.*

References

1. Dussan V.,E.B., Davis,S.H.: *On the motion of a fluid-fluid interface along a solid surface.* J. Fluid Mech. 65, 71-95(1974)
2. Kondrat'ev, V.A.: *Boundary value problems for elliptic equations in domains with conical or angular points.* Trudy Moskov. Mat. Obshch. 16, 209-292(1967)
3. Kröner, D.: *Asymptotische Entwicklungen für Strömungen von Flüssigkeiten mit freiem Rand und dynamischem Kontaktwinkel.* SFB-72 Preprint 809, Bonn, 1-105(1986)
4. Kröner, D.: *Asymptotic expansions for a flow with a dynamic contact angle.* In: The Navier-Stokes equations. Theory and numerical methods/ed. by J. Heywood, K. Masuda, R. Rautmann and V.A. Solonnikov. Lecture Notes Math. 1431, Springer-V., 1990, 49-59
5. Lowndes, J.: *The numerical simulation of the steady movement of a fluid meniscus in a capillary tube.* J. Fluid Mech. 101, 631-646 (1980)
6. Pileckas, K.: *Solvability of a problem of plane motion of a viscous incompressible fluid with noncompact free boundary.* Diff. Equ. Appl. Inst. of Math. Cybern. Acad. Sci. Lit. SSR 30, 57-96 1981 (in Russian)
7. Silliman, W.J., Scriven, L.E.: *Separating flow near a static contact line: Slip at a wall and shape of a free surface.* J. Comput.

Phys. 34, 287-313(1980)

8. Socolowsky, J.: *Mathematische Untersuchungen freier Randwertaufgaben der Hydrodynamik viskoser Flüssigkeiten.* Habilitation thesis, Merseburg 1989
9. Socolowsky, J.: *Solvability of a stationary problem on the plane motion of two viscous incompressible liquids with non-compact free boundaries.* Z. Angew. Math. Mech. (ZAMM) 71 (1991), 9 (in press), Zap. Nauchn. Sem. LOMI 163, 146-153(1987)(in Russian)
10. Socolowsky, J.: *The solvability of a free boundary problem for the Navier-Stokes equations with a dynamic contact line.* (1991, to be submitted)
11. Socolowsky, J.: *Elementary investigations of the equation for a free surface.* Techn. Mechanik 12 (1991), 4 (in press)
12. Solonnikov, V.A.: *Solvability of the problem of plane motion of a heavy viscous incompressible capillary liquid partially filling a container.* Izv. Akad. Nauk SSSR 43, 203-236(1979) (in Russian), [English transl.: Math. USSR-Izv. 14, 193-221(1980)]
13. Solonnikov, V.A.: *On the Stokes equation in domains with nonsmooth boundaries and on a viscous incompressible flow with a free surface.* Nonlinear partial diff. equations and their applications, College de France Seminar 3, 340-423, 1980/1981
14. Verfürth, R.: *Finite element approximation of incompressible Navier-Stokes equations with slip boundary condition.* Numer. Math. 50, 697-721(1987)

EVOLUTION FREE BOUNDARY PROBLEM FOR EQUATIONS OF MOTION OF VISCOUS COMPRESSIBLE BAROTROPIC LIQUID

V.A. SOLONNIKOV
St. Petersburg Branch of V.A. Steklov Mathematical Institute of the Russian Academy of Sciences,
Fontanka 27, 191011 St. Petersburg, Russia
A. TANI
Department of Mathematics, Faculty of science and technology,
Keio University, 3-14-1, Hiyoshi, Kohokuku, Yokohama 223, Japan

1. Introduction.

This paper is concerned with evolution free boundary problem for the Navier-Stokes equations governing the motion of an isolated mass of a viscous compressible barotropic liquid bounded by a free surface. It is required to find a bounded domain $\Omega_t \subset \mathbf{R}^3, t > 0$,the velocity vector field $v(x,t) = (v_1, v_2, v_3)$, and a scalar positive density $\rho(x,t)$ satisfying the following equations, initial and boundary conditions:

$$\rho(v_t + (v\cdot\nabla)v) - \nabla T(v,\rho) = 0, \quad \rho_t + \nabla\cdot\rho v = 0 \quad (x\epsilon\Omega_t, t>0),$$

$$v|_{t=0} = v_0(x), \quad \rho|_{t=0} = \rho_0(x) \quad (x\epsilon\Omega_0)$$

$$Tn - \sigma H n = 0 \quad (x\epsilon\Gamma_t \equiv \partial\Omega_t, t>0). \tag{1.1}$$

Here $\nabla = (\frac{\partial}{\partial x_1}, \frac{\partial}{\partial x_2}, \frac{\partial}{\partial x_3}), T = (-p(\rho) + \mu'\nabla\cdot v)I + \mu S(v)$ is the stress tensor, $S(v)$ is the strain tensor with the elements $S_{ij} = \frac{\partial v_i}{\partial x_j} + \frac{\partial v_j}{\partial x_i}$, I is the unit 3×3 - matrix, μ, μ' and σ are constant coeffitiens satisfying the conditions $\sigma > 0, \mu > 0, \mu' + \frac{2}{3}\mu \geq 0$, n is the exterior unit normal to Γ_t, and $H(x,t)$ is the twice mean curvature of Γ_t which is negative if Ω_t is convex in a neighbourhood of x; finally, the pressure $p(\rho)$ is a smooth increasing function of $\rho > 0$. The domain Ω_0 is given. For $t > 0, \Omega_t$ may be defined as the set of points $x = x(t,\xi)$ where $x(\tau,\xi)$ is a solution of the Cauchy problem

$$\frac{\partial x}{\partial \tau} = v(x(\tau,\xi)\tau), \quad (0 \leq \tau \leq t), \quad x(0,\xi) = \xi, \forall\xi\epsilon\Omega_0 \tag{1.2}$$

This condition (it is called a kinematic boundary condition) may be written in a different way. For instance, if Γ_t is determined by the equation

$$|x - x_0| = R(\omega, t), \quad \omega = \frac{x - x_0}{|x - x_0|} \tag{1.3}$$

on the unit sphere $S_1 : |\omega| = 1$, then this condition reduces to

$$R_t(\omega,t) = v_r - \frac{1}{R} v_\omega \cdot \nabla_\omega R, \quad \omega\epsilon S_1, \tag{1.4}$$

where $\nabla_\omega R$ is the surface gradient of R on S_1, and $v_r = v\cdot\omega, \quad v_\omega = v - \omega v_r$ are radial and angular components of v, respectively.

For a viscous incompressible liquid, the free boundary problem governing the unsteady motion of an isolated liquid mass was studied in a series of papers of V.A. Solonnikov

[12-16]. For a viscous compressible liquid (barotropic or general) this problem was investigated by A. Valli, P. Secchi and the present authors [6-8, 17,18,20]. In [8] and [20] it is proved in the case $\sigma = 0$ that this free boundary problem has a unique solution in a certain finite time interval (the solution was found in some anisotropic Hölder or Sobolev spaces, respectively). For $\sigma > 0$ the same kind of result is obtained in [6,7,17,18] (papers [17,18] are concerned with the barotropic liquid, i.e. exactly with problem (1.1), (1.2)). In the present paper we prove that the solution of this problem is defined for all $t \geq 0$, provided that initial data are close to the equilibrium rest state, i.e. v_0 is small, ρ_0 is close to a constant and Ω_0 is close to a ball. We also investigate the behavior of the solution as $t \to \infty$ and find a limiting regime. For the incompressible liquid this result was obtained in [13,15]. In the case of compressible liquid, one of the main estimates obtained in [13,15] (for instance, inequality (3.6) in [13]) fails, since localization in time is impossible. The estimate of solution of problem (1.1), (1.3), (1.4) given here (see (5.8)) is similar to the estimate of A. Matsumura and T. Nishida for the problem with adherence boundary condition $v|_\Gamma = 0$ [5] but it is obtained in other functional spaces and by a different technics. According to a private communication of Professor W. Zajaczkowski, he also proved the solvability of problem (1.1),(1.3),(1.4) in an infinite time interval for initial data close to the equilibrium rest state.
Let us say a few words about local solvability of problem (1.1),(1.2). The theorem established in [18] reads as follows.
THEOREM 1.1
Let $\Gamma_0 \epsilon W_2^{\ell+5/2}$ with $\ell \epsilon (1/2, 1)$. For arbitrary $v_0 \epsilon W_2^{\ell+1}(\Omega_0)$ and $\rho_0 \epsilon W_2^{\ell+1}(\Omega_0), \rho_0 \geq a > 0$, satisfying the compatibility conditions

$$T(v_0, \rho_0)n - \sigma H(\xi, 0) n|_{\xi \epsilon \Gamma_o} = 0$$

problem (1.1),(1.2) has a solution in a finite time interval $(0, T_1)$ whose magnitude T_1 depends on the norms of $v_o, \rho_0, H(\xi, 0)$. As functions of the Lagrangean coordinates (ξ, t), v and ρ have the following differentiability properties:

$$v \epsilon W_2^{\ell+2, \ell/2+1}(Q_{T_1}), \quad \nabla \rho \epsilon W_2^{\ell, \ell/2}(Q_{T_1}), \quad (Q_{T_1} = \Omega_0 \times (0, T_1)),$$

$$\rho \epsilon W_2^{\ell+1}(\Omega_o), \quad \rho_t \epsilon W_2^{\ell}(\Omega_0), \quad \forall t \epsilon (0, T_1).$$

In this class the solution is unique.

Here by $W_2^\ell(\Omega), (\ell > 0, \Omega \subset \mathbf{R}^n)$ we mean spaces of S.L. Sobolev - L.N. Slobodetskii (see [10,9]) denoted very often by $H_\ell(\Omega)$. They consist of functions which are square integrable in Ω together with their generalized (weak) derivatives up to order $[\ell]$, and the norm in $W_2^\ell(\Omega)$ is defined by

$$||u||^2_{W_2^\ell(\Omega)} = \sum_{|j| \leq \ell} \int_\Omega |D^j u(x)|^2 dx$$

if ℓ is an integer (i.e. $\ell = [\ell]$), and by

$$||u||^2_{W_2^\ell(\Omega)} = ||u||^2_{W_2^{[\ell]}(\Omega)} + \sum_{|j|=[\ell]} \int_\Omega \int_\Omega \frac{|D^j u(x) - D^j u(y)|^2}{|x-y|^{n+2(\ell-[\ell])}} dx dy$$

in the opposite case. By $W_2^{\ell,\ell/2}(Q_T)$, $Q_T = \Omega \times (0,T)$, we mean anisotropic spaces of functions depending on $(x,t)\epsilon Q_T$ which may be identified with $L_2(0,T; W_2^\ell(\Omega)) \cap L_2(\Omega; W_2^{\ell/2}(0,T))$.. One of the equivalent norms in $W_2^{\ell,\ell/2}(Q_T)$ may be defined as

$$||u||^2_{W_2^{\ell,\ell/2}(Q_T)} = \int_0^T ||u||^2_{W_2^\ell(\Omega)} dt + \int_\Omega ||u||^2_{W_2^{\ell/2}(0,T)} dx$$

The main idea of the proof of a local existence theorem is the linearization of problem (1.1),(1.2) written in the Lagrangean coordinates. In virtue of (1.2), the passage to the Lagrangean coordinates makes it possible to write the free boundary problem (1.1),(1.2) as a nonlinear initial-boundary value problem for v, ρ in a given domain $\Omega_0 \equiv \Omega$. Let $\hat{v}(\xi,t) = v(x(\xi,t),t)$, $\hat{\rho}(\xi,t) = \rho(x(\xi,t),t)$. The connection between Eulerian and Lagrangean coordinates is given by

$$x(\xi,t) = \xi + \int_0^t \hat{v}(\xi,\tau)d\tau, \tag{1.5}$$

and problem (1.1), (1.2) reduces to

$$\begin{aligned} &\hat{\rho}\hat{v}_t - \hat{\nabla}\hat{T} = 0, \ \ \hat{\rho}_t + \hat{\rho}\hat{\nabla}\cdot\hat{v} = 0 \ \ (\xi\epsilon\Omega, t > 0) \\ &\hat{v}|_{t=0} = v_0(\xi), \ \ \hat{\rho}|_{t=0} = \rho_0(\xi) \ \ (\xi\epsilon\Omega) \\ &\hat{T}n - \sigma\Delta(t)x(\xi,t) = 0 \ \ (\xi\epsilon\Gamma_0 \equiv \Gamma). \end{aligned} \tag{1.6}$$

Here

$$\hat{\nabla} = (\sum_{i=1}^3 \frac{\partial\xi_i}{\partial x_k}\frac{\partial}{\partial\xi_i})_{k=1,2,3} = \frac{A}{J}\nabla, \ \ J = det(\frac{\partial x_i}{\partial\xi_k}) = exp(\int_0^t \hat{\nabla}\cdot\hat{v}d\tau) = 1 + \int_0^t A\nabla\cdot\hat{v}d\tau,$$

A is a matrix whose elements A_{ij} are cofactors to

$$\frac{\partial x_i}{\partial\xi_j} = \delta_{ij} + \int_0^t \frac{\partial\hat{v}_i}{\partial\xi_j}d\tau, \ \ \hat{T} = (-p(\hat{\rho}) + \mu'\hat{\nabla}\cdot\hat{v}) + \mu\hat{S}(\hat{v}),$$

$$\hat{S}_{ij} = \frac{1}{J}\sum_{k=1}^3 (A_{ik}\frac{\partial\hat{v}_j}{\partial\xi_k} + A_{jk}\frac{\partial\hat{v}_y}{\partial\xi_k}),$$

and $\Delta(t)$ is the Laplace-Beltrami operator on the surface Γ_t that can be considered as the image of Γ under the transformation (1.5). The equation for $\hat{\rho}$ in (1.6) is easily integrated:

$$\hat{\rho}(\xi,t) = \rho_o(\xi)exp(-\int_0^t \hat{\nabla}\cdot\hat{v}d\tau) = \rho_0(\xi)J^{-1}(\xi,\tau)$$

and (1.6) reduces to initial-boundary value problem for $\hat{v}$, namely,

$$\begin{aligned} &\hat{v}_t - \rho_0^{-1}(\xi)A\hat{\nabla}\hat{T}'(\hat{v}) + \rho_0^{-1}A\hat{\nabla}p(\rho_0 J^{-1}) = 0, \ \ \hat{v}|_{t=0} = v_0(\xi), \\ &-p(\rho_0 J^{-1})n + \hat{T}'(\hat{v})n - \sigma\Delta(t)x(\xi,t)|_{\xi\epsilon\Gamma} = 0 \end{aligned} \tag{1.7}$$

where

$$\hat{T}' = \hat{T} - p(\hat{\rho})I = \mu'\hat{\nabla} \cdot \hat{v}I + \mu\hat{S}(\hat{v}).$$

Along with (1.7), the following linear problem is considered

$$u_t - \rho_0^{-1}(\xi)\nabla T'(u) = f(\xi, t) \quad (\xi \epsilon \Omega, t \epsilon (0,T)) \quad u|_{t=0} = u_0(\xi)$$

$$T'(u)n_0 - \sigma n_0(n_0 \cdot \Delta \int_0^t u d\tau)|_{\xi\epsilon\Gamma} = b(\xi, t) \tag{1.8}$$

Here $T'(u) = \mu'\nabla \cdot u + \mu S(u)$, n_0 is a unit exterior normal to Γ, Δ is the Laplace-Beltrami operator on Γ, f, u_0 and b are given vector-valued functions.
The following theorem is proved in [18].

Theorem 1.2
Let

$$\Gamma \subset W_2^{\ell+3/2}, \ell\epsilon(1/2, 1), \rho_o \epsilon W_2^{\ell+1}(\Omega), \rho_0 \geq a > 0, u_0 \epsilon W_2^{\ell+1}(\Omega), f \epsilon W_2^{\ell,\ell/2}(Q_T),$$

$$b_\tau = b - n_o(b \cdot n_o) \epsilon W_2^{\ell+1, \frac{\ell+1}{2}}(G_T), \quad G_T = \Gamma \times (0,T), \quad T < \infty$$

and suppose that $b \cdot n_0$ can be written as a sum

$$b \cdot n_0 = b' + \int_o^t B d\tau \text{ with } b' \epsilon W_2^{\ell+1, \frac{\ell+1}{2}}(G_T), \quad B \epsilon W_2^{\ell-1/2, \ell/2-1/4}(G_T).$$

Assume also that the compatibility condition holds

$$T'(u_o)n_o|_{\xi\epsilon\Gamma} = b|_{t=0}.$$

Then problem (1.8) has a unique solution $u \epsilon W_2^{\ell+2, \ell/2+1}(Q_T)$, and this solution satisfies the inequality

$$||u||^2_{W_2^{\ell+2,\ell/2+1}(Q_T)} \leq c_1(T)(||u_o||^2_{W_2^{\ell+1}(\Omega)} + ||f||^2_{W_2^{\ell,\ell/2}(Q_T)} +$$

$$||b_\tau||^2_{W_2^{\ell+1, \frac{\ell+1}{2}}(G_T)} + ||b'||^2_{W_2^{\ell+1, \frac{\ell+1}{2}}(G_T)} + ||B||^2_{W_2^{\ell-1/2, l/2-1/4}(G_T)}. \tag{1.9}$$

On the basis of this theorem, one can esablish the solvability of problem (1.7) in a certain finite time interval (0,T_1) and prove theorem 1.1. Moreover, just as in the case of incompressible liquid (see [13], theorem 6) one can show that the solution of problem (1.6) (or (1.7)) possesses some additional regularity with respect to t, when $t > 0$. In particular, we shall need the following result.

Theorem 1.3
Under the hypotheses of theorem 1.1, the norm $||\hat{v}||_{W_2^{\ell+2}(\Omega)}$ is uniformly bounded in the interval $t\epsilon(t_0, T_1)$ with arbitrary positive t_0, and

$$||\hat{v}(\cdot, t)||^2_{W_2^{\ell+2}(\Omega)} \leq c_2(t_o)||\hat{v}||^2_{W_2^{\ell+2,\ell/2+1}(Q_T)}.$$

We refer the reader to the paper [18] for the details and focus our attention on the following theorem on the solvatibility of problem (1.1),(1.3),(1.4) in an infinite time interval $t > 0$.

Theorem 1.4
Suppose that the hypotheses of theorem 1.1 are satisfied and that Γ_0 is defined by equation (1.3) with $R(\cdot,0)\epsilon W_2^{\ell+5/2}(S_1)$. Assume also that

$$||v_o||^2_{W_2^{\ell+1}(\Omega_0)} + ||\rho_o - \bar{\rho}||^2_{w_2^{\ell+1}(\Omega_0)} + ||R(\cdot,0) - \bar{R}||^2_{W_2^{\ell+5/2}(S_1)} \leq \varepsilon < 1 \qquad (1.10)$$

where $\bar{\rho}$ and $\bar{R}$ are two positive constants satisfying the relation

$$p(\bar{\rho}) = \frac{2\sigma}{\bar{R}} \qquad (1.11)$$

Then, T_1 is a montone increasing function of $1/\varepsilon$ which tends to infinity as $\varepsilon \rightarrow 0$. Moreover,if ε is sufficiently small, then the solution may be extended into an infinite time interval $t > 0$. The free surface Γ_t is determined by equation (1.3) with

$$x_o = Vt, \;\; V = (\int_{\Omega_o} \rho_o dx)^{-1} \int_{\Omega_o} \rho_o v_o dx, \;\; R(\cdot,t)\epsilon W_2^{\ell+5/2}(S_1), \;\; \forall t > 0.$$

As $t \rightarrow \infty$, the solution tends to a quasi-stationary solution of (1.1) corresponding to the rotation of a liquid as a rigid body about an axis which is parallel to the vector $\int_{\Omega_o} \rho_o[v_o \times x]dx$ and which is moving uniformly with a constant speed V.

2. Conservation laws and their consequences.
The proof of theorem 1.4 is based on a-priori estimates for solutions of a problem (1.1), (1.2) in an arbitrary time interval $t\epsilon(0,T)$. Conservation laws yield a first preliminary estimate of this type.

Theorem 2.1
For a solution (v,ρ) of problem (1.1),(1.2) the following relations hold:

$$\frac{d}{dt}\int_{\Omega_t} \rho dx = 0 \;\; \text{(conservation of mass)} \qquad (2.1)$$

$$\frac{d}{dt}\int_{\Omega_t} \rho v \cdot \eta dx = 0 \;\; \text{(conservation of momentum)} \qquad (2.2)$$

$$\frac{d}{dt}(\frac{1}{2}\int_{\Omega_t} \rho|v|^2 dx + \sigma|\Gamma_t| + \int_{\Omega_t} q(\rho)dx) + \mathrm{E}(v) = 0 \;\; \text{(conservation of energy)} \qquad (2.3)$$

Here η is an arbitrary vector field of the form $\eta = a + b \times x$ with constant a and b, $q(\rho) = \rho \int_{\rho_1}^{\rho} \frac{p(s)}{s^2} ds, \rho_1 \geq 0, |\Gamma_t|$ is the surface area of Γ_t and

$$\mathrm{E}(v) = \int_{\Omega'} (\mu'|\nabla \cdot v|^2 + \frac{\mu}{2}|S(v)|^2)dx$$

Proof.
(2.1) follows from the well-known formula

$$\frac{d}{dt}\int_{\Omega_t} f(x,t)dx = \int_{\Omega_t} (f_t + \nabla \cdot vf)dx \qquad (2.4)$$

and from the continuity equation $\rho_t + \nabla \cdot \rho v = 0$. (2.2) is established by the same arguments (see also theorem 5 in [13] for the incompressible liquid). Finally,

$$\frac{d}{dt}\int\limits_{\Omega_t} \frac{1}{2}\rho|v|^2 dx = \int\limits_{\Omega_t} \rho(v_t + (v\cdot\nabla)v)dx = -\mathrm{E}(v) + \sigma\int\limits_{\Gamma_t} Hv\cdot ndS + \int\limits_{\Omega_t} p\cdot divvdx$$

As shown in [13], $\int\limits_{\Gamma_t} Hv \cdot ndS = \frac{d|\Gamma_t|}{dt}$. Since $q'(\rho) = \frac{p(\rho)+q(\rho)}{\rho}$, we can easily conclude from (2.4) and from the continuity equation that

$$\frac{d}{dt}\int\limits_{\Omega_t} q(\rho)dx = \int\limits_{\Omega_t} q'(\rho)(\rho_t + v\cdot\nabla\rho)dx + \int qdivvdx = -\int\limits_{\Omega} pdivvdx$$

and (2.3) is proved.
Let us write (2.1) and (2.2) in an equivalent form

$$\int\limits_{\Omega_t} \rho(x,t)dx = \int\limits_{\Omega_0} \rho_o(x)dx \equiv m,$$

$$\int\limits_{\Omega_t} \rho v dx = \int\limits_{\Omega_0} \rho_o(x)v_o(x)dx \equiv mV$$

$$\int\limits_{\Omega_t} \rho[v\times x]dx = \int\limits_{\Omega_o} \rho_o[v_o\times x]dx \equiv M$$

where m is the mass of the liquid, $V = (V_1, V_2, V_3)$ is the velocity of the barycenter of Ω_t and M is the total angular momentum. Without restriction of generality we may suppose that $V = 0$ and $M = (0, 0, \gamma)$, since we can arrive at this case after the transforamtion

$$x' = x - Vt, v' = v - V$$

and the rotation of coordinate axes. When we now place the origine at the barycenter, we obtain

$$\int\limits_{\Omega_t} \rho x_i dx = 0, \quad i = 1,2,3 \tag{2.5}$$

Theorem 2.2
Suppose that

$$p'(\bar\rho) - \frac{p(\bar\rho)}{3\bar\rho} \equiv b > 0 \tag{2.6}$$

$$|R(w,\tau) - \bar R| + |\nabla_\omega R(\omega,\tau)| \le \delta\bar R, \quad 0 \le \tau \le t, \tag{2.7}$$

$$|\rho(x,\tau) - \bar\rho| \le \delta_1\bar\rho, \quad 0 \le \tau \le t \tag{2.8}$$

where δ and δ_1 are small positive numbers and $\nabla_\omega R$ is the gradient of R on the unit sphere. Then there exist such positive constants $c_1 - c_5$ independent of t that

$$\frac{1}{2}\int\limits_{Q_t} \rho|\vec v|^2 dx + c_1\sigma\int\limits_{S_1} (|R-\bar R|^2 + |\nabla_\omega R|^2)d\omega + c_2\int\limits_{\Omega_t} (\rho-\bar\rho)^2 dx +$$

$$\int_0^t \mathrm{E}(\vec{v})d\tau \le c_3 \int_{\Omega_t} \rho_0|\vec{v}_0|^2dx + c_4\sigma \int_{S_1} (|R-\bar{R}|^2 + |\nabla_\omega R|^2)d\omega + c_5 \int_{\Omega_0} (\rho_0-\bar{\rho})^2dx \qquad (2.9)$$

where $|\bar{\Omega}| = \frac{4}{3}\pi\bar{R}^3$.

Proof.

After elementary transformations we can write (2.3) in the form

$$\frac{d}{dt}\Big(\frac{1}{2}\int_{\Omega_t} \rho|v|^2dx + \sigma(|\Gamma_t| - 4\pi R_t^2) + 4\pi R_t^2\sigma - p(\bar{\rho})|\Omega_t| + \int_{\Omega_t} \hat{q}(\rho)dx\Big) + \mathrm{E}(v) = 0 \qquad (2.10)$$

where $\hat{q}(\rho) = q(\rho) - q(\bar{\rho}) - q'(\bar{\rho})(\rho-\bar{\rho})$ and $R_t = (\frac{3|\Omega_t|}{4\pi})^{1/3}$ is the radius of a ball whose volume equals $|\Omega_t|$. In virtue of (1.11), we have

$$4\pi R_t^2\sigma - p(\bar{\rho})|\Omega_t| = \frac{\sigma|\bar{\Omega}|}{\bar{R}}[3(\frac{|\Omega_t|}{\bar{\Omega}})^{2/3} - 2\frac{|\Omega_t|}{|\bar{\Omega}|} - 1] + \frac{\sigma|\bar{\Omega}|}{\bar{R}} \qquad (2.11)$$

Since the function $\varphi(x) = 3x^{2/3} - 2x - 1$ satisfies the conditions $\varphi(0) = 0, \varphi'(0) = 0,\ \varphi''(0) = -\frac{2}{3} < 0$, the condition (2.7) implies

$$(-\frac{1}{3} - c_6\sigma)(\frac{|\Omega_t|}{|\bar{\Omega}|} - 1)^2 \le \varphi(\frac{|\Omega_t|}{|\bar{\Omega}|}) \le (-\frac{1}{3} + c_6\delta)(\frac{|\Omega_t|}{|\bar{\Omega}|} - 1)^2 \qquad (2.12)$$

with a certain $c_6 > 0$. Moreover, it follows from $q''(\bar{\rho}) = \frac{p'(\bar{\rho})}{\rho} > 0$ and from (2.8) that

$$(\frac{1}{2}\frac{p'(\bar{\rho})}{\bar{\rho}} - c_7\delta_1)(\rho-\bar{\rho})^2 \le \hat{q}(\rho) \le (\frac{1}{2}\frac{p'(\bar{\rho})}{\bar{\rho}} + c_7\delta_1)(\rho-\bar{\rho})^2$$

Let us estimate the sum

$$\frac{1}{2}\frac{p'(\bar{\rho})}{\bar{\rho}}\int_{\Omega_t}(\rho-\bar{\rho})^2dx - \frac{\sigma|\bar{\Omega}|}{3\bar{R}}(\frac{|\Omega_t|}{|\bar{\Omega}|} - 1)^2 \equiv S(t)$$

Clearly,

$$\int_{\Omega_t}(\rho-\bar{\rho})dx = m - \bar{\rho}|\Omega_t| = \bar{\rho}(|\bar{\Omega}| - |\Omega_t|) + m - \bar{m},$$

$$|\Omega_t| - |\Omega| = \frac{1}{3}\int_{S_1}(R^3(\omega,t) - \bar{R}^3)d\omega \qquad (2.13)$$

where $\bar{m} = |\bar{\Omega}|\bar{\rho}$ is the mass of the liquid in the equilibrium rest state. It follows that

$$(|\bar{\Omega} - |\Omega_t|)^2 \le \frac{1+\varepsilon}{\bar{\rho}^2}|\Omega_t|\int_{\Omega_t}(\rho-\bar{\rho})^2dx + \frac{1+\varepsilon^{-1}}{\bar{\rho}^2}(m-\bar{m})^2, \ \forall\varepsilon > 0,$$

and, as a consequence,

$$S(t) \ge \frac{1}{2}(\frac{p'(\bar{\rho})}{\bar{\rho}} - \frac{(1+\varepsilon)p(\bar{\rho})}{3\bar{\rho}^2}\frac{|\Omega_t|}{|\bar{\Omega}|})\int_{\Omega_t}(\rho-\bar{\rho})^2dx - \frac{(1+\varepsilon^{-1})p(\bar{\rho})}{6\bar{\rho}^2|\bar{\Omega}|}(m-\bar{m})^2 \qquad (2.14)$$

Now, when we integrate (2.10) with respect to t and take (2.11) - (2.14) into account, we obtain

$$\frac{1}{2}\int\limits_{\Omega_t}\rho|v|^2dx+\sigma(|\Gamma_t|-4\pi R_t^2)+(\frac{1}{2}b-c_8\varepsilon-c_9\delta_1)\int\limits_{\Omega_t}(\rho-\bar{\rho})^2dx+$$

$$+\int_0^t \mathrm{E}(v)dx\le\int\limits_{\Omega_0}\rho_0|v_0|^2dx+\sigma(\Gamma_0|-4\pi R_0^2)+$$

$$c_{10}\int\limits_{\Omega_0}(\rho-\rho_0)^2dx+c_{11}(\varepsilon)(m-\bar{m})^2.$$

We shall assume that δ_1 is so small that $\frac{1}{2}b-c_9\delta_1>0$ and we take ε so small that $\frac{1}{2}b-c_9\delta_1-c_8\varepsilon\ge c_{12}>0$.
Consider the expression $|\Gamma_t|-4\pi R_t^2$. Clearly, $|\Gamma_t|-4\pi R_t^2\ge 0$; moreover, in [12] (see theorem 3) this expression is estimated from below and from above by const $\int\limits_{S_1}(|\tilde{R}-\bar{R}|^2+|\nabla_w\tilde{R}|^2)d\omega$ where $\tilde{R}$ is a distance between $x\epsilon\Gamma_t$ and center of masses of Ω_t in the case $\rho=const$. In virtue of (2.5), this point has coordinates $a_i(t)=\frac{1}{\bar{\rho}|\Omega_t|}\int\limits_{\Omega_t}(\bar{\rho}-\rho)x_idx$.
Since $\tilde{R}^2=R^2+|a|^2-2a\cdot\omega R$, it is not hard to verify that

$$\int\limits_{S_1}(|\tilde{R}-R_t|^2+|\nabla_\omega\tilde{R}|^2)d\omega\le c_{13}\int\limits_{S_1}(|R-R_t|^2+|\nabla_w R|^2)d\omega+c_{14}|a(t)|^2,$$

$$\int\limits_{S_1}(|R-R_t|^2+|\nabla_\omega R|^2)d\omega\le c_{13}\int\limits_{S_1}(|\tilde{R}-R_t|^2+|\nabla_w\tilde{R}|^2)d\omega+c_{14}|a(t)|^2 \tag{2.15}$$

Taking also (2.13) into account we obtain

$$\frac{1}{2}\int\limits_{\Omega_t}\rho|v|^2dx+c_{12}\int\limits_{\Omega_t}(\rho-\bar{\rho})^2dx+\int_0^t\mathrm{E}(v)d\tau\le$$

$$\frac{1}{2}\int\limits_{\Omega_0}\rho_0|v_0|^2dx+c_{15}\int\limits_{\Omega_0}(\rho_0-\bar{\rho})^2dx+c_{16}\int\limits_{\Omega}(|R(\omega,0)-\bar{R}|^2+|\nabla_\omega R(\omega,0)|^2)d\omega \tag{2.16}$$

In virtue of (2.15), (2.13) the norm $\int\limits_{S_1}(R-\bar{R}|^2+|\nabla_\omega R|^2)d\omega$ also can be estimated by the right-hand side of (2.16), and we arrive at (2.9). The theorem is proved.

3. Equilibrium figures

This section is concerned with the solution of equations (1.1_1) that is independent of time and corresponds to the rotation of a finite mass of a compressible liquid as a rigid body about x_3-axis with a given angular momentum M. We suppose that the barycenter coincides with the origine of the coordinate system $\{x\}$. Let v_∞ be the velocity vector field corresponding to such a rotation. It has a form

$$v_\infty(x)=\alpha(x_2,-x_1,0)=\alpha[e_3\times x],\ \ e_{3i}=\delta_{3i}, \tag{3.1}$$

and the density ρ_∞ satisfies the equations

$$v_\infty\cdot\nabla\rho_\infty=0,\ \ \nabla p(\rho_\infty)=\rho_\infty\frac{\alpha^2}{2}\nabla|x'|^2\ \ (x'=(x_1,x_2,0) \tag{3.2}$$

The last equation is equivalent to

$$P(\rho_\infty) = \frac{\alpha^2}{2}|x'|^2 + N$$

with N = const and $P(\rho) = \int\limits_{\rho_1}^{\rho} \frac{p'(s)}{s} ds, \rho_1 \geq 0$. As $P'(\rho) = \frac{p'(\rho)}{\rho} > 0$, the inverse function $P^{-1}(\rho)$ is well defined and

$$\rho_\infty(x) = P^{-1}(\frac{\alpha^2}{2}|x'| + N). \tag{3.3}$$

Let Ω_∞ be domain occupied by a rotating liquid and let its boundary Γ_∞ be given by the equation $|x| = R_\infty(\frac{x}{|x|})$. The boundary condition (1.1_3) at Γ_∞ reduces to

$$\sigma H_\infty + p(P^{-1}(\frac{\alpha^2}{2}|x'|^2 + N)) = 0 \tag{3.4}$$

where H_∞ is the twice mean curvature of Γ_∞. The equation for angular momentum

$$\int\limits_{\Omega_\infty} \rho_\infty[v_\infty \times x]dx = M$$

implies

$$\alpha \int\limits_{\Omega_\infty} |x'|^2 \rho_\infty(x)dx = \gamma \tag{3.5}$$

Finally, we prescribe the total mass of the liquid, i.e.

$$\int\limits_{\Omega_\infty} \rho_\infty(x)dx = m \tag{3.6}$$

(we shall suppose later that $m = \int\limits_{\Omega_0} \rho_0 dx$). Given γ and m, we should find α, N, R_∞ satisfying (3.4) - (3.6).

Let $C^s(S_1)$ be a Hölder space of function defined on a unit sphere S_1 with a standard norm

$$|u|_{S_1}^{(s)} = \max_{\sigma_k c S_1}(\sum_{|j|<s} \sup_{\omega \epsilon \sigma_k} |D^j u(\omega)| +$$

$$+ \sum_{|j|=[s]} \sup_{\omega,\omega'} |\omega - \omega'|^{-s+[s]} |D^j u(\omega) - D^j u(\omega')|)$$

where s is a positive non-integer and $D^j u$ is a $|j| - th$ derivative of $u(\omega)$ computed in local coordinates on a submanifold

$$\sigma_k \subset S_1, \quad (\bigcup_{k=1}^{N} \sigma_k = S_1).$$

By $\tilde{C}^s(S_1)$ we shall mean the subspace of all rotationally symmetric functions from $C^s(S_1)$ (i.e. functions depending on $|x'|$ and x_3)which are even in x_3.

We can prove the following theorem.

Theorem 3.1.

Let $p\epsilon C^{1+\beta}(I_\rho)$ where $\beta\epsilon(0,1)$ and $I_{\bar\rho} = \{|\rho - \bar\rho| \le \delta_1, \bar\rho\}$, and let $b = p'(\hat\rho) - \frac{1}{3}\frac{p(\hat\rho)}{\hat\rho} = 0$.
For arbitrary γ and m, such that

$$|\gamma| + |m - \bar m| \le \varepsilon'$$

with a small $\varepsilon' > 0$, there exists a solution $(R_\infty, N, \alpha)\epsilon\tilde C^{2+\beta}(S_1) \times \mathrm{I\!R} \times \mathrm{I\!R}$ of (3.4) - (3.6) satisfying the inequality

$$|R_\infty - \bar R|_{S_1}^{(2+\beta)} + |p^{-1}(N) - \bar\rho| + |\alpha| \le c_1(|\gamma| + |m - \bar m|)$$

The solution is unique in the ball

$$|R_\infty - \bar R|_{S_1}^{(2+\beta)} \ + |p^{-1}(N) - \bar\rho| + |\alpha| \le c_1\varepsilon$$

of the space $\tilde C^{2+\beta}(S_1) \times \mathrm{I\!R} \times \mathrm{I\!R}$. Moreover, if $p\epsilon C^{1+k+\beta}(I_{\bar\rho})$, then $R_\infty \epsilon C^{2+k+\beta}(S_1)$ and the following estimates hold;

$$|R - R_\infty|_{S_1}^{(2+k+\beta)} \le c_2(|k|)(\gamma^2 + |m - \bar m|), \ \ |\alpha| \le c_3|\gamma|$$

The proof consists in the linearization of equations (3.4) - (3.6) and application of the implicit function theorem. The details will be given in a forthcoming paper [19].
Equilibrium figures of incompressible self-gravitating fluid rotating with a given small velocity or momentum and subjected to capillary forces at the boundary are constructed in [3.14]. It is also necessary to mention the papers [1,2] of J. Bemelmans devoted to stationary free boundary problems of an incompressible viscous flow in the presence of a given symmetrical force field. In [1] it is assumed that $\sigma > 0$; in [2] the case $\sigma = 0$ is considered, and the selfgravitation forces are taken into account.

4.Estimates of $v - v_\infty$ in L_2-norms.

Further estimates are obtained for $w = v - v_\infty, \tau = \rho - \rho_\infty$ (we suppose that the functions v_∞, ρ_∞ are defined by (3.1) and (3.3) in the whole space R^3). With the aid of (3.2) it is easy to show that these functions satisfy in Ω_t the equations

$$\tau_t + (v\cdot\nabla)\tau + \rho\cdot\nabla\cdot w + (w\cdot\nabla)\rho_\infty = 0,$$

$$\rho(w_t + (v\cdot\nabla)w + (w\cdot\nabla)v_\infty) - \nabla T'(w) + \nabla p(\rho) - \nabla p(\rho_\infty) =$$

$$= (\rho - \rho_\infty)\frac{\alpha^2}{2}\nabla|x'|^2 \tag{4.1}$$

where $T'(w) = \mu'\nabla\cdot wI + \mu S(w) = T'(v)$. The initial and boundary conditions for w, τ have the form

$$w|_{t=0} = v_0 - v_\infty \equiv w_0(x), \ \ \tau|_{t=0} = \rho_0 - \rho_\infty = \tau_0(x) \ \ (x\epsilon\Omega_o), \tag{4.2}$$

$$-p(\rho)n + T'(w)n - \sigma Hn|_{\Gamma_t} = 0$$

Taking (3.4) into accont we obtain

$$T'(w)n - (p(\rho) - p(\rho_\infty))n = \sigma(H - H_\infty)n + hn \tag{4.3}$$

and moreover

$$Rn\cdot T'(w)n - (Rp(\rho) - R_\infty p(\rho_\infty)) = \sigma(RH - R_\infty H_\infty) + R_\infty h \tag{4.4}$$

where $h = p(\rho_\infty(R\omega)) - p(\rho_\infty(R_\infty\omega)), x = R(\omega,t)\omega, |\omega| = 1$. In this section we obtain intermediate estimates in L_2-norms for $w, \tau, R - R_\infty$. We begin with an auxiliary proposition.

Lemma 4.1

Suppose that condition (2.7) holds with $\delta\epsilon(0,1/2)$. Then arbitrary

$$u\epsilon W_2^1(\Omega_t)$$

satisfies the inequality

$$||u||_{L_2(\Omega_t)} \le c_1|L(u)| + c_2||\nabla u||_{L_2(\Omega_t)} \tag{4.5}$$

where $L(u)$ is a linear functional in $W_2^1(\Omega_t)$ such that $L(1) \neq 0$.Here

$$c_1 = |L(1)|^{-1}|\Omega_t|^{1/2}, \quad c_2 = C + c_1||L||(1+C^2)^{1/2}, ||L|| = \inf_{v\epsilon W_2^1(\Omega_t)} |L(v)|||v||^{-1}_{W_2^1(\Omega_t)}$$

is the norm of L, and C is the constant in the inequality of Poincare.

Proof.

Inequality (4.5) is well known (see for instance [10]); we find bounds for the constants c_1, c_2.

We start with an evident inequality

$$||u||_{L_2(\Omega_t)} \le ||u - \bar{u}||_{L_2(\Omega_t)} + |\bar{u}||\Omega_t|^{1/2}$$

where $\bar{u} = |\Omega_t|^{-1} \int_{\Omega_t} u dx$. Since $L(\bar{u}) = \bar{u}L(1)$, we have

$$|\bar{u}||\Omega_t|^{1/2} \le c_1|L(\bar{u})| \le c_1|L(u)| + c_1|L(u-\bar{u})| \le$$

$$\le c_1|L(u)| + c_1||L||||u-\bar{u}||_{W_2^1(\Omega_t)} =$$

$$= c_1|L(u)| + c_1|||L||(||u-\bar{u}||^2_{L_2(\Omega_t)} + ||\nabla u||^2_{L_2(\Omega_t)})^{1/2}$$

It remains to apply the Poincare inequality

$$||u - \bar{u}||_{L_2(\Omega_t)} \le C||\nabla||_{L_2(\Omega_t)}$$

(clearly, C can be taken independent of t because of (2.7)), and the lemma is proved. The main result of this section is the following theorem.

Theorem 4.1

Suppose that (2.7),(2.8) hold with small δ, δ_1 and that γ is also small. Then

$$||w||_{L_2(\Omega_t)} \le c_3||S(v)||_{L_2(\Omega_t)} + c_4|\gamma|(\int_{S_1} |R - R_\infty|d\omega + \int_{\Omega_t} |\rho - \rho_\infty|dx) \tag{4.6}$$

$$||\rho - \rho_\infty||_{W_2^1(\Omega_t)} + ||R - R_\infty||_{W_2^1(S_1)} \le$$

$$\le c_5(||w_t||_{L_2(\Omega_t)} + ||w||_{W_2^2(\Omega_t)} + ||(w\cdot\nabla)w||_{L_2(\Omega_t)}) \tag{4.7}$$

with constants $c_3 - c_5$ independent of γ and t.

Proof.

We make use of the Korn's inequality in the form presented in [14], namely,

$$||w||_{L_2(\Omega_t)} \le c_6||S(w)||_{L_2(\Omega_t)} + c_7(|\int_{\Omega_t} wdx| + |\int_{\Omega_t}[w\times x]dx|) \tag{4.8}$$

Since $\int\limits_{\Omega_t} \rho v dx = 0$ and $\int\limits_{\Omega_t} \rho x dx = 0$, we have $\int\limits_{\Omega_t} \rho w dx = 0$ and

$$|\int\limits_{\Omega_t} w dx| = |\int\limits_{\Omega_t} (1 - \frac{\rho}{\bar{\rho}}) w dx| \leq \delta_1 |\Omega_t| ||w||_{L_2(\Omega_t)}. \tag{4.9}$$

Moreover, from the relations

$$\int\limits_{\Omega_t} \rho[w \times x] dx = \int\limits_{\Omega_t} \rho([v \times x] - [v_\infty \times x]) dx =$$

$$\int\limits_{\Omega_t} (\rho_\infty - \rho)[v_\infty \times x] dx - \int\limits_{\Omega_t} \rho_\infty [v_\infty \times x] dx + \int\limits_{\Omega_\infty} \rho_\infty [v_\infty \times x] dx,$$

and

$$\int\limits_{\Omega_t} u(x) dx - \int\limits_{\Omega_\infty} u(x) dx = \int\limits_{S_1} d\omega \int\limits_{R_\infty(\omega)}^{R(\omega)} u(r\omega) r^2 dr$$

applied to $u(x) = \rho_\infty [v_\infty \times x]$ it follows that

$$|\int\limits_{\Omega_t} [w \times x] dx| \leq |\int\limits_{\Omega_t} (1 - \frac{\rho}{\bar{\rho}})[w \times x] dx| + |\int\limits_{\Omega_t} \frac{\rho}{\bar{\rho}} [w \times x] dx| \leq \tag{4.10}$$

$$c_8 \alpha^2 \left(\int\limits_{\Omega_t} |\rho_\infty - \rho| dx + \int\limits_{S_1} |R - R_\infty| ds \right) + \delta_1 (1 + \delta) R |\Omega_t| ||w||_{L_2(\Omega_t)}$$

If δ_1 is so small that $c_7 |\Omega_t| (1 + R(1 + \delta)) \delta_1 < 1/2$, then (4.7) follows form (4.8) - (4.10), since $S(v) = S(w)$.
To prove (4.7), we estimate a certain functional $L(\rho - \rho_\infty)$ satisfying the condition of Lemma 4.1. After the integration of (4.4) over S_1 we obtain

$$\begin{aligned}
&\int_{S_1} R(p(\rho) - p(\rho_\infty))_{x=R(\omega)\omega} d\omega \\
= \quad &\int_{S_1} (R_\infty(\omega) - R(\omega)) p(\rho_\infty(R\omega)) d\omega \\
- \quad &\int_{S_1} R_\infty(\omega)[p(\rho_\infty(R\omega)) - p(\rho_\infty(R_\infty\omega))] d\omega \\
+ \quad &\int_{S_1} Rn \cdot T'(w) n|_{x=R\omega} d\omega \\
- \quad &\sigma \int_{S_1} (RH[R] - R_\infty H[R_\infty]) d\omega.
\end{aligned} \tag{4.11}$$

Applying again the formula $RH[R] = \nabla_\omega \frac{\nabla_\omega R}{\sqrt{R^2 + |\nabla_\omega R|^2}} - \frac{2R}{\sqrt{R^2 + |\nabla_\omega R|^2}}$, we easily show that

$$\begin{aligned}
&\int_{S_1} (RH[R] - R_\infty H[R_\infty]) d\omega \\
= \quad &2 \int_{S_1} \left(\frac{|\nabla_\omega R|^2}{K(R, \nabla_\omega R)} - \frac{|\nabla_\omega R_\infty|^2}{K(R_\infty, \nabla_\omega R_\infty)} \right) d\omega
\end{aligned}$$

with $K(R,\nabla_\omega R) = \sqrt{R^2+|\nabla_\omega R|^2}(R+\sqrt{R^2+|\nabla_\omega R|^2})$.This integral can be estimated exactly as in [15],§6, namely

$$\begin{aligned} |I| &\leq 2\int_{S_1} \frac{||\nabla_\omega R|^2 - |\nabla_\omega R_\infty|^2|}{K(R,\nabla_\omega R} d\omega \\ &+ 2\int_{S_1} |\nabla_\omega R_\infty|^2 \frac{|K(R,\nabla R) - K(R_\infty \nabla R_\infty)|}{K(R,\nabla R)K(R_\infty \nabla R_\infty)} d\omega \\ &\leq c_9(\gamma+\delta_1)\int_{S_1}(|R-R_\infty| + |\nabla(R-R_\infty)|)d\omega \end{aligned}$$

Further, since $|\nabla\rho_\infty(x)| \leq c_{10}\alpha^2$, we have

$$|\int_{S_1} R_\infty(\omega)[p(\rho_\infty(R\omega)) - p(\rho_\infty(R_\infty\omega))]d\omega| \leq c_{11}\gamma^2 \int_{S_1} |R-R_\infty|d\omega$$

Other integrals in (4.11) may be represented in the form

$$\begin{aligned} \int_{S_1}(R-R_\infty)p(\rho_\infty(R\omega))d\omega &= \frac{p(\overline{\rho})}{\overline{\rho}} \int_{S_1} \rho_\infty(R\omega)(R_\infty(\omega) - R(\omega,t))d\omega + I_2, \\ \int_{S_1} R(p(\rho) - p(\rho_\infty))d\omega &= p'(\overline{\rho})\overline{R}\int_S(\rho(R\omega) - \rho_\infty(R\omega))d\omega + I_3 \end{aligned}$$

where

$$\begin{aligned} |I_2| &= |\int_{S_1}((\frac{p\rho_\infty}{\rho_\infty}) - (\frac{p\overline{\rho}}{\overline{\rho}}))\rho_\infty(R\omega)(R_\infty - R)d\omega| \\ &\leq c_{12}(\gamma^2 + |m-\overline{m}|)\int_{S_1}|R_\infty - R|d\omega, \\ |I_3| &= |\int_{S_1}\big([R(p(\rho) - p(\rho_\infty) - p'(\rho_\infty)(\rho-\rho_\infty))] + (\rho-\rho_\infty)[Rp'(\rho_\infty) - \overline{R}p'(\overline{\rho})]\big)\,d\omega| \\ &\leq c_{13}(\delta+\delta_1+\gamma^2+|m-\overline{m}|)\int_{S_1}|\rho - \rho_\infty(R\omega)|d\omega. \end{aligned}$$

Finally, from the identity

$$\begin{aligned} 0 &= \int_{\Omega_t}\rho dx - \int_{\Omega_\infty}\rho_\infty dx \\ &= \int_{\Omega_t}(\rho-\rho_\infty)dx + \int_{S_1}d\omega\int_{R_\infty(\omega)}^{R(\omega,t)}\rho_\infty(s\omega)s^2 ds \\ &= \int_{\Omega_t}(\rho-\rho_\infty)dx + \frac{1}{3}\int_{S_1}\rho_\infty(R\omega)(R^3(\omega,t) - R^3_\infty(\omega))d\omega \\ &+ \int_{S_1}d\omega\int_{R^3_\infty(\omega)}^{R(\omega,t)}[\rho_\infty(s\omega) - \rho_\infty(R\omega)]s^2 ds \end{aligned} \tag{4.12}$$

we find that

$$\left|\int_{\Omega_t}(\rho-\rho_\infty)dx + \overline{R}^2\int_{S_1}(R-R_\infty)\rho_\infty(R\omega)d\omega\right| \leq c_{14}|\gamma^2+\delta|\int_{S_1}|R-R_\infty|d\omega$$

Collecting all the terms, we arrive at the inequality

$$
\begin{gathered}
|\mathcal{L}(\rho-\rho_\infty)| \le \\
c_{15}(\delta+\delta_1+\gamma^2+|m-\overline{m}|) \\
\left(\int_{S_1}|\rho-\rho_\infty|d\omega+\int_{S_1}(|R-R_\infty|+|\nabla(R-R_\infty)|)d\omega\right) \\
+\int_{S_1} R|n\cdot T'(w)n|d\omega
\end{gathered}
$$

for the functional

$$
\mathcal{L}(\rho-\rho_\infty)=p'(\overline{\rho})\int_{S_1}(\rho-\rho_\infty(R\omega))d\omega-\frac{p(\overline{\rho})}{\overline{\rho}R^3}\int_{\Omega_t}(\rho-\rho_\infty)dx.
$$

Since

$$
\mathcal{L}(1)=4\pi b-\frac{p(\overline{\rho})}{\overline{\rho}R^3}(|\Omega_t|-|\overline{\Omega}|)
$$

and

$$
|\mathcal{L}(1)|\ge 4\pi|b|-\frac{p(\overline{\rho})}{\overline{\rho}R^3}(|\Omega_t|-|\overline{\Omega}|)\ge 4\pi(|b|-\frac{p(\overline{\rho})}{3\overline{\rho}}\delta)>0
$$

for small δ, we can apply inequality (4.5) which gives

$$
\begin{aligned}
\|\rho-\rho_\infty\|_{W_2^1(\Omega_t)} \;\le\;& c_{16}\|\nabla(\rho-\rho_\infty)\|_{L_2(\Omega_t)} \\
&+c_{17}\int_{S_1} R|n\cdot T'(w)n|d\omega \\
&+c_{18}(\gamma^2+|m-\overline{m}|+\delta+\delta_1)\|R-R_\infty\|_{W_2^1(S_1)}\,,
\end{aligned}
$$

if $\mu^2+|m-\overline{m}|+\delta+\delta_1$ is small. Since

$$
\nabla(p(\rho)-p(\rho_\infty))=p'(\rho)\nabla(\rho-\rho_\infty)+(p'(\rho)-p'(\rho_\infty))\nabla\rho_\infty,
$$

$\nabla(\rho-\rho_\infty)$ may be expressed in terms of derivates of w from (4.1) and we obtain the estimate

$$
\begin{aligned}
&\|\rho-\rho_\infty\|_{W_2^1(\Omega_t)} \\
\le\;& c_{19}(\|w_t\|_{L_2(\Omega_t)}+\|w\|_{W_2^2(\Omega_t)}+\|(w\cdot\nabla)w\|_{L_2(\Omega_t)}) \\
&+c_{20}(\mu^2+|m-\overline{m}|+\delta+\delta_1)\|R-R_\infty\|_{W_2^1(S_1)}.
\end{aligned}
\tag{4.13}
$$

Let us now turn to the estimates of $R-R_\infty$. The main idea of our arguments will be the same as in [15], §6. We multiply (1.4) by $(R-R_\infty)$ and integrate over S_1. This leads to

$$
\begin{aligned}
&\sigma\int_{S_1}\left(\nabla_\omega\cdot\frac{\nabla_\omega R}{\sqrt{R^2+|\nabla_\omega R|^2}}-\nabla_\omega\cdot\frac{\nabla_\omega R_\infty}{\sqrt{R_\infty^2+|\nabla_\omega R_\infty|^2}}\right)(R-R_\infty)d\omega \\
&+\int_{S_1}p(\overline{\rho})(R-R_\infty)^2d\omega \\
=\;& 2\int_{S_1}\left(\frac{|\nabla_\omega R|^2}{K(R,\nabla_\omega R)}-\frac{|\nabla_\omega R_\infty|^2}{K(R_\infty,\nabla_\omega R_\infty)}\right)(R-R_\infty)d\omega-\int_{S_1}R_\infty h(R-R_\infty)d\omega \\
&-\int_{S_1}(Rp(\rho)-R_\infty p(\rho_\infty)-p(\overline{\rho})(R-R_\infty))\,(R-R_\infty)d\omega \\
&+\int_{S_1}Rn\cdot T'n(R-R_\infty)d\omega
\end{aligned}
$$

The left-hand side can be estimated exactly as in [15], and we obtain

$$\begin{aligned} &\tfrac{2\sigma}{R} \textstyle\int_{S_1} (|\nabla\omega(R-R_\infty)|^2 - 2|R-R_\infty|^2)\,d\omega \\ = &-\textstyle\int_{S_1} Rn\cdot T'(w)n(R-R_\infty)d\omega + \int_{S_1} R(p(\rho)-p(\rho_\infty))(R-R_\infty)d\omega + D\,, \quad (4.14) \\ &|D| \le c_{21}(|\gamma|^2+\delta_1)\|R-R_\infty\|^2_{W_2^1(S_1)}. \end{aligned}$$

As shown in [15] (see (2.34)),

$$\begin{aligned} &\textstyle\int_{S_1} (|\nabla_\omega(R-R_\infty|^2 - 2|R-R_\infty|^2)d\omega \\ \ge\ &\tfrac{4}{7}\|R-R_\infty\|^2_{W_2^1(S_1)} - c_{22}(\textstyle\int_{S_1}(R-R_\infty)d\omega)^2 - c_{23}\sum_i(\int_{S_1}(R-R_\infty)\omega_i d\omega)^2 \quad (4.15) \end{aligned}$$

From the identity (4.12) and from

$$\begin{aligned} 0 &= \int_{\Omega_t} \rho(x,t)x_i dx - \int_{\Omega_\infty} \rho_\infty(x)x_i dx \\ &= \int_{\Omega_t} (\rho(x,t)-\rho_\infty(x))x_i dx \\ &+ \int_{S_1} \omega_i d\omega \int_{R_\infty}^{R} (\rho_\infty(s\omega)-\overline{\rho})s^3 ds \\ &+ \overline{\rho}\int_{S_1} \omega_i \frac{R^4-R_\infty^4}{4} d\omega \quad i=1,2,3 \end{aligned}$$

it follows that the right-hand side of (4.15) is not less than

$$\frac{4}{7}\|R-R_\infty\|^2_{W_2^1(S_1)} - c_{24}(\gamma^2+|m-\overline{m}|)\|R-R_\infty\|^2_{L_2(S_1)} - c_{25}\|\rho-\rho_\infty\|^2_{L_2(\Omega_t)}$$

so at the end we get

$$\begin{aligned} &\tfrac{4\sigma}{7R}\|R-R_\infty\|^2_{W_2^1(S_1)} \\ &\le \tfrac{c_{26}\sigma}{R}(\gamma^2+|m-\overline{m}|)\|R-R_\infty\|_{L_2(S_1)} \\ &+\tfrac{c_{27}\sigma}{R}\|\rho-\rho_\infty\|^2_{L_2(\Omega_t)} + \|R^2 n\cdot T'n\|_{L_2(S_1)}\|R-R_\infty\|_{L_2(S_1)} \\ &+\|R^2(p(\rho)-p(\rho_\infty))\|_{L_2(S_1)}\|R-R_\infty\|_{L_2(S_1)} + |D| \end{aligned}$$

This inequalities, (4.14) and (4.15) imply (4.7) and the theorem is proved.

5 Estimates of $w, \tau, R - R_\infty$ in Sobolev spaces and proof of theorem 1.4

In this section we obtain estimates of w, τ and $R - R_\infty$ in W_2^l-norms. It is convenient to make beforehand a coordinate transformation mapping Ω_t onto the ball $B_{\overline{R}} : |x| \leq \overline{R}$. Consider first a general transformation

$$x = x(y,t) \tag{5.1}$$

and let $\tilde{f}(y,t) = f(x(y,t),t)$. We have

$$\frac{\partial f}{\partial x_k} = \sum_{k,m=1}^{3} \mathcal{J}_{mk} \frac{\partial \tilde{f}}{\partial y_m}, \quad \frac{\partial \tilde{f}}{\partial t} = \tilde{f}_t + \sum_{k=1}^{3} \frac{\partial f}{\partial x_k} \dot{x}_k,$$

$$\tilde{f}_t = \frac{\partial \tilde{f}}{\partial t} - \sum_{m=1}^{3} \dot{x}_k \, \mathcal{J}_{mk} \frac{\partial \tilde{f}}{\partial y_m}$$

where $\dot{x}_k = \frac{\partial x_k}{\partial t}$ and $\mathcal{J}_{mk} = \frac{\partial y_m}{\partial x_k}$;, hence, equations (4.1) take a form

$$\tilde{\tau}_t - \sum_{k,m} \mathcal{J}_{mk} \, \dot{x}_k \, \tilde{\tau}_{y_m} + (\tilde{v} \cdot \tilde{\nabla})\tilde{\tau} + \tilde{\rho} \tilde{\nabla} \cdot \tilde{w} + (\tilde{w} \cdot \tilde{\nabla})\tilde{\rho}_\infty = 0,$$

$$\tilde{\rho}(\tilde{w}_t - \sum_{k,m} \mathcal{J}_{mk} \, \dot{x}_k \, \tilde{w}_{y_m} + (\tilde{v} \cdot \tilde{\nabla})\tilde{w} + (\tilde{w} \cdot \tilde{\nabla})\tilde{v}_\infty) - \tilde{\nabla}\tilde{T}'(\tilde{w}) \tag{5.2}$$

$$+ \tilde{\nabla}(p(\tilde{\rho}) - p(\tilde{\rho}_\infty)) - (\tilde{\rho} - \tilde{\rho}_\infty)\frac{\alpha^2}{2}\tilde{\nabla}|\tilde{x}'|^2 = 0$$

Here we have set $\tilde{\nabla} = \mathsf{J}^* \nabla = (\sum_m \mathcal{J}_{mk} \frac{\partial}{\partial y_m})_{k=1,2,3}$ (J is a matrix with elements $\mathcal{J}_{mk}$) and

$$\tilde{T}'_{ik}(\tilde{w}) = \mu' \delta_{ik} \tilde{\nabla} \cdot \tilde{w} + \mu \tilde{S}_{ik}(\tilde{w}),$$

$$\tilde{S}_{ik}(\tilde{w}) = \sum_{m=1}^{3} (\mathcal{J}_{mi} \frac{\partial \tilde{w}_k}{\partial y_m} + \mathcal{J}_{mk} \frac{\partial \tilde{w}_i}{\partial y_m}).$$

Transformed initial and boundary conditions read

$$\tilde{w}|_{t=0} = \tilde{w}_0(z), \quad \tilde{\tau}|_{t=0} = \tilde{\tau}_0(z), \tag{5.3}$$

$$\tilde{T}'(\tilde{w})\mathsf{J}^*\tilde{n} - (p(\tilde{\rho}) - p(\tilde{\rho}_\infty))\mathsf{J}^*\tilde{n} = \sigma(\tilde{H} - \tilde{H}_\infty)\mathsf{J}^*\tilde{n} + \tilde{h}\mathsf{J}^*\tilde{n} \tag{5.4}$$

where $\tilde{n}$ is the exterior normal to the image of Γ_t under the transformation $y = y(x)$. Finally, condition (1.4) may be written in the form

$$RR_t = Rw_r - w_\omega \cdot \nabla_\omega R - v_{\infty\omega} \cdot \nabla_\omega (R - R_\infty), \quad |\omega| = 1, \tag{5.5}$$

since $v_{\infty r} = 0$ and $v_\infty \cdot \nabla_\omega R_\infty = 0$.

We define $x(y,t)$ by the formula

$$x(y,t) = y(1 + \Phi(y,t)), \quad y \in B_{\overline{R}}, \tag{5.6}$$

where $\Phi(y,t) = \frac{1}{\overline{R}}(R(\frac{y}{|y|},t) - \overline{R})$ on the sphere $S_{\overline{R}} : |y| = \overline{R}$ and $\Phi(y,t)$, $y \in B_{\overline{R}}$ is an extension of this function made in such a way that

$$\|\Phi(\cdot,t)\|_{W_2^{l+3}(B_{\overline{R}})} \le C_1 \|\Phi(\cdot,t)\|_{W_2^{l+5/2}(S_{\overline{R}})} \le C_2 \|R(\cdot,t) - \overline{R}\|_{W_2^{l+5/2}(S_1)}$$

The Jacobi matrix of the transformation (5.1), (5.6) equals

$$\begin{pmatrix} 1+\Phi+y_1\Phi_{y_1} & y_1\Phi_{y_2} & y_1\Phi_{y_3} \\ y_2\Phi_{y_1} & 1+\Phi+y_2\Phi_{y_2} & y_2\Phi_{y_3} \\ y_3\Phi_{y_1} & y_3\Phi_{y_2} & 1+\Phi+y_3\Phi_{y_3} \end{pmatrix} = [(1+\Phi)\delta_{ik} + y_i\Phi_{y_k}]_{i,k=1,2,3}$$

and its determinant is equal to $(1+\Phi)^2(1+\Phi+\sum_k y_k\Phi_{y_k}) = (1+\Phi)^2(1+\Phi+r\Phi_r)$, $r = |y|$. The transformation is invertible, if

$$|\Phi| + r|\Phi_r| < 1 \tag{5.7}$$

The functions $\mathcal{J}_{mk}$ are elements of the inverse matrix, i.e.

$$\mathcal{J}_{mk} = \frac{\delta_{mk}}{1+\Phi} - \frac{y_m\Phi_{y_k}}{(1+\Phi)(1+\Phi+r\Phi_r)}$$

Under the condition (5.7) (which is fulfilled if $R(\omega,t)$ satisfies (2.7) with a small δ) we have

$$\|\mathcal{J}_{mk} - \delta_{mk}\|_{W_2^{l+2}(B_{\overline{R}})} \le C_3 \|R - \overline{R}\|_{W_2^{l+5/2}(S_1)}.$$

We can now formulate the main a-priori estimate for the solution of our free boundary problem.

<u>**Theorem 5.1**</u>

Let $(w,\tau,R-R_\infty)$ be defined for $0 < t < T$ and let

$$\begin{aligned} N_{l,T}^2[\tilde{w},\tilde{\tau},R-R_\infty] \equiv{}& \|\tilde{\tau}\|^2_{W_2^{l+2,l/2+1}(\tilde{Q}_T)} + \int_0^T \|\tilde{\tau}\|^2_{W_2^{l+1}(\tilde{Q}_T)}\,dt + \sup_{t\in(0,T)} \|\tilde{w}\|^2_{W_2^{l+1}(B_{\overline{R}})} \\ &+ \sup_{t\in(0,T)} \|\tilde{\tau}\|^2_{W_2^{l+1}(B_{\overline{R}})} + \sup_{t\in(0,T)} \|\tilde{\tau}_t\|^2_{W_2^{l}(B_{\overline{R}})} \\ &+ \sup_{t\in(0,T)} \|R-R_\infty\|^2_{W_2^{l+2}(S_1)} + \|R-R_\infty\|^2_{L_2(0,T;\,W_2^{l+5/2}(S_1))} \le \delta_2 \end{aligned}$$

where $\tilde{Q}_T = B_{\overline{R}} \times (0,T)$ and δ_2 is a certain small positive number ($\delta_2 \le 1$, see also (5.13) below). Suppose also that the numbers γ, δ, δ_1 are sufficiently small. Then

$$\begin{aligned} N_{l,T}^2[\tilde{w},\tilde{\tau},R-R_\infty] \le{}& C_4\Big(\|v_0\|^2_{W_2^{l+1}(\Omega_0)} + \|\rho_0 - \overline{\rho}\|^2_{W_2^{l+1}(\Omega_0)} \\ &+ \|R(\cdot,0) - \overline{R}\|^2_{W_2^{l+5/2}(S_1)}\Big) \end{aligned} \tag{5.8}$$

with a constant C_4 independent of T.

Scheme of the proof

The proof of (5.8) is carried out in several steps.

Step 1 We establish the inequality

$$\begin{aligned}
&\int_0^T \|\nabla\tilde{w}\|^2_{W_2^{l+1}(B_{\overline{R}})}\,dt + \sup_{t\in(0,T)} \|\tilde{\tau}\|^2_{W_2^{l+1}(B_{\overline{R}})} + \int_0^T \|\nabla\tilde{\tau}\|^2_{W_2^{l}(B_{\overline{R}})}\,dt \\
&\quad + \sup_{t\in(0,T)} \|R-R_\infty\|^2_{W_2^{l+2}(S_1)} \\
&\le C_5\,(\|\tilde{w}_0\|^2_{W_2^{l+1}(B_{\overline{R}})} + \|\tilde{\tau}_0\|^2_{W_2^{l+1}(B_{\overline{R}})} + \|R(\cdot,0)-\overline{R}\|^2_{W_2^{l+2}(S_1)} \qquad (5.9)\\
&\quad + (\delta_2+\varepsilon)N^2_{l,T}[\tilde{w},\tilde{\tau},R-R_\infty] + N^2_{l-\lambda,T}[\tilde{w},\tilde{\tau},R-R_\infty] + \|\tilde{w}_t\|^2_{L_2(0,T;\,W_2^{l}(B_{\overline{R}}))}),
\end{aligned}$$

where $\lambda\in(0,l)$ and ε is the same as in (1.10).

Step 2 The difference $R-R_\infty$ is estimated in the norm $L_2(0,T;\;W_2^{l+5/2}(S_1))$:

$$\int_0^T \|R-R_\infty\|^2_{W_2^{l+5/2}(S_1)}\,dt \le C_6 \int_0^T (\|\tilde{\tau}\|^2_{W_2^{l+1/2}(\Gamma)} + \|\nabla\tilde{w}\|^2_{W_2^{l+1/2}(\Gamma)})\,dt \qquad (5.10)$$

Step 3 The derivative $\tilde{w}_t$ is estimated in $L_2(B_{\overline{R}};\;W_2^{l/2}(0,T))$-norm:

$$\begin{aligned}
&\int_{B_{\overline{R}}} \|\tilde{w}_t\|^2_{W_2^{l/2}(0,T)}\,dz \le C_7\,(\|v_0\|^2_{W_2^{l+1}(\Omega_0)} + \|\rho_0-\overline{\rho}\|^2_{W_2^{l+1}(\Omega_0)}) \\
&\quad + C_8((\delta_2+\varepsilon)N^2_{l,T}[\tilde{w},\tilde{\tau},R-R_\infty] + N^2_{l-\lambda,T}[\tilde{w},\tilde{\tau},R-R_\infty]) \qquad (5.11)
\end{aligned}$$

Step 4 For the norm of $\tilde{w}_t$ in the right hand side of (5.9) the following interpolation inequality is established:

$$\begin{aligned}
\int_0^T \|\tilde{w}_t\|^2_{W_2^{l}(B_{\overline{R}})}\,dt \;&\le\; \varepsilon_1 \int_0^T \|\tilde{w}\|^2_{W_2^{l+2}(B_{\overline{R}})}\,dt \qquad (5.12)\\
&\quad + C_9(\varepsilon_1)\int_{B_{\overline{R}}} \|\tilde{w}\|^2_{W_2^{l/2}(0,T)}\,dy, \quad \textit{for all } \varepsilon_1\in(0,1).
\end{aligned}$$

Let us show that estimate (5.8) follows from (5.9)–(5.12) and from the results of §2 and §4. For sufficiently small ε_1, (5.9)–(5.12) imply

$$\begin{aligned}
N^2_{l,T} \;&\equiv\; N^2_{l,T}[\tilde{w},\tilde{\tau},R-R_\infty] \le C_{10}\,(\|v_0\|^2_{W_2^{l+1}(\Omega_0)} + \|\rho_0-\overline{\rho}\|^2_{W_2^{l+1}(\Omega_0)} \\
&\quad + \|R(\cdot,0)-\overline{R}\|^2_{W_2^{l+5/2}(S_1)} + (\delta_2+\varepsilon)N^2_{l,T} + N^2_{l-\lambda,T})
\end{aligned}$$

If

$$(\varepsilon+\delta_2)\,C_{10} \le \frac{1}{2} \qquad (5.13)$$

then

$$\begin{aligned}
N^2_{l,T} \;&\le\; 2C_{10}\,(\|v_0\|^2_{W_2^{l+1}(\Omega_0)} + \|\rho_0-\overline{\rho}\|^2_{W_2^{l+1}(\Omega_0)} \\
&\quad + \|R(\cdot,0)-\overline{R}\|^2_{W_2^{l+5/2}(S_1)} + N^2_{l-\lambda,T})
\end{aligned}$$

The norm $N_{l-\lambda,T}$ may be evaluated with the aid of interpolation inequality

$$N^2_{l-\lambda,T} \le \varepsilon_2 N^2_{l,T} + C_{11}(\varepsilon_2)(\|\tilde{w}\|^2_{L_2(\tilde{Q}_T)} + \|\tilde{\tau}\|^2_{L_2(\tilde{Q}_T)} + \|R - R_\infty\|^2_{L_2(S_1\times(0,T))}),$$

so, chosing $\varepsilon_2 \le \frac{1}{4C_{10}}$ we obtain

$$\begin{aligned} N^2_{l,T} \;\le\; & 4C_{10}(\|v_0\|^2_{W_2^{l+1}(\Omega_0)} + \|\rho_0 - \overline{\rho}\|^2_{W_2^{l+1}(\Omega_0)} + \|R(\cdot,0) - \overline{R}\|^2_{W_2^{l+5/2}(S_1)}) \\ & +4C_{10}C_{11}(\|\tilde{w}\|^2_{L_2(\tilde{Q}_T)} + \|\tilde{\tau}\|^2_{L_2(\tilde{Q}_T)} + \|R - R_\infty\|^2_{L_2(S_1\times(0,T))}) \end{aligned} \tag{5.14}$$

Now we make use of (4.6) and (4.7) to estimate the sum of L_2-norms in the right-hand side. We have

$$\begin{aligned} & \|\tilde{w}\|^2_{L_2(\tilde{Q}_T)} + \|\tilde{\tau}\|^2_{L_2(\tilde{Q}_T)} + \|R - R_\infty\|^2_{L_2(S_1\times(0,T))} \\ \le\; & C_{12}(\|\tilde{w}\|^2_{W_2^{2,1}(\tilde{Q}_T)} + \int_0^T \|S(v)\|^2_{L_2(\Omega_t)}\,dt) \\ & +C_{13}\gamma^2(\|\tilde{\tau}\|^2_{L_2(\tilde{Q}_T)} + \|R - R_\infty\|^2_{L_2(S_1\times(0,T))}) \\ \le\; & \varepsilon_3 N^2_{l,T} + C_{14}(\varepsilon_3)\|\tilde{w}\|^2_{L_2(\tilde{Q}_T)} + C_{12}\int_0^T \|S(v)\|^2_{L_2(\Omega_t)}\,dt) \\ & +C_{13}\gamma^2(\|\tilde{\tau}\|^2_{L_2(\tilde{Q}_T)} + \|R - R_\infty\|^2_{L_2(S_1\times(0,T))}) \end{aligned} \tag{5.15}$$

Applying again (4.6) we show that the right-hand side does not exceed

$$\begin{aligned} & \varepsilon_3 N^2_{l,T} + (C_{14}(\varepsilon_3) + 1)C_{12}\int_0^T \|S(v)\|^2_{L_2(\Omega_t)}\,dt \\ & +C_{13}(C_{14}(\varepsilon_3) + 1)\gamma^2(\|\tilde{\tau}\|^2_{L_2(\tilde{Q}_T)} + \|R - R_\infty\|^2_{L_2(S_1\times(0,T))}) \end{aligned}$$

We fix the parameter $\varepsilon_3 \le \frac{1}{16C_{10}C_{11}}$ and impose one more restruction on γ :

$$(C_{14} + 1)C_{13}\gamma^2 \le \frac{1}{2}.$$

Under these conditions (5.15) yields

$$\begin{aligned} & \|\tilde{w}\|^2_{L_2(\tilde{Q}_T)} + \|\tilde{\tau}\|^2_{L_2(\tilde{Q}_T)} + \|R - R_\infty\|^2_{L_2(S_1\times(0,T))} \\ \le\; & \frac{1}{8C_{10}C_{11}} N^2_{l,T} + 2C_{12}(C_{14} + 1)\int_0^T \|S(v)\|^2_{L_2(\Omega_t)}\,dt \end{aligned}$$

and we obtain

$$\begin{aligned} N^2_{l,T} \;\le\; & 8C_{10}(\|v_0\|^2_{W_2^{l+1}(\Omega_0)} + \|\rho_0 - \overline{\rho}\|^2_{W_2^{l+1}(\Omega_0)} + \|R(\cdot,0) - \overline{R}\|^2_{W_2^{l+5/2}(S_1)}) \\ & +16C_{10}C_{11}C_{12}(C_{14} + 1)\int_0^T \|S(v)\|^2_{L_2(\Omega_t)}\,dt \end{aligned}$$

It remains to estimate the last term by inequality (2.9), and (5.8) is proved. The above assumptions conserning the magnitude of δ, δ_1, γ follow from (5.8), if the constant ε in (1.9) is sufficiently small.
Let us come back to inequalities (5.9)-(5.12). The latter is a variant of an estimate for

mixed derivates and is known for a large class of spaces (see[11]). The shortest way to prove (5.12) is to extend $\tilde{w}(z,t)$ into the whole space $I\!R^4$ and to make the Fourier transform; then (5.12) reduces to a simple algebraic inequality. (5.10) is a consequence of a coersive estimate for a secont order elliptic equation on the spere S_1. This equation can be obtained from (4.4), since $RH[R] - R_\infty H[R_\infty]$ may be written as second order differential operator applied to $R - R_\infty$ with the principal part

$$\frac{\Delta_\omega(R-R_\infty)}{\sqrt{R^2+|\nabla_\omega R|^2}} - \frac{1}{2}\nabla R \cdot \frac{\nabla_\omega[\nabla_\omega(R+R_\infty)\cdot\nabla_\omega(R-R_\infty)]}{(R^2+|\nabla_\omega R|^2)^{\frac{3}{2}}}$$

The estimate has a form

$$\begin{aligned}\|R-R_\infty\|^2_{W_2^{l+5/2}(S_1)} &\le C_{15}(\|Rn\cdot T'(w)n\|^2_{W_2^{l+1/2}(\Gamma_t)} \\ &+ \|\rho-\rho_\infty\|^2_{W_2^{l+1/2}(S_1)} + \|R-R_\infty\|^2_{L_2(S_1)}).\end{aligned}$$

If we integrate it with respect to t and make use of (4.7) we arrive at (5.10).
The proof of (5.9),(5.11) is too technical and can not presented here with all the details. The general scheme is the same as for the problem with boundary condition $v|_\Gamma = 0$ (see[5,16]). Inequality (5.9) is proved by local considerations. Let $y_0 \in \Omega$, $\zeta \in C_0^\infty(I\!R^3)$, $\zeta(y) = 1$ for $|y-y_0| \le d$, $\zeta(y) = 0$ for $|y-y_0| \ge 2d$, $d > 0$.
Suppose that $B_{2d}(y_0) \equiv \{\|y-y_0| \le 2d\} \subset B_{\overline{R}}$. Denote by $\Delta^s(z)f$ the s-th finite difference of the function f, i.e.

$$\Delta^s(z)f(y) = \sum_{k=0}^{s} C_s^k(-1)^{s-k}f(x+kz),$$

$$C_s^k = \binom{s}{k} = \frac{s!}{k!(s-k)!}.$$

From (5.2) we obtain

$$\Delta^s(z)\tilde{\tau}_t - \sum_{k,m=1}^{3} \mathcal{J}_{mk}\,\dot{x}_k\,\Delta^s(z)\tilde{\tau}_{y_m} + \tilde{v}\cdot\tilde{\nabla}\Delta^s(z)\tilde{\tau} + \overline{\rho}\hat{\nabla}\cdot\Delta^s(z)\hat{w} = H_s \qquad (5.16)$$

$$\begin{aligned}&\tilde{\rho}(\Delta^s(z)\tilde{w}_t - \textstyle\sum_{k,m=1}^3 \mathcal{J}_{mk}\,\dot{x}_k\,\Delta^s(z)\tilde{w}_{y_m} \\ &+(\tilde{v}\cdot\tilde{\nabla})\Delta^s(z)\tilde{w}) - \tilde{\nabla}\Delta^s(z)T'(\hat{w}) + p'(\overline{\rho})\tilde{\nabla}\Delta^s(z)\tilde{\tau} = F_s \qquad (5.17)\end{aligned}$$

$$\begin{aligned}\text{with } H_s \;=\; &\left[\Delta^s(z)(\sum_{k,m=1}^{3}\mathcal{J}_{mk}\,\dot{x}_k\,\tilde{\tau}_{y_m}) - \sum_{k,m=1}^{3}\mathcal{J}_{mk}\,\dot{x}_k\,\Delta^s(z)\tilde{\tau}_{y_m}\right] \\ &+\left[\tilde{v}\cdot\tilde{\nabla}\Delta^s(z)\tilde{\tau} - \Delta^s(z)(\tilde{v}\cdot\tilde{\nabla}\tilde{\tau})\right] \\ &+\overline{\rho}\left[\tilde{\nabla}\Delta^s(z)\tilde{w} - \Delta^s(z)(\tilde{\nabla}\cdot\tilde{w})\right] \\ &-\Delta^s(z)(\tilde{\rho}-\overline{\rho})\tilde{\nabla}\cdot\tilde{w} - \Delta^s(z)(\tilde{w}\cdot\tilde{\nabla})\tilde{\rho}_\infty,\end{aligned}$$

$$
\begin{aligned}
F_s &= [\hat{\rho}\Delta^s(z)\tilde{w}_t - \Delta^s(z)(\tilde{\rho}\tilde{w}_t)] \\
&+ \left[\Delta^s(z)(\sum_{k,m=1}^{3} \mathcal{J}_{mk}\,\dot{x}_k\,\tilde{\tau}_{y_m}) - \sum_{k,m=1}^{3} \mathcal{J}_{mk}\,\dot{x}_k\,\Delta^s(z)\tilde{\tau}_{y_m}\right] \\
&+ \left[(\tilde{v}\cdot\tilde{\nabla})\Delta^s(z)\tilde{w}) - \Delta^s(z)(\tilde{v}\cdot\tilde{\nabla})\tilde{w}\right] \\
&+ \left[\Delta^s(z)\tilde{\nabla}\tilde{T}'(\hat{w}) - \tilde{\nabla}\Delta^s(z)\tilde{T}'(\tilde{w}\right] \\
&+ \left[p'(\overline{\rho})\tilde{\nabla}\Delta^s(z)\tilde{\tau} - \Delta^s(z)\tilde{\nabla}(p(\tilde{\rho}) - p(\tilde{\rho}_\infty))\right].
\end{aligned}
$$

Let $s > l+1$ (for instance, s=2) and $\mathcal{J} = det\left(\frac{\partial y_m}{\partial x_k}\right) = \frac{1}{(1+\Phi)^2(1+\Phi+r\Phi_r)}$. We multiply (5.16) by

$$\zeta^2\frac{p'(\overline{\rho})}{\overline{\rho}\mathcal{J}}|z|^{-3-2(l+1)}\Delta^s(z)\tilde{\tau},$$

and (5.17) by

$$\zeta^2\frac{1}{\mathcal{J}}|z|^{-3-2(l+1)}\Delta^s(z)\tilde{w},$$

and integrate with respect to y, t and $z \in B_{d/s}$ making use of the formula

$$
\begin{gathered}
\int_{B_d(y_0)}(f_t + (\dot{v}\cdot\tilde{\nabla})f - \sum_{k,m}\mathcal{J}_{mk}\,\dot{x}_k\,f_{y_m})f\tfrac{\zeta^2}{\mathcal{J}}\,dy = \\
\tfrac{1}{2}\tfrac{d}{dt}\int_{B_d(y_0)} f^2\tfrac{\zeta^2\,dy}{\mathcal{J}} - \tfrac{1}{2}\int_{B_d(y_0)}\left[(\tilde{\nabla}\cdot(\dot{v}\zeta^2) - \sum_{k,m}\mathcal{J}_{mk}\,\dot{x}_k\,\tfrac{\partial\zeta^2}{\partial x_m}\right] f^2\tfrac{dy}{\mathcal{J}}.
\end{gathered}
$$

After elementary but lenghty calculations connected mainly with estimates of H_s and F_s we arrive at the inequality

$$
\begin{aligned}
&\frac{1}{2}\sup_{t\in(0,T)}\int_{|z|\le d/s}\frac{dz}{|z|^{3+2(l+1)}}\int_{B_d(y_0)}\left(\tilde{\rho}|\Delta^s(z)\tilde{w}|^2 + \frac{p'(\overline{\rho}}{\overline{\rho})}|\Delta^s(z)\tilde{\tau}|^2\right)\zeta^2\frac{dy}{\mathcal{J}} \qquad (5.18)\\
&+\int_0^T dt\int_{|z|\le d/s}\frac{dz}{|z|^{3+2(l+1)}}\int_{B_d(y_0)}\left(\frac{\mu}{2}|\tilde{S}(\Delta^s(z)\tilde{w})|^2 + \mu'|\tilde{\nabla}\cdot\Delta^s(z)\tilde{w}|^2\right)\zeta^2\frac{dy}{\mathcal{J}} \\
\le\ &\int_{|z|\le d/s}\frac{dz}{|z|^{3+2(l+1)}}\int_{B_d(y_0)}\left(\tilde{\rho}_0|\Delta^s(z)\tilde{w}_0|^2 + \frac{p'(\overline{\rho}}{\overline{\rho}}|\Delta^s(z)\tilde{\tau}_0|^2\right)\zeta^2\frac{dy}{\mathcal{J}(y,0)} \\
&+C_{16}(N^3_{l,T,B_{2d}} + N^2_{l-\lambda,T,B_{2d}})
\end{aligned}
$$

where N_{l,T,B_d} is the norm $N_{l,T}$ computed in the domain $B_d \subset B_{\overline{R}}$.
It is well known (see for instance [3]) that the expression

$$\left(\int_{|z|\le d/s}|z|^{-3-2(l+1)}dz\int_{\mathbb{R}^3}|\Delta^s(z)(\tilde{w}\zeta)|^2dy\right)^{1/2}$$

is eqivalent to the principal part of the norm $\|\tilde{w}\zeta\|_{W_2^{l+1}(\mathbb{R}^3)}$ hence, (5.18) implies

$$
\begin{aligned}
&\sup_{t\in(0,T)}\left(\|\tilde{w}\|^2_{W_2^{l+1}(B_d(y_0))} + (\|\tilde{\tau}\|^2_{W_2^{l+1}(B_d(y_0))}\right) + \int_0^T\|\nabla\tilde{w}\|^2_{W_2^{l+1}(B_d(y_0))}dt \\
&\le C_{13}\left((\|\tilde{w}_0\|^2_{W_2^{l+1}(B_{2d}(y_0))} + (\|\tilde{\tau}_0\|^2_{W_2^{l+1}(B_{2d}(y_0))}\right) + C_{14}(N^3_{l,T,B_{2d}} + N^2_{l-\lambda,T,B_{2d}})
\end{aligned}
$$

The derivates $\frac{\partial}{\partial y_i}(\tilde{\rho}-\tilde{\rho}_\infty)$ may be found from (5.2) which gives immediatly

$$\begin{aligned}
&\int_0^T \|\nabla(\tilde{\rho}-\tilde{\rho}_\infty)\|^2_{W_2^l(B_d(y_0))}dt \\
\leq\ & C_{15}\left(\int_0^T (\|\tilde{w}\|^2_{W_2^{l+2}(B_d(y_0))} + (\|\tilde{w}_t\|^2_{W_2^l(B_d))})dy\right. \\
+\ & \left.\gamma^2\int_0^T \|\nabla(\tilde{\rho}-\tilde{\rho}_\infty)\|^2_{W_2^l(B_d))}dt\right)
\end{aligned}$$

so, taking accoount of (4.7) we arrive at the "local estimate" of the type (5.9)

$$\begin{aligned}
&\sup_{t\in(0,T)}\left(\|\tilde{w}\|^2_{W_2^{l+1}(B_d(y_0))} + \|\tilde{\tau}\|^2_{W_2^{l+1}(B_d(y_0))}\right) + \int_0^T (\|\nabla\tilde{w}\|^2_{W_2^{l+1}(B_d)} + \|\nabla\tilde{\tau}\|^2_{W_2^l(B_d))})dy \\
&\leq C_{16}\left(\|\tilde{w}_0\|^2_{W_2^{l+1}(B_{2d})} + \|\tilde{\tau}_0\|^2_{W_2^{l+1}(B_{2d})}\right) + C_{17}(N^3_{l,T,B_{2d}} + N^2_{l-\lambda,T,B_{2d}})
\end{aligned}$$

Analogous estimates can be obained in the neighbourhood of a point $x_0 \in S_{\overline{R}} = \partial B_{\overline{R}}$. In this case one should make additional coordinate transformation mapping this neighbourhood onto a semi-ball $|\xi| \leq d_1\,, \xi_3 > 0$. Let (z_1, z_2, z_3) be cartesian coordinates in $I\!R^3$ with the origine at x_0 and with z_3-axis directed along an interior normal $-\hat{n}(x_0)$. The sphere $S_{\overline{R}}$ is determind by the equation

$$z_3 = \overline{R} - \sqrt{\overline{R}^2 - |z'|^2} = \phi(z')\,, z' = (z_1, z_2),\ |z'| \leq d_1 < \overline{R}.$$

We define new coordinates (ξ_1, ξ_2, ξ_3) by $\xi_1 = z_1,\ \xi_2 = z_2,\ \xi_3 = z_3 - \phi(z_1, z_2)$.
After this transformation functions $\mathcal{J}_{mk}$ in equation (5.2) should be replaced by $\mathcal{J}'_{sk} = \sum_m \mathcal{J}''_{sm}\mathcal{J}_{mk}$ where $\mathcal{J}''_{sm}$ are elements of the Jacobian matrix of the transformation $\xi = \xi(y)$. Estimate of the type (5.18) is established for finite differences $\Delta^s(z')\hat{w}$ and $\Delta^s(z')\tilde{\tau}$ with $z' = (z_1, z_2, 0)$; the differences $\Delta^s(z_3)\tilde{w}_{z_3z_3}$ and $\Delta^s(z_3)\tilde{\tau}_{z_3}$ are then found from (5.2). Due to the boundary conditions (5.4), additional positive term equivalent to $\sup_{t\in(0,T)}\|R - R_\infty\|^2_{W_2^{l+2}(S_1)}$ appears in the left-hand side of (5.9).
Finally, we say a few words about inequality (5.11). Let

$$\Delta_t^k(h)f(x,t) = \sum_{j=0}^{k} C_k^j(-1)^{k-j}f(x, t+jh)\,, k > l.$$

Clearly,

$$\tilde{\rho}\Delta_t^k(h)\tilde{w}_t - \tilde{\nabla}\tilde{T}'(\Delta_t^k(h)\tilde{w}) = G_k \tag{5.19}$$

with

$$\begin{aligned}
G_k =\ & \left[\tilde{\rho}\Delta_t^k(h)\tilde{w}_t - \Delta_t^k(h)\tilde{\rho}\tilde{w}_t\right] \\
&+ \left[\Delta_t^k(h)\tilde{\nabla}\tilde{T}'(\tilde{w}) - \tilde{\nabla}\tilde{T}'(\Delta_t^k(h)\tilde{w})\right] \\
&+ \Delta_t^k(h)\tilde{\rho}\left[\sum_{k,m}\mathcal{J}_{mk}\ \dot{x}_k\ \tilde{w}_{y_m} - (\tilde{v}\cdot\tilde{\nabla})\tilde{w} - (\tilde{w}\cdot\tilde{\nabla})\tilde{v}_\infty\right] \\
&- \Delta_t^k(h)\tilde{\nabla}(p(\tilde{\rho}) - p(\tilde{\rho}_\infty)) - \Delta_t^k(h)(\tilde{\rho}-\tilde{\rho}_\infty)\frac{\alpha^2}{2}\tilde{\nabla}|\tilde{x}'|^2.
\end{aligned}$$

From (5.19) it follows that

$$\int_0^{h_0} \frac{dh}{h^{1+l}} \int_0^{T-kh_0} \psi(t)dt \int_{B_{\overline{R}}} (\tilde{\rho}|\Delta_t^k(h)\tilde{w}_t|^2 - \tilde{\nabla}\tilde{T}'(\Delta_t^k(h)\tilde{w}) \cdot \Delta_t^k(h)\tilde{w}_t)\frac{dy}{\mathcal{J}} =$$
$$\int_0^{h_0} \frac{dh}{h^{1+l}} \int_0^{T-kh_0} \psi(t)dt \int_{B_{\overline{R}}} G_k \cdot \Delta_t^k(h)\tilde{w}_t \frac{dy}{\mathcal{J}} \tag{5.20}$$

where $\psi(t)$ is a smooth function vanishing fot $t \le t_0$ and equal to 1 for $t \ge 2t_0$ (we suppose that $T > max(kh_0, 2t_0)$). After elementary calculations it can be shown that

$$\begin{aligned} -\int_0^{T-kh_0} \psi(t)dt \int_{B_{\overline{R}}} (\tilde{T}'(\Delta_t^k(h)\tilde{w}) \cdot \Delta_t^k(h)\tilde{w}_t \frac{dy}{\mathcal{J}} = \quad (5.21)\\ \frac{1}{2}\int_{S_{\overline{R}}} \left(\tfrac{\mu}{2}|S(\Delta_t^k(h)\tilde{w})|^2 + \mu'|\tilde{\nabla}\Delta_t^k(h)\tilde{w}|^2\right) \frac{dy}{\mathcal{J}}\Big|_{t=T-kh_0} \\ -\frac{1}{2}\int_0^{T-kh_0} \psi'(t)dt \int_{B_{\overline{R}}} \left(\tfrac{\mu}{2}|\tilde{S}(\Delta_t^k(h)\tilde{w})|^2 + \mu'|\tilde{\nabla}\Delta_t^k(h)\tilde{w}|^2\right) \frac{dy}{\mathcal{J}} \\ -\int_0^{T-kh_0} \psi(t)dt \int_{S_{\overline{R}}} \tilde{T}'(\Delta_t^k(h)\tilde{w})\mathcal{J}^*\tilde{n} \cdot \Delta_t^k(h)\tilde{w}_t \frac{dS}{\mathcal{J}} + I, \\ |I| \le C_{18}\int_0^{T-kh_0} \psi(t)dt \int_{B_{\overline{R}}} (|\Phi_t| + |\nabla\Phi_t|)\,|\Delta_t^k(h)\tilde{w}|^2 \frac{dy}{\mathcal{J}}, \end{aligned}$$

hence (5.20) and (5.21) imply

$$\begin{aligned} &\int_0^{h_0} \frac{dh}{h^{1+l}} \int_0^{T-kh_0} \psi(t)dt \int_{B_{\overline{R}}} \tilde{\rho}|\Delta_t^k(h)\tilde{w}_t|^2 \frac{dy}{\mathcal{J}} \\ \le\ &\int_0^{h_0} \frac{dh}{h^{1+l}} \int_0^{T-kh_0} \psi(t)dt \left(\Big| \int_{B_{\overline{R}}} G_k \cdot \Delta_t^k(h)\tilde{w}_t \frac{dy}{\mathcal{J}} \Big| + |I| \right. \quad (5.22) \\ &\left. + \Big| \int_{S_{\overline{R}}} \tilde{T}'(\Delta_t^k(h)\tilde{w}_t)\mathcal{J}^*\tilde{n} \cdot \Delta_t^k(h)\tilde{w} \frac{dS}{\mathcal{J}} \Big| \right) \\ &+ \int_0^{h_0} \frac{dh}{h^{1+l}} \int_{t_0}^{2t_0} \psi'(t)dt \int_{B_{\overline{R}}} \left(\frac{\mu}{2}|S(\Delta_t^k(h)\tilde{w})|^2 + \mu'|\tilde{\nabla}\Delta_t^k(h)\tilde{w}|^2 \right) \frac{dy}{\mathcal{J}} \end{aligned}$$

The contribution of the last term may be evaluated with the aid of an local existence theorem. Other terms in (5.22) are estimated after some lengthy calculations.
As a result, we obtain (5.11). We restrict ourselves with these remarks concerning (5.9)-(5.12) and prove theorem 1.4 at the conclusion.

Proof of theorem 1.4 The boundary condition in (1.7) may be written in the form

$$\begin{aligned} &\tilde{T}'(\tilde{v})\tilde{n} - \sigma\tilde{n}(\tilde{n} \cdot \Delta(t)\textstyle\int_0^t \tilde{v}d\tau) \\ =\ &\left[\sigma(H_0 + \tfrac{2}{\overline{R}}) + p(\rho_0\mathcal{J}^{-1}) - p(\overline{\rho})\right]\tilde{n} + \sigma\tilde{n}(\tilde{n} \cdot \textstyle\int_0^t \frac{\partial\Delta(\tau)}{\partial\tau}\xi d\tau) \end{aligned}$$

($H_0 = H(\xi, 0)$) and from the estimate (1.9) applied to problem (1.7) it follows that

$$\begin{aligned} \|\tilde{v}\|^2_{W_2^{2+l,1+l/2}(Q_{T_1})} &\le C_{19}(T_1)(\|v_0\|^2_{W_2^{l+1}(\Omega)} + \|\rho_0 - \overline{\rho}\|^2_{W_2^{l+1}(\Omega)} \\ &\quad + \|R(\cdot, 0) - \overline{R}\|^2_{W_2^{l+5/2}(S_1)}) \\ &\le C_{19}(T_1)\varepsilon \end{aligned}$$

where $C_{19}(T_1)$ is an increasing function of T_1. This proves the first statement of the theorem. For ε sufficiently small ($\varepsilon \le \varepsilon_0$) we have proved the estimate (5.8) with the constant C_4 independent of T_1. In virtue of theorem 1.3, the norm $\|\tilde{v}(\cdot,t)\|_{W_2^{l+2}(\Omega)}$ is uniformly bounded for $t \geq \frac{T}{2}$,hence, $\|R(\cdot,t) - \overline{R}\|_{W_2^{l+5/2}(S_1)}$ is also uniformly bounded. It is possible therefore to apply theorem 1.1 once more (in new Lagrangean coordinates $\xi^{(1)} \in \Omega_{T_1}$) and to extend the solution into the interval $t \in (T_1, 2T_1)$. The estimate (5.8) holds for $t \leq 2T_1$. Clearly, we can repeat this procedure infinity many times and establish the boundedness of the norm $N_{l,\infty}[\tilde{w}, \tilde{\tau}, R - R_\infty]$. It follows that $\tilde{w}, \tilde{\tau}$ and $R - R_\infty$ tend to zero as $t \to \infty$ and theorem 1.4 is proved.

References

[1] Bemelmans, J., Gleichgewichtsfiguren mit Oberflächenspannung. Analysis, 1 (1981), 241-282

[2] Bemelmans, J., On a free boundary problem for the sationary Navier-Stokes equations. Ann. Inst. H.Poincarè (6) vol 4. (1987), p. 517-547

[3] Golovkin, K.K, On equivalent norms for fractional spaces. Proc. Inst. Math. Steklov, 66 (1962), 364-383; Amer. Math. Soc. Translations, (2) 81, 257-280

[4] Hölder, E., Gleichgewichtsfiguren rotierender Flüssigkeiten mit Oberflächenspannung. Math.Zeitschrift, 25 (1926), 188-208.

[5] Matsumura, A., Nishida, T., Initial-boundary value problems for the equations of motion of general fluids. Proc. V Intern. Symp. on Computing methods, Versailles, Dec. 14-18 (1981), 389-406.

[6] Secchi, P., On the motion of gaseous stars in the presence of radiation. Comm. PDE, 15 (1990), 185 - 204.

[7] Secchi, P., On the unequeness of motion of viscous gaseous stars. Math. Meth. Appl. Sci., 13 (1990), 391-404

[8] Secchi, P, Valli, A. A free boundary problem for compessible viscous flouid, Journ. Reine Angew. Math., 341 (1983), 1-31.

[9] Slobodetskii, L.N., Generalized S.L Sobolev spaces and their application to boundary value problems for partial differential equations, Sci. comm. Lening. Ped.Inst.Herzen, 197 (1958), 54-112, Amer. Math. Soc. Translations, (2) 57, 207-275

[10] Sobolev, S.L, Applications of functional analysis in mathematical physics, Leningrad 1950, 255p.

[11] Solonnikov, V.A., A priori estimates for second order parabolic equations, Proc. Math. Inst. Steklov, 70 (1964), 133-212; Amer. Math. Soc. Translations, (2) 65, 51-137

[12] Solonnikov, V.A., Solvability of a problem on a motion of a viscous incompressible liquid bounded by a free surface, Izvestia Acad. Sci. USSR., 41 (1977), 1388-1424.

[13] Solonnikov, V.A., On a nonstationary motion of a finite liquid mass bounded by a free surface. Zapiski nuchn. Semin. LOMI, 152 (1986),137-157

[14] Solonnikov, V.A., On a nonstationary motion of a isolated mass of a viscous incompressible fluid. Izvestia Acad. Sci. USSR, 51 (1987), 1065-1087.

[15] Solonnikov, V.A., On a nonstationary motion of a finite isolated mass of self-gravitating fluid. Algebra and analysis, 1 (1989), 207-246.

[16] Solonnikov, V.A., Solvability of initial-boudary value problem for equations of motion of a viscous compressible barotropic fluid in the spaces $W_2^{2+l,1+l/2}(Q_T)$, Zapiski nauchn. semin. LOMI, to appear

[17] Solonnikov, V.A., Tani, A., Free boundary problem for the compressible Navier-Stokes equations with the surface tension. Zapiski nauchn. semin. LOMI, (182) (1990), 142-148

[18] Solonnikov, V.A., Tani, A., Free boundary problem for a viscous compressible flow with the surface tension. Constantin Caratheodory: an international tribute, Th.M.Rassias (editor), World Sci. Publ., 1991, 1270-1303.

[19] Soloonikov, V.A., Tani, A., Equilibrium figures of slowly rotating viscous compressible barotropic capillary liquid. Advances in math. sciences, to appear.

[20] Tani, A., On the free boundray problem for compressible vicous fluid motion. Journ. Math. Kyoto Univ., 21 (1981), 839-859

Heat-conducting fluids with free surface in the case of slip-condition on the walls

Michael Wolff
Humboldt-Universität Berlin
Fachbereich Mathematik
Unter den Linden 6, O-1086 Berlin

1.Statement of the problem

There are given two concentrically placed cylinders, which are set up on a horizontal plane. The space between these cylinders is partially filled by a viscous heat-conducting fluid. The cylinders and the plane can rotate with different speeds ω_1, ω_2 and ω_3, respectively. This brings the fluid in motion due to the slip-condition on the material walls. Caused by the interaction of the surface tension, central forces, internal stresses of the fluid, gravity, heat-conduction and atmospherical pressure a free (i.e. apriori unknown) surface is formed as an upper boundary of the fluid's volume. Moreover, we assume that the wetting angles β_1 and β_2, which are formed by the free surface and the cylindrical walls, as well as the values of the temperature T_0, T_1, T_2, T_3 on the walls are given (see pict.).

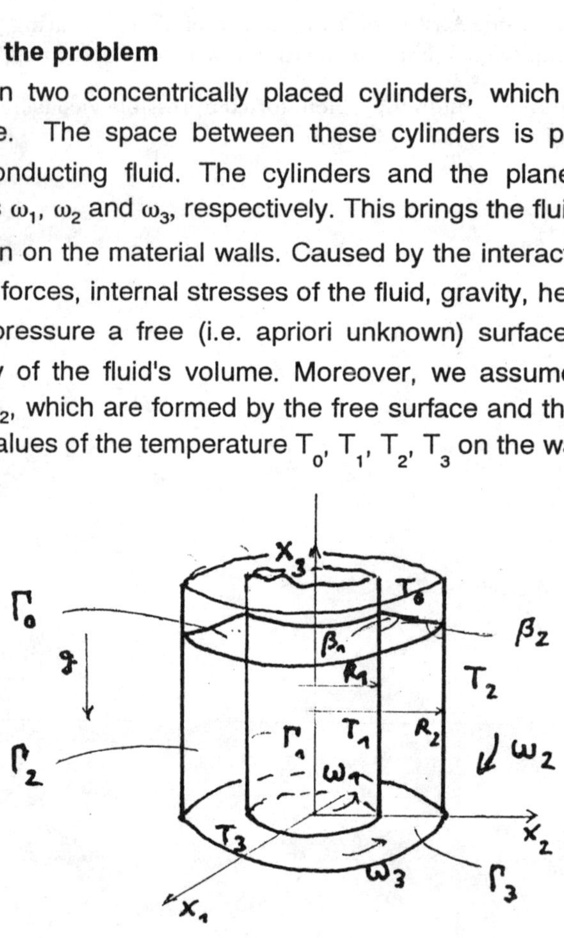

Here $Q:=\{(x_1, x_2, x_3)\in \mathbb{R}^3\colon R_1^2<x_1^2+x_2^2<R_2^2\,,\ 0<x_3<h(\sqrt{x_1^2+x_2^2})\}$

the fluid's volume, Γ_0 the (unknown) free surface, Γ_1, Γ_2, Γ_3 the contact surfaces where the fluid meets the cylinder's walls.

It is our aim to obtain a local existence and uniqueness result for velocity, presssure, temperature and the free surface of the complete stationary problem formulated below. We shall apply the implicit function theorem acting in weighted Hölder spaces. The present paper completes the author's studies [9]. There we have only considered non-slip-conditions (adherence) on the walls. Our work is based on the approach of Kondratjev [1] in order to study boundary value problems for PDEs in non-smooth domains. Further, our studies are related to [3, 4, 7]. We remark that in [2, 5] the slip-condition is considered in connection with dynamical contact points. Of course, there are many other contributions to free surface problems. See the literature cited in the references.
Let us introduce a threedimensional Cartesian coordinate system $\{x_1, x_2, x_3\}$, so that the x_3 - axis coincides with the cylinders' axis and is directed contrary to the gravity. We assume an axisymmmetric steady fluid's flow. The free surface is assumed to have the shape $x_3 = h(r)$, ($r=\sqrt{x_1^2 + x_2^2}$), where the function h is apriori unknown.
Let us formulate the problem in Cartesian coordinates:

(1.1) $$-S_{ij},_j + \rho v_j v_i,_j + \Phi,_i = 0 \quad \text{in } Q,$$

(1.2) $$v_i,_i = 0 \quad \text{in } Q,$$

(1.3) $$S_{ij} n_i \tau_j + \alpha_k v_i \tau_i = \alpha_k\, a_i(\omega_k)\tau_i \quad \text{on } \Gamma_k,\ k = 1,2,3;$$

slip-condition (if $\alpha_k = \infty$ we get adherence),

(1.4) $$v_i n_i = 0 \quad \text{on } \partial Q,$$

(1.5) $$S_{ij} n_i \tau_j = \sigma'(T)\, T,_i \tau_i \quad \text{on } \Gamma_0,$$

(1.6) $$\sigma(T)J(h) + \rho g h - \Phi + S_{ij} n_i n_j + p_a = 0 \quad \text{on } \Gamma_0,$$

(1.7) $$\frac{h'(R_i)}{\sqrt{1+h'^2(R_i)}} = (-1)^i \cos \beta_i := \alpha_i;\quad i = 1,2;$$

(1.8) $$\int_{R_1}^{R_2} r h(r)\, dr = c_h$$

(1.9) $$\rho v_i \varepsilon,_i + q_i,_i = S_{ij} v_i,_j \quad \text{in } Q, \text{ heat-conduction},$$

(1.10) $$q_i n_i = \beta_k (T - T_k) \quad \text{on } \Gamma_k,\ k = 0,1,2,3;$$

(1.11) $$S = \mu(T)D, \quad \varepsilon = \varepsilon(T), \quad q_i = -k(T)T,_i \quad \text{in } Q, \quad \text{material lows.}$$

The meaning of the above notation is as follows:

$v=\{v_1, v_2, v_3\}$, T - velocity, temperature; p_a, p - atmospheric pressure, pressure, $\Phi := p+\rho g x_3$ - reduced pressure; ρ, g, α_k (k=1,2,3) - density, gravity, slip coefficient, all are positive constants;
$n=\{n_1, n_2, n_3\}$, $\tau=\{\tau_1, \tau_2, \tau_3\}$ - unit outward normal, (arbitrary) unit tangent vector;
$a=\{a_1, a_2, a_3\}$ - boundary value of velocity determined by the given ω_1, ω_2, ω_3;
$\sigma=\sigma(T)>0$ - surface tension, $\mu=\mu(T)>0$ - viscosity, $D:=\frac{1}{2}(v_{i,j}+v_{j,i})$ - stetching tensor,
S - extra stress tensor,
$J(h):=\frac{1}{r}\frac{d}{dr}\frac{rh'}{\sqrt{1+h'^2}}$ - mean curvature of the free surface. By f' we denote the derivative of f, if f depends only on one variable. The spatial derivative of v with respect to x_i is denoted by $v_{,i}$. We suppose $\mathrm{tr}S:= S_{ii} = 0$, and the pressure's additive constant is determined by condition (1.8). Further, $k=k(T)>0$ is the heat conductivity, $\varepsilon=\varepsilon(T)$ is specific internal energy, $q=\{q_i\}$ is the heat flow vector, β_k are positive constants (k=0,1,2,3);
$c_h =$ const >0, where $V=2\pi c_h$ and V is the fluid's volume (filling the domain Q).
We are not interested in the numerical values of the positive constants ρ, g, α_k, β_k and so we put them equal to one.

2. Transfomation of the problem; Operator formulation

We transform the system (1.1) - (1.11) into cylindrical coordinates

$$\begin{cases} x_1=r\cos\Theta \\ x_2=r\sin\Theta \\ x_3=z \end{cases}.$$

Taking into account the axisymmetry, i.e. the independence of Θ and denoting $\Omega:=Q\cap\{\Theta=\text{const}\}$ as well as $\gamma_k:=\Gamma_k\cap\{\Theta=\text{const}\}$ we get

$$-[\frac{\partial}{\partial r}(\mu(T)\frac{\partial v^1}{\partial r})+\frac{1}{2}\frac{\partial}{\partial z}(\mu(T)(\frac{\partial v^1}{\partial z}+\frac{\partial v^3}{\partial r}))-\frac{\mu(T)v^1}{r^2}+\frac{\mu(T)}{r}\frac{\partial v^1}{\partial r}]+v^1\frac{\partial v^1}{\partial r}-r(v^2)^2+v^3\frac{\partial v^1}{\partial z}+\frac{\partial\Phi}{\partial r}=0 \text{ in } \Omega$$

$$(2.1)\qquad -[\frac{\partial}{\partial r}(\frac{\mu(T)}{2}\frac{\partial v^2}{\partial r})+\frac{\partial}{\partial z}(\frac{\mu(T)}{2}\frac{\partial v^2}{\partial z})+\frac{3\mu(T)}{2r}\frac{\partial v^2}{\partial r}]+v^1\frac{\partial v^2}{\partial r}+\frac{2v^1v^2}{r}+v^3\frac{\partial v^2}{\partial z}=0 \text{ in } \Omega;$$

$$-[\frac{\partial}{\partial r}(\frac{\mu(T)}{2}(\frac{\partial v^1}{\partial z}+\frac{\partial v^3}{\partial r}))+\frac{\partial}{\partial z}(\mu(T)\frac{\partial v^3}{\partial z})+\frac{\mu(T)}{2r}(\frac{\partial v^1}{\partial z}+\frac{\partial v^3}{\partial r})]+v^1\frac{\partial v^3}{\partial r}+v^3\frac{\partial v^3}{\partial z}+\frac{\partial\Phi}{\partial z}=0 \text{ in } \Omega;$$

$$(2.2)\qquad \frac{\partial}{\partial r}(rv^1)+\frac{\partial}{\partial z}(rv^3)=0 \quad \text{in } \Omega,$$

$$(2.3)\qquad -h'v^1+v^3=0 \quad \text{on } \gamma_0,$$

$$(2.4)\qquad v^1=0 \quad \text{on } \gamma_1 \text{ and } \gamma_2,$$

$$(2.5)\qquad v^3=0 \quad \text{on } \gamma_3,$$

(2.6) $$-\frac{\mu(T)}{2}\frac{\partial v^3}{\partial r} + v^3 = 0 \quad \text{on } \gamma_1;$$

(2.7) $$-\frac{\mu(T)}{2}\frac{\partial v^2}{\partial r} + v^2 - \omega_1 = 0 \quad \text{on } \gamma_1;$$

(2.8) $$\frac{\mu(T)}{2}\frac{\partial v^3}{\partial r} + v^3 = 0 \quad \text{on } \gamma_2;$$

(2.9) $$\frac{\mu(T)}{2}\frac{\partial v^2}{\partial r} + v^2 - \omega_2 = 0 \quad \text{on } \gamma_2;$$

(2.10) $$-\frac{\mu(T)}{2}\frac{\partial v^1}{\partial z} + v^1 = 0 \quad \text{on } \gamma_3;$$

(2.11) $$-\frac{\mu(T)}{2}\frac{\partial v^2}{\partial z} + v^2 - \omega_3 = 0 \quad \text{on } \gamma_3;$$

(2.12) $$\mu(T)\Big(-h'\frac{\partial v^1}{\partial r}+\frac{1}{2}\Big(\frac{\partial v^1}{\partial z}+\frac{\partial v^3}{\partial r}\Big)(1-h'^2)+h'\frac{\partial v^3}{\partial z}\Big) - \sigma'(T)\sqrt{1+h'^2}(T_{,1}+h'T_{,3})=0 \quad \text{on } \gamma_0;$$

(2.13) $$\mu(T)\Big(h'\frac{\partial v^2}{\partial r} - \frac{\partial v^2}{\partial z}\Big) = 0 \quad \text{on } \gamma_0;$$

(2.14) $$-\sigma(T)J(h)+h - \Phi+p_a+\frac{\mu(T)}{1+h'^2}\Big(h'^2\frac{\partial v^1}{\partial r} - h'\Big(\frac{\partial v^1}{\partial z}+\frac{\partial v^3}{\partial r}\Big)+\frac{\partial v^3}{\partial z}\Big)=0 \quad \text{on } \gamma_0;$$

(2.15) $$\frac{h'(R_i)}{\sqrt{1+h'^2(R_i)}} = \alpha_i; \quad i=1,2;$$

(2.16) $$\int_{R_1}^{R_2} rh(r)\,dr = c_h,$$

(2.17) $$\varepsilon'(T)\Big(v^1\frac{\partial T}{\partial r} + v^3\frac{\partial T}{\partial z}\Big) - \frac{\partial}{\partial r}\Big(k(T)\frac{\partial T}{\partial r}\Big) - \frac{\partial}{\partial z}\Big(k(T)\frac{\partial T}{\partial z}\Big) - \frac{k(T)}{r}\frac{\partial T}{\partial r} -$$
$$\frac{\mu(T)}{2}\Big\{2\Big(\frac{\partial v^1}{\partial r}\Big)^2 + 2\frac{\partial v^3}{\partial r}\frac{\partial v^1}{\partial z} + \Big(\frac{\partial v^1}{\partial z}\Big)^2 + \Big(\frac{\partial v^3}{\partial r}\Big)^2 + 2\Big(\frac{\partial v^3}{\partial z}\Big)^2 + r^2\Big(\frac{\partial v^2}{\partial r}\Big)^2 + r^2\Big(\frac{\partial v^2}{\partial z}\Big)^2\Big\}=0 \quad \text{in } \Omega$$

(2.18) $$\frac{k(T)}{\sqrt{1+h'^2}}(h'\,T_{,1}-T_{,3}) - \beta_0(T-T_0) = 0 \quad \text{on } \gamma_0;$$

(2.19) $$k(T)T_{,1} - \beta_1(T-T_1) = 0 \quad \text{on } \gamma_1,$$

(2.20) $$-k(T)T_{,1} - \beta_2(T-T_2) = 0 \quad \text{on } \gamma_2,$$

(2.21) $$k(T)T_{,3} - \beta_3(T-T_3) = 0 \quad \text{on } \gamma_3.$$

Note, that v^1, v^2, v^3 are the contravariant components of the velocity vector according to the cylindrical coordinates. To proceed further, we map the (apriori unknown) domain Ω onto the fixed domain Ω_0 determined by the rest state. This rest state is characterized by

(2.22) $$v=0,\ \Phi=\Phi_0,\ h=h_0,\ T=\hat{T},\ (\hat{T} \text{ is a fixed reference temperature}),$$
$$\omega_1=\omega_2=\omega_3=0, \quad T_i=\hat{T},\ i=0,1,2,3.$$

We can calculate Φ_0 and h_0 in the following way. From (2.14) and (2.22) we get

$$(2.23)\qquad -\frac{\sigma(\hat{T})}{r}\frac{d}{dr}\left(\frac{r\frac{dh_0}{dr}}{\sqrt{1+(\frac{dh_0}{dr})^2}}\right)+h_0-\Phi_0+p_a=0 \text{ for } r\in I:=(R_1,R_2)\,;$$

Multiplying this equation by r, integrating over I and using (2.16) we obtain

$$\Phi_0 = p_a-(R_1^2-R_2^2)^{-1}(\sigma(\hat{T})(R_2\alpha_2-R_1\alpha_1)+c_h).$$

And so we can derive from (2.23) the following equation

$$-\frac{1}{r}\frac{d}{dr}\left(\frac{r\frac{dh_0}{dr}}{\sqrt{1+(\frac{dh_0}{dr})^2}}\right)+\frac{h_0}{\sigma(\hat{T})}+\frac{(\sigma(\hat{T})(R_2\alpha_2-R_1\alpha_1)+c_h)}{\sigma(\hat{T})(R_2^2-R_1^2)}=0\,.$$

Together with the boundary conditions (2.15) this last equation forms a nonlinear elliptic problem having a unique solution $h_0\in C^\infty([R_1,R_2])$ [8], provided

$$(2.24)\quad |\cos \beta_1|<1\,,\quad |\cos \beta_2|<\frac{R_1}{R_2}\,;$$

As in [7, 9] we define a weighted Hölder space $C_s^l(\Omega)$ ($l>0$, $s<l$, both non-integer reals in our case) as the set of all $[l]$-times continuously differentiable functions in Ω satisfying

$$\|u\|^{(l)}_{s,\Omega_0} := \|u\|^{(s)}_{\Omega_0}+\sum_{s<|\alpha|<l}\sup_{x\in\Omega_0} d^{|\alpha|-s}(x)\,|D^\alpha u(x)| +$$

$$+\sum_{|\alpha|=[l]}\sup_{x,y\in\Omega_0} \mathit{min}(d^{l-s}(x),d^{l-s}(y))\,\frac{|D^\alpha u(x)-D^\alpha u(y)|}{|x-y|^{l-[l]}}<\infty,$$

where $d(x) := \mathit{min}\,(\|x-x^{(1)}\|,\|x-x^{(2)}\|)$, $x^{(i)}:=(R_i,h_0(R_i))$, $i=1,2$ and

$$\|u\|^{(s)}_{\Omega_0} := \sum_{|\alpha|<s}\sup_{x\in\Omega_0}|D^\alpha u(x)| + \sum_{|\alpha|=[s]}\sup_{x,y\in\Omega_0}|x-y|^{[s]-s}\,|D^\alpha u(x)-D^\alpha u(y)|,\qquad \text{if } s>0$$

and $\qquad \|u\|^{(s)}_{\Omega_0} := 0,\qquad$ if $s<0$.

The space $C_s^l(I)$ ($I:=(R_1,R_2)$) is defined analogously setting $x^{(i)}:=R_i$, $i=1,2$. Of course, C_s^l are Banach spaces.

We note the following result due to Solonnikov [7]. Let
$h\in C^{l+3}_{s+1}(I)$ be arbitrary with $\|h-h_0\|_{C^{l+3}_{s+1}(I)} \le \delta_0$ being sufficiently small. Then there exists a special one-to-one mapping
$X : \Omega_0 \leftrightarrow \Omega$ of the class $C^{l+3}_{s+1}(\Omega_0)$.

As an advantage of this transform we get a decoupling of the system for the auxillary problem after taking the partial Frechét derivative of the non-linear operator (cf. below). Instead of h we use $g:=h-h_0$ and the functions H_0, H, G are siutable continuations of h_0, h, g, respectively, from $I=[R_1, R_2]$ into $I\times IR^1_+$ (cf. [7] for details). We denote by $x=(x_1,x_2):=(r,z)\in\Omega$ the old coordinates and by $y=(y_1,y_2)\in\Omega_0$ the new coordinates. Then the Jacobi matrix of the transform X reads as

$$(B_{ij}) := \left[\frac{\partial y_i}{\partial x_j}\right] = \begin{bmatrix} 1, & 0 \\ \dfrac{G_{,2}H_{0,1}-G_{,1}H_{0,2}+G_{,1}}{G_{,2}+H_{0,2}-1}, & \dfrac{H_{0,2}-1}{G_{,2}+H_{0,2}-1} \end{bmatrix}$$

and we note the matrices

$$\left[A^k_{ij}\right] := \frac{\partial}{\partial x_i}\left[\frac{\partial y_k}{\partial x_j}\right], \qquad \text{of course, we get } A^1_{ij}=0 \ \forall i,j=1,2.$$

Further, the unknown functions v^i, Φ, T can be expressed in the new coordinates by

$$u^i(y_1, y_2) := v^i(r,z) = v^i(x_1, x_2), \qquad \Psi(y_1, y_2) := \Phi(r,z) = \Phi(x_1, x_2),$$
$$\vartheta(y_1, y_2) := T(r,z) = T(x_1, x_2).$$

Now we write the system (2.1) - (2.21) in the new coordinates:

$$- \frac{\partial}{\partial y_1}\left[\mu(\vartheta)\frac{\partial u^1}{\partial y_1}\right] - B_{21}\frac{\partial}{\partial y_2}\left[\mu(\vartheta)\frac{\partial u^1}{\partial y_1}\right] - B_{21}\frac{\partial}{\partial y_1}\left[\mu(\vartheta)\frac{\partial u^1}{\partial y_2}\right] -$$
$$(B_{21})^2\frac{\partial}{\partial y_2}\left[\mu(\vartheta)\frac{\partial u^1}{\partial y_2}\right] - \mu(\vartheta)\frac{\partial u^1}{\partial y_2}A^2_{11} - ½(B_{22})^2\frac{\partial}{\partial y_2}\left[\mu(\vartheta)\frac{\partial u^1}{\partial y_2}\right] - ½\mu(\vartheta)\frac{\partial u^1}{\partial y_2}A^2_{22}$$
$$(2.25) \qquad - ½\frac{\partial}{\partial y_2}\left[\mu(\vartheta)\frac{\partial u^3}{\partial y_1}\right] - ½\frac{\partial}{\partial y_2}\left[\mu(\vartheta)\frac{\partial u^3}{\partial y_2}\right]B_{21}B_{22} - ½\mu(\vartheta)\frac{\partial u^3}{\partial y_2}A^2_{21}$$
$$- \frac{\mu(\vartheta)}{y_1^2}u^1 - \frac{\mu(\vartheta)}{y_1}\left[\frac{\partial u^1}{\partial y_1} + B_{21}\frac{\partial u^1}{\partial y_2}\right] - u^1B_{21}\frac{\partial u^1}{\partial y_2} - y_1(u^2)^2 +$$
$$+ u^3B_{22}\frac{\partial u^1}{\partial y_2} + \frac{\partial \Psi}{\partial y_1} + B_{21}\frac{\partial \Psi}{\partial y_2} = 0 \ \text{ in } \Omega_0,$$

$$
\text{(2.26)}\quad -\frac{\partial}{\partial y_1}\left[\frac{\mu(\vartheta)}{2}\frac{\partial u^2}{\partial y_1}\right] - B_{21}\frac{\partial}{\partial y_2}\left[\frac{\mu(\vartheta)}{2}\frac{\partial u^2}{\partial y_1}\right] - B_{21}\frac{\partial}{\partial y_1}\left[\frac{\mu(\vartheta)}{2}\frac{\partial u^2}{\partial y_2}\right] -
$$

$$
-(B_{21})^2\frac{\partial}{\partial y_2}\left[\frac{\mu(\vartheta)}{2}\frac{\partial u^2}{\partial y_2}\right] - \tfrac{1}{2}\mu(\vartheta)\frac{\partial u^2}{\partial y_2}A_{11}^2 - \tfrac{1}{2}(B_{22})^2\frac{\partial}{\partial y_2}\left[\mu(\vartheta)\frac{\partial u^2}{\partial y_2}\right] -
$$

$$
\frac{3\mu(\vartheta)}{2y_1}\left[\frac{\partial u^2}{\partial y_1} + B_{21}\frac{\partial u^2}{\partial y_2}\right] + u^1\left[\frac{\partial u^2}{\partial y_1} + B_{21}\frac{\partial u^2}{\partial y_2}\right] + \frac{2}{y_1}u^1u^2 + u^3\frac{\partial u^2}{\partial y_2}B_{22} = 0 \ \text{ in } \Omega_0,
$$

$$
-\frac{\partial}{\partial y_1}\left[\frac{\mu(\vartheta)}{2}\frac{\partial u^1}{\partial y_1}\right]B_{22} - \frac{\partial}{\partial y_2}\left[\frac{\mu(\vartheta)}{2}\frac{\partial u^1}{\partial y_2}\right]B_{21}B_{22} - A_{12}^2\frac{\mu(\vartheta)}{2}\frac{\partial u^1}{\partial y_2} -
$$

$$
\text{(2.27)}\quad \frac{\partial}{\partial y_1}\left[\frac{\mu(\vartheta)}{2}\frac{\partial u^3}{\partial y_2}\right] - \frac{\partial}{\partial y_2}\left[\frac{\mu(\vartheta)}{2}\frac{\partial u^3}{\partial y_1}\right]B_{21} - \frac{\partial}{\partial y_1}\left[\frac{\mu(\vartheta)}{2}\frac{\partial u^3}{\partial y_2}\right]B_{21} - \frac{\partial}{\partial y_2}\left[\frac{\mu(\vartheta)}{2}\frac{\partial u^3}{\partial y_2}\right]B_{21}^2
$$

$$
-\tfrac{1}{2}\mu(\vartheta)\frac{\partial u^3}{\partial y_2}A_{11}^2 - \frac{\partial}{\partial y_2}\left[\mu(\vartheta)\frac{\partial u^3}{\partial y_2}\right]B_{22}^2 - \mu(\vartheta)\frac{\partial u^3}{\partial y_2}A_{22}^2 - \frac{\mu(\vartheta)}{2y_1}\left[B_{22}\frac{\partial u^1}{\partial y_2}\right] -
$$

$$
\frac{\mu(\vartheta)}{2y_1}\left[\frac{\partial u^3}{\partial y_1} + B_{21}\frac{\partial u^1}{\partial y_2}\right] + u^1\left[B_{21}\frac{\partial u^3}{\partial y_2} + \frac{\partial u^3}{\partial y_1}\right] + u^3B_{22}\frac{\partial u^3}{\partial y_2} + B_{22}\frac{\partial \Psi}{\partial y_2} = 0 \ \text{ in } \Omega_0,
$$

$$
\text{(2.28)}\qquad \nabla\cdot\left(\,(\det B\,)^{-1}B\,(y,u)\,\right) = 0 \quad \text{in } \Omega_0,
$$

$$
\text{(2.29)}\qquad -\,(h_0' + g')\,u^1 + u^3 = 0 \ \text{ on } \gamma_0,
$$

$$
\text{(2.30)}\quad \mu(\vartheta)\left[-(h_0' + g')\left[\frac{\partial u^1}{\partial y_1} + B_{21}\frac{\partial u^1}{\partial y_2}\right] + \tfrac{1}{2}\left[B_{22}\frac{\partial u^1}{\partial y_2} + \frac{\partial u^3}{\partial y_1} + B_{21}\frac{\partial u^3}{\partial y_2}\right]\left(1-(h_0' + g')^2\right)\right]
$$

$$
+\mu(\vartheta)(h_0' + g')\,B_{22}\frac{\partial u^3}{\partial y_2} - \sigma'(\vartheta)\sqrt{1+(h_0' + g')^2}\left[\frac{\partial \vartheta}{\partial y_1} + B_{21}\frac{\partial \vartheta}{\partial y_2} + (h_0' + g')B_{22}\frac{\partial \vartheta}{\partial y_2}\right] = 0 \text{ on } \gamma_0,
$$

$$
\text{(2.31)}\qquad \mu(\vartheta)\left[(h_0' + g')\left[\frac{\partial u^2}{\partial y_1} + B_{21}\frac{\partial u^2}{\partial y_2}\right] - B_{22}\frac{\partial u^2}{\partial y_2}\right] = 0 \text{ on } \gamma_0,
$$

$$
\text{(2.32)}\qquad -\sigma(\vartheta)J(h_0' + g') + h_0 + g - \Psi + p_a + \frac{\mu(\vartheta)}{1+(h_0' + g')^2}\left[(h_0' + g')^2\left\{\frac{\partial u^1}{\partial y_1} + B_{21}\frac{\partial u^1}{\partial y_2}\right\}\right.
$$

$$
\left. -\,(h_0' + g')\left\{B_{22}\frac{\partial u^1}{\partial y_2} + \frac{\partial u^3}{\partial y_1} + B_{21}\frac{\partial u^3}{\partial y_2}\right\} + B_{22}\frac{\partial u^3}{\partial y_2}\right] = 0 \qquad \text{on } \gamma_0,
$$

$$
\text{(2.33)}\qquad u^1 = 0 \qquad \text{on } \gamma_1 \text{ and } \gamma_2,
$$

$$
\text{(2.34)}\qquad u^3 = 0 \ \text{ on } \gamma_3,
$$

$$
\text{(2.35)}\qquad -\tfrac{1}{2}\mu(\vartheta)\left[\frac{\partial u^3}{\partial y_1} + B_{21}\frac{\partial u^3}{\partial y_2}\right] + u^3 = 0 \text{ on } \gamma_1,
$$

$$
\text{(2.36)}\qquad -\tfrac{1}{2}\mu(\vartheta)\left[\frac{\partial u^2}{\partial y_1} + B_{21}\frac{\partial u^2}{\partial y_2}\right] + u^2 - \omega_1 = 0 \text{ on } \gamma_1,
$$

(2.37) $\frac{1}{2}\mu(\vartheta)\left[\frac{\partial u^3}{\partial y_1} + B_{21}\frac{\partial u^3}{\partial y_2}\right] + u^3 = 0$ on γ_2,

(2.38) $\frac{1}{2}\mu(\vartheta)\left[\frac{\partial u^2}{\partial y_1} + B_{21}\frac{\partial u^2}{\partial y_2}\right] + u^2 - \omega_2 = 0$ on γ_2,

(2.39) $-\frac{1}{2}\mu(\vartheta)\frac{\partial u^1}{\partial y_2} + u^1 = 0$ on γ_3,

(2.40) $-\frac{1}{2}\mu(\vartheta)\frac{\partial u^3}{\partial y_2} + u^2 - \omega_3 = 0$ on γ_3,

$$\sigma'(\vartheta)\left\{u^1\frac{\partial\vartheta}{\partial y_1} + u^1B_{21}\frac{\partial\vartheta}{\partial y_2} + u^3B_{22}\frac{\partial\vartheta}{\partial y_2}\right\} - \frac{\partial}{\partial y_1}\left[k(\vartheta)\frac{\partial\vartheta}{\partial y_1}\right] - \frac{\partial}{\partial y_2}\left[k(\vartheta)\frac{\partial\vartheta}{\partial y_1}\right]B_{21} -$$

$$\frac{\partial}{\partial y_1}\left[k(\vartheta)\frac{\partial\vartheta}{\partial y_2}\right]B_{21} - \frac{\partial}{\partial y_2}\left[k(\vartheta)\frac{\partial\vartheta}{\partial y_2}(B_{21})^2\right] - \frac{\partial}{\partial y_2}\left[k(\vartheta)\frac{\partial\vartheta}{\partial y_2}(B_{22})^2\right] - k(\vartheta)\frac{\partial\vartheta}{\partial y_2}A^2_{11} -$$

(2.41) $$k(\vartheta)\frac{\partial\vartheta}{\partial y_2}A^2_{22} - \frac{k(\vartheta)}{y_1}\left[\frac{\partial\vartheta}{\partial y_1} + B_{21}\frac{\partial\vartheta}{\partial y_2}\right] - \mu(\vartheta)\left[\frac{\partial u^1}{\partial y_1} + B_{21}\frac{\partial u^1}{\partial y_2}\right]^2 -$$

$$\mu(\vartheta)\left[\frac{\partial u^3}{\partial y_1} + B_{21}\frac{\partial u^3}{\partial y_2}\right]B_{22}\frac{\partial u^1}{\partial y_2} - \frac{\mu(\vartheta)}{2}\left[B_{22}\frac{\partial u^1}{\partial y_2}\right]^2 - \frac{\mu(\vartheta)}{2}\left[\frac{\partial u^3}{\partial y_1} + B_{21}\frac{\partial u^3}{\partial y_2}\right]^2 -$$

$$(\mu(\vartheta)\left[B_{22}\frac{\partial u^3}{\partial y_2}\right]^2 - \frac{\mu(\vartheta)y_1^2}{2}\left[\frac{\partial u^2}{\partial y_1} + B_{21}\frac{\partial u^2}{\partial y_2}\right]^2 - \frac{\mu(\vartheta)y_1^2}{2}\left[B_{22}\frac{\partial u^2}{\partial y_2}\right]^2 = 0 \text{ in } \Omega_0,$$

(2.42) $$-\frac{k(\vartheta)}{\sqrt{1+(h_0'+g')^2}}\left\{(h_0'+g')\left[\frac{\partial\vartheta}{\partial y_1} + B_{21}\frac{\partial\vartheta}{\partial y_2}\right] - B_{22}\frac{\partial\vartheta}{\partial y_2}\right\} - \beta_0(\vartheta - T_0) = 0 \text{ on } \gamma_0,$$

(2.43) $k(\vartheta)\left[\frac{\partial\vartheta}{\partial y_1} + B_{21}\frac{\partial\vartheta}{\partial y_2}\right] - \beta_1(\vartheta - T_1) = 0$ on γ_1,

(2.44) $-k(\vartheta)\left[\frac{\partial\vartheta}{\partial y_1} + B_{21}\frac{\partial\vartheta}{\partial y_2}\right] - \beta_2(\vartheta - T_2) = 0$ on γ_2,

(2.45) $k(\vartheta)\frac{\partial\vartheta}{\partial y_2} - \beta_3(\vartheta - T_3) = 0$ on γ_3,

As in [7] we regard the system (2.25) - (2.45) as an operator equation

(2.46) $\mathcal{P}(e, \lambda)=0$,

where $e:=((u^1,u^2,u^3),\Psi,g,\vartheta)$ and $\lambda:=(\omega_1, \omega_2, \omega_3, T_0, T_1, T_2, T_3)$ are the unknown functions and the variable parameters, respectively. Of course, it holds

$$\mathcal{P}(e_0, \lambda_0) = 0,$$

where the rest state is represented by

$$e_0:=(0, \Phi_0, 0, \hat{T}) \text{ and } \lambda_0:=(0, 0, 0, \hat{T}, \hat{T}, \hat{T}, \hat{T}), \text{ (see above).}$$

We want to solve (2.46) in a neighbourhood of (e_0,λ_0) applying the implicit function theorem and getting $e=e(\lambda)$. Let us define the operator $\mathcal{P}$:

$\mathcal{P}_1$:= the left-hand sides of (2.25), (2.26) and (2.27)

$\mathcal{P}_2 := div\,((\det B)^{-1}B(y_1,u))$

$\mathcal{P}_3, \ldots, \mathcal{P}_6$:= left-hand sides of (2.29) - (2.32), respectively,

$\mathcal{P}_7, \ldots, \mathcal{P}_{17}$:= the left-hand sides of (2.35) - (2.45), respectively.

We can show that

$$\mathcal{P} : E \times \Lambda \to F,$$

where

$E := E_1 \times E_2 \times E_3 \times E_4$, $\quad F := F_1 \times \ldots \times F_{17}$ with the condition

$$(2.47) \qquad \int_{\Omega_0} a\,dy_1 dy_2 = \int_{\gamma_0} \frac{b y_1}{\sqrt{1+h_0'^2}}\,ds \quad \text{for all } a \in F_2 \text{ and } b \in F_3,$$

$$\Lambda := C^{1+s}_{s-1}(\gamma_1) \times C^{1+s}_{s-1}(\gamma_2) \times C^{1+s}_{s-1}(\gamma_3) \times F_{14} \times F_{15} \times F_{16} \times F_{17}.$$

And the spaces E_i and F_j are defined in the following way:

$E_1 := \{v \in [C^{2+s}_s(\Omega_0)]^3 : v^1 = 0$ on γ_1 and γ_2, $v_3 = 0$ on $\gamma_3\}$, $\quad E_2 := C^{1+s}_{s-1}(\Omega_0)$

$E_3 := \{g \in C^{3+s}_{s+1}(I) : \int_{R_1}^{R_2} rg\,dr = 0,\ g'(R_i) = 0,\ i=1,2\}$, $\quad E_4 := C^{2+s}_s(\Omega_0)$,

$F_1 := [C^s_{s-2}(\Omega_0)]^3$, $\quad F_2 := C^{1+s}_{s-1}(\Omega_0)$, $\quad F_3 := C^{2+s}_s(I)$, $\quad F_4 = F_5 := C^{1+s}_{s-1}(I)$,

$F_6 = F_7 := C^{1+s}_{s-1}(\gamma_1)$, $\quad F_8 = F_9 := C^{1+s}_{s-1}(\gamma_2)$, $\quad F_{10} = F_{11} := C^{1+s}_{s-1}(\gamma_3)$, $\quad F_{12} = F_{14} := F_4$,

$F_{13} := C^s_{s-2}(\Omega_0)$, $\quad F_{15} := F_6$, $\quad F_{16} := F_8$, $F_{17} := F_{10}$.

At first let the weight exponent s be arbitrary with $0<s<1$, it will be determined later when proving the lemmas. We remark that the domain Ω_0 has four corners and so each corner may have it own weight coefficient. For convenience we put the smallest one, denote it by s and bound it by $0<s<1$.

Further, we suppose $\|g\|_{E_3} \le \delta_0$ (cf. page 6, above) and we will see that the implicit function theorem is applicable. The main part of this procedure consists in proving the existence of a bounded inverse of the partial Frechét derivative $\mathcal{P}_E(e_0, \lambda_0)$. We return to this matter in the next section.

3. Formulation of the result. Auxillary problems

Theorem 3.1: Let the condition (2.24) be given. Further, we assume

$$\mu, \varepsilon, k, \sigma \in C^3(\mathbb{R}), \quad h_0(r)>0 \text{ in } I, \quad \mu(\hat{T})>0, \quad k(\hat{T})>0, \quad \sigma(\hat{T})>0.$$

Then there exist a weight exponent s with $0<s<1$ and two positive reals d_1 and d_2, such that for each $\lambda=(\omega_1, \omega_2, \omega_3, T_0, T_1, T_2, T_3)\in B(\lambda_0,d_2)\subset\Lambda$ a unique solution $e=e(\lambda)=(v, \Phi, g, T)\in B(e_0,d_1)\subset E$ to equation (2.46) exists. In addition, $e=e(\lambda)$ is continuous in $B(\lambda_0,d_2)$ and analytic, if μ, ε, k, σ are analytic too.

The proof of this theorem follows from the lemmas below.

The Theorem 3.1 is a special application of the well-known implicit function theorem. According to this we have to verify the following assertions

(A) $\mathcal{P} : U(e_0) \subset E \times \Lambda \to F$, ($U(e_0)$ is a neighbourhood of e_0).

(B) i) $\mathcal{P}$ is continuous on $U(e_0) \times \Lambda$.

ii) $\mathcal{P}$ is partially with respect to E Fréchét differentiable on $U(e_0) \times \Lambda$.

iii) This partial derivative $\mathcal{P}_E$ is continuous at the point (e_0,λ_0).

(C) For each $f\in F$ there exists uniquely one $e\in E$ as a solution of the (linear) equation

(3.1) $\quad \mathcal{P}_E(e_0,\lambda_0)e = f$

satisfying the following estimate

(3.2) $\quad ||e||_E < c||f||_F$.

Clearly, it is not difficult to prove the assertions (A) and (B) under the given assumptions. The substancial part of the proof of Theorem 3.1 is the verification of (C). After calculating the partial derivative $\mathcal{P}_E(e_0,\lambda_0)$ we write the equation (3.1) in the following expanded form

$$-\mu_0\frac{\partial^2 u^1}{\partial y_1^2} - \frac{\mu_0}{2}\frac{\partial^2 u^1}{\partial y_2^2} - \frac{\mu_0}{2}\frac{\partial^2 u^3}{\partial y_1 \partial y_2} - \frac{\mu_0 u^1}{y_1^2} - \frac{\mu_0}{y_1}\frac{\partial u^1}{\partial y_1} + \frac{\partial \Psi}{\partial y_1} = f_1 \text{ in } \Omega_0, \tag{3.3}$$

$$-\frac{\mu_0}{2}\frac{\partial^2 u^2}{\partial y_1^2} - \frac{\mu_0}{2}\frac{\partial^2 u^2}{\partial y_2^2} - \frac{3\mu_0}{2y_1}\frac{\partial u^2}{\partial y_1} = f_2 \text{ in } \Omega_0, \tag{3.4}$$

$$-\frac{\mu_0}{2}\frac{\partial^2 u^3}{\partial y_1^2} - \frac{\mu_0}{2}\frac{\partial^2 u^1}{\partial y_1 \partial y_2} - \mu_0\frac{\partial^2 u^3}{\partial y_2^2} - \frac{\mu_0}{2y_1}\left\{\frac{\partial u^1}{\partial y_2} + \frac{\partial u^3}{\partial y_1}\right\} + \frac{\partial \Psi}{\partial y_2} = f_3 \text{ in } \Omega_0, \tag{3.5}$$

$$\frac{\partial}{\partial y_1}(y_1 u^1) + \frac{\partial}{\partial y_3}(y_1 u^3) = m \quad \text{in } \Omega_0, \tag{3.6}$$

(3.7) $$- \frac{y_1}{(1+h_0'^2)^{1/2}} (-h_0' u^1 + u^3) = f_4 \text{ on } \gamma_0,$$

(3.8) $$\mu_0\left\{-h_0' \frac{\partial u^1}{\partial y_1} + \tfrac{1}{2} \left(\frac{\partial u^1}{\partial y_2} + \frac{\partial u^3}{\partial y_1}\right)(1-h_0'^2) + h_0' \frac{\partial u^3}{\partial y_2}\right\} + \\ - \sigma'(\hat{T}) (1+h_0'^2)^{1/2} \left[\frac{\partial \vartheta}{\partial y_1} + h_0' \frac{\partial \vartheta}{\partial y_3}\right] = f_5 \text{ on } \gamma_0,$$

(3.9) $$\mu_0\left(h_0' \frac{\partial u^2}{\partial y_1} - \frac{\partial u^2}{\partial y_2}\right) = f_6 \text{ on } \gamma_0,$$

(3.10) $$- \sigma'(\hat{T})\vartheta\, y_1^{-1} \frac{\partial}{\partial y_1} (y_1 h_0' (1+h_0'^2)^{-1/2}) - \sigma_0 y_1^{-1} \frac{\partial}{\partial y_1} (y_1 g' (1+h_0'^2)^{\frac{-3}{2}}) + \\ + g - \Psi + \frac{\mu_0}{1+h_0'^2} \left\{h_0'^2 \frac{\partial u^1}{\partial y_1} - h_0' \left(\frac{\partial u^1}{\partial y_2} + \frac{\partial u^3}{\partial y_1}\right) + \frac{\partial u^3}{\partial y_2}\right\} = f_7 \text{ on } \gamma_0,$$

(3.11) $$- \frac{\mu_0}{2} \frac{\partial u^3}{\partial y_1} + u^3 = f_8 \text{ on } \gamma_1,$$

(3.12) $$- \frac{\mu_0}{2} \frac{\partial u^2}{\partial y_1} + u^2 = f_9 \text{ on } \gamma_1,$$

(3.13) $$\frac{\mu_0}{2} \frac{\partial u^3}{\partial y_1} + u^3 = f_{10} \text{ on } \gamma_2,$$

(3.14) $$\frac{\mu_0}{2} \frac{\partial u^2}{\partial y_1} + u^2 = f_{11} \text{ on } \gamma_2,$$

(3.15) $$- \frac{\mu_0}{2} \frac{\partial u^1}{\partial y_1} + u^1 = f_{12} \text{ on } \gamma_3,$$

(3.16) $$- \frac{\mu_0}{2} \frac{\partial u^2}{\partial y_1} + u^2 = f_{13} \text{ on } \gamma_3,$$

(3.17) $$- k_0 \frac{\partial^2 \vartheta}{\partial y_1^2} - k_0 \frac{\partial^2 \vartheta}{\partial y_2^2} - \frac{k_0}{y_1} \frac{\partial \vartheta}{\partial y_1} = f_{14} \quad \text{in } \Omega_0,$$

(3.18) $$k_0\left\{h_0' (1+h_0'^2)^{-1/2} \frac{\partial \vartheta}{\partial y_1} - (1+h_0'^2)^{-1/2} \frac{\partial \vartheta}{\partial y_2}\right\} - \beta_0 \vartheta = f_{15} \text{ on } \gamma_0,$$

(3.19) $$k_0 \frac{\partial \vartheta}{\partial y_1} - \beta_1 \vartheta = f_{16} \text{ on } \gamma_1,$$

(3.20) $$- k_0 \frac{\partial \vartheta}{\partial y_1} - \beta_2 \vartheta = f_{17} \text{ on } \gamma_2,$$

(3.21) $$k_0 \frac{\partial \vartheta}{\partial y_2} - \beta_3 \vartheta = f_{18} \text{ on } \gamma_3,$$

Here the positive constants μ_0, k_0 and σ_0 are the abbrivations for $\mu(\hat{T})$, $k(\hat{T})$ and $\sigma(\hat{T})$, respectively.

Of course, the equations (3.3) - (3.21) form a system of linear elliptic PDEs and of an ODE completed by (linear) boundary conditions. Fortunately, this system is

partially decoupled and so we can solve the auxillary problem (3.17) - (3.21) for ϑ at first. After this the remaining auxillary problems (3.4), (3.9), (3.12), (3.14) and (3.16) for u^2, (3.3), (3.5), (3.6), (3.7), (3.8), (3.11), (3.13) and (3.15) for $\{u^1, u^3, \Psi\}$ and (3.10) for g will be solved. Of course, some boundary conditions are also involved in the definitions of the spaces E_i. Applying the approach due to Kondratjev [1], we have to study related problems to these auxillary problems for T, v^2 and $\{v^1, v^3, \Phi\}$, respectively, in an unbounded sector (cf. [3, 6, 7] for details). But in difference to [3, 7] we have another boundary conditions as a consequence of the slip-condition.
Now we will deal with the auxillary problems.

Lemma 3.1: There exists a real number s_1 with $0<s_1<1$, so that for all $s\in(0,s_1]$ and for all $f_{14}\in F_{13}$, $f_{15}\in F_{14}$, $f_{16}\in F_{15}$, $f_{17}\in F_{16}$ and $f_{18}\in F_{17}$ exactly one solution $\vartheta\in E_4$ exists to the problem (3.17) - (3.21). Moreover, the following estimates holds

(3.22) $$\|\vartheta\|_{E_4} \leq c(\|f_{14}\|_{F_{13}} + \|f_{15}\|_{F_{14}} + \|f_{16}\|_{F_{15}} + \|f_{17}\|_{F_{16}} + \|f_{18}\|_{F_{17}})$$

(c is independent of ϑ).

Proof: As in [3] the approach due to Kondratjev [1, 7] is applicable to the problem (3.17) - (3.21). At first the existence of a weak solution is proved. Step by step the needed regularity can be established. The cruzial point is the investigation of a model problem in an unbounded sector (cf. [6] for details of this approach). The Mellin transform of this model problem yields to an ODE depending on a parameter. The smallest eigenvalue of this eigenvalue problem determines the weight exponent of the weighted Hölder space E4 and the estimate (3.22) can be established. The main difference to Lagunova [3] consists in the boundary conditions. Because of we have slip-conditions on the cylinder walls, we get the auxillary problem in the sector with Neumann type conditions on both sides of the angle and the characteristic equation is

$\sin \sigma\beta = 0$ (β stands for β_1 and β_2, respectevely)

with the smallest positive root $\sigma_0 = \frac{\pi}{\beta}$. After some calculation we see that the weight s_1 must satisfy the estimate:

(3.23) $$0 < s_1 < \min\{1, \frac{\pi}{\beta_1}, \frac{\pi}{\beta_2}\}$$

(We take s_1 smaller then 1 in order to avoid further compatibility conditions). ■

Lemma 3.2: There exists a real number s_2 with $0<s_2<1$, so that for all $s\in(0,s_2]$ and for all $f_2\in C^s_{s-2}(\Omega_0)$, $f_6\in F_5$, $f_9\in F_7$, $f_{11}\in F_9$ and $f_{13}\in F_{10}$ exactly one solution $u^2\in E_4$ exists to the problem (3.4), (3.9), (3.12), (3.14) and (3.16). Moreover, the following estimates holds

(3.24) $$||u^2||_{E_4} \leq c(||f_2||_{C^s_{s-2}(\Omega_0)} + ||f_6||_{F_5} + ||f_9||_{F_7} + ||f_{11}||_{F_9} + ||f_{13}||_{F_{10}})$$

(c is independent of u^2).

Proof: We remark that the auxillary problems for u^1 and ϑ have the same structure. Clearly, the proofs are equal, too. Further, s_2 satisfies the condition (3.23). ■

Lemma 3.3: Let the assumptions of Lemma 3.1 be given. Then there exists a real number s_3 with $0<s_3\leq s_1$, so that for all $s\in(0,s_3]$ and for all f_1, $f_3\in C^s_{s-2}(\Omega_0)$, $m\in F_2$, $f_4\in F_3$, (with m and f_4 satisfying the compatibility condition (2.)) $f_5\in F_5$, $f_8\in F_7$, $f_{10}\in F_9$ and $f_{12}\in F_{10}$ exactly one solution

$\{u^1, 0, u^3, \Psi\} \in E_1 x E_2$ with $\int_{\Omega_0} \Psi dy = 0$ exists to the problem (3.3), (3.5), (3.6), (3.7), (3.8), (3.11), (3.13) and (3.15). Moreover, the following estimates holds

(3.24) $$||u^1,0,u^3||_{E_1} + ||\Psi||_{E_2} \leq c(||f_1,f_3||_{[C^s_{s-2}(\Omega_0)]^2} + ||m||_{F_2} + ||f_4||_{F_3} + ||f_5||_{F_5} +$$

$$+ ||f_8||_{F_7} + ||f_{10}||_{F_9} + ||f_{12}||_{F_{10}} + ||f_{14}||_{F_{13}} + ||f_{15}||_{F_{14}} + ||f_{16}||_{F_{15}} + ||f_{17}||_{F_{16}} + ||f_{18}||_{F_{17}})$$

(c is independent of $\{u^2, 0, u^3, \Psi\}$).

Proof: The problem in Lemma 3.3 is a Stokes problem with mixed Dirichlet and Neumann boundary condition for the components u^1 and u^3. The proof is similar to the proof of the corresponding problems in [3, 7] and it has the general scheme as discribted above in the proof of Lemma 3.1. In difference to those authors we deal with slip-conditions on the rigid walls. As a consequence of this the model problem in the sector involves another boundary conditions and the inspection of the corresponding eigenvalue problem (cf. [6, 7]) yields the characteristic equation

$\sigma^2 \cos 2\sigma\beta - \sigma^2 \cos 2\beta = 0$ (β stands for β_1 and β_2, respectevely).

By some calculation (cf. [6]) we get the following estimate

$$0<s_3<\begin{cases} 1, & \text{if } \beta_1, \beta_2<\frac{\pi}{2} \\ \min\{\frac{\pi}{\beta_1}-1, \frac{\pi}{\beta_2}-1\}, & \text{if } \frac{\pi}{2}<\beta_1, \beta_2<\pi \end{cases} \quad . \quad \blacksquare$$

To the end, we deal with the problem (3.10). Clearly, this is a linear Sturm-Liouville problem.

Lemma 3.4: Let the full data $f\in F$ be given and let s be the minimum of s_2 and s_3. Then there exists a unique solution $g\in E_3$ to the problem (3.10) satisfying the estimate

(3.25) $$\|g\|_{E_3} \le c\|f\|_F$$

(c is independent of f).

Proof: After calculation of u^1, u^2, u^3, Ψ and ϑ the equation (3.10) is a pure Sturm-Liouville problem for g. In order to satisfy the integral condition in the definition of the space E_3 we take $\Psi+a$ instead of Ψ (a being a real constant). Then a is determined by (3.10) and the the boundary conditions of g. As in [3, 7, 9] the Sturm-Liouville problem can be solved and the needed regularity will be obtained.

Of course, we can go back to Cartesian coordinates and the original problem (1.1) - (1.11) has a unique solution for small data, i.e., for small rotary speeds and small deviations from the reference temperature. ■

The author is indebted to Professor Solonnikov and Dr. Socolowsky for fruitful discussions when preparing this paper.
This work will not be published at other places.

References:

[1] Kondratjev, V.A.: Boundary value problems for elliptic equations in domains with conical and regular points. - Trudy Moskov. Mat. Obsc. 16(1967), 209-292 (in Russian).

[2] Kröner, D.: Asymptotische Entwicklungen von Flüssigkeiten mit freiem Rand und dynamischem Kontaktwinkel. - Preprint no. 806(1986), SFB 72, Universität Bonn.

[3] Lagunova, M.V.: On solvability of the plane problem of thermo-capillary convektion. - Probl. Mat. Ana. 10(1986), 33-47(in Russian), Leningrad State university.

[4] Sattinger, D.H.: On the free surface of a viscous fluid motion. - Proc. R. Soc. London A 349(1976), 183-204.

[5] Socolowsky, J.: Solvebility of stationary problems of the plane motion of two viscous incompressible liquids with non-compact free boudary. - to appear in ZAMM 71(9)(1991).

[6] Socolowsky, J, Wolff, M.: On some model problems in flow problems with contact angles.- to appear in ZAMM

[7] Solonnikov, V.A.: On the Stokes equations in domains with non-smooth boundaries and on viscous incompressible flow with a free surface. - in "Nonlinear partial differential equations and their applications. College de France, Seminar",vol. 3(1983), 340-423, Ed.: H. Brezis, J.L. Lions.

[8] Wolff, M.: Mathematische Untersuchungen gewisser Klassen einfacher Flüssigkeiten mit teilweise freier Oberfläche. - Berlin, Humboldt-Universität, Sektion Mathematik, Diss. A.

[9] Wolff, M.: Stationary flow of heat-conducting fluids with free surface between concentric cylinders. - Berlin, Humboldt -Universität, Sektion Mathematik, Seminarbericht Nr. 99(1988).

On some coercive estimates for the Stokes problem in unbounded domains

Wolfgang BORCHERS and Tetsuro MIYAKAWA

1. Statement of the results

Let Ω be an exterior domain in $\mathbb{R}^n$, $n \geq 3$, with smooth boundary Γ. We denote by $\mathbf{X}_r$ the $\mathbf{L}^r$-closure of $\mathbf{C}_{0,\sigma}^\infty(\Omega)$, the set of smooth solenoidal vector fields with compact support in Ω. Using the Helmholtz decomposition [18]:

$$\mathbf{L}^r(\Omega) = \mathbf{X}_r \oplus \mathbf{G}_r \quad (1 < r < \infty) \tag{1.1}$$

with

$$\mathbf{G}_r = \{\nabla p \in \mathbf{L}^r(\Omega);\ p \in L^r_{loc}(\overline{\Omega})\}$$

and the associated bounded projector $P = P_r$ onto $\mathbf{X}_r$, we define the Stokes operator $A = A_r$ as

$$A_r v = -P_r \Delta v, \quad for \quad v \in D(A_r) = \mathbf{W}^{2,r}(\Omega) \cap \mathbf{W}_0^{1,r}(\Omega) \cap \mathbf{X}_r.$$

Here and in what follows we use the standard notation for the Sobolev spaces [1]. The norm in $\mathbf{W}^{m,r}(\Omega)$ and in $W^{m,r}(\Omega)$ is denoted by $\|\cdot\|_{m,r} = \|\cdot\|_{m,r,\Omega}$, where bold face letters always indicate spaces of $\mathbb{R}^n$-valued functions. As is well known, the operator A_r is a densely defined closed linear operator in $\mathbf{X}_r$ and solving the equation $A_r v = f$ for given $f \in \mathbf{X}_r$ amounts to solving the exterior stationary Stokes problem

$$\begin{aligned} -\Delta v + \nabla p &= f, \ (x \in \Omega) \\ \nabla \cdot v &= 0, \ (x \in \Omega) \\ v|_\Gamma &= 0, \end{aligned} \tag{S}$$

in $\mathbf{L}^r$ space. Since $-A_r$ generates a bounded analytic semigroup in $\mathbf{X}_r$ [6], the fractional powers $A_r^\alpha, \alpha \geq 0$, are well defined [5,6,11].

In this paper we study the estimates of the following form:

$$\|\partial^2 v\|_r \leq C\|Av\|_r, \quad v \in D(A_r), \tag{1.2}$$

$$\|\partial v\|_r \leq C\|A^{\frac{1}{2}} v\|_r, \quad v \in D(A_r^{1/2}), \tag{1.3}$$

$$\|\partial v\|_{r^*} \leq C\|Av\|_r, \qquad v \in D(A_r),\ 1 < r < n,\ 1/r^* = 1/r - 1/n, \tag{1.4}$$

where $\|\cdot\|_r = \|\cdot\|_{0,r}$ is the usual $\mathbf{L}^r$-norm and $\partial^m v$ stands for general m-th order derivatives of the function v. The inequalities reverse to (1.2) and (1.3) are now well known. Indeed, the definition of the Stokes operator A_r immediately implies

$$\|Av\|_r \leq C\|\partial^2 v\|_r \quad for\ all \quad 1 < r < \infty. \tag{1.5}$$

Furthermore, it is shown in [5] that

$$\|A^{1/2} v\|_r \leq C\|\partial v\|_r \quad for\ all\ 1 < r < \infty. \tag{1.6}$$

In contrast to these two estimates, (1.2) and (1.3) have been established, for all $1 < r < \infty$, only in the cases of the whole space $\mathbb{R}^n$, the halfspace $\mathbb{R}^n_+$, and bounded domains in $\mathbb{R}^n$, $n \geq 2$. For the whole spaces, they are easily obtained from the Calderón-Zygmund theory on singular integrals [24]; the case of the halfspaces is discussed in [4]. For the case of bounded domains we refer to [10]. Using these results, we can deduce the so-called $\mathbf{L}^q$-$\mathbf{L}^r$ estimates in the most general form for the semigroup generated by $-A$ on $\mathbb{R}^n$, $\mathbb{R}^n_+$, and bounded domains; see for instance [4,11]. As for the case of the exterior problem in $\mathbb{R}^n$, $n \geq 3$, estimate (1.2) is established in [6] only for $1 < r < n/2$, and (1.3) is proved in [5] only for $1 < r < n$. Consequently, the resulting $\mathbf{L}^q$ -$\mathbf{L}^r$ estimates for the corresponding semigroup are not so satisfactory as compared with the above-mentioned cases (see [5,6,12,14]). It is therefore desirable to establish (1.1) and (1.2) for a broader range of r. In this paper, however, we give a negative answer to the problem stated above. Namely, we shall prove

THEOREM 1.1. . *Let $n \geq 3$ and consider the Stokes operator over an exterior domain in $\mathbb{R}^n$ with smooth boundary. Then :*

(i) Estimate (1.2) is not valid for any r such that $n/2 \leq r < \infty$.

(ii) Estimate (1.3) is not valid for any r such that $n \leq r < \infty$.

(iii) Estimate (1.4) holds if and only if $1 < r < n/2$.

Assertion (i) is also investigated in [8,9,17] (compare also the remark in [5, p. 207]) and will be reproved here since our method immediately implies Assertion (ii) above. Estimate (1.2) is found in Shinbrot [20] in the case of bounded domains. In view of the result (i), it would be impossible to deduce estimates (1.2) and (1.4) in the two-dimensional case. On the other hand, estimate (1.3) is trivial in case $n = r = 2$, because A_2 is the positive self-adjoint operator in $\mathbf{X}_2$ associated to the bilinear form $(\nabla u, \nabla v)$ defined on $\mathbf{X}_2 \cap \mathbf{W}_0^{1,2}(\Omega)$.

Our next theorem gives a positive answer for a class of domains with noncompact boundary. Let $D \subset \mathbb{R}^n$ $(n \geq 2)$ be a compactly perturbed halfspace with a smooth boundary, i.e. there exists a ball $B_{\rho_0} = \{|x| \geq \rho_0\} \subset \mathbb{R}^n$ $(\rho_0 \geq 0)$ such that

$$D \backslash B_{\rho_0} = \mathbb{R}^n_+ \backslash B_{\rho_0}, \tag{1.7}$$

where $\mathbb{R}^n_+ = \{x = (x_1, \ldots, x_n) \in \mathbb{R}^n : x_n > 0\}$. Recall that the Poincaré inequality is not valid for D.

THEOREM 1.2. . *Let $n \geq 2$, $1 < r < n$ and let $D \subset \mathbb{R}^n$ satisfy (1.7). Then*

(i) For $v \in \mathbf{W}^{2,r}(D) \cap \mathbf{W}_0^{1,r}(D)$ with div $v = 0$ and $p \in W^{1,r}(D)$ there holds the estimate

$$\|\partial^2 v\|_r + \|\nabla p\|_r \leq C\| - \Delta v + \nabla p\|_r, \tag{1.8}$$

with a constant $C > 0$ not depending on v, p.

(ii) Estimate (1.8) is not valid for any r such that $n \leq r < \infty$, unless $D = \mathbb{R}^n_+$.

Note that the case $n = r = 2$ in (1.8) is excluded, so that even the Hilbert space structure can not improve the range of r. Using the projector P_r one can show that (1.8) is equivalent to (1.2). However, to avoid here the construction of P_r for the domain D, we have used the equivalent formulation (1.8) instead of (1.2) (compare [21]).

Throughout, $C > 0$ is a constant which may vary from line to line.

2. Proof of Theorem 1.1

To show Theorem 1.1, we need the following result of Bogovski [3]:

THEOREM 2.1. *Let Ω_b be a bounded domain in $\mathbb{R}^n$, $n \geq 2$, with locally Lipschitz boundary. Then there is a linear operator S_{Ω_b} from $C_0^\infty(\Omega_b)$ to $\mathbf{C}_0^\infty(\Omega_b)$ such that, for $1 < r < \infty$ and $m = 0, 1, 2, \ldots$,*

$$\|\partial^{m+1} S_{\Omega_b} f\|_r \leq C \|\partial^m f\|_r$$

with a constant $C = C(m, r, \Omega_b)$, and

$$\nabla \cdot S_{\Omega_b} f = f \text{ holds provided } \int_{\Omega_b} f\, dx = 0.$$

A full proof of Theorem 2.2 is given in [7]. As a consequence of its proof we obtain

COROLLARY 2.2. *Let $n \geq 2$, $y \in \mathbb{R}^n$, $0 \neq t \in \mathbb{R}$, and let*

$$\Omega_b(y,t) = \{(1-t)y + tx\, ;\ x \in \Omega_b\}.$$

Then, the constant $C(m, r, \Omega_b(y,t))$ associated to the operator $S_{\Omega_b(y,t)}$ is independent of y and t.

See [7, Theorem 2.10]. We now begin the proof of Theorem 1.1. Without loss of generality, we may assume that $0 \notin \overline{\Omega}$.

(a) Suppose first $n/2 < r < \infty$. We show that estimate (1.2) is not valid. To do so, consider the Stokes problem:

$$\begin{aligned} -\Delta v + \nabla p &= 0, && (x \in \Omega) \\ \nabla \cdot v &= 0, && (x \in \Omega) \\ v|_\Gamma = 0; \quad v &\to c \text{ as } |x| \to \infty. && \end{aligned} \tag{2.1}$$

for any given constant vector $c \neq 0$. By the substitution $v = w + c$, Problem (2.1) is transformed into

$$\begin{aligned} -\Delta w + \nabla p &= 0, && (x \in \Omega) \\ \nabla \cdot w &= 0, && (x \in \Omega) \\ w|_\Gamma = -c; \quad w &\to 0 \text{ as } |x| \to \infty. && \end{aligned} \tag{2.2}$$

Problem (2.2) is uniquely solved with the aid of the theory of hydrodynamical potentials as given in [16, Chap.3], and the solution $v = w + c$ of Problem (2.2) and the associated pressure p satisfy

$$|v(x) - c| = \mathcal{O}(|x|^{2-n}), \ \ |\partial v(x)| = \mathcal{O}(|x|^{1-n}), \ \ |\partial^2 v(x)| = \mathcal{O}(|x|^{-n}), \tag{2.3}$$

and

$$|p(x)| = O(|x|^{1-n}), \ \ |\partial p(x)| = O(|x|^{-n}) \tag{2.4}$$

as $|x| \to \infty$. Although [16, Chap.3] discusses only the case $n = 3$, one can easily generalize the result to higher dimensions, starting from the expression of the fundamental solution of the Stokes equations as given in [19]. We thus conclude that $0 < \|\partial^2 v\|_q < \infty$ for all $1 < q < \infty$. Now fix a function $\psi \in C_0^\infty(I\!R^n)$ so that $0 \le \psi \le 1$ and

$$\psi(x) = \begin{cases} 1 & (|x| \le 1) \\ 0 & (|x| \ge 2) \end{cases}$$

and set

$$\psi_N(x) = \psi(x/N) \ \ for \ \ N = 1, 2, \ldots .$$

Using these functions, we define

$$v_n = v\psi_N - S_N(v \cdot \nabla\psi_N) \tag{2.5}$$

where S_N is the operator given in Theorem 2.1 with respect to the bounded domain

$$\Omega_{b,N} = \{x\,;\ N < |x| < 2N\}.$$

Since $S_N(v \cdot \nabla\psi_N) \in \mathbf{W}_0^{2,r}(\Omega_{b,N})$, we can regard $S_N(v \cdot \nabla\psi_N) \in \mathbf{W}^{2,r}(I\!R^n)$ by defining it to be 0 outside $\Omega_{b,N}$. Obviously, $v_N \in \mathbf{W}^{2,r}(\Omega) \cap \mathbf{W}_0^{1,r}(\Omega)$ for large N. By the relation $v \cdot \nabla\psi_N = \nabla \cdot (v\psi_N)$ and the divergence theorem we see that, if ν denotes the unit outward normal to the boundary $\partial\Omega_{b,N}$, then for large N,

$$\begin{aligned} \int_{\Omega_{b,N}} v \cdot \nabla\psi_N \, dx &= \int_{\Omega_{b,N}} \nabla \cdot (v\psi_N)\, dx = \int_{\{|x|=N\}} v \cdot \nu \, dS \\ &= -\int_{\Omega \cap \{|x|<N\}} \nabla \cdot v \, dx = 0. \end{aligned}$$

Theorem 2.1 thus implies

$$\nabla \cdot S_N(v \cdot \nabla\psi_N) = v \cdot \nabla\psi_N$$

and therefore $v_N \in D(A_r)$ for large N. We now show that

$$\|\partial^2(v - v_N)\|_r \to 0 \ \ as \ \ N \to \infty. \tag{2.6}$$

To this end we write

$$\begin{aligned} \|\partial^2(v - v_N)\|_r &\le C\left(\|(1-\psi_N)\partial^2 v\|_r + \|v\partial^2\psi_N\|_r + \|\partial\psi_N \cdot \partial v\|_r\right) \\ &\quad + \|\partial^2 S_N(v \cdot \nabla\psi_N)\|_r \\ &\equiv I_1(N) + I_2(N) + I_3(N) + I_4(N). \end{aligned}$$

We easily see that $I_1(N) \to 0$. For $I_2(N)$ we have

$$I_2(N) \leq CN^{-2} \left[\int_{N \leq |x| \leq 2N} dx \right]^{\frac{1}{r}} \leq CN^{-2+n/r} \to 0.$$

Since (2.3) implies $|\partial v| \in L^r$,

$$I_3(N) \leq CN^{-1} \left[\int_{N \leq |x| \leq 2N} |\partial v|^r \, dx \right]^{\frac{1}{r}} \leq CN^{-2+n/r} \to 0.$$

By Theorem 2.1 and Corollary 2.2,

$$I_4(N) \leq C\|\partial(v \cdot \nabla\psi_N)\|_r \leq C[I_2(N) + I_3(N)] \to 0$$

with C independent of N. Observe that, by Corollary 2.2, the constant $C(m, r, N)$ associated to the operator S_N is independent of N. This proves (2.6). In particular, we have

$$\|\partial^2 v_N\|_r \to \|\partial^2 v\|_r \neq 0; \quad \Delta v_N \to \Delta v \quad \text{in } \mathbf{W}^r(\Omega). \tag{2.7}$$

Since $\Delta v = \nabla p \in \mathbf{L}^r$, we get $P\Delta v = P(\nabla p) = 0$; so (2.7) yields

$$\|Av_N\|_r = \|P\Delta v_N\|_r \to \|P\Delta v\|_r = 0. \tag{2.8}$$

From (2.7) and (2.8) we conclude that (1.2) is not valid for $n/2 < r < \infty$.

(b) Suppose next that $r = n/2$. In this case, the foregoing calculation shows that $I_j(N) \to 0$ for $j = 1, 3$, and $I_2(N)$ and $I_4(N)$ are bounded. Hence $\|\partial^2 v_N\|_{n/2}$ are bounded. But, since the set $\Omega_{b,N}$, which contains the support of $S_N(v \cdot \nabla\psi_N)$, goes to the spatial infinity as $N \to \infty$, the definition (2.5) of v_N implies

$$(\partial^2 v_N, \varphi) \to (\partial^2 v, \varphi) \quad as \quad N \to \infty,$$

for all $\varphi \in C_0^\infty(\Omega)^{n^3}$. Hence $\partial^2 v_N$ converges weakly to $\partial^2 v$ in $L^{n/2}$. We can now apply Mazur's theorem [25, p.120] on weak convergence to see that, if we chooose a suitable convex combination $\hat{v}_N$ of $\{v_k;\ k \geq N\}$ for each N, then (2.6) and (2.7) hold with r and v_N replaced by $n/2$ and $\hat{v}_N$, respectively. Since $\Delta v = \nabla p \in \mathbf{L}^{n/2}$, we conclude that

$$\|A\hat{v}_N\|_{n/2} \to \|P\Delta v\|_{n/2} = 0$$

contradicting (1.1). This proves (i).

We next prove (ii). Let $n < r < \infty$. Since $\partial(v_n - v)$ identically vanishes in a neighborhood of the boundary S, the Sobolev inequality applies to yield

$$\begin{aligned} \|\partial(v_N - v)\|_r &\leq C\|\partial^2(v_N - v)\|_s \to 0 \\ \|\partial(\hat{v}_N - v)\|_n &\leq C\|\partial^2(\hat{v}_N - v)\|_{n/2} \to 0, \end{aligned} \tag{2.9}$$

where $1/s = 1/r + 1/n$. We thus have

$$\|\partial v_N\|_r \to \|\partial v\|_r \neq 0; \ \|\partial\hat{v}_N\|_n \to \|\partial v\|_n \neq 0. \tag{2.10}$$

From (2.9) and (1.6) it follows that

$$A^{1/2}v_N \to u \ \text{in } \mathbf{X}_r\,; \quad A^{1/2}\hat{v}_N \to w \ \text{in } \mathbf{X}_n$$

But, we already know that

$$\lim_{N\to\infty} \|Av_N\|_r = 0 \ \text{ and } \limsup_{N\to\infty} \|A\hat{v}_N\|_n \le \limsup_{N\to\infty} \|Av_N\|_n = 0.$$

Since the operators $A_r^{1/2}$ are all closed, we see that $u \in D(A_r^{1/2}), w \in D(A_n^{1/2})$ and

$$A_r^{1/2}u = 0; \quad A_n^{1/2}w = 0.$$

Since $A_r^{1/2}$ are injective (see [5]), we obtain $u = w = 0$; so we conclude that

$$\|A^{1/2}v_N\|_r \to 0; \quad \|A^{1/2}\hat{v}_n\|_n \to 0. \tag{2.11}$$

Combining (2.10) and (2.11) gives assertion (ii).

We finally prove Assertion (iii). When $1 < r < n/2$, estimate (1.2) is valid. So, by a Sobolev-type inequality as given in [6,7] we obtain

$$\|\partial u\|_{r^*} \le C\|\partial^2 u\|_r \le C\|Au\|_r$$

which shows (1.4) for $1 < r < n/2$. On the other hand, if $n/2 < r < n$, then $n < r^* < \infty$. Thus, the foregoing argument shows

$$\|\partial v_N\|_{r^*} \to \|\partial v\|_{r^*} \ne 0; \quad \|Av_n\|_r \to 0,$$

and

$$\|\partial \hat{v}_N\|_n \to \|\partial v\|_n \ne 0; \quad \|A\hat{v}_N\|_{n/2} \to 0.$$

Hence, (1.4) is not valid for $n/2 \le r < n$. The proof is complete.

3. Proof of Theorem 1.2

As is clear from the preceding investigation, the main part of the proof will be a uniqueness statement for solutions v, p of (S) having finite second and first order L^r-Dirichlet integral respectively.

In the following we need the homogeneous Sobolev space $\hat{W}^{1,r}(D)\,(1 < r < n)$, which is the completion of $W^{1,r}(D)$ with respect to the first order L^r-Dirichlet norm $\|\nabla \cdot\|_r$. Furthermore, let $\hat{\mathbf{W}}_A^r(D)$ be the completion of the set

$$\mathbf{W}_A^r(D) := \{u \in \mathbf{W}^{2,r}(D) : u = 0 \ on \ \partial D, \nabla \cdot u = 0 \ in \ D\}$$

with respect to the second order Dirichlet norm $\|\partial^2 \cdot\|_r$. For any $\rho > \rho_0$ let $S_\rho = S_\rho(D)$ be the strip-like domain $S_\rho = \{x = (x_1, \dots, x_n) \in D : x_n < \rho\}$ and denote by $Ru = u|_{S_\rho}$ the restriction to S_ρ of a function u defined on D.

LEMMA 3.1. *Let $n \geq 2, 1 < r < n, 1/s = 1/r - 1/n$. Then*
(i) $\hat{W}^{1,r}(D) \subset L^s(D)$,
(ii) $R\,\hat{\mathbf{W}}^r_A(D) \subset \mathbf{L}^s(S_\rho)$, $\rho > \rho_0$,
with continuous imbeddings.

PROOF. Let $\tilde{D}_\rho$ denote the reflection of $D_\rho = D \backslash B_\rho$ along the hyperplane $x_n = 0$. Then $\Omega := D_\rho \cup \tilde{D}_\rho$ is the domain exterior to the ball B_ρ. For $u \in W^{1,r}(D)$ the function

$$\tilde{u} = \begin{cases} \hat{u} = & reflection\ of\ u\ in\ \tilde{D}_\rho \\ u & in\ D_\rho \end{cases}$$

belongs to $W^{1,r}(\Omega)$, hence the continuous imbedding $\hat{W}^{1,r}(\Omega) \subset L^s(\Omega)$ [7, Corollary 4.5, (c)] yields

$$\|u\|_{s,D_\rho} = \frac{1}{2}\|\tilde{u}\|_{s,\Omega} \leq C\,\|\nabla u\|_{r,D_\rho}. \tag{3.1}$$

In particular, since $s > r$, (3.1) implies

$$\|u\|_{1,r,B_\rho \cap B_{\rho_1}} \leq C\,\|\nabla u\|_{r,D_\rho},$$

for any $\rho_1 > \rho > \rho_0$. Now, recalling that

$$u \to \|\nabla u\|_{r,D\cap B_{\rho_1}} + \|u\|_{1,r,B_\rho \cap B_{\rho_1}}$$

defines an equivalent norm on $W^{1,r}(D \cap B_{\rho_1})$, Assertion (i) follows from the well known Sobolev imbedding $W^{1,r}(D \cap B_{\rho_1}) \subset L^s(D \cap B_{\rho_1})$. Finally, Assertion (ii) is a consequence of (i) and the Poincaré inequality $\|u\|_{r,S_\rho} \leq C\,\|\nabla u\|_{r,S_\rho}$. This proves Lemma 3.1.

COROLLARY 3.2. *Let $v \in \hat{\mathbf{W}}^r_A(\mathbb{R}^n_+)$, $p \in \hat{W}^{1,r}(\mathbb{R}^n_+)$ $(n \geq 2,\ 1 < r < n)$ be a solution of (S) in $\mathbb{R}^n_+$ with $f \equiv 0$. Then $v \equiv \nabla p \equiv 0$.*

PROOF. By definition there exist sequences $(v_j) \subset \mathbf{W}^r_A(\mathbb{R}^n_+)$ and $(p_j) \subset W^{1,r}(\mathbb{R}^n_+)$ such that for $j \to \infty$

$$\begin{aligned} v_j &\to v \quad in \quad \hat{\mathbf{W}}^r_A(\mathbb{R}^n_+) \\ p_j &\to p \quad in \quad \hat{W}^{1,r}(\mathbb{R}^n_+). \end{aligned}$$

Hence we also get $f_j = -\Delta v_j + \nabla p_j \to 0$ in $\mathbf{L}^r(\mathbb{R}^n_+)$ as $j \to \infty$. Denoting by $A_r = -P_r\Delta$ the Stokes operator in $\mathbf{X}_r = \mathbf{X}_r(\mathbb{R}^n_+)$ (see [4]), we have $A_r v_j = P_r f_j$ and $\nabla p_j = (1 - P_r)f_j + A_r v_j + \Delta v_j$. Theorem 3.6 (i) in [4] therefore yields

$$\begin{aligned} \|\partial^2 v\|_r = \lim_{j\to\infty} \|\partial^2 v_j\|_r &\leq C \lim_{j\to\infty} \|A_r v_j\| = 0 \\ \|\nabla p\|_r = \lim_{j\to\infty} \|\nabla p_j\|_r &\leq C \lim_{j\to\infty} (\|f_j\|_r + \|\Delta v_j\|_r) = 0. \end{aligned}$$

Consequently v is a solenoidal affin function vanishing for $x_n = 0$, i.e. $v = \beta x_n$ with $\beta = (\beta_1, \ldots, \beta_n) \in \mathbb{R}^n$, $\beta_n = 0$. From Lemma 3.1 we conclude $\beta = 0$ as asserted.

We are now in a position to prove the main uniqueness statement in this section.

THEOREM 3.3. *Corollary 3.2 remains valid if $I\!R^n_+$ is replaced by D.*

PROOF. Let $v \in \hat{\mathbf{W}}^r_A(D)$, $p \in \hat{W}^{1,r}(D)$ be a homogeneous solution of Problem (S) in D. In the first step we show that v has a finite L^2-Dirichlet integral. For fixed $N_0 \geq \rho_0$ set

$$u = v \cdot \varphi_{N_0} - S^+_{N_0}(v \cdot \nabla\varphi_{N_0}), \quad q = p \cdot \varphi_{N_0}, \tag{3.2}$$

with $\varphi_{N_0} = 1 - \psi_{N_0}$ (ψ_N from Section 2) and $S^+_{N_0}$ is the operator from Theorem 2.1 with respect to the domain $\Omega^+_{b,N_0} = B_{N_0} \cap B_{2N_0} \cap I\!R^n_+$. For $N_0 \geq \rho_0$, we find $\nabla \cdot u = 0$ in $I\!R^n_+$ by the same reasoning as in the preceding section. Consequently the pair u, q solves

$$\begin{aligned} -\Delta u + \nabla q &= f_{N_0} \quad (x \in I\!R^n_+) \\ \nabla \cdot u &= 0 \quad (x \in I\!R^n_+) \\ u &= 0 \quad (x_n = 0), \end{aligned} \tag{3.3}$$

where $f_{N_0} = -\nabla v \nabla\varphi_{N_0} - v\Delta\varphi_{N_0} + \Delta S^+_{N_0}(v \cdot \nabla\varphi_{N_0}) + p\nabla\varphi_{N_0}$ for $x \in \Omega^+_{b,N_0}$ and $f_{N_0} = 0$ otherwise. Since the Stokes system (S) is elliptic in the sense of [2], it follows that the homogeneous solution v, p is locally smooth up to the boundary. Therefore we see that $v \cdot \nabla\varphi_{N_0}$ is smooth and vanishes on $\partial(\Omega^+_{b,N_0})$, hence $v \cdot \nabla\varphi_{N_0} \in W^{1,s}_0(\Omega^+_{b,N_0})$ and by Theorem 2.1 $S_{N_0}(v \cdot \nabla\varphi_{N_0}) \in \mathbf{W}^{2,s}_0(\Omega^+_{b,N_0})$. Extending $S_{N_0}(v \cdot \nabla\varphi_{N_0})$ by zero, this shows that f_{N_0} is well defined, has a compact support and

$$f_{N_0} \in \mathbf{L}^s(I\!R^n_+) \quad \textit{for each } 1 \leq s < \infty. \tag{3.4}$$

We consider the sequence

$$u_k = (k^{-1} + A_r)^{-1} P_r f_{N_0}, \quad k = 1, 2, \ldots, \tag{3.5}$$

which is well defined since $P_r f_{N_0} \in \mathbf{X}_r(I\!R^n_+)$ by (3.4). Theorem 3.6 (i) in [4] implies for $0 < k \leq j$

$$\begin{aligned} \|\partial^2(u_j - u_k)\|_r &\leq C\,\|(k^{-1} - j^{-1})A_r(j^{-1} + A_r)^{-1}(k^{-1} + A_r)^{-1} P_r f_{N_0}\|_r \\ &\leq C\,\|A_r(j^{-1} + A_r)^{-1}\|\;\|k^{-1}(k^{-1} + A_r)^{-1} P_r f_{N_0}\|_r, \end{aligned} \tag{3.6}$$

where $\|\cdot\|$ denotes the operator norm. Now by Corollary 3.4 in [4] we have the uniform bounds

$$\|k^{-1}(k^{-1} + A_r)^{-1}\| \leq M_r \tag{3.7}$$

$$\|A_r(k^{-1} + A_r)^{-1}\| \leq \tilde{M}_r \tag{3.8}$$

In particular, (3.7) implies that the range $R(A_r)$ of A_r is dense in $\mathbf{X}_r$ [15, Theorem 3.1]. It follows

$$k^{-1}(k^{-1} + A_r)^{-1} \to 0 \quad \textit{strongly in } \mathbf{X}_r \textit{ as } k \to \infty. \tag{3.9}$$

From (3.5)-(3.9) we conclude that (u_k) is a Cauchy sequence in $\hat{\mathbf{W}}^r_A(I\!R^n_+)$. Next, rewriting (3.5) as

$$k^{-1}u_r - \Delta u_k + \nabla q_k = f_{N_0}, \tag{3.10}$$

we find $\nabla q_k = (1 - P_r)(f_{N_0} + \Delta u_k)$, so that q_k is a Cauchy sequence in $\hat{W}^{1,r}(I\!R^n_+)$. We denote by u_0 and q_0 the limit of u_k and q_k respectively. In view of Lemma 3.1 it is easily seen that $(k^{-1}u_k, w) \to 0$ as $k \to \infty$, for each $w \in \mathbf{C}_0^\infty(\mathbb{R}_+^n)$. Hence the pair u_0, q_0 solves (3.3) and Corollary 3.2 yields $u_0 = u$, $\nabla q_0 = \nabla q$. On the other hand, using the explicit representation of the Stokes resolvent (see [4, Eqn. (3.8), p. 144]), we have

$$(k^{-1} + A_r)^{-1} = (k^{-1} + A_s)^{-1} \; on \; \mathbf{X}_r \cap \mathbf{X}_s,$$

and since by (3.4) $P_r f_{N_0} = P_s f_{N_0}$, the same arguments as above yield

$$u = u_0 \in \hat{\mathbf{W}}_A^s(\mathbb{R}_+^n), \quad q = q_0 \in \hat{W}^{1,s}(\mathbb{R}_+^n), \quad 1 < s < n. \tag{3.11}$$

If $n \geq 3$ we may choose $s = 2n/(n+2)$ and Lemma 3.1 implies $u \in \hat{\mathbf{W}}^{1,2}(\mathbb{R}_+^n)$. Recalling that u and v coincide outside of a compact set, this shows that v is a homogeneous solution with finite L^2-Dirichlet integral.

Next we consider the case $n = 2$. In this case we have to argue differently. Let $w \in \mathbf{V}(\mathbb{R}_+^2)$ denote the unique weak solution of the variational problem

$$(\nabla w, \nabla z) = (f_{N_0}, z) \quad for \; all \; z \in \mathbf{V}(\mathbb{R}_+^2), \tag{3.12}$$

where $(\cdot,\cdot)$ is the scalar product in $\mathbf{L}^2(\mathbb{R}_+^2)$, $(\nabla w, \nabla z) = (\nabla w_1, \nabla z_1) + (\nabla w_2, \nabla z_2)$ and $\mathbf{V}(\mathbb{R}_+^2)$ is the completion of $\mathbf{C}_{0,\sigma}^\infty(\mathbb{R}_+^2)$ with respect to the norm $\|\nabla \cdot\|_2$. Note that $z \to (f_{N_0}, z)$ extends uniquely to a bounded linear functional on $\mathbf{V}(\mathbb{R}_+^2)$. Indeed, since *supp* f_{N_0} is compact, this follows from the continuous imbedding $\mathbf{V}(\mathbb{R}_+^2) \subset \mathbf{L}_{loc}^2(\overline{\mathbb{R}_+^2})$ as a consequence of the Poincaré inequality in S_ρ.

Let $\pi \in L^2(\mathbb{R}_+^2)$ denote the associated pressure (see [22, Theorem 5], [23, Lemma 2.1]). Clearly, we have $w \in \mathbf{W}_{loc}^{2,s}(\overline{\mathbb{R}_+^2})$, $\pi \in W_{loc}^{1,s}(\overline{\mathbb{R}_+^2})$ $(1 < s < \infty)$. For large N we define

$$\begin{aligned} w_N &= w \cdot \psi_N - S_N^+(w \cdot \nabla\psi_N), \quad \pi_N = \pi\psi_N \\ f_N &= f_{N_0}\psi_N - 2\nabla w \nabla\psi_n - w\Delta\psi_N + \Delta S_N^+(w \cdot \nabla\psi_N) + \pi\nabla\psi_N, \end{aligned} \tag{3.13}$$

then obviously $w_N \in D(A_r)$ and $A_r w_N = P_r f_N$. Direct computation gives in view of Theorem 3.6 (i) in [4] and Corollary 2.1

$$\|\partial^2 w_N\|_r + \|\nabla\pi_N\|_r \leq C(\|f_{N_0}\|_r + \|\partial w \cdot \partial\psi_N\|_r + \|w\partial^2\psi_N\|_r + \|\pi\nabla\psi_N\|_r), \tag{3.14}$$

where $C > 0$ does not depend on N, π, w, f_{N_0}.

Next, applying the Hölder inequality for the domain $\Omega_{b,N}^+ = B_N \cap B_{2N} \cap \mathbb{R}_+^2$ yields

$$\|\partial w \cdot \partial\psi_N\|_r \leq \|\nabla\psi_N\|_{2r/(2-r)} \left[\int_{\Omega_{b,N}^+} |\nabla w|^2 \, dx\right]^{1/2} \leq C\, N^{-2+2/r}\|\nabla w\|_2, \tag{3.15}$$

similarly

$$\|\pi\nabla\psi_N\|_r \leq C\, N^{-2+2/r}\|\pi\|_2. \tag{3.16}$$

Moreover, using in addition the Poincaré inequality for $\Omega_{b,N}^+$ gives

$$\|w\partial^2\psi_N\|_r \leq C\, N^{-3+2/r} \left[\int_{\Omega_{b,N}^+} |w|^2 \, dx\right]^{1/2} \leq C\, N^{-2+2/r}\|\nabla w\|_2. \tag{3.17}$$

Inserting (3.15)-(3.17) into (3.14) leads to the estimate

$$\|\partial^2 w_N\|_r + \|\nabla \pi_N\|_r \leq C\,(\|f_{N_0}\|_r + \|\nabla w\|_2 + \|\pi\|_2), \tag{3.18}$$

where $C > 0$ does not depend on N, w, π and f_{N_0}. Thus, there exist subsequences (again denoted by w_N, π_N) such that w_N, π_N converge weakly in $\hat{\mathbf{W}}^r_A(I\!R^2_+)$ and $\hat{W}^{1,r}(I\!R^2_+)$ respectively. On the other hand

$$w_N \to w, \quad \pi_N \to \pi \quad in\ L^2_{loc}(\overline{I\!R^2_+}),$$

as is easily seen from the definition. Therefore we conclude $w \in \hat{\mathbf{W}}^r_A(I\!R^2_+),\ \pi \in \hat{W}^{1,r}(I\!R^2_+)$, hence $u = w,\ q = \pi$ by Corollary 3.2.

Finally, to show that v and thus ∇p vanish in D, we consider the identity

$$(\nabla v, \nabla v_N) = 0 \quad with\ v_N = v\psi_n - S^+_N(v \cdot \nabla \psi_N). \tag{3.19}$$

Making repeated use of Corollary 2.2 and the Poincaré inequality for $\Omega^+_{b,N}$, we obtain

$$\begin{aligned}\|\nabla(v - v_N)\|_2 &\leq \|(1-\psi_N)\nabla v\|_2 + C\|v\partial\psi_N\|_2 \\ &\leq C\left[\int_{I\!R^n_+\cap\{|x|\geq N\}} |\nabla v|^2\,dx\right]^{1/2} \to 0 \quad as\ N \to \infty,\end{aligned}$$

and Theorem 3.3 follows by letting $N \to \infty$ in (3.19).

Proof of Theorem 1.2

(i) Let $v \in W^r_A(D),\ p \in W^{1,r}(D)$ and $f = -\Delta v + \nabla p$. We argue as in [6]. Multiplying v, p by $\varphi_1 = \psi_{N_0}$ and $\varphi_2 = 1 - \psi_{N_0}$ and setting

$$v_i = v \cdot \varphi_i - S^+_{N_0}(v \cdot \nabla\varphi_i), \quad p_i = p \cdot \varphi_i \quad (i = 1, 2),$$

we see that v_i, p_i satisfy

$$-\Delta v_i + \nabla p_i = f_i, \quad \nabla \cdot v_i = 0 \quad in\ D_i, \quad v_{i|\partial D_i} = 0,$$

where $D_2 = I\!R^n_+$ and $D_1 \subset I\!R^n_+$ is a smooth and bounded domain containing $B_{2N_0} \cap I\!R^n_+$. The right hand sides are given by

$$f_i = f\varphi_i - 2\nabla v \nabla\varphi_i - v\Delta\varphi_i + \Delta S^+_{N_0}(v \cdot \nabla\varphi_i) + p\nabla\varphi_i.$$

Applying (1.8) for D_1, D_2 and summing the resulting inequalities we obtain (see [6] for more details)

$$\|\partial^2 v\|_r + \|\nabla p\|_r \leq C\,(\|f\|_r + \|v\|_{1,r,D_1} + \|p\|_{r,D_1}). \tag{3.20}$$

Assume now on the contrary that (1.8) does not hold in D. In this case there exist sequences $v_k \in W^r_A(D),\ p_k \in W^{1,r}(D)$ such that

$$\|\partial^2 v_k\|_r + \|\nabla p_k\|_r = 1\ and\ \|-\Delta v_k + \nabla p_k\|_r \to 0\ as\ k \to \infty. \tag{3.21}$$

Lemma 3.1 implies that $\|v_k\|_{2,r,D_1} + \|p_k\|_{1,r,D_1}$ remains bounded and so by compactness we can assume that $v_k \to v$, $p_k \to p$ weakly in $\hat{\mathbf{W}}^r_A(D)$, $\hat{W}^{1,r}(D)$ and strongly in $\mathbf{W}^{1,r}_{loc}(\overline{D})$, $L^r_{loc}(\overline{D})$ respectively. By the weak convergence it is then easy to see that v, p form a homogeneous solution and Theorem 3.3 implies $v = \nabla p = 0$. But then from (3.20) we conclude $\|\partial^2 v_k\|_r + \|\nabla p_k\|_r \to 0$ as $k \to \infty$, contradicting (3.21). This proves (i).

(ii) We consider the function

$$u(x) = \beta x_n \cdot \varphi_{N_0} - S^+_{N_0}(x_n \beta \cdot \nabla \varphi_{N_0})$$

with $0 \neq \beta = (\beta_1, \dots \beta_n) \in I\!R^n$, $\beta_n = 0$. Computing $-\Delta u$ yields

$$-\Delta u = -\beta x_n \Delta \varphi_{N_0} - 2\beta \frac{\partial}{\partial x_n} \varphi_{N_0} + \Delta S^+_{N_0}(x_n \beta \cdot \nabla \varphi_{N_0}) =: f_{N_0}.$$

Obviously, *supp* $f_{N_0} \subset D \cap I\!R^n_+$ is compact and (3.4) is fullfilled.

Let $v \in \mathbf{V}(D)$ be the unique solution of

$$(\nabla w, \nabla z) = (f_{N_0}, z) \ \textit{for all} \ z \in \mathbf{V}(D),$$

with associated pressure $\pi \in L^2(D)$. Since f_{N_0} has a compact support in $\Omega^+_{b.N}$, we have

$$\partial^\alpha w \in \mathbf{L}^2(D \backslash \overline{B}_{3N_0}), \ \partial^\gamma \pi \in L^2(D \backslash \overline{B}_{3N_0}), \tag{3.22}$$

for any $\alpha, \gamma \in I\!N^n_0$, $|\alpha| \geq 1$ (see [13]). After modification on a set of measure zero, (3.22) shows that $\partial^\alpha w$ and $\partial^\gamma \pi$ are uniformly bounded in $D \backslash B_{3N_0}$ in the pointwise sense. In particular, w is uniformly Lipschitz continuous, hence w and thus $v = u - w$ can grow at most linearly as $|x| \to \infty$. Using this, direct computation yields that

$$v_N = v \cdot \psi_N - S^+_N(v \cdot \nabla \psi_N), \quad \pi_N = \pi \psi_N$$

are bounded in $\hat{\mathbf{W}}^r_A(D)$ resp. $\hat{W}^{1,r}(D)$ $(n \leq r < \infty)$ uniformly with respect to N. Indeed, Corollary 2.2 implies

$$\begin{aligned} \|\partial^2 v_N\|_r &\leq C\,[\|\psi_N \partial^2 v\|_r + \|\partial v \cdot \partial \psi_N\|_r + \|v \partial^2 \psi_N\|_r \\ &= C\,[I_1(N) + I_2(N) + I_3(N)]. \end{aligned}$$

Recalling $\partial^2 v = -\partial^2 w$ for $|x| \geq 2N_0$ and $\partial^2 w \in \mathbf{L}^\infty(D \backslash B_{3N_0})$, we easily see $\partial^2 v \in \mathbf{L}^r(D)$ and $I_1(N) \to \|\partial^2 v\|_r$ as $N \to \infty$. Moreover, using $(1 + |x|)^{-1} v$, $\partial v \in \mathbf{L}^\infty(D)$, we obtain

$$I_2(N) + I_3(N) \leq C\,N^{-1} \left[\int_{\Omega^+_{b,N}} dx \right]^{1/r} + C\,N^{-2} \left[\int_{\Omega^+_{b,N}} |x|^r \, dx \right]^{1/r} \leq C\,N^{-1+n/r}$$

with a similar estimate for $\|\partial \pi_N\|_r$. The same argument as in Section 2 gives

$$(\partial^2 v_N, z) \to (\partial^2 v, z), \ \ (\nabla \pi_N, z) \to (\nabla \pi, z) \ \ \textit{as} \ N \to \infty$$

for all $z \in \mathbf{C}^\infty_0(D)$. Therefore the uniform boundedness of $\partial^2 v_N$, $\nabla \pi_N$ in $\mathbf{L}^r(D)$ implies

$$\partial^2 v_N \to \partial^2 v, \ \ \nabla \pi_N \to \nabla \pi \ \textit{as} \ N \to \infty, \ \ \textit{weakly in} \ \mathbf{L}^r(D).$$

By Mazur's theorem we can thus find convex combinations, denoted by $\hat{v}_N$, $\hat{\pi}_N$, such that $\partial^2\hat{v}_N$, $\nabla\hat{\pi}_N$ converge strongly in $\mathbf{L}^r(D)$. Hence, as $N \to \infty$, we have

$$\|\partial^2\hat{v}_N\|_r \to \|\partial^2 v\|_r, \quad \|\nabla\hat{\pi}_N\|_r \to \|\nabla\pi\|_r, \tag{3.23}$$

and

$$\| - \Delta\hat{v}_N + \nabla\hat{\pi}_N\|_r \to 0,$$

since by construction the pair v, π solves the homogeneous system (S) in D.

Now suppose $\|\partial^2 v\|_r = 0$, then $v(x) = \tilde{\beta}x_n$ for some $\tilde{\beta} \in I\!R^n$, $\tilde{\beta}_n = 0$. Since $u = 0$ for $|x| \leq N_0$ it follows $w(x) = -\tilde{\beta}x_n$ for $|x| \leq N_0$. Therefore w can not vanish on ∂D unless $\tilde{\beta} = 0$ or $\partial D = \partial I\!R^n_+$. On the other hand we have $u = \beta x_n$ for $|x| \geq 2N_0$ and thus $w = u - v$ can not have finite Dirichlet integral unless $\beta = \tilde{\beta}$ and we conclude $\|\partial^2 v\|_r \neq 0$. This completes the proof.

References

[1] R. A. Adams, *Sobolev Spaces*, Academic Press, New York, 1975.

[2] S. Agmon, A. Douglis and L. Nirenberg, Estimates Near the Boundary for Solutions of Elliptic Partial Differential Equations satisfying general Boundary Conditions II. Pure Appl. Math. **17** (1964), 35–92.

[3] M. E. Bogovski, Solutions of the first boundary value problem for the equations of continuity of an incompressible medium, Soviet Math. Dokl. **20** (1979), 1094–1098.

[4] W. Borchers and T. Miyakawa, L^2 decay for the Navier-Stokes flow in halfspaces, Math. Ann. **282** (1988), 139–155.

[5] W. Borchers and T. Miyakawa, Algebraic L^2 decay for the Navier-Stokes flows in exterior domains, Acta Math. **165** (1990), 189–227.

[6] W. Borchers and H. Sohr, On the semigroup of the Stokes operator for exterior domains in L^q spaces, Math. Z. **196** (1987), 415–425.

[7] W. Borchers and H. Sohr, On the equations rot $v = g$ and div $u = f$ with zero boundary conditions, Hokkaido Math. J. **19** (1990), 67–87.

[8] P. Deuring, The Stokes-system in exterior domains: existence, uniqueness and regularity of solutions in L^p-spaces, Com. Part. Diff. Equa. **16** (1991), 1513-1528.

[9] G. P. Galdi and C. G. Simader, Existence, Uniqueness and L^q-Estimates for the Stokes-problem in an Exterior Domain, Arch. Rat. Mech. Anal. **112** (1990), 291–318.

[10] Y. Giga, Domains of fractional powers of the Stokes operator in L_r spaces, Arch. Rational Mech. Anal. **89** (1985), 251–265.

[11] Y. Giga, Solutions for semilinear parabolic equations in L^p and regularity of weak solutions of the Navier-Stokes system, J. Differential Equations **62** (1986), 186–212.

[12] Y. Giga and H. Sohr, On the Stokes operator in exterior domains, J. Fac. Sci. Univ. Tokyo, Sect. IA **36** (1989), 103–130.

[13] J. G. Heywood, On Uniqueness questions in the theory of viscous flow, Acta Math. **136** (1976) 61–102.

[14] H. Iwashita, $L_q - L_r$ estimates for solutions of nonstationary Stokes equations in an exterior domain and the Navier-Stokes initial value problem in L_q spaces, Math. Ann. **285** (1989), 265–288.

[15] H. Komatsu, Fractional Powers of Operators, Pac. J. Math. **19**, 285–346 (1966).

[16] O. A. Ladyzhenskaya, *The Mathematical Theory of Viscous Incompressible Flow*, Gorden & Breach, New York, 1969.

[17] P. Maremonti and V. A. Solonnikov, On estimates of solutions of the Stokes system in exterior domains, Zap. Nauch. Semin. LOMI **180** (1990), 105-120.

[18] T. Miyakawa, On nonstationary solutions of the Navier-Stokes equations in an exterior domain, Hiroshima Math. J. **12** (1982), 115–140.

[19] P. Secchi, On the stationary and nonstationary Navier-Stokes equations in $\mathbb{R}^n$, Annali di Mat. Pura e Appl. **153** (1988), 293–306.

[20] M. Shinbrot, *Lectures on Fluid Mechanics*, Gorden and Breach, New York, 1973.

[21] C. G. Simader and H. Sohr, A new Approach to the Helmholtz Decomposition and the Neumann Problem in L^q-spaces for Bounded and Exterior Domains. In: Mathematical Problems Relating to the Navier-Stokes Equation, G. P. Galdi (Ed.), World Scientific Publishers (to appear).

[22] V. A. Solonnikov, On the solvability of Boundary and Initial-Boundary Value Problems for the Navier-Stokes System in Domains with Noncompact Boundaries, Pac. J. Math. **93**, No. 2 (1981), 443–458.

[23] V. A. Solonnikov, Stokes and Navier-Stokes equations in domains with noncompact boundaries, in College de France Seminar, **4**, 240–349, Pitman, 1983.

[24] E. M. Stein, *Singular Integrals and Differentiability Properties of Functions*, Princeton Univ. Press, Princeton, 1970.

[25] K. Yosida, *Functional Analysis*, Springer-Verlag, Berlin, 1965.

Department of Mathematics,
University of Paderborn,
D-4790 Paderborn, Germany

Department of Applied Science,
Faculty of Engeneering,
Kyushu University
Hakozaki, Fukuoka 812, Japan

This note is in a final form and no similar paper has been or is being submitted elsewhere.

THE STEADY NAVIER–STOKES PROBLEM FOR LOW REYNOLDS NUMBER VISCOUS JETS INTO A HALF SPACE

Huakang Chang
University of British Columbia

§ 1. Introduction

In a half space $\Omega = \{ \, x \in \mathbb{R}^3 \; : \; x_3 > 0 \, \}$, the Dirichlet problem of the steady Navier–Stokes equations is posed as follow:

$$\left\{ \begin{array}{ll} \Delta u - \nabla p = \lambda \, u \cdot \nabla u \, , & \nabla \cdot u = 0 \quad \textit{for } x \in \Omega \, , \\ u(x) = b(x) \;\; \textit{for } x \in \partial\Omega \, , \quad \textit{and} & u(x) \to 0 \;\; \textit{as} \;\; |x| \to \infty \, . \end{array} \right. \tag{1}$$

Throughout this paper, it will be assumed that the boundary value b has a compact support and is smooth.

It is well known that the existence of smooth solutions with finite Dirichlet integrals can be proven by the methods of functional analysis. But, there is not much information available about the decay of these solutions at infinity. It is not even known if there exists at least one such solution with nice decay property at infinity. Because of this, the uniqueness of these solutions is not evident, although uniqueness is expected in the case of small data. In this paper, we construct a solution of (1) by potential theoretic methods. The solution that is obtained decays like $|x|^{-2}$ at infinity. Then, we show that the constructed solution is unique in the class of all functions which have finite Dirichlet integrals and decay like $|x|^{-1}$ at infinity. In other words, every solution which decays like $|x|^{-1}$ at infinity have to decay like $|x|^{-2}$ at infinity for small data.

§ 2. Stokes problem and Green's tensor

The corresponding Stokes problem of (1) is posed as follows:

$$\left\{ \begin{array}{ll} \Delta u - \nabla p = 0 \, , & \nabla \cdot u = 0 \quad \textit{for } x \in \Omega \, , \\ u(x) = b(x) \;\; \textit{for } x \in \partial\Omega \, , \quad \textit{and} & u(x) \to 0 \;\; \textit{as} \;\; |x| \to \infty \, . \end{array} \right. \tag{2}$$

It has been shown by [11], [1], [6], that there exists a unique solution of (2) with a finite Dirichlet integral. The solution has a potential representation

$$\left\{ \begin{array}{rcl} u_0(x) & = & \displaystyle\int_{\xi_3=0} b(\xi) \, K(x,\xi) \, d\xi_1 d\xi_2 \, , \\ p_0(x) & = & \displaystyle\int_{\xi_3=0} b(\xi) \, k(x,\xi) \, d\xi_1 d\xi_2 \, , \end{array} \right. \tag{3}$$

where $K(x,\xi)$ is a 3×3 matrix, and $k(x,\xi)$ a 1×3 vector, with components

$$K_{ij}(x, \xi) = \frac{3}{2\pi} \frac{(x_3 - \xi_3)(x_i - \xi_i)(x_j - \xi_j)}{|x - \xi|^5} \, , \qquad i, j = 1, 2, 3 \, , \tag{4}$$

$$k_i(x, \xi) = -\frac{1}{\pi} \frac{\partial}{\partial x_i} \left(\frac{x_3 - \xi_3}{|x - \xi|^3} \right) , \qquad i = 1, 2, 3 \, . \tag{5}$$

Lemma 1. *If $x_0 = (0,\,0,\,-1)$ and $b \in C_0^\infty(\partial\Omega)$, then there exists a positive constant a_0 such that the unique solution $\{u_0,\,p_0\}$ of (2) satisfies*

$$|u_0(x)| \le \frac{a_0}{|x-x_0|^2}\,, \qquad |\nabla u_0(x)|\,,\ |p_0(x)| \le \frac{a_0}{|x-x_0|^3}\,, \tag{6}$$

for all $x \in \Omega$.

For a pair $\{\,u\,,p\,\}$, consisting of a solenoidal vector function $u(x)$ and a scalar function $p(x)$, the related stress tensor Tu and its adjoint $T'u$ are defined to be the 3×3 matrix functions with entries

$$\begin{aligned} (Tu)_{ij}(x) &= -p(x)\,\delta_{ij} + \Big[D_{x_j}u_i(x) + D_{x_i}u_j(x)\Big]\,, \\ (T'u)_{ij}(x) &= p(x)\,\delta_{ij} + \Big[D_{x_j}u_i(x) + D_{x_i}u_j(x)\Big]\,, \end{aligned} \tag{7}$$

for $i\,,\,j = 1\,,\,2\,,\,3$, where δ_{ij} is the Kronecker delta notation.

By means of the divergence theorem, we have

$$\int_D [v\cdot(\Delta u - \nabla p) - u\cdot(\Delta v + \nabla q)]\,dx = \oint_{\partial D} [v\cdot Tu - u\cdot T'v]\cdot ds\,, \tag{8}$$

for any bounded domain D with smooth boundary ∂D, and any pairs $\{\,u\,,\,p\}$ and $\{\,v\,,\,q\,\}$, where $u\,,\,v$ are solenoidal vector functions, and $p\,,\,q$ are scalar functions. Here ds is interpreted as a directed infinitesimal surface element on the boundary.

In $I\!R^3$, the fundamental tensor for the Stokes equations is the pair $\{\,E\,,\,e\,\}$ (cf. [11], [5], [9]), where $E = [\,E_{ij}\,]$ is a 3×3 matrix, and $e = [\,e_i\,]$ is a 1×3 vector, with components

$$E_{ij}(x,y) = E_{ij}(x-y) = -\frac{1}{8\pi}\left[\frac{\delta_{ij}}{|x-y|} + \frac{(x_i-y_i)(x_j-y_j)}{|x-y|^3}\right], \tag{9}$$

$$e_i(x,y) = e_i(x-y) = -\frac{1}{4\,\pi}\,\frac{x_i-y_i}{|x-y|^3}\,, \tag{10}$$

for all $i,j = 1,2,3$. Clearly, $E(x,y) = E(x-y)$ is symmetric, that is, $E(x,y) = E(y,x)$, and $e(x,y) = -e(y,x)$. Of course, the fundamental tensor $\{\,E\,,\,e\,\}$ is constructed to satisfy

$$\begin{cases} \Delta_x E(x,y) - \nabla_x e(x,y) = \delta(x-y)\,I\,, \\ \Delta_y E(x,y) + \nabla_y e(x,y) = \delta(x-y)\,I\,, \end{cases} \tag{11}$$

and

$$\nabla\cdot E(x,y) = 0\,, \qquad x \ne y\,, \tag{12}$$

where $\delta(x,y) = \delta(x-y)$ is the dirac delta function, and $I = [\,\delta_{ij}\,]$ is the 3×3 identity matrix. By an easy calculation, one can show that the fundamental tensor satisfies

$$\begin{aligned} |\,E(x,y)\,| &\le \frac{\sqrt{6}}{8\,\pi\,|x-y|}\,, & |\,e(x,y)\,| &\le \frac{1}{4\,\pi\,|x-y|^2}\,, \\ |\,\nabla\,E(x,y)\,| &\le \frac{\sqrt{10}}{8\,\pi\,|x-y|^2}\,, & |\,\nabla e(x,\,y)| &\le \frac{\sqrt{6}}{4\,\pi\,|x-y|^3}\,, \end{aligned} \tag{13}$$

for all $x \neq y$ in $I\!R^3$.

Using the functional analysis method, it can be shown that the boundary value problem

$$\left\{ \begin{array}{ll} \triangle_y A(x, y) + \nabla_y \alpha(x, y) = 0 \,, \quad \nabla_y \cdot A(x, y) = 0 & \text{for } x, y \in \Omega \\ A(x, y) = E(x, y) & \text{on } \partial\Omega \end{array} \right. \tag{14}$$

has a unique solution with finite Dirichlet integral. Since E decays like r^{-1} at infinity, the solution $\{ A, \alpha \}$ has a potential representation

$$A(x, y) = \int_{\xi_3=0} E(x, \xi)\, K(y, \xi)\, d\xi_1 d\xi_2 \,, \tag{15}$$

$$\alpha(x, y) = -\int_{\xi_3=0} E(x, \xi)\, k(y, \xi)\, d\xi_1 d\xi_2 \,. \tag{16}$$

Lemma 2. *$A(x, y)$ has the symmetric properties*

$$A_{ij}(x, y) = A_{ij}(y, x) = A_{ji}(x, y) = A_{ji}(y, x) \,, \tag{17}$$

for all $i, j = 1, 2, 3$ and all $x, y \in \Omega$.

PROOF : From the fact that $E_{ij}(x, y) = E_{ij}(y, x) = E_{ji}(x, y) = E_{ji}(y, x)$, and $K_{ij}(x, y) = K_{ji}(x, y)$, it is not difficult to show that $A(x, y)$ is a symmetric matrix. We need only show that $A(x, y) = A(y, x)$. By a calculation, we find that

$$K(x, y)\Big|_{y_3=0} = 2\,T_y E(y, x) \cdot \mathbf{n}(y)\Big|_{y_3=0} = 2\,T_y' E(x, y) \cdot \mathbf{n}(y)\Big|_{y_3=0} \,. \tag{18}$$

Thus,

$$A(x, y) = \int_{\xi_3=0} E(x, \xi)\, K(y, \xi)\, d\xi_1 d\xi_2 = 2 \int_{\xi_3=0} E(x, \xi)\, T_\xi E(\xi, y) \cdot \mathbf{n}(\xi)\, d\xi_1 d\xi_2 \,.$$

Let Ω_R be a smoothly bounded domain which is only slightly different from the semiball of radius R. Let $\partial\Omega_R = S_R \cup \Gamma_R$ with $\Gamma_R \subset \partial\Omega$. We can assume that $\lim_{R\to\infty} \Gamma_R = \partial\Omega$, and that for any $x, y \in \Omega$, R can be chosen large enough so that $x, y \in \Omega_R$. Using the divergence formula (8) on the tensors $E(x, \xi)$ and $E(y, \xi)$ in the domain Ω_R, we have

$$\begin{aligned} & \oint_{\partial\Omega_R} \left\{ E(x, \xi) \cdot T_\xi E(\xi, y) - E(\xi, y) \cdot T_\xi' E(x, \xi) \right\} \cdot \mathbf{n}(\xi)\, ds_\xi \\ = & \int_{\Omega_R} \{ E(x, \xi)\, [\triangle_\xi E(\xi, y) - \nabla_\xi e(\xi, y)] - E(\xi, y)\, [\triangle_\xi E(x, \xi) + \nabla_\xi e(x, \xi)] \}\, d\xi \\ = & \int_{\Omega_R} [\, E(x, \xi)\, I\, \delta(\xi - y) - E(\xi, y)\, I\, \delta(\xi - x) \,]\, d\xi \\ = & \; E(x, y) - E(x, y) = 0 \,. \end{aligned}$$

This implies that

$$\oint_{\partial\Omega_R} E(x, \xi) \cdot T_\xi E(\xi, y) \cdot \mathbf{n}(\xi)\, ds_\xi = \oint_{\partial\Omega_R} E(\xi, y) \cdot T_\xi' E(x, \xi) \cdot \mathbf{n}(\xi)\, ds_\xi \,.$$

Taking the limit as $R \to \infty$, we easily see that

$$\lim_{R\to\infty} \int_{S_R} E(x,\xi)\cdot T_\xi E(\xi,y)\cdot \mathbf{n}(\xi)\, ds_\xi = \lim_{R\to\infty} \int_{S_R} E(\xi,y)\cdot T'_\xi E(x,\xi)\cdot \mathbf{n}(\xi)\, ds_\xi = 0\ .$$

Hence, we obtain

$$\int_{\partial\Omega} E(x,\,\xi)\cdot T_\xi E(\xi,\,y)\cdot \mathbf{n}(\xi)\, ds_\xi = \int_{\partial\Omega} E(\xi,\,y)\cdot T'_\xi E(x,\,\xi)\cdot \mathbf{n}(\xi)\, ds_\xi\ .$$

This implies (17), since

$$\int_{\xi_3=0} E(x,\,\xi)\,K(y,\,\xi)\,d\xi_1 d\xi_2 = \int_{\xi_3=0} E(\xi,\,y)\,K(x,\,\xi)\,d\xi_1 d\xi_2\ . \tag{19}$$

Q.E.D.

The tensor $A(x\,,\,y)$ can be written as

$$A(x,\,y) = \int_{\xi_3=0} E(\xi,\,y)\,K(x,\,\xi)\,d\xi_1 d\xi_2 = \int_{\xi_3=0} E(\xi,\,y)\,K(y,\,\xi)\,d\xi_1 d\xi_2\,. \tag{20}$$

The Green's tensor $\{\,G,\,g\,\}$ is obtained by setting

$$\begin{cases} \mathbf{G}(\,x,\,y\,) &=& E(\,x,\,y\,) - A(\,x,\,y\,)\,, \\ g(\,x,\,y\,) &=& e(\,x,\,y\,) - \alpha(\,x,\,y\,)\,. \end{cases} \tag{21}$$

It satisfies

$$\begin{cases} \Delta_x \mathbf{G}(x,\,y) - \nabla_x g(x,\,y) = \delta(x-y)\,I & \text{for } x,\,y\in\Omega \\ \Delta_y \mathbf{G}(x,\,y) + \nabla_y g(x,\,y) = \delta(x-y)\,I & \text{for } x,\,y\in\Omega \\ \nabla\cdot\mathbf{G}(x,\,y) = 0 & \text{for } x,\,y\in\Omega\,, \end{cases} \tag{22}$$

and has the properties

i) $G_{ij}(x,\,y) = G_{ij}(y,\,x) = G_{ji}(x,\,y) = G_{ji}(y,\,x)$ for all $x \neq y \in \Omega$ and $i,\,j = 1,2,3$;

ii) $\lim_{x_3\to 0+} \mathbf{G}(x,y) = \lim_{y_3\to 0+} \mathbf{G}(x,y) = 0$;

iii) $\mathbf{G}(x,\,y) \to 0$ as $|y| \to \infty$ for fixed x, and vice versa.

Note that

$$K(\,x,\,y\,)\Big|_{y_3=0} = 2\,T'_y E(\,x,\,y\,)\cdot\mathbf{n}(y)\Big|_{y_3=0} = T'_y \mathbf{G}(\,x,\,y\,)\cdot\mathbf{n}(y)\Big|_{y_3=0}\ . \tag{23}$$

Lemma 3. *There exists a positive constants C such that*

$$|\mathbf{G}(x,\,y)| \le \frac{C}{|x-y|}\,, \tag{24}$$

for all $x \neq y \in \bar{\Omega}$.

PROOF : From (21), it is clear that

$$|\mathbf{G}(x, y)| \le |E(x, y)| + |A(x, y)| . \tag{25}$$

Recalling (13), we only need to consider the auxiliary term $A(x, y)$. From (20), we have

$$\begin{aligned} |A|^2 &= \sum_{i,j=1}^{3} A_{ij}^2 = \sum_{i,j=1}^{3} \left[\sum_{l=1}^{3} \int_{\xi_3=0} K_{jl}(x, \xi)\, E_{il}(\xi, y)\, d\xi_1 d\xi_2 \right]^2 \\ &\le \left[\int_{\xi_3=0} \sum_{i,j=1}^{3} \left| \sum_{l=1}^{3} K_{jl}(x, \xi)\, E_{il}(\xi, y) \right| d\xi_1 d\xi_2 \right]^2 . \end{aligned}$$

Thus, by a calculation,

$$|A(x, y)| \le \frac{9\, x_3}{8\, \pi^2} \int_{\xi_3=0} \frac{d\xi_1 d\xi_2}{|x-\xi|^3\, |y-\xi|} .$$

From properties of harmonic functions on a half space (see [2]), we have

$$\frac{x_3}{2\,\pi} \int_{\xi_3=0} \frac{d\xi_1 d\xi_2}{|x-\xi|^3\, |y-\xi|} = \frac{1}{|x-y'|} \le \frac{1}{|x-y|} , \tag{26}$$

where $y' = (y_1, y_2, -y_3)$ whenever $y = (y_1, y_2, y_3) \in \Omega$. Therefore, we obtain

$$|A(x, y)| \le \frac{9}{4\,\pi\, |x-y'|} \le \frac{9}{4\,\pi\, |x-y|} . \tag{27}$$

By (13) and (27), the constant C can be chosen as $(36 + 2\sqrt{6})/16\,\pi < 0.82$.

Q.E.D.

The following is a result of Solonnikov [13]. An alternative proof is sketched below.

Lemma 4. *There exists a positive constant c such that the gradient of the Green's tensor $\mathbf{G}(x, y)$ satisfies*

$$|\nabla \mathbf{G}(x, y)| \le \frac{c}{|x-y|^2} , \tag{28}$$

for all $x, y \in \Omega$.

PROOF : By the definition (21) and the estimate (13) for the fundamental tensor, we need to consider $|\nabla A|$ only.

$$D_{y_h} A_{ij}(x, y) = \frac{3x_3}{16\pi^2} \int_{\xi_3=0} \sum_{l=1}^{3} \frac{(x_j - \xi_j)(x_l - \xi_l)}{|x-\xi|^5} F_{i,l,h}(\xi, y)\, d\xi_1 d\xi_2 , \tag{29}$$

where

$$F_{i,l,h}(\xi, y) = \frac{\xi_h - y_h}{|\xi - y|^3} \delta_{il} - \frac{\xi_i - y_i}{|\xi - y|^3} \delta_{lh} - \frac{\xi_l - y_l}{|\xi - y|^3} \delta_{ih} + \frac{3(\xi_i - y_i)(\xi_l - y_l)(\xi_h - y_h)}{|\xi - y|^5} .$$

From (26), we see that for $h = 1\,,\,2\,,$

$$-\frac{y_h - x_h}{|x-y'|^3} = \frac{\partial}{\partial y_h}\left(\frac{1}{|x-y'|}\right) = \frac{x_3}{2\,\pi}\int_{\xi_3=0}\frac{(\xi_h - y_h)}{|x-\xi|^3\,|y-\xi|^3}\,d\xi_1 d\xi_2\,,$$

$$-\frac{y_3 + x_3}{|x-y'|^3} = \frac{\partial}{\partial y_3}\left(\frac{1}{|x-y'|}\right) = \frac{x_3}{2\,\pi}\int_{\xi_3=0}\frac{(\xi_3 - y_3)}{|x-\xi|^3\,|y-\xi|^3}\,d\xi_1 d\xi_2\,.$$

By comparing these with each term in (29), we see that there exists a constant c_1 such that

$$|\nabla_y A(x,\,y)| \;\le\; \frac{c_1}{|x-y|^2}\,.$$

Thus, the result claimed in the lemma has been proved.

Q.E.D.

Theorem 1. *There exists a constant M such that*

$$\int_\Omega \frac{1}{|y-x_0|^4}\,|\,\nabla_y \mathbf{G}(x,\,y)\,|\;dy \le \frac{M}{|x-x_0|^2}\,, \tag{30}$$

for all $x \in \Omega$ and fixed $x_0 = (0,\,0,\,-1) \notin \bar{\Omega}$.

PROOF : First of all, we see that $dist(x_0\,,\,\partial\Omega) \ge 1$ for all $x \in \Omega$. By *Lemma 4*, we have

$$\int_\Omega \frac{1}{|y-x_0|^4}\,|\,\nabla_y \mathbf{G}(x,\,y)\,|\;dy \le c\int_\Omega \frac{dy}{|y-x_0|^4\,|x-y|^2}\,.$$

Using a domain decomposition, the integral on the right can be estimated as follows:

$$\int_\Omega \frac{dy}{|y-x_0|^4\,|y-x|^2} \equiv I_1 + I_2 + I_3 + I_4\,,$$

where

$$\begin{aligned}
I_1 &\equiv \int_{\Omega_{|y-x|\le\frac{1}{2}|x-x_0|}} \frac{dy}{|y-x|^2\,|y-x_0|^4}\\
&\le \int_{\Omega_{|y-x|\le\frac{1}{2}|x-x_0|}} \frac{dy}{|y-x|^2\,|y-x_0|^3}\\
&\le \frac{8}{|x-x_0|^3}\int_{|y-x|\le\frac{1}{2}|x-x_0|} \frac{dy}{|y-x|^2}\\
&= \frac{16\,\pi}{|x-x_0|^2}\,,\\
I_2 &\equiv \int_{\Omega_{|y-x_0|\le\frac{1}{2}|x-x_0|}} \frac{dy}{|y-x|^2\,|y-x_0|^4}\\
&\le \frac{4}{|x-x_0|^2}\int_{\Omega_{|y-x_0|\le\frac{1}{2}|x-x_0|}} \frac{dy}{|y-x_0|^4}\\
&\le \frac{4}{|x-x_0|^2}\int_{1\le|y-x_0|\le\frac{1}{2}|x-x_0|} \frac{dy}{|y-x_0|^4}\\
&\le \frac{16\,\pi}{|x-x_0|^2}\,,
\end{aligned}$$

$$
\begin{aligned}
I_3 &\equiv \int_{\Omega_{|y-x_0|,|y-x|\geq\frac{1}{2}|x-x_0|,|y-x_0|\geq|y-x|}} \frac{dy}{|y-x_0|^4\,|y-x|^2} \\
&\leq \int_{\frac{1}{2}|x-x_0|\leq|y-x|<\infty} \frac{dy}{|y-x|^6} \\
&\leq \frac{16\,\pi}{|x-x_0|^2}\,, \\
I_4 &\equiv \int_{\Omega_{|y-x_0|,|y-x|\geq\frac{1}{2}|x-x_0|,|y-x_0|\leq|y-x|}} \frac{dy}{|y-x_0|^4\,|y-x|^2} \\
&\leq \int_{\frac{1}{2}|x-x_0|\leq|y-x_0|<\infty} \frac{dy}{|y-x_0|^6} \\
&\leq \frac{16\,\pi}{|x-x_0|^2}\,.
\end{aligned}
$$

Combining these, we obtain the estimate (30).

Q.E.D.

§ 3. Solvability of the Navier–Stokes problem in a half space

The main result in this section is the following:

Theorem 2. *Under the assumptions of the Lemma 1, for λ satisfying*

$$
|\,\lambda\,| < \frac{1}{4\,a_0\,M}\,, \tag{31}
$$

there exists at least one vector function $u \in \mathcal{C}^\infty(\Omega) \cap \mathcal{C}(\bar{\Omega})$ solving the problem (1) and satisfying

$$
|u(x)| \leq \frac{2\,a_0}{|x-x_0|^2}\,, \qquad |\nabla u(x)| \leq \frac{2\,a_0}{|x-x_0|^2}\,, \qquad \text{for } \ x \in \Omega\,, \tag{32}
$$

where a_0 is as in Lemma 1. Thus, u has a finite Dirichlet integral.

Under appropriate assumptions about the regularity and decay of f, the Dirichlet problem for the Stokes equations

$$
\begin{cases}
\Delta u - \nabla p = f\,, & \nabla \cdot u = 0 \quad \text{for } x \in \Omega\,, \\
u(x) = b(x) \ \text{for } x \in \partial\Omega\,, & \text{and} \quad u(x) \to 0 \ \text{as } |x| \to \infty\,.
\end{cases}
$$

has a unique solution

$$
u(x) = \int_\Omega f(y)\,\mathbf{G}(x,y)\,dy + \int_{y_3=0} u(y)\,K(x,y)\,dy_1 dy_2\,, \tag{33}
$$

with an associated pressure

$$
p(x) = \int_\Omega f(y)\,g(x,y)\,dy + \int_{y_3=0} u(y)\,k(x,y)\,dy_1 dy_2\,. \tag{34}
$$

Using (33) and (34), the problem (1) is formally equivalent to the integral equation,

$$u(x) = \lambda \int_\Omega u(y) \cdot \nabla u(y)\, \mathbf{G}(x,y)\, dy + u_0(x) \ , \tag{35}$$

with an associated pressure

$$p(x) = \lambda \int_\Omega u(y) \cdot \nabla u(y) g(x,y)\, dy + p_0(x) \ , \tag{36}$$

where $\{ u_0 \, , \, p_0 \}$ is the solution of corresponding Stokes problem (2). Furthermore, integrating by parts, we see that (35) is formally equivalent to

$$u(x) = -\lambda \int_\Omega u(y) \cdot \nabla_y \mathbf{G}(x,y)\, u(y)\, dy + u_0(x) \ . \tag{37}$$

Lemma 5. *Under the assumptions of the Theorem 2, for λ satisfying (31), there exists at least one vector function $u \in C^1(\Omega) \cap C(\bar{\Omega})$ solving the integral equation (37) and satisfying (32).*

Proof : We seek a solution of (37) in the form

$$u(x) \equiv \sum_{n=0}^{\infty} u_n(x)\, \lambda^n \, , \tag{38}$$

where u_0 is given by (3). To find u, we formally substitute (38) into the integral equation (37) obtaining

$$\sum_{n=0}^{\infty} u_n(x)\lambda^n = -\lambda \sum_{n=0}^{\infty} \lambda^n \int_\Omega \sum_{k=0}^{n} u_k(y) \cdot \nabla_y \mathbf{G}(x,y)\, u_{n-k}(y)\, dy + u_0(x) \, ,$$

or

$$\sum_{n=0}^{\infty} u_{n+1}(x)\, \lambda^{n+1} = -\sum_{n=0}^{\infty} \lambda^{n+1} \int_\Omega \sum_{k=0}^{n} u_k(y) \cdot \nabla_y \mathbf{G}(x,y)\, u_{n-k}(y)\, dy \, .$$

By comparing the coefficients of λ^n on the both sides, we obtain

$$u_{n+1}(x) = -\int_\Omega \sum_{k=0}^{n} u_k(y) \cdot \nabla_y \mathbf{G}(x,y)\, u_{n-k}(y)\, dy \ . \tag{39}$$

We divide the proof into the following steps:

i) For all $n \geq 0$,

$$|x - x_0|^2 \, |u_n(x)| \leq a_n \ , \tag{40}$$

where $\{ a_n \}$ is defined by the recurrence formula

$$a_{n+1} = M \sum_{k=0}^{n} a_k\, a_{n-k} \ , \qquad \text{for } n = 0, 1, 2, 3, \cdots \ , \tag{41}$$

where a_0 is as in *Lemma 1* and M is as in *Theorem 1*..

This can be shown by induction. First of all, (40) holds for $n = 0$ by *Lemma 1.* Suppose that (40) holds for all $k \leq n$, the recurrence relation (39) and *Theorem 1* imply

$$\begin{aligned} |u_{n+1}(x)| &\leq \int_\Omega \frac{1}{|y-x_0|^4} |\nabla_y G(x\,,\,y)| \sum_{k=0}^{n} a_k\, a_{n-k} dy \\ &\leq \frac{M}{|x-x_0|^2} \sum_{k=0}^{n} a_k\, a_{n-k} = \frac{a_{n+1}}{|x-x_0|^2} . \end{aligned}$$

ii) For all $n \geq 0$, u_n defined by (39) belong to $\mathcal{C}^\infty(\Omega) \cap \mathcal{C}(\bar{\Omega})$.

This can also be shown by induction. First, we know that $u_0 \in \mathcal{C}^\infty(\Omega) \cap \mathcal{C}(\bar{\Omega})$ by *Lemma 1.* Assume that the statement is true for all $k \leq n$. Using integration by part and the formulas (39), (40), we have

$$u_{n+1}(x) = \int_\Omega \sum_{k=0}^{n} u_k(y) \cdot \nabla u_{n-k}(y) \mathbf{G}(x,y)\, dy\ , \tag{42}$$

which implies that $u_{n+1} \in \mathcal{C}^\infty(\Omega) \cap \mathcal{C}(\bar{\Omega})$.

iii) For λ satisfying (31), the series $\sum_{n=0}^\infty a_n \lambda^n$ is convergent and satisfies

$$\sum_{n=0}^{\infty} a_n\, |\lambda^n| \leq 2\, a_0\ . \tag{43}$$

Hence, $u(x) = \sum_{n=0}^\infty u_n(x)\, \lambda^n$ is uniformly and absolutely convergent in variable x. It is obvious that u belongs to $\mathcal{C}(\bar{\Omega})$ and satisfies the first estimate in (32).

From the recurrence relation (41), we can calculate that

$$\sum_{n=0}^{\infty} a_n\, \lambda^n = \begin{cases} \dfrac{1}{2\,M\,\lambda} \left[1 - \sqrt{1 - 4\, a_0\, M\, \lambda}\right] & \text{if} \quad \lambda \neq 0 \\ a_0 & \text{if} \quad \lambda = 0\, , \end{cases} \tag{44}$$

which is analytic and bounded by $2\, a_0$ for λ satisfying (31).

iv) For all $n \geq 0$,

$$|\nabla u_n(x)| \leq \frac{a_n}{|x-x_0|^2} . \tag{45}$$

Thus, $u \in \mathcal{C}^1(\Omega) \cap \mathcal{C}(\bar{\Omega})$ and satisfies the second estimate in (32).

We can prove the estimate (45) by induction. *Lemma 1* implies that (45) holds for $n = 0$. Assume (45) holds for all $k \leq n$. From (42), we obtain

$$\nabla u_{n+1}(x) = \int_\Omega \sum_{k=0}^{n} u_k(y) \cdot \nabla u_{n-k}(y)\ \nabla_x \mathbf{G}(x,y)\, dy\ .$$

It implies that

$$\begin{aligned} |\nabla u_{n+1}(x)| &\leq \sum_{k=0}^{n} a_k\, a_{n-k} \int_\Omega \frac{1}{|y-x_0|^4} |\nabla_x \mathbf{G}(x,y)|\, dy \\ &\leq \frac{M}{|x-x_0|^2} \sum_{k=0}^{n} a_k\, a_{n-k} = \frac{a_{n+1}}{|x-x_0|^2} . \end{aligned}$$

By the same reason we mentioned before, the series $\sum_{n=0}^{\infty} \nabla u_n(x)\,\lambda^n$ is uniformly and absolutely convergent in the variable x for λ satisfying (31). It implies that $u \in \mathcal{C}^1(\Omega)$ and

$$\nabla u(x) = \sum_{n=0}^{\infty} \nabla u_n(x)\,\lambda^n \ .$$

The second estimate in (32) follows immediately.

The construction of u implies that u solves the integral equation (37).

Q.E.D.

PROOF of *Theorem 2* : We only need to show that the constructed u in *Lemma 5*, solves the problem (1) and belongs to $\mathcal{C}^\infty(\Omega) \cap \mathcal{C}(\bar{\Omega})$.

Since $u \in \mathcal{C}^1(\Omega) \cap \mathcal{C}(\bar{\Omega})$ and satisfies (32) and (37), we have that $u \in \mathcal{C}^{1+\alpha}(\Omega)$ by potential theory. Integrating by parts, we see that u satisfies (35). A potential bootstrapping argument provides that $u \in \mathcal{C}^\infty(\Omega) \cap \mathcal{C}(\bar{\Omega})$. The properties of the Green's tensor imply that u solves the Dirichlet problem (1).

Q.E.D.

§ 4. Generalized solutions and uniqueness

Let $\mathcal{D}(\Omega) = \{\, \varphi \in \mathcal{C}_0^\infty(\Omega) \ : \ \nabla \cdot \varphi = 0 \,\}$, and let $\mathcal{J}_0(\Omega)$ be the completion of $\mathcal{D}(\Omega)$ in the Dirichlet norm $\|\nabla \varphi\|$. Obviously, u_0, obtained in *Lemma 1*, is a solenoidal extension of b over $\bar{\Omega}$ and satisfies i) u_0 has a finite Dirichlet integral, ii) $u_0(x) \to 0$ as $|x| \to \infty$. We will use b to denote a solenoidal extension of itself satisfying the above properties in this section.

Definition 1. *u is a generalized solution of the Navier–Stokes problem* (1) *provided that $u - b \in \mathcal{J}_0(\Omega)$, and*

$$\int_\Omega \nabla u : \nabla \varphi \, dx = -\,\lambda \int_\Omega (u \cdot \nabla u) \cdot \varphi \, dx \,, \tag{46}$$

for all $\varphi \in \mathcal{D}(\Omega)$.

In the definition, the test functions φ cannot be increased from the space $\mathcal{D}(\Omega)$ to its completion $\mathcal{J}_0(\Omega)$, since the right side of (46) may not be well defined for all $\varphi \in \mathcal{J}_0(\Omega)$. By a theorem in [6], we have the following

Proposition 1. *If u and v are two generalized solutions of the problem* (1), *then the difference $w = u - v$ belongs to $\mathcal{J}_0(\Omega)$.*

If $\{u\,,\,p\}$ is a classical solution of the problem (1) and u has a finite Dirichlet integral, then u is a generalized solution. Obviously, the solution u obtained in *Theorem 2* has a finite Dirichlet integral, and is thus a generalized solution.

Theorem 3. *For λ satisfying* (31), *let u be the solution obtained in Theorem 2 and v be any generalized solution of* (1) *which decays like $|x|^{-1}$ at infinity. Then $v = u$ on Ω.*

Proof : Let $w = v - u$. By *Proposition 1*, $w \in \mathcal{J}_0(\Omega)$. From the definition of generalized solution, it is easy to verify that w satisfies

$$\int_\Omega \nabla w : \nabla\varphi \, dx = - \int_\Omega (w \cdot \nabla w + w \cdot \nabla u + u \cdot \nabla w) \, \varphi \, dx \, ,$$

for all $\varphi \in \mathcal{D}(\Omega)$. By the Hardy inequality, the decay properties of u and v imply that the right side defines a continuous linear functional on $\mathcal{D}(\Omega)$, and hence on its completion $\mathcal{J}_0(\Omega)$. Substituting w for φ, we have

$$\int_\Omega |\nabla w|^2 \, dx = - \int_\Omega (w \cdot \nabla w + w \cdot \nabla u + u \cdot \nabla w) \cdot w \, dx \, .$$

Integrating by parts, one can easily verify that

$$\int_\Omega u \cdot \nabla w \cdot w = \int_\Omega w \cdot \nabla w \cdot w = 0 \, ,$$

and

$$- \int_\Omega w \cdot \nabla u \cdot w \, dx = \int_\Omega w \cdot \nabla w \cdot u \, dx \, .$$

Thus, using the Schwarz and Hardy inequalities, and (32) one has

$$\begin{aligned} \|\nabla w\|^2 &= \lambda \int_\Omega w \cdot \nabla w \cdot u \ \le \ |\lambda| \, \|\nabla w\| \sqrt{\int_\Omega u^2 \, w^2} \\ &\le \ 2\, a_0 \, |\lambda| \, \|\nabla w\| \sqrt{\int_\Omega \frac{w^2}{|x - x_0|^2}} \ \le \ 4\, a_0 \, |\lambda| \, \|\nabla w\|^2 \, . \end{aligned}$$

The assumption (31) implies that $4\, a_0 \, |\lambda| < 1$, and hence we conclude that $\|\nabla w\| = 0$. Therefore, $w = 0$.

Q.E.D.

References

[1] L. Cattabriga, Su un problema al contorno relativo al sistema di equazioni di Stokes, *Rend. Sem. Mat. Univ. Padova* **31**(1961), 308-340.

[2] R. Courant & D. Hilbert, Method of Mathematical Physics *Volume II* Interscience, New York, 1962.

[3] R. Finn, On the steady state solutions of the Navier-Stokes equations III, *Acta Math.* **105**(1961), 197-244.

[4] R. Finn, On the exterior stationary solution of the Navier-Stokes equations, and associated perturbation problems, *Arch. Rat. Mech. Anal.* **19**(1965), 363-406.

[5] H. Fujita, On the existence and regularity of the steady-state solutions of the Navier-Stokes equations, *J. Fac. Sci. Univ. Tokyo, Sec. I A Math* **9**(1961), 59-102.

[6] J.G. HEYWOOD, On uniqueness questions in the theory of viscous flow, *Acta Math.* **136**(1976), 61-102.

[7] J.G. HEYWOOD, The Navier-Stokes equations: On the existence, regularity and decay of solutions, *Indiana Univ. Math. J.* **29**(1980), 639-681.

[8] L.V. KAPITANSKII & K.I. PILETSKAS, On spaces of solenoidal vector fields and boundary value problems for the Navier-Stokes equations in domains with noncompact boundaries, *Proc. Steklov Inst. Math.* **Issue 2**(1984), 3-34.

[9] O.A. LADYZHENSKAYA, *The Mathematical Theory of Viscous Incompressible Flow,* Second Edition, Gordon and Breach, New York, 1969.

[10] J. LERAY, Étude de diverse équations intégrales non linéaires et de quelques problémes que pose l'hydrodynamique, *J. Math. Pures Appl.* **12**(1933), 1-82.

[11] F.K.G. ODQVIST, Die Randwertaufgaben der Hydrodynamik zäher Flüssigkeiten, *Stockholm: P.A.Norstedt & Söner* 1928, see also *Math. Z.* **32**(1930), 329-375.

[12] V.A. SOLONNIKOV, On differential properties of the solutions of the first boundary–value problem for nonstationary systems of Navier–Stokes equations, *Trudy. Mat. Inst. Steklov,* **73**(1964), 221–191.

[13] V.A. SOLONNIKOV, On Green's matrices for elliptic boundary value problems I *Proc. Steklov Inst. Math.* **110**(1970), 123-170

An Approach to Resolvent Estimates for the Stokes Equations in L^q-spaces

Reinhard Farwig
and
Hermann Sohr

Fachbereich Mathematik-Informatik,
Universität-GH Paderborn
Warburger Straße 100, D-4790 Paderborn

1 Introduction

In the present paper we give a selfcontained rather elementary approach to resolvent estimates of the Stokes equations for a class of unbounded domains $\Omega \subseteq \mathbb{R}^n, n \geq 2$. First we develop an independent proof for the whole space $\mathbb{R}^n$ and the halfspace $\mathbb{R}^n_+$. In contrast to the literature our approach admits nonzero divergence; this is important since later on we use transformation and localization arguments which naturally lead to nonvanishing divergence terms. Our proof for halfspaces is based directly on the multiplier technique and works also for div $u \neq 0$. In Section 4 we consider a class of domains with noncompact boundaries obtained by a deformation of the halfspace (perturbed halfspace H_ω with some function ω describing the deformation). Our resolvent estimates for this class play a key role for proving results on more general domains via localization procedures.

For complex $\lambda \neq 0$, $\lambda \notin \mathbb{R}_-$, we consider the following *generalized Stokes system*

$$\begin{aligned} \lambda u - \Delta u + \nabla p &= f \quad \text{in } \Omega \\ \operatorname{div} u &= g \quad \text{in } \Omega \\ u &= 0 \quad \text{on } \partial\Omega \end{aligned} \tag{1.1}$$

which leads in the special case $g = 0$ to the usual resolvent system for solenoidal vector fields $u = (u_1, \dots, u_n)$. In the special case $g = 0$ we define the Stokes operator A_q as follows:

For $1 < q < \infty$ let $L^q_\sigma(\Omega)$ denote the closure in $L^q(\Omega)^n = L^q(\Omega) \times \dots \times L^q(\Omega)$ of the space $C^\infty_{0,\sigma}(\Omega)$ which consists of all smooth solenoidal vector fields having a compact support in Ω. There is a bounded projection operator $P_q : L^q(\Omega)^n \to$

$L^q_\sigma(\Omega)$ vanishing on the space $G^q(\Omega) = \{\nabla p \in L^q(\Omega)^n : p \in W^{1,q}_{loc}(\Omega)\}$ of gradients $\nabla p = (\partial_1, \ldots, \partial_n)p$ where $\partial_i = \partial/\partial x_i$, $i = 1, \ldots, n$; see [4], [6], [7], [9], [10], [11], [12], [13] and [16] for details. The Stokes operator A_q is defined by $A_q = -P_q\Delta$ with domain of definition

$$D(A_q) = L^q_\sigma(\Omega) \cap W^{2,q}(\Omega)^n \cap W^{1,q}_0(\Omega)^n.$$

Applying P_q to (1.1) when $g = 0$ we obtain the Stokes resolvent problem

$$(\lambda + A_q)u = f, \quad u \in D(A_q) \tag{1.2}$$

in the space $L^q_\sigma(\Omega)$.

Consider some given $0 < \varepsilon < \pi$ and the sector

$$S_\varepsilon = \{z \in \mathbb{C} : z \neq 0, \ |\arg z| < \pi - \varepsilon\}$$

in the complex plane $\mathbb{C}$. Concerning $g = 0$ in (1.1) we are mainly interested in the *resolvent estimate*

$$\|(\lambda u, \ A_q u)\|_q \leq C_\varepsilon \|f\|_q \tag{1.3}$$

for all $\lambda \in S_\varepsilon$, $f = \lambda u + A_q u \in L^q_\sigma(\Omega)$, where $C_\varepsilon > 0$ has to be independent of λ, f. Further we used the notation

$$\|(f_1, \ldots, f_k)\|_q = \|f_1\|_q + \ldots + \|f_k\|_q$$

where $\|\cdot\|_q = \|\cdot\|_{L^q(\Omega)}$ denotes the L^q-norm on Ω. Estimate (1.3) is basic for the nonstationary theory; it leads to

$$\|(\lambda + A_q)^{-1} f\|_q \leq C_\varepsilon |\lambda|^{-1} \ \|f\|_q \tag{1.4}$$

for all $\lambda \in S_\varepsilon$, $f \in L^q_\sigma(\Omega)$. In the case $0 < \varepsilon < \pi/2$ the inequality (1.4) implies that $-A_q$ generates an *analytic semigroup* e^{-tA_q}, $t \geq 0$, which is uniformly bounded for $t \geq 0$, i.e.

$$\|e^{-tA_q} f\|_q \leq M \|f\|_q, \ \ t \geq 0, \ \ f \in L^q_\sigma(\Omega)\,, \tag{1.5}$$

where $M > 0$ is independent of t, f. For $\Omega = \mathbb{R}^n$ or $\mathbb{R}^n_+$, and under some restrictions also for $\Omega = H_\omega$, we can prove the more general *extended resolvent estimate*

$$\|(\lambda u, \sqrt{|\lambda|}\, \nabla u, \nabla^2 u)\|_q \leq C_\varepsilon \|f\|_q \tag{1.6}$$

for all $\lambda \in S_\varepsilon$, $f = \lambda u + A_q u \in L^q_\sigma(\Omega)$, including all second order partial derivatives $\nabla^2 u = (\partial_i \partial_j u)_{i,j=1,\ldots,n}$ of u. Letting $\lambda \to 0$ we obtain from (1.6) in particular

$$\|\nabla^2 u\|_q \leq C_\varepsilon \|A_q u\|_q\,, \quad u \in D(A_q). \tag{1.7}$$

Let us give a short survey on the literature concerning the resolvent estimate (1.4). Solonnikov [13] proved (1.4) for bounded and exterior domains $\Omega \subseteq \mathbb{R}^3$ with $\partial\Omega \in C^2$, where $\lambda \in S_\varepsilon$ is restricted to $|\lambda| \geq \delta > 0$ if Ω is an exterior domain and where $0 < \varepsilon < \pi/2$ was not given arbitrarily. Giga [8] proved (1.4) for bounded domains $\Omega \subseteq \mathbb{R}^n$ $(n \geq 2)$ with $\partial\Omega \in C^{2+\mu}$ $(0 < \mu < 1)$ where $0 < \varepsilon < \pi$ was arbitrarily given. For exterior domains he proved (1.4) under the restriction that C_ε may depend additionally on a positive lower bound of $|\lambda|$. This restriction has been removed by Borchers-Sohr [3] if $n \geq 3$; in particular this proves (1.5) for exterior domains $\Omega \subseteq \mathbb{R}^n$, $n \geq 3$; see also Deuring [5]. The estimates (1.4), (1.6) are well known for the whole space $\mathbb{R}^n$ and the halfspace $\mathbb{R}^n_+$, $n \geq 2$; see [4], [9], [10].

Our generalized resolvent estimate of (1.1) for $\Omega = \mathbb{R}^n$, $\mathbb{R}^n_+$ or H_ω with nonzero divergence div $u = g$ is of the form

$$\|(\lambda u, \sqrt{|\lambda|}\nabla u, \nabla^2 u, \nabla p)\|_q \leq C_\varepsilon(\|(f, \nabla g)\|_q + \|\lambda g\|_{-1,q,*}) \tag{1.8}$$

for all $\lambda \in S_\varepsilon$, $f \in L^q(\Omega)^n$ and $g \in W^{1,q}(\Omega) \cap \hat{W}^{-1,q,*}(\Omega)$. In order to explain the space $\hat{W}^{-1,q,*}(\Omega)$ and its norm $\|\cdot\|_{-1,q,*}$ let q' be the conjugate exponent of q and let $\hat{W}^{1,q'}(\Omega)$ be the completion of the space

$$C_0^\infty(\overline{\Omega}) = \{v|_\Omega : v \in C_0^\infty(\mathbb{R})\}$$

under the Dirichlet norm $\|\nabla \cdot\|_{q'}$. Then $\hat{W}^{-1,q,*}(\Omega)$ is defined as the dual space of $\hat{W}^{1,q'}(\Omega)$, and for $g \in \hat{W}^{-1,q,*}(\Omega)$

$$\|g\|_{-1,q,*} = \sup\{| < g, \varphi > | \ : \ \varphi \in \hat{W}^{1,q'}(\Omega), \ \|\nabla\varphi\|_{q'} = 1\},$$

where $< g, \varphi >$ means the value of g at φ. Note that when $\Omega = \mathbb{R}^n$, the space $\hat{W}^{1,q'}(\mathbb{R}^n)$ coincides with $\hat{W}_0^{1,q'}(\mathbb{R}^n) = \overline{C_0^\infty(\mathbb{R}^n)}^{\|\nabla\cdot\|_{q'}}$ and $\hat{W}^{-1,q,*}(\mathbb{R}^n)$ coincides with $\hat{W}^{-1,q}(\mathbb{R}^n) = \hat{W}_0^{1,q'}(\mathbb{R}^n)^*$, the dual space of $\hat{W}_0^{1,q'}(\mathbb{R}^n)$; then we write $\|\cdot\|_{-1,q}$ instead of $\|\cdot\|_{-1,q,*}$.

This paper is organized as follows. In Section 2 we solve (1.1) for $\Omega = \mathbb{R}^n$ and prove (1.8). The case $\Omega = \mathbb{R}^n_+$ in Section 3 is much more complicated. To solve (1.1) in this case we first reduce this system to the case $f = 0$, $g = 0$, $u \neq 0$ on $\partial\mathbb{R}^n_+$ and then we remove the pressure term ∇p by transforming it to a fourth order partial differential equation for the nth component u_n. This system is solved by applying the partial Fourier transform with respect to the variables $x' = (x_1, \ldots, x_{n-1})$; this leads to an explicit formula for the solution. The estimate (1.8) is proved by multiplier techniques; see e.g. [14].

In Section 4 we solve (1.1) for the perturbed halfspace H_ω by transforming H_ω to $\mathbb{R}^n_+$. In Section 5 we shortly explain the localization method for proving the resolvent estimate (1.4) for more general domains via the results on $\mathbb{R}^n$, $\mathbb{R}^n_+$ and H_ω.

2 Generalized resolvent estimates for $\mathbb{R}^n$

In the whole space $\mathbb{R}^n$ we have to omit the boundary condition in (1.1) and consider for $\lambda \in S_\varepsilon$ the following generalized resolvent problem:

$$\begin{aligned} \lambda u - \Delta u + \nabla p &= f \\ \operatorname{div} u &= g. \end{aligned} \tag{2.1}$$

We solve (2.1) for $f \in L^q(\mathbb{R}^n)^n$ and $g \in W^{1,q}(\mathbb{R}^n) \cap \hat{W}^{-1,q}(\mathbb{R}^n)$ where $\hat{W}^{-1,q}(\mathbb{R}^n) = \hat{W}^{1,q'}(\mathbb{R}^n)^*$ is the space defined in the Introduction. Note that g is identified with the functional $\varphi \to < g, \varphi > = \int_{\mathbb{R}^n} g(x)\,\varphi(x)\,dx$ and that

$$\begin{aligned} \|g\|_{-1,q} &= \sup\{|< g,\varphi >| \; : \; \varphi \in C_0^\infty(\mathbb{R}^n), \|\nabla\varphi\|_{q'} = 1\} \\ &= \sup\{|\int_{\mathbb{R}^n} u \cdot (\nabla\varphi)\,dx| \; : \; \varphi \in C_0^\infty(\mathbb{R}^n), \|\nabla\varphi\|_{q'} = 1\} \\ &\leq \|u\|_q. \end{aligned}$$

Actually, the elements of $\hat{W}^{1,q}(\mathbb{R}^n)$ are uniquely determined only modulo constants the same being true for the space $\hat{W}^{1,q}(\Omega)$ with an arbitrary domain $\Omega \subset \mathbb{R}^n$.

Our result on $\mathbb{R}^n$ now reads as follows.

Theorem 2.1 *Let $n \geq 2$, $1 < q < \infty$ and $0 < \varepsilon < \pi$. Then for every $f \in L^q(\mathbb{R}^n)^n$, $g \in W^{1,q}(\mathbb{R}^n) \cap \hat{W}^{-1,q}(\mathbb{R}^n)$ and $\lambda \in S_\varepsilon$ there is a unique pair $(u, \nabla p) \in W^{2,q}(\mathbb{R}^n)^n \times L^q(\mathbb{R}^n)^n$ solving (2.1). Moreover $(u, \nabla p)$ is subject to the inequality*

$$\|(\lambda u, \sqrt{|\lambda|}\,\nabla u, \nabla^2 u, \nabla p)\|_q \leq C_\varepsilon(\|(f, \nabla g)\|_q + \|\lambda g\|_{-1,q}) \tag{2.2}$$

where $C_\varepsilon = C_\varepsilon(q,n) > 0$ is a constant independent of f, g and $\lambda \in S_\varepsilon$.

Proof. Let Δ be the Laplacian in $\mathbb{R}^n$ with domain $D(\Delta) = W^{2,q}(\mathbb{R}^n)^n$, range $R(\Delta) \subseteq L^q(\mathbb{R}^n)^n$ and let $(-\Delta)^{-1}$ be the inverse of $-\Delta$. Since by assumption $g \in W^{1,q}(\mathbb{R}^n) \cap \hat{W}^{-1,q}(\mathbb{R}^n)$ we conclude that $\nabla g \in R(\Delta)$ and that $u_g = -(-\Delta)^{-1}\nabla g \in D(\Delta)$ is well defined. Using the well known fundamental solution of $-\Delta$, the Calderon-Zygmund estimate implies that

$$\|u_g\|_q \leq C\|g\|_{-1,q}, \quad \|\nabla u_g\|_q \leq C\|g\|_q, \quad \|\nabla^2 u_g\|_q \leq C\|\nabla g\|_q.$$

In the same way we find a unique $\nabla p \in L^q(\mathbb{R}^n)^n$ satisfying

$$< \nabla p, \nabla\varphi > = < f, \nabla\varphi > + < \lambda g, \varphi > + < \nabla g, \nabla\varphi > \tag{2.3}$$

for all $\varphi \in C_0^\infty(\mathbb{R}^n)$. This means

$$-\Delta p = -\operatorname{div} f + (\lambda - \Delta) g$$

in the sense of distributions. We obtain the estimate

$$\|\nabla p\|_q \leq C(\|f\|_q + \|\nabla g\|_q + \|\lambda g\|_{-1,q}).$$

Then we solve the equation

$$(\lambda - \Delta)v = f - (\lambda - \Delta)u_g - \nabla p$$

by using the Fourier transform and get that $(u, \nabla p) = (v + u_g, \nabla p)$ is the desired solution pair of (2.1). By applying the multiplier theorem in [14] we get the estimate

$$\begin{aligned} \|(\lambda v, \sqrt{|\lambda|}\,\nabla v, \nabla^2 v)\|_q &\le C_1\|f - (\lambda - \Delta)u_g - \nabla p\|_q \\ &\le C_2(\|f\|_q + \|\lambda g\|_{-1,q} + \|\nabla g\|_q + \|\nabla p\|_q) \\ &\le C_3(\|f\|_q + \|\lambda g\|_{-1,q} + \|\nabla g\|_q) \end{aligned}$$

which leads to (2.2). The uniqueness of $(u, \nabla p)$ is obvious. This proves Theorem 2.1.

Remark 2.1 Write $f = (f', f_n)$ with $f' = (f_1, \dots, f_{n-1})$ and correspondingly $u = (u', u_n)$ in Theorem 2.1 and assume that f' and g are even functions with respect to x_n where $x = (x', x_n) = (x_1, \dots, x_n) \in \mathbb{R}^n$ are the variables; furthermore assume that f_n is an odd function with respect to x_n. Then an easy symmetry consideration implies that u' and p are even in x_n while u_n is odd in x_n. In particular, $u_n(x', 0) = 0$ for all $x' \in \mathbb{R}^{n-1}$.

The even extension h_e of some function h on $\mathbb{R}^n_+$ to all of $\mathbb{R}^n$ is defined by

$$h_e(x) = \begin{cases} h(x', x_n) & \text{for } x_n > 0 \\ h(x', -x_n) & \text{for } x_n < 0, \end{cases}$$

while the odd extension h_o is defined by

$$h_o(x) = \begin{cases} h(x', x_n) & \text{for } x_n > 0 \\ -h(x', -x_n) & \text{for } x_n < 0. \end{cases}$$

3 Generalized resolvent estimates for the halfspace $\mathbb{R}^n_+$

Consider the halfspace

$$\mathbb{R}^n_+ = \{x = (x_1, \dots, x_n) = (x', x_n) \in \mathbb{R}^n : x_n > 0\}$$

and for all $\lambda \in S_\varepsilon$, $0 < \varepsilon < \pi$, the generalized resolvent problem

$$\begin{aligned} \lambda u - \Delta u + \nabla p &= f \quad \text{in } \mathbb{R}^n_+ \\ \operatorname{div} u &= g \quad \text{in } \mathbb{R}^n_+ \\ u &= 0 \quad \text{on } \Gamma = \partial\mathbb{R}^n_+ = \{x \in \mathbb{R}^n : x_n = 0\}. \end{aligned} \tag{3.1}$$

The case $g \equiv 0$ is well known in the literature; see [4], [10], [15]. The crucial case $g \not\equiv 0$ is important for our approach, since by this generalization we can avoid the use of Bogovskii's theory [2]; see [3] and [9]. Recall that the assumption $g \in W^{1,q}(\mathbb{R}^n_+) \cap \hat{W}^{-1,q,*}(\mathbb{R}^n_+)$ in the theorem below is quite natural if $u \in W^{2,q}(\mathbb{R}^n_+)^n \cap W_0^{1,q}(\mathbb{R}^n_+)^n$ and $g = \operatorname{div} u$. Actually, since for every $\varphi \in C_0^\infty(\overline{\mathbb{R}^n_+})$

$$|<g,\varphi>| = |\int g\varphi\, dx| = |-\int u\cdot\nabla\varphi\, dx| \leq \|u\|_q \|\nabla\varphi\|_{q'},$$

we have $g \in \hat{W}^{-1,q,*}(\mathbb{R}^n_+)$ and $\|g\|_{-1,q,*} \leq \|u\|_q$.

Our main result on (3.1) reads as follows.

Theorem 3.1 *Let $n \geq 2$, $1 < q < \infty$ and $0 < \varepsilon < \pi$. Then for every $\lambda \in S_\varepsilon$, $f \in L^q(\mathbb{R}^n_+)^n$ and $g \in W^{1,q}(\mathbb{R}^n_+) \cap \hat{W}^{-1,q,*}(\mathbb{R}^n_+)$ there is a unique $(u, \nabla p) \in W^{2,q}(\mathbb{R}^n_+)^n \times L^q(\mathbb{R}^n)^n$ solving the system (3.1). Moreover, $(u,\nabla p)$ is subject to the inequality*

$$\|(\lambda u, \sqrt{|\lambda|}\,\nabla u, \nabla^2 u, \nabla p)\|_q \leq C_\varepsilon(\|(f,\nabla g)\|_q + \|\lambda g\|_{-1,q,*}) , \tag{3.2}$$

where $C_\varepsilon = C_\varepsilon(q,n) > 0$ is independent of f, g and λ.

Remark 3.1 Consider g in Theorem 3.1. Then it is easy to give sufficient conditions for $g \in \hat{W}^{-1,q,*}(\mathbb{R}^n_+)$. Let g satisfy one of the following conditions:

i) $g \in L^q(\mathbb{R}^n_+)$, supp g *compact*, $\int_{\mathbb{R}^n_+} g\, dx = 0$, $1 < q < \infty$.

ii) $g \in L^s(\mathbb{R}^n_+)$, *where s is defined by* $\frac{1}{n} + \frac{1}{q} = \frac{1}{s}$ *and* $q > \frac{n}{n-1}$.

iii) $g \in L^q_{loc}(\mathbb{R}^n_+)$ *and* $|x|g \in L^q(\mathbb{R}^n_+)$ *if* $q > \frac{n}{n-1}$.

iv) $g \in L^q_{loc}(\mathbb{R}^n_+)$, $|x|\,g \in L^q(\mathbb{R}^n_+)$ *and* $\int_{\mathbb{R}^n_+} g(x)\, dx = 0$ *if* $q < \frac{n}{n-1}$

(note that $|x|\,g \in L^q(\mathbb{R}^n_+)$ yields $g \in L^1(\mathbb{R}^n_+)$).

Then $g \in \hat{W}^{-1,q,*}(\mathbb{R}^n_+)$ and $\|g\|_{-1,q,*}$ may be estimated by $\|g\|_q$, $\|g\|_s$ and $\||x|\,g\|_q$ in the cases i), ii) and iii), iv), respectively. Roughly speaking, the conditions i) – iv) impose a decay of g at infinity.

Proof of Theorem 3.1. First observe that it is sufficient to prove Theorem 3.1 including (3.2) for all $\lambda \in S_\varepsilon$ with $|\lambda| = 1$. Then a well known scaling procedure (see, e.g. [4]) shows the validity of the general result. So assume $|\lambda| = 1$ in the following. We prove the theorem in several steps.

a) In the first step we reduce the system (3.1) to the case $f = 0$, $g = 0$, $u_n = 0$ and $u' = \Phi \neq 0$ on Γ where $u = (u', u_n)$. For this purpose write $f = (f', f_n)$ and let f'_e, g_e be the even extensions of f', g, respectively, while f_{no} denotes the odd extension of f_n defined in the preceding section. Next we choose a solution pair $(U, \nabla P)$ for $F = (f'_e, f_{no})$, $G = g_e$ according to Theorem 2.1 on $\mathbb{R}^n$. Observe that $G \in W^{1,q}(\mathbb{R}^n)$ since g_e is the even extension of g. Let $U = (U', U_n)$ and $\Phi' = -U'(x', 0)$. Then from (2.2) we conclude using $|\lambda| = 1$ that

$$\begin{aligned}\|\Phi'\|_{W^{2-\frac{1}{q},q}(\mathbb{R}^{n-1})} &\leq C_1\|U\|_{W^{2,q}(\mathbb{R}^n)}\\ &\leq C_2(\|(F,\nabla G)\|_{L^q(\mathbb{R}^n)} + \|G\|_{\hat{W}^{-1,q}(\mathbb{R}^n)})\\ &\leq C_3(\|(f,\nabla g)\|_{L^q(\mathbb{R}^n_+)} + \|g\|_{-1,q,*}).\end{aligned} \tag{3.3}$$

b) In the next step we reduce the system (3.1) by subtracting $(U, \nabla P)$ to the following system with $f = 0$, $g = 0$ and $u = (u', u_n) = (\Phi', 0)$ on Γ,

$$\begin{aligned} \lambda u - \Delta u + \nabla p &= 0 \\ \text{div } u &= 0 \\ u' &= \Phi \quad \text{on } \Gamma \\ u_n &= 0 \quad \text{on } \Gamma. \end{aligned} \tag{3.4}$$

Here we assume that $\Phi \in W^{2-\frac{1}{q},q}(\mathbb{R}^{n-1})^{n-1}$, $\lambda \in S_\varepsilon$, $|\lambda| = 1$, and we prove the existence of a unique $(u, \nabla p) \in W^{2,q}(\mathbb{R}^n_+)^n \times L^q(\mathbb{R}^n_+)^n$ satisfying (3.4); moreover we show

$$\|(u, \nabla u, \nabla^2 u, \nabla p)\|_q \le C \|\Phi'\|_{W^{2-\frac{1}{q},q}(\mathbb{R}^{n-1})}. \tag{3.5}$$

Starting with $(U, \nabla P)$ in a), setting $\Phi'(x') = -U'(x', 0)$ and solving (3.4) with (3.5) we obtain the desired solution of (3.1) for $|\lambda| = 1$ in the form $(u + U,\ \nabla p + \nabla P)$. Thus (3.2) follows from (2.2) and (3.5). The uniqueness for (3.1) follows from the uniqueness for (3.4). Therefore, in the following we may restrict ourselves to (3.4) with (3.5).

c) In order to remove the pressure term ∇p we transform (3.4) in a fourth order partial differential equation for u_n where $u = (u', u_n)$, $u' = (u_1, \ldots, u_{n-1})$. Let ∇', Δ' and div$'$ denote the gradient, the Laplacian and the divergence with respect to x' only. Further let $\partial_n = \partial/\partial_{x_n}$. Then apply $\partial_n \text{div}'$ to the first $n-1$ equations (3.4) and $-\Delta'$ to the nth equation (3.4); adding both scalar equations and inserting $\text{div}'\, u' = -\partial_n u_n$, we obtain from (3.4) the following system

$$\begin{aligned} -\Delta(\lambda - \Delta) u_n &= 0 \quad \text{in } \mathbb{R}^n_+ \\ u_n &= 0 \quad \text{on } \Gamma \\ \partial_n u_n &= -\text{div}'\, \Phi' \quad \text{on } \Gamma. \end{aligned} \tag{3.6}$$

d) Next we solve (3.6) by introducing the partial Fourier transform $\hat{} = F'$ with respect to the variables x' in the sense of distributions,

$$\hat{u}(\xi', x_n) = (F'u)(\xi', x_n) = (2\pi)^{-\frac{n-1}{2}} \int_{\mathbb{R}^{n-1}} u(x', x_n)\, e^{-ix'\cdot\xi'}\, dx', \quad \xi' \in \mathbb{R}^{n-1}.$$

Then (3.6) is transformed into the fourth order differential equation for $\hat{u}(\xi', x_n)$,

$$\begin{aligned} (|\xi'|^2 - \partial_n^2)(\lambda + |\xi'|^2 - \partial_n^2)\hat{u}_n(\xi', x_n) &= 0 \\ \hat{u}_n(\xi', 0) &= 0 \\ \partial_n \hat{u}_n(\xi', 0) &= -i\,\xi' \cdot \hat{\Phi}'. \end{aligned} \tag{3.7}$$

Here $|\xi'|$ is the Euclidean norm of ξ'. We abbreviate $s = |\xi'|$. For fixed $\xi' \in \mathbb{R}^{n-1}$, $\lambda \in S_\varepsilon$, $|\lambda| = 1$, there is a unique bounded solution $\hat{u}_n$ of (3.7) which is a linear combination of $e^{-s x_n}$ and $e^{-\sqrt{\lambda + s^2}\, x_n}$. Thus we get

$$\hat{u}_n(\xi', x_n) = i\,\xi'\, m_0(s, x_n)\, \hat{\Phi}'(\xi') \tag{3.8}$$

with

$$m_0(s, x_n) = \frac{e^{-\sqrt{\lambda+s^2}\,x_n} - e^{-s\,x_n}}{\sqrt{\lambda+s^2} - s}$$

which leads to u_n.

e) Now we construct u' and p in (3.4) where $u = (u', u_n)$. For this purpose define $h(x', x_n)$ by its Fourier transform $\hat{h}(\xi', x_n) = e^{-\sqrt{\lambda+s^2}\,x_n}\,\hat{\Phi}'(\xi')$ and put

$$\hat{u}'(\xi', x_n) = \frac{i\,\xi'}{s^2}\partial_n\,\hat{u}_n(\xi', x_n) + (1 - \frac{\xi'\xi'}{s^2})\;\hat{h}(\xi', x_n) \tag{3.9}$$

where $\xi'\xi'$ denotes the dyadic product of ξ' with itself. From $\hat{u}'$ we obtain u' and get $\operatorname{div}'\,u' = -\partial_n\,u_n$ which yields $\operatorname{div}\,u = 0$ for $u = (u', u_n)$. Furthermore we define p by its Fourier transform

$$\hat{p}(\xi', x_n) = -\frac{1}{s^2}(\lambda + s^2 - \partial_n^2)\partial_n\,\hat{u}_n. \tag{3.10}$$

Then a direct calculation shows that $u = (u', u_n)$ and p is a solution of (3.4).

f) It remains to show that $u \in W^{2,q}(\mathbb{R}^n_+)^n$, $\nabla p \in L^q(\mathbb{R}^n_+)^n$ and that (3.5) is satisfied. The uniqueness of $(u, \nabla p)$ in (3.4) follows easily from the above consideration. To prove (3.5) we use the well known multiplier technique; see [14]. Recall the multiplier condition for a function $m(\cdot) \in C^\infty(\mathbb{R}^{n-1} \setminus \{0\})$,

$$s^k|(\nabla')^k m(\xi')| \le c_k, \quad k = 0, 1, \ldots, 1 + [\tfrac{n-1}{2}], \tag{3.11}$$

where the constants $c_k > 0$ are independent of $\xi' \in \mathbb{R}^{n-1}$. By a direct calculation we get the following facts:

i) For every $x_n > 0$, $m_1(\xi') = s\,m_0(s, x_n)$ with m_0 from (3.8) satisfies the condition (3.11) with $c_k(x_n) = C(1 + x_n)^{-1}$ where $C > 0$ is independent of x_n and $\lambda \in S_\varepsilon$, $|\lambda| = 1$. There is a $\delta > 0$ such that even $m_2(\xi') = e^{\delta\,s\,x_n} s\,m_0(s, x_n)$ satisfies (3.11) with $c_k(x_n) = C(1 + x_n)^{-1}$ and C as above.

ii) Let $m(\cdot) \in C^\infty(\mathbb{R}^{n-1} \setminus \{0\})$ be a multiplier satisfying (3.11). Then the result of i) holds true for $m_1(\xi') = m(\xi')s\,m_0(s, x_n)$ as well as for $m_2(\xi') = e^{-\sqrt{\lambda+s^2}x_n}$ and $m_3(\xi') = \partial_n\,m_0(s, x_n)$. Observe that the constant C above depends on ε in S_ε. First we consider $u_n(x', x_n)$ in step d). Since m_1 in i) is a multiplier and $\Phi' \in W^{2-\frac{1}{q},q}(\Gamma)^{n-1} \subseteq L^q(\Gamma)^{n-1}$, we obtain from (3.8) using the multiplier theorem that $u_n(\cdot, x_n) \in L^q(\mathbb{R}^{n-1})$ and $\|u_n(\cdot, x_n)\|_q \le C(1 + x_n)^{-1}\|\Phi'\|_q$. This yields $u_n \in L^q(\mathbb{R}^n_+)$ and $\|u_n\|_{L^q(\mathbb{R}^n_+)} \le C\|\Phi'\|_{L^q(\mathbb{R}^{n-1})}$. To estimate ∇u_n and $\nabla^2\,u_n$, we introduce $v(x', x_n)$ by its Fourier transform $\hat{v}(\xi', x_n) = e^{-\delta\,s\,x_n}\hat{\Phi}'(\xi')$ with $\delta > 0$ as above in i). From the theory of the Poisson semigroup (see [1], [14]) it is known that $\nabla v \in L^q(\mathbb{R}^n_+)^{n(n-1)}$, $|\nabla^2 v| \in L^q(\mathbb{R}^n_+)$ and

$$\|(\nabla v,\, \nabla^2 v)\|_{L^q(\mathbb{R}^n_+)} \le C\|\Phi'\|_{W^{2-\frac{1}{q},q}(\mathbb{R}^{n-1})}. \tag{3.12}$$

Writing $\hat{u}_n(\xi', x_n) \;=\; i\,\xi'\,m_0(s, x_n)e^{\delta\,s\,x_n}\;\hat{v}(\xi', x_n)$ and

$$\begin{aligned}\widehat{\nabla' u_n}\,(\xi', x_n) &= i\,\xi'\,m_0(s, x_n)\cdot e^{\delta\,s\,x_n}\widehat{\nabla' v}\,(\xi', x_n),\\ \widehat{\partial_n\,u_n}\,(\xi', x_n) &= (\partial_n m_0(s, x_n))\,e^{\delta\,s\,x_n}\widehat{\operatorname{div}' v}\,(\xi', x_n),\end{aligned}$$

applying i), ii) just as before and using (3.12) we obtain

$$\|(\nabla' u_n, \partial_n u_n)\|_{L^q(\mathbb{R}^n_+)} \le C\|(\nabla' v, \operatorname{div} v)\|_{L^q(\mathbb{R}^n_+)}.$$

This yields $\|\nabla u_n\|_{L^q(\mathbb{R}^n_+)} \le C\|\Phi'\|_{W^{2-\frac{1}{q},q}(\mathbb{R}^{n-1})}$. In the same way we get $\|\nabla^2 u_n\|_{L^q(\mathbb{R}^n_+)} \le C\|\Phi'\|_{W^{2-\frac{1}{q},q}(\mathbb{R}^{n-1})}$. The estimate of u' and p' defined by (3.9), (3.10) is quite similar. Consequently $u' \in W^{2,q}(\mathbb{R}^n_+)^{n-1}$, $\nabla p \in L^q(\mathbb{R}^n_+)^n$ and $\|(u', \nabla u', \nabla^2 u', \nabla p)\|_{L^q(\mathbb{R}^n_+)} \le C\|\Phi'\|_{W^{2-\frac{1}{q},q}(\mathbb{R}^{n-1})}$. This shows (3.5) and the proof of Theorem 3.1 is complete.

4 Generalized resolvent estimates for the perturbed halfspace

Here we prove resolvent estimates for a class of domains $\mathbb{R}^n$, $n \ge 2$, having a noncompact boundary. By a transformation and a perturbation argument this case can be reduced to the case $\mathbb{R}^n_+$. A further generalization can be obtained by the localization method (see [3], [9]). Our class is defined as follows.

Consider some continuous function $\omega : \mathbb{R}^{n-1} \to \mathbb{R}$ such that

$$\partial_i \omega \in L^\infty(\mathbb{R}^{n-1}),\ \partial_i \partial_j \omega \in L^1_{loc}(\mathbb{R}^{n-1}),\ i,j = 1,\ldots,n-1. \tag{4.1}$$

Thus $\omega \in C^{0,1}(\mathbb{R}^{n-1}) \cap W^{2,1}_{loc}(\mathbb{R}^{n-1})$. Then the perturbed halfspace H_ω is defined by

$$H_\omega = \{x = (x_1,\ldots,x_n) \in \mathbb{R}^n \ : x_n > \omega(x_1,\ldots,x_{n-1})\}. \tag{4.2}$$

For $\omega = 0$ we come back to the halfspace $H_0 = \mathbb{R}^n_+$. Under some smallness assumptions on ω and some restriction on q , we can prove for H_ω essentially the same resolvent estimate as for the halfspace $\mathbb{R}^n_+$. Recall that the element $g \in W^{1,q}(H_\omega)$ defines the functional $\varphi \to < g, \varphi > = \int_{H_\omega} g\varphi\, dx$ for all $\varphi \in C_0^\infty(\overline{H}_\omega)$, and $g \in \hat{W}^{-1,q,*}(H_\omega)$ means that

$$\|g\|_{-1,q,*} = \sup\{| < g, \varphi > | : \varphi \in C_0^\infty(\overline{H}_\omega), \|\nabla\varphi\|_{q'} = 1\}$$

is finite . Then for all $\lambda \in S_\varepsilon$, $\varepsilon > 0$, we consider the generalized resolvent problem

$$\begin{aligned} \lambda u - \Delta u + \nabla p &= f && \text{in } H_\omega \\ \operatorname{div} u &= g && \text{in } H_\omega \\ u &= 0 && \text{on } \Gamma_\omega = \partial H_\omega\,. \end{aligned} \tag{4.3}$$

Recall the notations $x = (x', x_n), \nabla = (\nabla', \partial_n), \nabla'^2 = (\partial_i\partial_j)_{i,j=1,\ldots,n-1}$. Then our result on (4.3) reads as follows.

Theorem 4.1 *Let $n \ge 2$, $1 < q < \infty$, $0 < \varepsilon < \pi$ and $\omega \in C^{0,1}(\mathbb{R}^{n-1}) \cap W^{2,1}_{loc}(\mathbb{R}^{n-1})$. Further consider $f \in L^q(H_\omega)^n$ and $g \in W^{1,q}(H_\omega) \cap \hat{W}^{-1,q,*}(H_\omega)$. Then there are constants $K = K(n,q,\varepsilon) > 0, \lambda_0 = \lambda_0(\omega,n,q,\varepsilon) > 0$ independent of f, g with the following properties :*

a) If $\|\nabla'\omega\|_\infty \le K$ *and* $\|\nabla'^2\omega\|_\infty < \infty$, $\lambda \in S_\varepsilon$ *and* $|\lambda| \ge \lambda_0$, *then there is a unique* $(u, \nabla p) \in W^{2,q}(H_\omega)^n \times L^q(H_\omega)^n$ *solving* (4.3); $(u, \nabla p)$ *is subject to*

$$\|(\lambda u, \sqrt{|\lambda|}\,\nabla u, \nabla^2 u, \nabla p)\|_{L^q(H_\omega)} \le C_\varepsilon(\|(f, \nabla g)\|_{L^q(H_\omega)} + \|\lambda g\|_{-1,q,*}), \tag{4.4}$$

where $C_\varepsilon = C_\varepsilon(q, n, \omega) > 0$ *is independent of* f, g, $\lambda \in S_\varepsilon$, $|\lambda| \ge \lambda_0$.

b) Let $1 < q < n-1$, $n \ge 3$ *and* $\|\nabla'\omega\|_\infty \le K$. *If* $\|\nabla'^2\omega\|_{L^{n-1}(\mathbb{R}^{n-1})} \le K$ *or if* $\||x'|\nabla'^2\omega\|_{L^\infty(\mathbb{R}^{n-1})} \le K$, *then for all* $\lambda \in S_\varepsilon$ *there is a unique* $(u, \nabla p) \in W^{2,q}(H_\omega)^n \times L^q(H_\omega)^n$ *solving* (4.3), *and* (4.4) *holds with* C_ε *independent of* f, g *and* $\lambda \in S_\varepsilon$.

Proof. We reduce the problem (4.3) to the halfspace by a transformation. In the first step we explain it and prove some properties.

i) Consider the function $\Phi : H_\omega \to \mathbb{R}^n_+$ defined by $\Phi(x) = (x', x_n - \omega(x'))$. Then we set $\tilde{x} = (\tilde{x}', \tilde{x}_n) = \Phi(x)$, $\tilde{x}' = x'$, $\tilde{x}_n = x_n - \omega(x')$, and we write $\tilde{\partial}_i = \partial/\partial\tilde{x}_i$, $i = 1, \dots, n$, $\tilde{\nabla} = (\tilde{\partial}_1, \dots, \tilde{\partial}_n) = (\tilde{\nabla}', \tilde{\partial}_n)$; analogously we define $\tilde{\Delta}$ and $\widetilde{\operatorname{div}}$. Obviously Φ is a bijection with Jacobian equal to 1. Each function u on H_ω is transformed to $\tilde{u}$ on $\mathbb{R}^n_+$ by the definition $u(x) = \tilde{u}(\tilde{x})$, $\tilde{x} = \Phi(x)$, $x \in H_\omega$. Then we have $\partial_i u = (\tilde{\partial}_i - \omega_i\tilde{\partial}_n)\tilde{u}_n$ where $\omega_i = \partial_i\omega$, $i = 1, \dots, n-1$, and $\partial_n = \tilde{\partial}_n$. It follows

$$\begin{aligned}
\Delta u(x) &= [\tilde{\Delta} + |\nabla'\omega|^2\tilde{\partial}_n^2 - 2(\nabla'\omega)\cdot(\tilde{\nabla}\tilde{\partial}_n) - (\Delta'\omega)\tilde{\partial}_n]\,\tilde{u}(\Phi(x)),\\
\nabla p(x) &= [\tilde{\nabla} - (\nabla'\omega)\tilde{\partial}_n]\,\tilde{p}(\Phi(x)),\\
\operatorname{div} u(x) &= [\widetilde{\operatorname{div}} - \nabla'\omega\cdot\tilde{\partial}_n]\,\tilde{u}(\Phi(x)).
\end{aligned} \tag{4.5}$$

For $u \in W^{2,q}(H_\omega)$ this leads to $\|u\|_{L^q(H_\omega)} = \|\tilde{u}\|_{L^q(\mathbb{R}^n_+)}$ and the estimates $\|\nabla u\|_{L^q(H_\omega)} \le C(1 + \|\nabla'\omega\|_\infty)\,\|\tilde{\nabla}\tilde{u}\|_{L^q(\mathbb{R}^n_+)}$,

$$\|\nabla^2 u\|_{L^q(H_\omega)} \le c(1 + \|\nabla'\omega\|_\infty^2)\,\|\tilde{\nabla}^2\tilde{u}\|_{L^q(\mathbb{R}^n_+)} + c\|(\nabla'^2\omega)\,\tilde{\partial}_n\,\tilde{u}\|_{L^q(\mathbb{R}^n_+)}. \tag{4.6}$$

Finally, since for $g \in W^{1,q}(H_\omega) \cap \hat{W}^{-1,q,*}(H_\omega)$

$$\int_{H_\omega} g(x)\,\varphi(x)\,dx = \int_{\mathbb{R}^n_+} \tilde{g}(\tilde{x})\,\tilde{\varphi}(\tilde{x})\,d\tilde{x}, \quad \varphi \in C_0^\infty(\overline{H}_\omega),$$

we obtain

$$\|g\|_{-1,q,*} \le C\|\tilde{g}\|_{\hat{W}^{-1,q,*}(\mathbb{R}^n_+)}. \tag{4.7}$$

If $1 < q < n-1$, $n \ge 3$, and $|\nabla'^2\omega| \in L^{n-1}(\mathbb{R}^{n-1})$, we define s by $\frac{1}{n-1} + \frac{1}{s} = \frac{1}{q}$ and use for each $x_n > 0$ the inequality

$$\|\tilde{\partial}_n\,\tilde{u}(\cdot, x_n)\|_{L^s(\mathbb{R}^{n-1})} \le C\|\tilde{\nabla}'\tilde{\partial}_n\tilde{u}(\cdot, x_n)\|_{L^q(\mathbb{R}^{n-1})},$$

the constant C being independent of x_n. Then Hölder's inequality yields

$$\begin{aligned}
\|(\nabla'^2\omega)\tilde{\partial}_n\tilde{u}\|^q_{L^q(\mathbb{R}^n_+)} &\le C\int_0^\infty dx_n \int_{\mathbb{R}^{n-1}} dx'\,|\nabla'^2\omega|^q\,|\tilde{\partial}_n\tilde{u}(x', x_n)|^q\\
&\le C\,\|\nabla'^2\omega\|^q_{L^{n-1}(\mathbb{R}^{n-1})}\,\|\tilde{\nabla}'\tilde{\partial}_n\tilde{u}\|^q_{L^q(\mathbb{R}^n_+)}.
\end{aligned}$$

Therefore we get

$$\|\nabla^2 u\|_{L^q(H_\omega)} \le C(1 + \|\nabla'\omega\|_\infty^2 + \|\nabla'^2\omega\|_{L^{n-1}(\mathbb{R}^{n-1})})\|\tilde{\nabla}^2\tilde{u}\|_{L^q(\mathbb{R}^n_+)}. \tag{4.8}$$

However, if $|x'|\,|\nabla'^2\omega| \in L^\infty(\mathbb{R}^{n-1})$, we use the weighted inequality

$$\||x'|^{-1}\tilde{\partial}_n\tilde{u}(\cdot, x_n)\|_{L^q(\mathbb{R}^{n-1})} \le C\|\tilde{\nabla}'\tilde{\partial}_n\tilde{u}(\cdot, x_n)\|_{L^q(\mathbb{R}^{n-1})}$$

and obtain

$$\|\nabla^2 u\|_{L^q(H_\omega)} \le C(1 + \|\nabla'\omega\|_\infty^2 + \||x'|\nabla'^2\omega\|_{L^\infty(\mathbb{R}^{n-1})})\|\tilde{\nabla}^2\tilde{u}\|_{L^q(\mathbb{R}^n_+)}. \tag{4.9}$$

ii) Now we solve (4.3) by a perturbation argument. For this purpose let us consider the Banach spaces $X = (W^{2,q}(H_\omega) \cap W_0^{1,q}(H_\omega))^n \times L^q(H_\omega)^n$ and $Y = L^q(H_\omega)^n \times (W^{1,q}(H_\omega) \cap \hat{W}^{-1,q,*}(H_\omega))$ with norms $\|(u, \nabla p)\|_X =$ $\|(\lambda u, \sqrt{|\lambda|}\,\nabla u, \nabla^2 u, \nabla p)\|_{L^q(H_\omega)}$ and $\|(f,g)\|_Y = \|(f, \nabla g)\|_{L^q(H_\omega)} + \|\lambda g\|_{-1,q,*}$, respectively. Then we define the operator $A : X \to Y$ for the system (4.3) by $A(u, \nabla p) = (\lambda u - \Delta u + \nabla p, \operatorname{div} u)$. Let $\tilde{X}, \tilde{Y}, \|(\tilde{u}, \tilde{\nabla}\tilde{p})\|_{\tilde{X}}$, $\|(\tilde{f}, \tilde{g})\|_{\tilde{Y}}$ and $\tilde{A}$ be the corresponding expressions when $H_\omega, u, p, f, g, \Delta, \nabla, \operatorname{div}$ are replaced by $\mathbb{R}^n_+, \tilde{u}, \tilde{p}, \tilde{f}, \tilde{g}, \tilde{\Delta}, \tilde{\nabla}, \widetilde{\operatorname{div}}$, respectively. A calculation using (4.5) shows that $[A(u, \nabla p)](x) = [\tilde{A}(\tilde{u}, \tilde{\nabla}\tilde{p})](\tilde{x}) + [B(\tilde{u}, \tilde{\nabla}\tilde{p})](\tilde{x})$ where $B(\tilde{u}, \tilde{\nabla}\tilde{p}) = (-|\nabla'\omega|^2\tilde{\partial}_n^2\tilde{u} + 2(\nabla'\omega)\cdot(\tilde{\nabla}\tilde{\partial}_n\tilde{u}) + (\Delta'\omega)\tilde{\partial}_n\tilde{u} - (\nabla'\omega)\tilde{\partial}_n\tilde{p},\ -\nabla'\omega\cdot\tilde{\partial}_n\tilde{u})$ is treated as a perturbation of $\tilde{A}(\tilde{u}, \tilde{\nabla}\tilde{p})$. Consider the assertion in 4.1 a). Then for sufficiently small $\|\nabla'\omega\|_\infty$, sufficiently large λ_0 and $|\lambda| \ge \lambda_0$, a direct calculation using (3.2),(4.6) and (4.7) shows that there is a constant C_0 with $0 < C_0 < 1$ such that

$$\|B(\tilde{u}, \tilde{\nabla}\tilde{p})\|_{\tilde{Y}} \le C_0\|\tilde{A}(\tilde{u}, \tilde{\nabla}\tilde{p})\|_{\tilde{Y}}.$$

So we may apply Kato's perturbation criterion; we conclude that A is an isomorphism from X onto Y and it holds

$$\begin{aligned} \|(u, \nabla p)\|_X &\le C_1\|(\tilde{u}, \tilde{\nabla}\tilde{p})\|_{\tilde{X}} \le C_2\|\tilde{A}(\tilde{u}, \tilde{\nabla}\tilde{p})\|_{\tilde{Y}} \\ &\le C_3\|\tilde{A}(\tilde{u}, \tilde{\nabla}\tilde{p}) + B(\tilde{u}, \tilde{\nabla}\tilde{p})\|_{\tilde{Y}} \le C_4\|A(u, \nabla p)\|_Y \end{aligned}$$

which proves (4.4). In the case 4.1 b) we get the same estimates even for all $\lambda \in S_\varepsilon$ if we additionally use (4.8) or (4.9). Thus Theorem 4.1 is proved.

Remark 4.1 Let us consider some examples of perturbed halfspaces H_ω.

a) Let $\omega \in C^2(\mathbb{R}^{n-1})$, $n \ge 2$, such that $\omega(x') = \alpha|x'|$ for large $|x'|$. Thus H_ω coincides for large $|x'|$ with a cone of opening angle $2\arctan\frac{1}{\alpha}$. By Theorem 4.1 a) the generalized resolvent estimate (4.4) holds true in $L^q(H_\omega)$, $1 < q < \infty$, for large $|\lambda|$, provided that $\|\nabla'\omega\|_\infty$ and in particular that $|\alpha|$ is sufficiently small, i.e. the opening angle of the cone is close to π.

b) Theorem 4.1 a) also applies to perturbed halfspaces H_ω with oscillating boundaries given by $\omega(x') = \sin(\alpha|x'|)$, such that neither the frequency nor the amplitude are decreasing.

c) Using the condition $\||x'|\nabla'^2\omega\|_\infty \le K$, Theorem 4.1 b) applies to the cone $H_\omega = \{(x', x_n);\ x_n > \alpha|x'|\} \subset \mathbb{R}^n$, $n \ge 3$, if $|\alpha|$ is sufficiently small.

d) Analogously, the generalized resolvent estimate holds true for "oscillating" halfspaces $H_\omega \subset \mathbb{R}^n$, $n \geq 3$, such that either the amplitude is decreasing ($\omega(x') \sim |x'|^{-1-\varepsilon}\sin(\alpha|x'|)$, $\varepsilon \geq 0$, for large $|x'|$) or such that the frequency is decreasing ($\omega(x') = \sin(\alpha|x'|^{\frac{1}{2}-\varepsilon})$, $\varepsilon \geq 0$).

5 Further consequences

Our generalized resolvent estimates for $\mathbb{R}^n, \mathbb{R}^n_+$ and H_ω enable us to give a simple proof of the basic estimate

$$\|(\lambda + A_q)^{-1} f\|_q \leq C(\Omega, \varepsilon, q)\,|\lambda|^{-1}\,\|f\|_q, \quad f \in L^q_\sigma(\Omega), |\arg\lambda| < \pi - \varepsilon \tag{5.1}$$

for the well known case of bounded and exterior domains $\Omega \subseteq \mathbb{R}^n, n \geq 2$, ([8], [9], [11], [13]), as well as for a new more general class of domains with noncompact boundaries. For bounded and exterior domains $\Omega \subseteq \mathbb{R}^n$ with $\partial\Omega \in C^{2+\mu}(0 < \mu < 1)$, (5.1) is known for each given $0 < \varepsilon < \pi$ excepting the case $n = 2$ where $C(\Omega, \varepsilon, q) > 0$ in (5.1) may depend additionally on a positive lower bound of $|\lambda|$, $\lambda \in S_\varepsilon$. By our method, we get the same restriction in this case but we only need $\partial\Omega \in C^{1,1}$. The same method also yields (5.1) for a class of $C^{1,1}$-domains obtained from $\mathbb{R}^n_+$ by an arbitrary not necessarily small deformation of $\mathbb{R}^n_+$ within a bounded region.

Our method to prove (5.1) rests on well known localization arguments which reduce the general case $\Omega \subseteq \mathbb{R}^n$ to the special cases $\mathbb{R}^n$, $\mathbb{R}^n_+$, H_ω; see [3],[9] for details concerning this procedure. To explain this method we start with the equations

$$\begin{aligned} \lambda u - \Delta u + \nabla p &= f && \text{in } \Omega \\ \operatorname{div} u &= 0 && \text{in } \Omega \\ u &= 0 && \text{on } \partial\Omega \end{aligned} \tag{5.2}$$

for $u \in W^{2,q}(\Omega)^n, \nabla p \in L^q(\Omega)^n$. Consider an open covering $\overline{\Omega} \subseteq \bigcup_{i=0}^m \Omega_i$ where Ω_0 is unbounded if Ω is an exterior domain and $\Omega_1, \ldots, \Omega_m$ are sufficiently small open balls. Considering cut-off functions $\varphi_0 \in C_0^\infty$, $\varphi_i \in C_0^\infty$, $i = 1, \ldots, n$, with $\sum_{i=0}^m \varphi_i = 1$ on Ω, $\varphi_0 = 0$ in a neighborhood of $\partial\Omega_0$, $0 \leq \varphi_i \leq 1$, and multiplying (5.2) with φ_i we obtain the local equations

$$\begin{aligned} \lambda(\varphi_i u) - \Delta(\varphi_i u) + \nabla(\varphi_i p) &= \varphi_i f - 2(\nabla\varphi_i)(\nabla u) - (\Delta\varphi_i)u + (\nabla\varphi_i)p \\ \operatorname{div}(\varphi_i u) &= (\nabla\varphi_i)u \qquad\qquad \text{in } \Omega_i \cap \Omega \\ \varphi_i u &= 0 \qquad\qquad\qquad\quad \text{on } \partial(\Omega_i \cap \Omega) \end{aligned} \tag{5.3}$$

which are treated as equations in $\mathbb{R}^n$ or H_ω. The conditions on ω in Theorem 4.1 are satisfied if $\Omega_1, \ldots, \Omega_m$ are sufficiently small; here we have to use b) in 4.1 for small $|\lambda|$ under the restriction on q. So we may apply our generalized resolvent estimates to (5.3); we obtain now an estimate for (5.2) which contains additional lower order terms. These can be removed by a compactness argument. Observe that $\operatorname{div}(\varphi_i u) \neq 0$; this shows the importance of our generalized resolvent estimates. The compactness argument requires a uniqueness assertion which is available from the L^2-theory and some regularity property. The restriction on q in 4.1 b) can be removed by duality and interpolation arguments as in [3], [9].

Consider the three-dimensional convex or concave cone

$$H_\omega = \{(x', x_3) \in \mathbb{R}^3 : x_3 > \alpha|x'|\}$$

with $|\alpha|$ sufficiently small, see Remark 4.1 c). Assuming the existence of the continuous projection $P_q : L^q(H_\omega)^n \to L^q_\sigma(H_\omega)$, Theorem 4.1 b) implies that the Stokes operator $A_q = -P_q\Delta$ generates a bounded analytic semigroup e^{-tA_q}, $t \geq 0$, in $L^q_\sigma(H_\omega)$, $1 < q < 2$. By duality arguments we get the same results in $L^q_\sigma(H_\omega)$, $2 < q < \infty$. In particular $e^{-tA_q}u_0$ tends to zero as $t \to \infty$ for each $u_0 \in L^q_\sigma(H_\omega)$, $1 < q < \infty$. In the limit $\lambda = 0$ we conclude the unique solvability of the Stokes equation (5.2) in the three-dimensional cone H_ω with opening angle close to π in each L^q–space, $1 < q < 2$.

References

[1] Agmon, S., Douglis, A., Nirenberg, L.: Estimates near the boundary for solutions of elliptic partial differential equations satisfying general boundary conditions. I, Comm. Pure Appl. Math., 12 (1959), 623–727.

[2] Bogovskii, M.E.: Solution of the first boundary value problem for the equation of continuity of an incompressible medium, Soviet Math. Dokl., 20 (1979), 1094–1098.

[3] Borchers, W., Sohr, H.: On the semigroup of the Stokes operator for exterior domains in L_q-spaces, Math. Z. 196 (1987), 415–425.

[4] Borchers, W., Miyakawa, T.: L^2-decay for the Navier-Stokes equations in halfspaces, Math. Ann. 282 (1988), 139–155.

[5] Deuring, P.: The resolvent problem for the Stokes sytem in exterior domains: an elementary approach, Math. Meth. in the Appl. Sci. 13 (1990), 335–349.

[6] Fujiwara, D., Morimoto, H.: An L_r-theorem of the Helmholtz decomposition of vector fields, J. Fac. Sci. Univ. Tokyo, Sect. I A, 24 (1977), 685–700.

[7] Galdi, G.P., Simader, C.G.: Existence, uniqueness and L^q-estimates for the Stokes problem in an exterior domain, Arch. Rat. Mech. Anal. 112 (1990), 291–318.

[8] Giga, Y.: Analyticity of the semigroup generated by the Stokes operator in L_r spaces, Math. Z. 178 (1981), 297–329.

[9] Giga, Y., Sohr, H.: On the Stokes operator in exterior domains, J. Fac. Sci. Univ. Tokyo, Sect. I A, 36 (1989), 103–130.

[10] Mc Cracken, M.: The resolvent problem for the Stokes equations on halfspace in L_p, SIAM J. Math. Anal. 12 (1981), 221–228.

[11] Miyakawa, T.: On nonstationary solutions of the Navier-Stokes equations in an exterior domain, Hiroshima Math. J. 12 (1982), 115–140.

[12] Simader, C.G., Sohr, H.: A new approach to the Helmholtz decomposition in L_q-spaces for bounded and exterior domains, Series on Advances Mathematics for Applied Science, Vol. 11, World Scientific 1992

[13] Solonnikov, V.A.: Estimates for solutions of nonstationary Navier- Stokes equations, J. Soviet Math. 8 (1977), 467–529.

[14] Triebel, H.: Interpolation theory, function spaces, differential operators, North-Holland, Amsterdam 1978.

[15] Ukai, S.: A solution formula for the Stokes equation in $\mathbb{R}^n_+$ Comm. Pure Appl. Math. 40 (1987), 611–621.

[16] Wahl, von W.: Vorlesungen über das Außenraumproblem für die instationären Gleichungen von Navier-Stokes, SFB 256 Nichtlineare partielle Differentialgleichungen, Vorlesungsreihe Nr. 11, Universität Bonn 1989.

ON THE OSEEN BOUNDARY-VALUE PROBLEM IN EXTERIOR DOMAINS

Giovanni P. Galdi

Istituto di Ingegneria dell'Universita'
Via Scandiana 21, 44100 Ferrara

Introduction.

During the last years, a remarkable interest has been directed towards the resolution of the linearized equations of stationary flows past an obstacle $\mathcal{O}$ (say) of a viscous incompressible fluid and the derivation of corresponding estimates both in L^q and weighted Sobolev spaces, *cf.*, *e.g.*, Maremonti & Solonnikov (1985,1986), Specovius-Neugebauer (1986), Galdi & Simader (1990), Maslennikova (1990), Kozono & Sohr (1991), Varnhorn (*cf.* these Proceedings). Such a work is sometimes aimed to the resolution of nonlinear problems. For example, one may use these results to obtain sharp bounds on time decay rate of solutions to the *initial* boundary value problem for the full nonlinear Navier-Stokes equations, *cf. e.g.*, Borchers & Miyakawa (1990). All the mentioned papers deal with the *Stokes equations*. However, as is well-known, if $\mathcal{O}$ is moving with a non-zero velocity -a situation which is very appealing from the physical viewpoint- a more appropriate linearization of the Navier-Stokes equations is that described by the *Oseen equations*. Mathematically, the boundary-value problem is described as follows:

$$\left.\begin{aligned} &\Delta \mathbf{v} - \mathcal{R}e\frac{\partial \mathbf{v}}{\partial x_1} = \nabla p + \mathbf{f} \\ &\nabla\cdot\mathbf{v} = 0 \end{aligned}\right\} \quad \text{in } \Omega \tag{0.1}$$

with

$$\begin{aligned} &\mathbf{v} = \mathbf{v}_* \ \text{ at } \partial\Omega \\ &\lim_{|x|\to\infty} \mathbf{v}(x) = \mathbf{v}_\infty \end{aligned} \tag{0.2}$$

where $\Omega\subset\mathbb{R}^n$ is the domain exterior to $\mathcal{O}$, $\mathcal{R}e$ is a non-negative dimensionless (Reynolds) number, $\mathbf{f}$ is the (opposite of the) body force acting on the fluid and $\mathbf{v}_*$ and $\mathbf{v}_\infty$ are prescribed field (on $\partial\Omega$) and vector (in $\mathbb{R}^n$), respectively. If we formally take $\mathcal{R}e=0$ we have the Stokes problem, while $\mathcal{R}e>0$ (which we shall assume throughout) describes the Oseen one. Because of the term $\mathcal{R}e\frac{\partial \mathbf{v}}{\partial x_1}$, one expects that the asymptotic properties of solutions to the Stokes and

---

This paper is in final form and no similar paper has been or is being submitted elswhere.

Oseen problems be quite different and, specifically, that they resemble the behaviour of the corresponding fundamental solutions, *cf.* Chang & Finn (1961). Concerning Oseen problem, we may wish to mention the contributions of Babenko (1973) who derives some L^q estimates for n=3 and of Salvi who proves existence of strong solutions and corresponding L^q estimates again for n=3 (*cf.* these Proceedings). We also recall that Farwig has performed a detailed analysis of the Oseen boundary-value problem in weighted Sobolev spaces of Hilbert type for n=3 and has used these results to show existence and asymptotic behaviour of solutions to the *nonlinear* problem (*cf.* Farwig (1991a), (1991b)).

In the present paper we furnish a complete study of the Oseen problem (0.1), (0.2) in arbitrary dimension $n\geq 2$. Specifically we prove existence, uniqueness and corresponding estimates for *weak* and *strong* solutions in L^q spaces, where the allowed values of q, as in the Stokes problem , depend on n. It is interesting to observe that, if n=2, the two components of the velocity field $v=(v_1,v_2)$ belong to *different* L^q spaces and, in particular, the component v_2 transverse to the wake exhibits a behaviour at infinity "better" than v_2. It is also expected (even though not yet proved) that, in a complete analogy with the Stokes problem (*cf.* Galdi & Simader (1990)), if q does not satisfy the restrictions imposed in this paper there should appear a non-empty null set and therefore all the above results should be reformulated in suitable quotient spaces. Applications of the results contained in this paper to nonlinear Navier-Stokes problem are given in Galdi (1991, 1992). Finally, we would like to observe that we have assumed that Ω is as smooth as required by the standard elliptic theory (*cf.* Cattabriga (1961)). However, this smoothness can be relaxed and, in particular, using the recent theory of Galdi, Simader & Sohr (1992) the results of this paper concerning *weak solutions* continue to hold also for lipschitz domains with not too sharp corners.

1. Existence, uniqueness and L^q estimates in the whole space.

The notations we shall use are rather standard. In particular, we indicate by $C_0^\infty(\Omega)$ the set of all indefinitely differentiable functions of compact support in Ω, by $W^{m,q}(\Omega)$ the Sobolev space on Ω with m order of differentiation and q exponent of summability and set

$$\|u\|^q_{m,q,\Omega} = \sum_{|\alpha|=0}^{m} \int_\Omega |D^\alpha u|^q ,$$

where the subscript Ω will be omitted if no confusion arises. We have $W^{0,q}(\Omega)=L^q(\Omega)$ and we set $\|u\|_{0,q,\Omega}\equiv\|u\|_{q,\Omega}$. Also, we denote by $D^{m,q}(\Omega)$, the class of functions u with

$$|u|_{m,q,\Omega} = \sum_{|\alpha|=m} (\int_\Omega |D^\alpha u|^q)^{1/q}$$

finite and by $D_0^{1,q}(\Omega)$ the completion of $C_0^\infty(\Omega)$ in the norm $|u|_{1,q,\Omega}$. Furthermore, $\mathcal{D}_0^{1,q}(\Omega)$ indicates the subspace of $D_0^{1,q}(\Omega)$ constitued by solenoidal functions. Finally, $D_0^{-1,q}(\Omega)$ is the dual space of $D_0^{1,q'}(\Omega)$ and the duality pairing between the two spaces is denoted by [,].

The aim of this section is to show existence, uniqueness and appropriate estimates of solutions v,p to the *non-homogeneous Oseen system*

$$\left.\begin{aligned} &\Delta v - \mathcal{R}e\frac{\partial v}{\partial x_1} - \nabla p = \mathcal{R}e f \\ &\nabla\cdot v = g \end{aligned}\right\} \quad \text{in } \mathbb{R}^n, \tag{1.1}$$

in Lebesgue spaces $L^q(\mathbb{R}^n)$. In establishing estimates for (1.1) it is important to single out the dependence of the constants entering the estimates on the dimensionless parameter $\mathcal{R}e$. We shall therefore consider the problem

$$\left.\begin{aligned} &\Delta v - \frac{\partial v}{\partial x_1} - \nabla p = f \\ &\nabla\cdot v = g \end{aligned}\right\} \quad \text{in } \mathbb{R}^n, \tag{1.2}$$

and establish corresponding estimates for its solutions. The analogous ones for solutions to (1.1) will be then obtained by the replacements

$$f \to f/\mathcal{R}e^2,\ g \to g/\mathcal{R}e\ ,\ p \to p/\mathcal{R}e\ ,\ x_1 \to \mathcal{R}e x_1\ . \tag{1.3}$$

To show the afore-said results, we shall make use of an appropriate tool due to P.I.Lizorkin (1963) and which we are going to describe. Denote by $\mathcal{S}(\mathbb{R}^n)$ the space of *functions of rapid decrease* consisting of elements u from $C^\infty(\mathbb{R}^n)$ such that

$$\sup_{x\in\mathbb{R}^n} |x_1|^{\alpha_1}\cdot\ldots\cdot|x_n|^{\alpha_n}|D^\beta u(x)| < \infty$$

for all $\alpha_1,\ldots,\alpha_n \ge 0$ and $|\beta|\ge 0$. For $u\in\mathcal{S}(\mathbb{R}^n)$ we denote by $\hat{u}$ its *Fourier transform*:

$$\hat{u}(\xi) = \frac{1}{(2\pi)^{n/2}}\int_{\mathbb{R}^n} e^{-ix\cdot\xi}u(x)dx\ .$$

It is well-known that $\hat{u}\in\mathcal{S}(\mathbb{R}^n)$. Given a function $\Phi:\mathbb{R}^n\to\mathbb{R}$, let us consider the *integral transform*

$$Tu \equiv h(x) = \frac{1}{(2\pi)^{n/2}} \int_{\mathbb{R}^n} e^{ix\cdot\xi}\Phi(\xi)\hat{u}(\xi)d\xi, \quad u\in\mathcal{S}(\mathbb{R}^n). \tag{1.4}$$

Lemma 1.1 (Lizorkin). *Let* $\Phi:\mathbb{R}^n\to\mathbb{R}$ *be continuous together with the derivative*

$$\frac{\partial^n\Phi}{\partial\xi_1\ldots\partial\xi_n}$$

and all preceding derivatives for $|\xi_i|>0$, $i=1,\ldots,n$. *Then if for some* $\beta\in[0,1)$ *and* $M>0$

$$|\xi_1|^{\kappa_1+\beta}\cdot\ldots\cdot|\xi_n|^{\kappa_n+\beta}\left|\frac{\partial^\kappa\Phi}{\partial\xi_1^{\kappa_1}\ldots\partial\xi_n^{\kappa_n}}\right| \le M ,$$

where κ_i *is zero or one and* $\kappa = \sum_{i=1}^n \kappa_i=0,1,\ldots n$, *the integral transform* (4.4) *defines a bounded linear operator from* $L^q(\mathbb{R}^n)$ *into* $L^r(\mathbb{R}^n)$, $1<q<\infty$, $1/r=1/q-\beta$, *and we have*

$$\|Tu\|_r\le c\|u\|_q \quad ,$$

with a constant c *depending only on* M, r *and* q.

With this result in hands, we shall now look for a solution to (1.2) corresponding to $f\in C_0^\infty(\mathbb{R}^n)$ and $g\equiv0$ of the form

$$\begin{aligned} v(x) &= \frac{1}{(2\pi)^{n/2}}\int_{\mathbb{R}^n} e^{ix\cdot\xi}\, V(\xi)d\xi \\ p(x) &= \frac{1}{(2\pi)^{n/2}}\int_{\mathbb{R}^n} e^{ix\cdot\xi}\, P(\xi)d\xi \end{aligned} \tag{1.5}$$

Replacing (1.5) into (1.2) provides the following algebraic system for **V**, **P**

$$\begin{aligned} (\xi^2+i\xi_1)V_m(\xi) + i\xi_m P(\xi) &= -\hat{f}_m(\xi) \\ i\xi_m V_m(\xi) &= 0 \end{aligned} \tag{1.6}$$

where m=1,...,n. Solving (1.6) for **V** and **P** delivers

$$\begin{aligned} V_m(\xi) &= \frac{\xi_m\xi_k - \xi^2\delta_{mk}}{\xi^2(\xi^2+i\xi_1)}\hat{f}_k(\xi) \\ P(\xi) &= -\frac{1}{i}\,\frac{\hat{f}_k(\xi)\xi_k}{\xi^2} \quad . \end{aligned} \tag{1.7}$$

Since $\hat{f}\in\mathcal{S}(\mathbb{R}^n)$, it is not hard to show that (1.5) and (1.7) define a C^∞ solution to (1.2) with $g\equiv0$. Let us now determine some L^q estimates for v and p. This will be done with the aid of Lemma 1.1. Setting

$$\phi_{mk}(\xi) = \frac{\xi_m\xi_k - \xi^2\delta_{mk}}{\xi^2(\xi^2 + i\xi_1)}, \tag{1.8}$$

we can easily prove the following lemma.

Lemma 1.2. *Let* $n\geq2$ *and let* ϕ_{mk} *be given by* (1.8) *with* m, k *ranging in* $\{1,\dots,n\}$. *Then, the assumptions of* Lemma 1.1 *are satisfied*:

(a) *by* ϕ_{mk} *with* $\beta=2/(n+1)$;

(b) *by* $\xi_\ell\phi_{mk}$ *with* $\beta=1/(n+1)$ *and* $\ell\in\{1,\dots,n\}$;

(c) *by* $\xi_1\phi_{mk}$ *with* $\beta=0$;

(d) *by* $\xi_s\xi_\ell\phi_{mk}$ *with* $\beta=0$ *and* $s,\ell\in\{1,\dots,n\}$.

Finally, if $n=2$, *for all* $\ell,k\in\{1,2\}$ *the assumptions of* Lemma 1.1 *are satisfied:*

(e) *by* ϕ_{2k} *with* $\beta=1/2$;

(f) *by* $\xi_\ell\phi_{2k}$ *with* $\beta=0$.

From (1.5) and $(1.7)_1$, with the help of Lemma 1.1 and Lemma 1.2 (c), (d), it follows at once

$$\left\|\frac{\partial v}{\partial x_1}\right\|_q \leq c\|f\|_q \tag{1.9}$$

and

$$|v|_{2,q} \leq c\|f\|_q. \tag{1.10}$$

Also, observing that for all $s,k=1,\dots,n$ the function $\xi_s\xi_k/\xi^2$ satisfies the assumptions of Lemma 1.1 with $\beta=0$ we have

$$|p|_{1,q} \leq c\|f\|_q. \tag{1.11}$$

From (1.9)-(1.11) we conclude

$$\left\|\frac{\partial v}{\partial x_1}\right\|_q + |v|_{2,q} + |p|_{1,q} \leq c\|f\|_q, \quad 1<q<\infty, \tag{1.12}$$

with $c=c(n,q)$. In the case $n=2$, we are able to obtain a sharper estimate on the component v_2 of the velocity field. Specifically, from (1.5), $(1.7)_1$, Lemma 1.1 and from Lemma 1.2 (f), we recover

$$|v_2|_{1,q} \leq c\|f\|_q, \quad 1<q<\infty, \tag{1.13}$$

which, along with (1.12), then furnishes

$$|v_2|_{1,q}+\left\|\frac{\partial v_1}{\partial x_1}\right\|_q+|v|_{2,q}+|p|_{1,q}\le c\|f\|_q\ ,\ 1<q<\infty. \tag{1.14}$$

Estimates (1.12), (1.14) can be generalized to derivatives of arbitrary order. Actually, noticing that

$$\widehat{D_\ell u} = i\xi_\ell \hat{u}\ ,$$

after differentiating (1.5) a finite number of times we may use the same reasonings leading to (1.12) and (1.14) we deduce for all $\ell\ge 0$

$$\left|\frac{\partial v}{\partial x_1}\right|_{\ell,q}+|v|_{\ell+2,q}+|p|_{\ell+1,q}\le c|f|_{\ell,q}\ ,\ 1<q<\infty\ , \tag{1.15}$$

and, for n=2,

$$|v_2|_{\ell+1,q}+\left|\frac{\partial v_1}{\partial x_1}\right|_{\ell,q}+|v|_{\ell+2,q}+|p|_{\ell+1,q}\le c|f|_{\ell,q}\ ,1<q<\infty \tag{1.16}$$

Other estimates can be obtained by suitably restricting the range of values of q. To this end, assume $1<q<n+1$; we then obtain from Lemma 1.1 and Lemma 1.2 (b), for all $\ell\ge 0$

$$|v|_{\ell+1,s_1}\le c|f|_{\ell,q}\ ,\ s_1=\frac{(n+1)q}{n+1-q}\ ,\ 1<q<n+1,$$

which, together with (1.15) implies, for all $\ell\ge 0$

$$|v|_{\ell+1,s_1}+\left|\frac{\partial v}{\partial x_1}\right|_{\ell,q}+|v|_{\ell+2,q}+|p|_{\ell+1,q}\le c|f|_{\ell,q}\ ,s_1=\frac{(n+1)q}{n+1-q}\ ,\ 1<q<n+1, \tag{1.17}$$

where $c=c(n,\ell,q)$. In addition to this inequality, for the case n=2, from Lemma 1.2 (e) we derive

$$|v_2|_{\ell,\frac{2q}{2-q}}\le c|f|_{\ell,q},\ 1<q<2,$$

which together with (1.16) and (1.17) furnishes for all $\ell\ge 0$

$$\begin{aligned}&|v_2|_{\ell,\frac{2q}{2-q}}+|v_2|_{\ell+1,q}+|v|_{\ell+1,\frac{3q}{3-q}}+\left|\frac{\partial v_1}{\partial x_1}\right|_{\ell,q}\\ &\quad+|v|_{\ell+2,q}+|p|_{\ell+1,q}\le c|f|_{\ell,q},\ 1<q<2.\end{aligned} \tag{1.18}$$

Let us now assume $1<q<(n+1)/2$. Again from (1.5), $(1.7)_1$, Lemma 1.1 and Lemma 1.2 (a) we obtain

$$|v|_{\ell,s_2}\le c|f|_{\ell,q}\ ,\ s_2=\frac{(n+1)q}{n+1-2q}\ ,\ 1<q<(n+1)/2.$$

This inequality along with (1.18) gives the following estimate for all $\ell\ge 0$:

$$|v|_{\ell,s_2}+|v|_{\ell+1,s_1}+\left|\frac{\partial v}{\partial x_1}\right|_{\ell,q}+|v|_{\ell+2,q}+|p|_{\ell+1,q}\le c|f|_{\ell,q}$$

$$s_1=\frac{(n+1)q}{n+1-q},\quad s_2=\frac{(n+1)q}{n+1-2q},\quad 1<q<(n+1)/2. \tag{1.19}$$

In particular, if n=2, (1.18) and (1.19) furnish

$$|v_2|_{\ell,\frac{2q}{2-q}}+|v_2|_{\ell+1,q}+|v|_{\ell,\frac{3q}{3-2q}}+|v|_{\ell+1,\frac{3q}{3-q}}$$

$$+\left|\frac{\partial v_1}{\partial x_1}\right|_{\ell,q}+|v|_{\ell+2,q}+|p|_{\ell+1,q}\le c|f|_{\ell,q},\quad 1<q<3/2. \tag{1.20}$$

We shall next consider existence of solutions to (1.2) when both $f,g\not\equiv 0$, with $f,g\in C_0^\infty(\mathbb{R}^n)$. We look for a solution of the form $v=w+h$ where w solves

$$\begin{aligned}\Delta w-\frac{\partial w}{\partial x_1}&=\nabla p+F\\ \nabla\cdot w&=0\end{aligned} \tag{1.21}$$

with

$$F=f-\Delta h+\frac{\partial h}{\partial x_1} \tag{1.22}$$

and

$$h=\nabla\Psi-e_1\Psi \tag{1.23}$$

with ψ solving

$$\Delta\Psi-\frac{\partial\Psi}{\partial x_1}=g. \tag{1.24}$$

Since

$$\Delta h-\frac{\partial h}{\partial x_1}=\nabla g-e_1 g\in C_0^\infty(\mathbb{R}^n),$$

w can be determined by means of (1.5), (1.7) with $f\equiv F$. As a consequence, w will obey estimates (1.15), (1.17) and (1.19) (for n=2: (1.16), (1.18) and (1.20)) with f replaced by F, that is, for all $\ell\ge 0$ we have, if $1<q<\infty$:

$$\left|\frac{\partial w}{\partial x_1}\right|_{\ell,q}+|w|_{\ell+2,q}+|p|_{\ell+1,q}\le c(|f|_{\ell,q}+|g|_{\ell+1,q}+|g|_{\ell,q}) \tag{1.25}$$

and, if $1<q<n+1$:

$$|w|_{\ell+1,s_1}+\left|\frac{\partial w}{\partial x_1}\right|_{\ell,q}+|w|_{\ell+2,q}+|p|_{\ell+1,q}\le c(|f|_{\ell,q}+|g|_{\ell+1,q}+|g|_{\ell,q}) \tag{1.26}$$

while, if $1<q<(n+1)/2$:

$$|w|_{\ell,s_2}+|w|_{\ell+1,s_1}+\left|\frac{\partial w}{\partial x_1}\right|_{\ell,q}+|w|_{\ell+2,q}+|p|_{\ell+1,q}\le c(|f|_{\ell,q}+|g|_{\ell+1,q}+|g|_{\ell,q}) \tag{1.27}$$

with s_1 and s_2 defined in (1.19). In the case n=2, from (1.16), (1.18) and

(1.20) we obtain, in addition, if $1<q<\infty$:

$$|w_2|_{\ell+1,q}+\left|\frac{\partial w_1}{\partial x_1}\right|_{\ell,q}+|w|_{\ell+2,q}+|p|_{\ell+1,q}\le c(|f|_{\ell,q}+|g|_{\ell+1,q}+|g|_{\ell,q}), \tag{1.28}$$

if $1<q<2$:

$$|w_2|_{\ell,\frac{2q}{2-q}}+|w_2|_{\ell+1,q}+|w|_{\ell+1,\frac{3q}{3-q}}+\left|\frac{\partial w_1}{\partial x_1}\right|_{\ell,q}$$
$$+|w|_{\ell+2,q}+|p|_{\ell+1,q}\le c(|f|_{\ell,q}+|g|_{\ell+1,q}+|g|_{\ell,q}), \tag{1.29}$$

while, if $1<q<3/2$:

$$|w_2|_{\ell,\frac{2q}{2-q}}+|w_2|_{\ell+1,q}+|w|_{\ell,\frac{3q}{3-2q}}+|w|_{\ell+1,\frac{3q}{3-q}}$$
$$+\left|\frac{\partial w_1}{\partial x_1}\right|_{\ell,q}+|w|_{\ell+2,q}+|p|_{\ell+1,q}\le c(|f|_{\ell,q}+|g|_{\ell+1,q}+|g|_{\ell,q}). \tag{1.30}$$

We have now to estimate the field **h**. To this end, we notice that the system (1.24) is the scalar version of the Oseen system and that for it we can directly show existence and corresponding L^q estimates by the same techniques used before. In particular, we prove for all $\ell\ge 0$, if $1<q<\infty$:

$$\left|\frac{\partial\Psi}{\partial x_1}\right|_{\ell,q}+|\Psi|_{\ell+2,q}\le c|g|_{\ell,q} \tag{1.31}$$

and, if $1<q<n+1$:

$$|\Psi|_{\ell+1,s_1}+\left|\frac{\partial\Psi}{\partial x_1}\right|_{\ell,q}+|\Psi|_{\ell+2,q}\le c|g|_{\ell,q} \tag{1.32}$$

while, if $1<q<(n+1)/2$:

$$|\Psi|_{\ell,s_2}+|\Psi|_{\ell+1,s_1}+\left|\frac{\partial\Psi}{\partial x_1}\right|_{\ell,q}+|\Psi|_{\ell+2,q}\le c|g|_{\ell,q}\,. \tag{1.33}$$

Consequently, for all $\ell\ge 0$, we derive the following bounds for **h**, if $1<q<\infty$:

$$|h|_{\ell+2,q}\le c_1(|\Psi|_{\ell+3,q}+|\Psi|_{\ell+2,q})\le c(|g|_{\ell+1,q}+|g|_{\ell,q}) \tag{1.34}$$

$$\left|\frac{\partial h}{\partial x_1}\right|_{\ell,q}\le c_1\left(\left|\frac{\partial\Psi}{\partial x_1}\right|_{\ell+1,q}+|\Psi|_{\ell,q}\right)\le c(|g|_{\ell+1,q}+|g|_{\ell,q}) \tag{1.35}$$

$$|h_2|_{\ell+1,q}\le c_1|\Psi|_{\ell+2,q}\le c|g|_{\ell,q}\,; \tag{1.36}$$

if $1<q<n+1$:

$$|h|_{\ell+1,s_1}\le c_1(|\Psi|_{\ell+2,q}+|\Psi|_{\ell,q})\le c(|g|_{\ell+1,q}+|g|_{\ell,q}) \tag{1.37}$$

while, if $1<q<(n+1)/2$:

$$|h|_{\ell,s_2} \le c_1(|\Psi|_{\ell+1,s_2} + |\Psi|_{\ell,s_2}) \le c(|g|_{\ell+1,q} + |g|_{\ell,q}). \tag{1.38}$$

Furthermore, if n=2, from inequality (II.5.9) it follows for $q \in (1,2)$

$$|\Psi|_{\ell+1,\frac{2q}{2-q}} \le c_1 |\Psi|_{\ell+2,q},$$

so that (1.23) and (1.31) give for n=2 and $1<q<2$

$$|h_2|_{\ell+1,\frac{2q}{2-q}} \le c|g|_{\ell,q}. \tag{1.39}$$

Recalling that **v**=**w**+**h**, we may collect (1.25)-(1.27) and (1.34)-(1.39) to conclude the validity of the following inequalities for all $\ell \ge 0$, if $1<q<\infty$:

$$\left|\frac{\partial \mathbf{v}}{\partial x_1}\right|_{\ell,q} + |\mathbf{v}|_{\ell+2,q} + |p|_{\ell+1,q} \le c(|\mathbf{f}|_{\ell,q} + |g|_{\ell+1,q} + |g|_{\ell,q}) \tag{1.40}$$

and, if $1<q<n+1$

$$|\mathbf{v}|_{\ell+1,s_1} + \left|\frac{\partial \mathbf{v}}{\partial x_1}\right|_{\ell,q} + |\mathbf{v}|_{\ell+2,q} + |p|_{\ell+1,q} \le c(|\mathbf{f}|_{\ell,q} + |g|_{\ell+1,q} + |g|_{\ell,q}) \tag{1.41}$$

while, if $1<q<(n+1)/2$

$$\begin{aligned}|\mathbf{v}|_{\ell,s_2} + |\mathbf{v}|_{\ell+1,s_1} + \left|\frac{\partial \mathbf{v}}{\partial x_1}\right|_{\ell,q} + |\mathbf{v}|_{\ell+2,q} + |p|_{\ell+1,q}\\ \le c(|\mathbf{f}|_{\ell,q} + |g|_{\ell+1,q} + |g|_{\ell,q})\end{aligned} \tag{1.42}$$

with s_1 and s_2 given in (1.19). In addition to the above inequalities, in the case n=2, from (1.28)-(1.30) and (1.34)-(1.39) we have, if $1<q<\infty$:

$$|v_2|_{\ell+1,q} + \left|\frac{\partial v_1}{\partial x_1}\right|_{\ell,q} + |\mathbf{v}|_{\ell+2,q} + |p|_{\ell+1,q} \le c(|\mathbf{f}|_{\ell,q} + |g|_{\ell+1,q} + |g|_{\ell,q}), \tag{1.43}$$

if $1<q<2$:

$$\begin{aligned}|v_2|_{\ell,\frac{2q}{2-q}} + |v_2|_{\ell+1,q} + |\mathbf{v}|_{\ell+1,\frac{3q}{3-q}} + \left|\frac{\partial v_1}{\partial x_1}\right|_{\ell,q}\\ + |\mathbf{v}|_{\ell+2,q} + |p|_{\ell+1,q} \le c(|\mathbf{f}|_{\ell,q} + |g|_{\ell+1,q} + |g|_{\ell,q}),\end{aligned} \tag{1.44}$$

while, if $1<q<3/2$:

$$\begin{aligned}|v_2|_{\ell,\frac{2q}{2-q}} + |v_2|_{\ell+1,q} + |\mathbf{v}|_{\ell,\frac{3q}{3-2q}} + |\mathbf{v}|_{\ell+1,\frac{3q}{3-q}}\\ + \left|\frac{\partial v_1}{\partial x_1}\right|_{\ell,q} + |\mathbf{v}|_{\ell+2,q} + |p|_{\ell+1,q} \le c(|\mathbf{f}|_{\ell,q} + |g|_{\ell+1,q} + |g|_{\ell,q}).\end{aligned} \tag{1.45}$$

Assume now **f** and g merely belonging to $W^{m,q}(\mathbb{R}^n)$ and $W^{m+1,q}(\mathbb{R}^n)$, respectively. We may use a standard density argument to establish, with the help of (1.40)–(1.45), existence of solutions to (1.2) and related estimates under the above-stated larger assumptions on **f** and g. Corresponding results for system (1.1) can be then obtained via the transformation (1.3). Finally, uniqueness of such solutions can be obtained from the results of Chang & Finn (1961). We have thus proved

Theorem 1.1. *Given* $f\in W^{m,q}(\mathbb{R}^n)$, $g\in W^{m+1,q}(\mathbb{R}^n)$, $m\geq 0$, $1<q<\infty$, *there exists a pair of functions* v, p *with* $v\in W^{m+2,q}(B_R)$, $p\in W^{m+1,q}(B_R)$, *for any* $R>0$, *and satisfying* a.e. *the non-homogeneous Oseen system* (1.1). *Moreover, for all integers* $\ell\in[0,m]$, *the quantities*

$$\left|\frac{\partial v}{\partial x_1}\right|_{\ell,q}, |v|_{\ell+2,q}, |p|_{\ell+1,q}$$

are finite and satisfy the estimate

$$\mathcal{R}e\left|\frac{\partial v}{\partial x_1}\right|_{\ell,q}+|v|_{\ell+2,q}+|p|_{\ell+1,q}\leq c(|f|_{\ell,q}+|g|_{\ell+1,q}+\mathcal{R}e|g|_{\ell,q}). \tag{1.46}$$

If $n=2$, *also* $|v_2|_{\ell+1,q}$ *is finite and it holds*

$$\mathcal{R}e|v_2|_{\ell+1,q}+\mathcal{R}e\left|\frac{\partial v_1}{\partial x_1}\right|_{\ell,q}+|v|_{\ell+2,q}+|p|_{\ell+1,q}\leq c(|f|_{\ell,q}+|g|_{\ell+1,q}+\mathcal{R}e|g|_{\ell,q}).$$

If $1<q<n+1$ $|v|_{\ell+1,s_1}$ *is finite,* $s_1=(n+1)q/(n+1-q)$, *and it is*

$$(\mathcal{R}e)^{1/(n+1)}|v|_{\ell+1,s_1}+\mathcal{R}e\left|\frac{\partial v}{\partial x}\right|_{\ell,q}+|v|_{\ell+2,q}+|p|_{\ell+1,q}$$
$$\leq c(|f|_{\ell,q}+|g|_{\ell+1,q}+\mathcal{R}e|g|_{\ell,q}).$$

If $n=2$ *and* $1<q<2$ *also* $|v_2|_{\ell,\frac{2q}{2-q}}$ *is finite and we have*

$$\mathcal{R}e|v_2|_{\ell,\frac{2q}{2-q}}+\mathcal{R}e|v_2|_{\ell+1,q}+(\mathcal{R}e)^{1/3}|v|_{\ell+1,\frac{3q}{3-q}}+\mathcal{R}e\left|\frac{\partial v_1}{\partial x_1}\right|_{\ell,q}$$
$$+|v|_{\ell+2,q}+|p|_{\ell+1,q}\leq c(|f|_{\ell,q}+|g|_{\ell+1,q}+\mathcal{R}e|g|_{\ell,q}). \tag{1.47}$$

Furthermore, if $1<q<(n+1)/2$, $|v|_{\ell,s_2}$ *is finite,* $s_2=(n+1)q/(n+1-2q)$, *and*

$$(\mathcal{R}e)^{2/(n+1)}|v|_{\ell,s_2}+(\mathcal{R}e)^{1/(n+1)}|v|_{\ell+1,s_1}+\mathcal{R}e\left|\frac{\partial v}{\partial x_1}\right|_{\ell,q}$$
$$+|v|_{\ell+2,q}+|p|_{\ell+1,q}\leq c(|f|_{\ell,q}+|g|_{\ell+1,q}+\mathcal{R}e|g|_{\ell,q}), \tag{1.48}$$

so that, in particular, if $n=2$, *from* (1.47) *and* (1.48) *it follows*

$$\mathcal{R}e|v_2|_{\ell,\frac{2q}{2-q}} + \mathcal{R}e|v_2|_{\ell+1,q} + (\mathcal{R}e)^{2/3}|v|_{\ell,\frac{3q}{3-2q}}$$

$$+(\mathcal{R}e)^{1/3}|v|_{\ell+1,\frac{3q}{3-q}} + \mathcal{R}e\left|\frac{\partial v_1}{\partial x_1}\right|_{\ell,q} + |v|_{\ell+2,q} + |p|_{\ell+1,q}$$

$$\le c(|f|_{\ell,q} + |g|_{\ell+1,q} + \mathcal{R}e|g|_{\ell,q}).$$

Here the constants c_i, $i=1,2,3$, *depend only on* n, ℓ *and* q. *Finally, if* w,τ *is another solution to* (1.2) *corresponding to the same data* f,g *with* $\left|\frac{\partial w}{\partial x_1}\right|_{\ell,q}$ *and* $|w|_{\ell+2,q}$ *finite, for some* $\ell\in[0,m]$, *then*

$$\left|\frac{\partial}{\partial x_1}(w-v)\right|_{\ell,q} \equiv |w-v|_{\ell+2,q} \equiv |\tau-p|_{\ell+1,q} \equiv 0.$$

The last part of this section is devoted to show existence and uniqueness of a *q-weak solution to* (1.2). By this we mean a function $v\in D_0^{1,q}(\mathbb{R}^n)$ satisfying the relations

$$(\nabla v,\nabla\psi) - (v,\frac{\partial\psi}{\partial x_1}) - (p,\nabla\cdot\psi) = -[f,\psi]\,, \qquad (v,\nabla\varphi) = -(g,\varphi) \tag{1.49}$$

for some $p\in L^q_{loc}(\mathbb{R}^n)$ and all $\psi,\varphi\in C_0^\infty(\mathbb{R}^n)$. As before, we begin to take $g=0$ and $f\in C_0^\infty(\mathbb{R}^n)$. Consequently, in such a case, (1.5), (1.7) is a C^∞ solution to (1.2). We now observe that f can be written in a divergence form, that is, $f_j(x) = D_\ell g_{\ell j}(x)$ where $g=g(x)$ is a second-order tensor field satisfying

$$|f|_{-1,q} \le \|g\|_q \le c_1|f|_{-1,q}\,, \text{ for all } q\in(1,\infty). \tag{1.50}$$

Actually, we may choose

$$g_{\ell j} = D_\ell(\mathcal{E}*f_j)\,,$$

where $\mathcal{E}$ is the Laplace fundamental solution in $\mathbb{R}^n$ and $*$ denotes convolution operation, and then use Calderón–Zygmund theorem. Since

$$\widehat{D_\ell g_{\ell j}} = i\sum_{\ell=1}^n \xi_\ell\,\hat{g}_{\ell j}\,,$$

from (1.5)–(1.7) we obtain

$$\begin{aligned} v(x) &= \frac{i}{(2\pi)^{n/2}}\sum_{\ell=1}^n\int_{\mathbb{R}^n} e^{ix\cdot\xi}\xi_\ell\phi_{mk}(\xi)\hat{g}_{\ell j}(\xi)d\xi \\ p(x) &= \frac{1}{(2\pi)^{n/2}}\sum_{\ell=1}^n\int_{\mathbb{R}^n} e^{ix\cdot\xi}\xi_\ell\xi_k\xi^{-2}\hat{g}_{\ell j}(\xi)d\xi\,, \end{aligned} \tag{1.51}$$

where ϕ_{mk} is defined in (1.8). With the aid of Lemmas 1.1, 1.2 (d), from

(1.51) it follows

$$|v|_{1,q}+\|p\|_q \le c_2\|g\|_q \;,\; 1<q<\infty,$$

and (1.50) implies

$$|v|_{1,q}+\|p\|_q \le c_3|f|_{-1,q} \;,\; 1<q<\infty. \tag{1.52}$$

In the case n=2, we can use Lemma 1.2 (f) to obtain an additional estimate for the component v_2 of the velocity field, namely,

$$\|v_2\|_q \le c_4\|g\|_q \;,\; 1<q<\infty.$$

Therefore, in such a case, this last relation along with (1.50) and (1.52) furnishes

$$\|v_2\|_{1,q}+|v_1|_{1,q}+\|p\|_q \le c_5|f|_{-1,q} \;,\; 1<q<\infty. \tag{1.53}$$

If we restrict the range of values of q, we may obtain another useful estimate. Specifically, from $(1.51)_1$, Lemmas 1.1 and 1.2 (b) it follows

$$\|v\|_{s_1} \le c_6\|g\|_q \;,\quad s_1=\frac{(n+1)q}{n+1-q} \;,\; 1<q<n+1,$$

so that (1.50) and (1.52) imply

$$\|v\|_{s_1}+|v|_{1,q}+\|p\|_q \le c_7|f|_{-1,q} \;,\; s_1=\frac{(n+1)q}{n+1-q} \;,\; 1<q<n+1. \tag{1.54}$$

Moreover, if n=2, inequalities (1.53) and (1.54) deliver

$$\|v_2\|_{1,q}+\|v\|_{\frac{3q}{3-q}}+|v_1|_{1,q}+\|p\|_q \le c_8|f|_{-1,q} \;,\; 1<q<3. \tag{1.55}$$

Starting from (1.52), (1.53), using a standard density argument we can prove that for any $f\in D_0^{-1,q}(\mathbb{R}^n)$ there exists a pair v,p with $v\in \mathcal{D}_0^{1,q}(\mathbb{R}^n)$, $p\in L^q(\mathbb{R}^n)$ satisfying equations (1.49) with $g\equiv 0$. Moreover, v,p verify (1.52), (1.53) and, if $1<q<n+1$, (1.54), (1.55).

It is easy to extend this result to the case $g\not\equiv 0$. Precisely, assume $g\in L^q(\mathbb{R}^n)\cap D_0^{-1,q}(\mathbb{R}^n)$ and let Ψ be a weak solution to (1.24), namely, Ψ solves

$$(\nabla\Psi,\nabla\varphi) - (\Psi,\frac{\partial\varphi}{\partial x_1}) = -[g,\varphi] \tag{1.56}$$

for all $\varphi\in C_0^\infty(\mathbb{R}^n)$. Employing just the same technique used for Theorem 1.1 and for problem (1.49) with $g\equiv 0$, we can show with no difficulty the existence of a solution $\Psi\in D^{2,q}(\mathbb{R}^n)\cap D_0^{1,q}(\mathbb{R}^n)$ to (1.56) such that

$$|\Psi|_{2,q}\le c\|g\|_q \;,\quad |\Psi|_{1,q}\le c|g|_{-1,q} \;, \tag{1.57}$$

and, if $1<q<n+1$,

$$\|\Psi\|_{s_1}+|\Psi|_{1,q}\le c|g|_{-1,q} \;,\; s_1=\frac{(n+1)q}{n+1-q} \;. \tag{1.58}$$

We then look for a q-weak solution to (1.2) of the form v=w+h, where

$w \in \mathcal{D}_0^{1,q}(\mathbb{R}^n)$ solves (1.56) with $f - \Delta h + \frac{\partial h}{\partial x_1}$ in place of f and $g=0$, while h is defined by (1.23) with $\Psi \in D_0^{2,q}(\mathbb{R}^n) \cap D_0^{1,q}(\mathbb{R}^n)$ solving (1.56)-(1.57). It holds, for $1<q<\infty$,

$$\|h_2\|_q \leq c_1 |g|_{-1,q} \quad , |h|_{1,q} \leq c_1 (\|g\|_q + |g|_{-1,q}) \tag{1.59}$$

and, if $1<q<n+1$,

$$\|h\|_{s_1} \leq c_2 (\|g\|_q + |g|_{-1,q}). \tag{1.60}$$

Now, by (1.52), (1.53), w, p satisfy the inequalities

$$\begin{aligned} |w|_{1,q} + \|p\|_q &\leq c \left| f - \Delta h + \frac{\partial h}{\partial x_1} \right|_{-1,q} \\ \|w_2\|_{1,q} + |w_1|_{1,q} + \|p\|_q &\leq c \left| f - \Delta h + \frac{\partial h}{\partial x_1} \right|_{-1,q} \quad \text{(if } n=2) \end{aligned} \tag{1.61}$$

and, if $1<q<n+1$

$$\begin{aligned} \|w\|_{s_1} + |w|_{1,q} + \|p\|_q &\leq c \left| f - \Delta h + \frac{\partial h}{\partial x_1} \right|_{-1,q} , \quad s_1 = \frac{(n+1)q}{n+1-q} \\ \|w_2\|_{1,q} + \|w\|_{\frac{3q}{3-q}} + |w_1|_{1,q} + \|p\|_q &\leq c |f|_{-1,q} , \quad \text{(if } n=2). \end{aligned} \tag{1.62}$$

In view of (1.23), (1.59)-(1.62) and of the denseness of $C_0^\infty(\Omega)$ in $D_0^{-1,q}(\Omega)$, *cf.* Galdi & Simader (1990), and using the replacements (1.3) we may conclude that, for given f and g with the regularity properties stated previously, there exists a q-weak solution to (1.1) satisfying together with the corresponding pressure p the following estimates: if $1<q<\infty$

$$\begin{aligned} |v|_{1,q} + \|p\|_q &\leq c(|f|_{-1,q} + \mathcal{R}e|g|_{-1,q} + \|g\|_q) \; . \\ \mathcal{R}e\|v_2\|_q + |v|_{1,q} + \|p\|_q &\leq c(|f|_{-1,q} + \mathcal{R}e|g|_{-1,q} + \|g\|_q) \quad \text{(if } n=2) \end{aligned} \tag{1.63}$$

and if $1<q<n+1$

$$\begin{aligned} &(\mathcal{R}e)^{1/(n+1)} \|v\|_{s_1} + |v|_{1,q} + \|p\|_q \\ &\quad \leq c(|f|_{-1,q} + \mathcal{R}e|g|_{-1,q} + \|g\|_q) \; , \quad s_1 = \frac{(n+1)q}{n+1-q} \\ &\mathcal{R}e\|v_2\|_q + (\mathcal{R}e)^{1/3}\|v\|_{\frac{3q}{3-q}} + |v|_{1,q} + \|p\|_q \\ &\quad \leq c(|f|_{-1,q} + \mathcal{R}e|g|_{-1,q} + \|g\|_q) \quad \text{(if } n=2) \end{aligned} \tag{1.64}$$

with $c=c(n,q)$. Uniqueness of solutions just obtained follows from the work of Chang & Finn (1961).

The above results are summarized in the following

Theorem 1.2. *Given* $f\in D_0^{-1,q}(\mathbb{R}^n)$, $g\in L^q(\mathbb{R}^n)\cap D_0^{-1,q}(\mathbb{R}^n)$, $1<q<\infty$, *there exist* $v\in D_0^{1,q}(\mathbb{R}^n)$, $p\in L^q(\mathbb{R}^n)$ *satisfying* (1.49) *for all* $\psi\in C_0^\infty(\mathbb{R}^n)$. *Moreover,* v, p *verify* (1.63) *and, if* $1<q<n+1$, *then*

$$v\in L^{s_1}(\mathbb{R}^n),\ s_1=\frac{(n+1)q}{n+1-q},$$

with

$$v_2\in L^q(\mathbb{R}^2)\ ,\ \textit{if}\ n=2,$$

and estimate (1.64) *holds. Finally, if* $w\in D_0^{1,q}(\mathbb{R}^n)$, $\tau\in L^q_{loc}(\mathbb{R}^n)$ *is another solution to* (1.49) *corresponding to the same data* f *and* g, *then* $w\equiv v$, $\tau\equiv p+\text{const}$.

2. Existence, uniqueness and L^q estimates in exterior domains.

The aim of the present section is to investigate to what extent the theorems proved in Section 1 in the whole space can be extended to the more general situation when the region of flow is an *exterior domain*. Let us begin to consider the Oseen problem (0.1), (0.2) in an exterior domain Ω of class C^{m+2}, $m\geq 0$, with data $f\in C_0^\infty(\bar{\Omega})$, $v_*\in W^{m+2-1/q,q}(\partial\Omega)$, $1<q<\infty$, and $v_\infty=0$. By the results of Chang & Finn (1961), Finn (1965) and Finn & Smith (1967) we then construct a solution v,p such that

$$v\in W^{m+2,q}_{loc}(\bar{\Omega})\cap C^\infty(\Omega),\ p\in W^{m+1,q}_{loc}(\bar{\Omega})\cap C^\infty(\Omega)$$

and which at large distances has the asymptotic structure of the fundamental Oseen solution. Denote next by ψ a "cut-off" function which equals one in $\Omega^{R/2}$ and zero in Ω_ρ, where $R/2>\rho>\delta(\Omega^c)$. Putting $u=\psi v$, $\pi=\psi p$, from (0.3) it follows that u,π satisfies the following Oseen problem in $\mathbb{R}^n$

$$\begin{aligned}\Delta u - \mathcal{R}e\frac{\partial u}{\partial x_1} - \nabla\pi &= F\\ \nabla\cdot u &= g\end{aligned}\tag{2.1}$$

where

$$F=\psi f+\mathcal{R}e\frac{\partial\psi}{\partial x_1}v+(2\nabla\psi\cdot\nabla v+\Delta\psi v-p\nabla\psi)\ ,g=\nabla\psi\cdot v\ .\tag{2.2}$$

Employing Theorem 1.1 we deduce the existence of a solution w,τ to (2.1), (2.2) satisfying, in particular, the following properties

$$w\in\bigcap_{\ell=0}^{m} D^{\ell+2,q}(\mathbb{R}^n),\quad \tau\in\bigcap_{\ell=0}^{m} D^{\ell+1,q}(\mathbb{R}^n),\ 1<q<\infty,$$

$$w\in\bigcap_{\ell=0}^{m} D^{\ell+1,s_1}(\mathbb{R}^n),\quad s_1=\frac{(n+1)q}{n+1-q},\ 1<q<n+1, \tag{2.3}$$

$$w\in\bigcap_{\ell=0}^{m} D^{\ell,s_2}(\mathbb{R}^n),\quad s_2=\frac{(n+1)q}{n+1-2q},\ 1<q<(n+1)/2.$$

By the uniqueness results of Chang & Finn (1961) we deduce $u\equiv w$, $\pi\equiv p$+const. Recalling (2.2) and that $v=u$, $p=\pi$ in $\Omega^{R/2}$, from Theorem 1.1 it follows that for all $\ell\in[0,m]$ and all $q\in(1,(n+1)/2)$ v,p obey the inequality

$$\begin{aligned}(\mathcal{R}e)^{2/(n+1)}|v|_{\ell,s_2,\Omega^{R/2}}&+(\mathcal{R}e)^{1/(n+1)}|v|_{\ell+1,s_1,\Omega^{R/2}}\\ +\mathcal{R}e\left|\frac{\partial v}{\partial x_1}\right|_{\ell,q,\Omega^{R/2}}&+|v|_{\ell+2,q,\Omega^{R/2}}+|p|_{\ell+1,q,\Omega^{R/2}}\\ &\le c(|f|_{\ell,q}+(1+\mathcal{R}e)|v|_{\ell+1,q,\Omega_R}+|p|_{\ell,q,\Omega_R})\end{aligned} \tag{2.4}$$

where $s_1=\dfrac{(n+1)q}{n+1-q}$, $s_2=\dfrac{(n+1)q}{n+1-2q}$ and, for $n=2$,

$$\begin{aligned}\mathcal{R}e|v_2|_{\ell,\frac{2q}{2-q}}&+\mathcal{R}e|v_2|_{\ell+1,q}+(\mathcal{R}e)^{2/3}|v|_{\ell,\frac{3q}{3-2q}}\\ +(\mathcal{R}e)^{1/3}|v|_{\ell+1,\frac{3q}{3-q}}&+\mathcal{R}e\left|\frac{\partial v_1}{\partial x_1}\right|_{\ell,q}+|v|_{\ell+2,q}+|p|_{\ell+1,q}\\ &\le c(|f|_{\ell,q}+(1+\mathcal{R}e)|v|_{\ell+1,q,\Omega_R}+|p|_{\ell,q,\Omega_R})\end{aligned} \tag{2.4)'$$

From Cattabriga (1961) we have

$$\begin{aligned}\|v\|_{m+2,q,\Omega_R}&+\|p\|_{m+1,q,\Omega_R}\\ \le c_2\Big\{\|f\|_{m,q,\Omega_R}&+\|v_*\|_{m+2-1/q,q(\partial\Omega)}+\mathcal{R}e\left\|\frac{\partial v}{\partial x_1}\right\|_{m,q,\Omega^R}\\ \|v\|_{m+2-1/q,q(\partial B_R)}&+\|v\|_{q,\Omega_R}+\|p\|_{q,\Omega_R}\Big\}\end{aligned} \tag{2.5}$$

where, as usual, the origin of coordinates has been taken in the interior of Ω^c. By the trace theorem, *cf.* Nečas (1967), we have

$$\|v\|_{m+2-1/q,q(\partial B_R)}\le c_3\{|v|_{m+2,q,\Omega^{R/2}}+\|v\|_{m+1,q,\Omega_R}\}. \tag{2.6}$$

Furthermore, by the Sobolev embedding theorem, it is

$$\begin{aligned}\|v\|_{m,s_2,\Omega_{R/2}}+\sum_{\ell=0}^{m}|v|_{\ell+1,s_1,\Omega_{R/2}}&\le c_4\|v\|_{m+2,q,\Omega_R}\\ \|v\|_{m,\frac{2q}{2-q},\Omega_{R/2}}\le c_4\|v\|_{m+1,q,\Omega_R}&,\ \text{if } n=2\end{aligned} \tag{2.7}$$

and so, collecting (2.4), (2.4)′–(2.7), we derive, in particular, for some

$c=c(n,q,\Omega,m)$ and all $q\in(1,(n+1)/2)$

$$\begin{aligned} &a_1\|v\|_{m,s_2,\Omega}+\mathcal{R}e\left\|\frac{\partial v}{\partial x_1}\right\|_{m,q,\Omega} \\ &+\sum_{\ell=0}^{m}\left\{a_2\,|v|_{\ell+1,s_1,\Omega}+|v|_{\ell+2,q,\Omega}+|p|_{\ell+1,q,\Omega}\right\} \qquad (2.8)\\ &\le c_5(\|f\|_{m,q,\Omega}+\|v_*\|_{m+2-1/q,q(\partial\Omega)}+(1+\mathcal{R}e)\|v\|_{m+1,q,\Omega_R}+\|p\|_{m,q,\Omega_R}) \end{aligned}$$

and, if $n=2$,

$$\begin{aligned} &\mathcal{R}e\left(\|v_2\|_{m,\frac{2q}{2-q},\Omega}+\|\nabla v_2\|_{m+1,q,\Omega}\right)+a_1\|v\|_{m,\frac{3q}{3-2q},\Omega}+\mathcal{R}e\left\|\frac{\partial v}{\partial x_1}\right\|_{m,q,\Omega} \\ &+\sum_{\ell=0}^{m}\left\{a_2\,|v|_{\ell+1,\frac{3q}{3-q},\Omega}+|v|_{\ell+2,q,\Omega}+|p|_{\ell+1,q,\Omega}\right\} \qquad (2.8)'\\ &\le c_5(\|f\|_{m,q,\Omega}+\|v_*\|_{m+2-1/q,q(\partial\Omega)}+(1+\mathcal{R}e)\|v\|_{m+1,q,\Omega_R}+\|p\|_{m,q,\Omega_R}) \end{aligned}$$

where

$$a_1=\min\{1,\mathcal{R}e^{2/(n+1)}\},\quad a_2=\min\{1,\mathcal{R}e^{1/(n+1)}\}$$
$$s_1=\frac{(n+1)q}{n+1-q},\quad s_2=\frac{(n+1)q}{n+1-2q} \qquad (2.9)$$

By a repeated use of Ehrling's inequality, for all $\varepsilon>0$ it follows

$$\|p\|_{m,q,\Omega_R}\le\varepsilon\|p\|_{m+1,q,\Omega_R}+c_6\|p\|_{q,\Omega_R} \qquad (2.10)$$

with $c_6=c_6(\varepsilon,n,m,q,\Omega_R)$. In addition, possibly modifying p by a suitable constant (what causes no loss of generality), from Cattabriga (1961) and Gagliardo (1957) we derive

$$\|p\|_{q,\Omega_R}\le c_7[(1+\mathcal{R}e)\|v\|_{1,q,\Omega_R}+\mathcal{R}e\|f\|_{q,\Omega}]. \qquad (2.11)$$

Inequalities (2.8), (2.8)', (2.10) and (2.11) then yield

$$\begin{aligned} &a_1\|v\|_{m,s_2,\Omega}+\mathcal{R}e\left\|\frac{\partial v}{\partial x_1}\right\|_{m,q,\Omega}+\sum_{\ell=0}^{m}\left\{a_2|v|_{\ell+1,s_1,\Omega}+|v|_{\ell+2,q,\Omega}+|p|_{\ell+1,q,\Omega}\right\} \qquad (2.12)\\ &\le c_8(\|f\|_{m,q,\Omega}+\|v_*\|_{m+2-1/q,q(\partial\Omega)}+(1+\mathcal{R}e)\|v\|_{m+1,q,\Omega_R}). \end{aligned}$$

and, if $n=2$,

$$\begin{aligned} &\mathcal{R}e\left(\|v_2\|_{m,\frac{2q}{2-q}}+\|\nabla v_2\|_{m+1,q}\right)+a_1\|v\|_{m,\frac{3q}{3-2q}}+\mathcal{R}e\left\|\frac{\partial v}{\partial x_1}\right\|_{m,q} \\ &+\sum_{\ell=0}^{m}\left\{a_2\,|v|_{\ell+1,\frac{3q}{3-q}}+|v|_{\ell+2,q}+|p|_{\ell+1,q}\right\} \qquad (2.12)'\\ &\le c_5(\|f\|_{m,q}+\|v_*\|_{m+2-1/q,q(\partial\Omega)}+(1+\mathcal{R}e)\|v\|_{m+1,q,\Omega_R}) \end{aligned}$$

We now look for an inequality of the type

$$\|v\|_{m+1,q,\Omega} \le c_9(\|f\|_{m,q,\Omega} + \|v_*\|_{m+2-1/q,q(\partial\Omega)}) \tag{2.13}$$

for a suitable constant independent of v, f and v_*. The proof of (2.13) can be obtained, as in Galdi & Simader (1990), by a contradiction argument based on compactness of the imbedding $W^{m+2,q}(\Omega)\to W^{m+1,q}_{loc}(\Omega)$ and on uniqueness of solutions to the Oseen problem belonging to $L^r(\Omega)$, for some r>1. Once (2.13) is proved, one can then show (we omit details, *cf.* Galdi (1992))

Theorem 2.1. *Let* Ω *be an exterior domain in* $\mathbb{R}^n$ *of class* C^{m+2}, $m\ge 0$. *Given* $f\in W^{m,q}(\Omega)$, $v_*\in W^{m+2-1/q,q}(\partial\Omega)$, $1<q<(n+1)/2$, $v_\infty\in\mathbb{R}^n$, *there exists one and only one corresponding solution* v,p *to the Oseen problem* (0.1), (0.2) *such that*

$$v-v_\infty\in W^{m,s_2}(\Omega)\cap\left\{\bigcap_{\ell=0}^{m}\left[D^{\ell+1,s_1}(\Omega)\cap D^{\ell+2,q}(\Omega)\right]\right\},\quad p\in\bigcap_{\ell=0}^{m}D^{\ell+1,q}(\Omega)$$

with $s_1=(n+1)q/(n+1-q)$, $s_2=(n+1)q/(n+1-2q)$. *If* n=2 *we also have*

$$v_2-v_{\infty 2}\in W^{m,2q/(2-q)}(\Omega)\cap\left(\bigcap_{\ell=0}^{m}D^{\ell+1,q}(\Omega)\right).$$

Moreover, v,p *verify*

$$a_1\|v-v_\infty\|_{m,s_2}+\mathcal{R}e\left\|\frac{\partial v}{\partial x_1}\right\|_{m,q}+\sum_{\ell=0}^{m}\left\{a_2\,|v|_{\ell+1,s_1}+|v|_{\ell+2,q}+|p|_{\ell+1,q}\right\}$$
$$\le c(\|f\|_{m,q}+\|v_*\|_{m+2-1/q,q(\partial\Omega)}+|v_\infty|)$$

and, if n=2,

$$\mathcal{R}e\left(\|v_2-v_{\infty 2}\|_{m,\frac{2q}{2-q}}+\|\nabla v_2\|_{m+1,q}\right)+a_1\|v-v_\infty\|_{m,\frac{3q}{3-2q}}+\mathcal{R}e\left\|\frac{\partial v}{\partial x_1}\right\|_{m,q}$$
$$+\sum_{\ell=0}^{m}\left\{a_2|v|_{\ell+1,\frac{3q}{3-q}}+|v|_{\ell+2,q}+|p|_{\ell+1,q}\right\}\le c(\|f\|_{m,q}+\|v_*\|_{m+2-1/q,q(\partial\Omega)}+|v_\infty|)$$

with a_1 *and* a_2 *given in* (2.9). c *depends on* m,q,n,Ω *and* $\mathcal{R}e$. *However, if* $q\in(1,n/2)$ *and* $\mathcal{R}e\in(0,B]$ *for some* B>0, c *depends solely on* m,q,n,Ω *and* B.

Our final objective is to extend Theorem 1.2 to the case of an exterior domain Ω of class C^2. We start with $f\in C_0^\infty(\bar\Omega)$, $v_*\in W^{1-1/q,q}(\partial\Omega)$ and $v_\infty=0$. As in Theorem 2.1, corresponding to these data there exists a solution v,p to (0.1), (0.2) with

$$v\in W^{1,q}_{loc}(\bar\Omega)\cap C^\infty(\Omega),\quad p\in L^q_{loc}(\bar\Omega)\cap C^\infty(\Omega)$$

satisfying the asymptotic behaviour of the type described in Theorem 2.2. Furthermore, $u\equiv\psi v$ and $\pi\equiv\psi p$ satisfy problem (2.1), (2.2) in $\mathbb{R}^n$, to which we apply the results stated in Theorem 1.2. We thus deduce the existence of a solution w,τ to (2.1), (2.2) enjoying the following properties

$$w\in D^{1,q}(\mathbb{R}^n)\cap L^{s_1}(\mathbb{R}^n),\quad \tau\in L^q(\mathbb{R}^n),\quad s_1=(n+1)q/(n+1-q),\quad 1<q<n+1$$
$$w_2\in L^q(\mathbb{R}^2),\ \text{if } n=2. \tag{2.14}$$

together with the estimates

$$(\mathcal{R}e)^{1/(n+1)}\|w\|_{s_1}+|w|_{1,q}+\|\tau\|_q\le c_1(|F|_{-1,q}+\mathcal{R}e|g|_{-1,q}+\|g\|_q) \tag{2.15}$$
$$\mathcal{R}e\|w_2\|_q+(\mathcal{R}e)^{1/3}\|w\|_{\frac{3q}{3-q}}+|w|_{1,q}+\|\tau\|_q\le c_1(|F|_{-1,q}+\mathcal{R}e|g|_{-1,q}+\|g\|_q)\ (\text{if } n=2)$$

As in Theorem 1.1, we prove $w\equiv u$, and (up to a constant) $\tau\equiv p$. Assuming $q>n/(n-1)$ and reasoning exactly as in Galdi & Simader (1990) we show

$$\mathcal{R}e|F|_{-1,q}+\mathcal{R}e|g|_{-1,q}+\|g\|_q\le c_2\{|f|_{-1,q}+(1+\mathcal{R}e)\|v\|_{q,\Omega_R}+\|p\|_{-1,q,\Omega_R}\} \tag{2.16}$$

and so, recalling that $u=v$, $\pi=p$ in $\Omega^{R/2}$, from (2.14)–(2.16) we deduce

$$(\mathcal{R}e)^{1/(n+1)}\|v\|_{s_1,\Omega^{R/2}}+|v|_{1,q,\Omega^{R/2}}+\|p\|_{q,\Omega^{R/2}}$$
$$\le c_3\{|f|_{-1,q}+(1+\mathcal{R}e)\|v\|_{q,\Omega_R}+\|p\|_{-1,q,\Omega_R}\}. \tag{2.17}$$

$$\mathcal{R}e\|v_2\|_{q,\Omega^{R/2}}+(\mathcal{R}e)^{1/3}\|v\|_{\frac{3q}{3-q},\Omega^{R/2}}\qquad (\text{if } n=2)$$
$$+|v|_{1,q,\Omega^{R/2}}+\|p\|_{q,\Omega^{R/2}}\le c_1(|F|_{-1,q}+\mathcal{R}e|g|_{-1,q}+\|g\|_q)$$

To obtain estimate "near" the boundary we apply inequality (5.10) of Galdi & Simader (1990), that is,

$$\|v\|_{1,q,\Omega_R}+\|p\|_{q,\Omega_R}\le c_4\Big(|f|_{-1,q}+\|v_*\|_{1-1/q,q(\partial\Omega)}+(1+\mathcal{R}e)\|v\|_{q,\Omega_R}$$
$$+\|p\|_{-1,q,\Omega_R}+\|v\|_{1-1/q,q(\partial B_R)}\Big) \tag{2.18}$$

where we used the obvious inequality $\|f\|_{-1,q,\Omega_R}\le|f|_{-1,q}$. Employing the trace theorem of Gagliardo (1957) at the boundary term on ∂B_R in (2.18) we deduce

$$\|v\|_{1,q,\Omega_R}+\|p\|_{q,\Omega_R}\le c_5\Big(|f|_{-1,q}+\|v_*\|_{1-1/q,q(\partial\Omega)}+(1+\mathcal{R}e)\|v\|_{q,\Omega_R}$$
$$+\|p\|_{-1,q,\Omega_R}+\|v\|_{1,1-1/q(\partial B_R)}\Big).$$

Taking into account that , by the Sobolev embedding theorem it is

$$\|v\|_{s_1,\Omega_{R/2}}\le c_6\|v\|_{1,q,\Omega_{R/2}}, \tag{2.19}$$

combining (2.17), (2.18) and (2.19) we conclude, for all $q\in(n/(n-1),n+1)$

$$a_2\|v\|_{s_1,\Omega}+|v|_{1,q,\Omega}+\|p\|_{q,\Omega}\le c_7\{|f|_{-1,q,\Omega}+\|v_*\|_{1-1/q,q(\partial\Omega)}$$
$$+(1+\mathcal{R}e)\|v\|_{q,\Omega_R}+\|p\|_{-1,q,\Omega_R}\}, \tag{2.20}$$

$$\mathcal{R}e\|v_2\|_{q,\Omega}+a_2\|v\|_{\frac{3q}{3-q},\Omega}+|v|_{1,q,\Omega}+\|p\|_{q,\Omega}\le c_7(|f|_{-1,q}+\|v_*\|_{1-1/q,q(\partial\Omega)}$$
$$+(1+\mathcal{R}e)\|v\|_{q,\Omega_R}+\|p\|_{-1,q,\Omega_R}\} \qquad \text{(if } n=2)$$

where a_2 is defined in (2.9). By a reasoning totally similar to that developed in the proof of Theorem 7.1 we can prove the existence of a constant c_8 independent of v,p,f and v_* such that

$$\|v\|_{q,\Omega_R}+\|p\|_{-1,q,\Omega_R}\le c_8\{|f|_{-1,q}+\|v_*\|_{1-1/q,q(\partial\Omega)}\}. \tag{2.21}$$

c_8 will depend, in general, also on $\mathcal{R}e$. However, if $q\in(n/(n-1),n)$, employing Sobolev inequality, *i.e.*,

$$\|v\|_{nq/(n-q)}\le c|v|_{1,q}$$

one shows, as in Theorem 7.1, that c_8 can be chosen independent of $\mathcal{R}e$ ranging in $(0,B]$, with an arbitrarily fixed B.

From what established, one can then show (we omit details, *cf.* Galdi (1992))

Theorem 2.2. *Let Ω be an exterior domain in $\mathbb{R}^n$ of class C^2. Given $f\in D_0^{-1,q}(\Omega)$, $v_*\in W^{1-1/q,q}(\partial\Omega)$, $n/(n-1)<q<n+1$, and $v_\infty\in\mathbb{R}^n$, there exists one and only one q-generalized solution v to (0.1), (0.2). Furthermore, it holds*

$$v-v_\infty\in L^{s_2}(\Omega),\quad s_2=(n+1)q/(n+1-q)$$
$$p\in L^q(\Omega)$$

with p pressure field associated to v by Lemma 1.1, and, if $n=2$, $v_2\in L^q(\Omega)$. Finally, v and p obey the following estimate

$$a_2\|v-v_\infty\|_{s_2}+|v|_{1,q}+|p|_q\le c\{|f|_{-1,q}+\|v_*\|_{1-1/q,q(\partial\Omega)}+|v_\infty|\}$$

and, if $n=2$,

$$\mathcal{R}e\|v_2\|_{q,\Omega}+a_2\|v\|_{\frac{3q}{3-q},\Omega}+|v|_{1,q,\Omega}+\|p\|_{q,\Omega}\le c\{|f|_{-1,q}+\|v_*\|_{1-1/q,q(\partial\Omega)}+|v_\infty|\}$$

where $c=c(n,q,\Omega,\mathcal{R}e)$. If $n>2$ and $q\in(n/(n-1),n)$, $\mathcal{R}e\in(0,B]$, with B arbitrarily fixed positive number, c depends solely on n,q,Ω and B.

Acknowledgment. This work was made under the auspices of GNFM of Italian C.N.R and with the financial support of M.P.I. 40% and 60% contracts.

Bibliography

Babenko, K.I., 1973, *On the Stationary Solutions of the Problem of Flow past a Body of a Viscous Incompressible Fluid,* Math. USSR Sbornik **20** (1), 1-24.

Borchers, W. & Miyakawa, T., 1990, *Algebraic* L^2 *Decay for the Navier-Stokes Flows in Exterior Domains,* Acta Math. **165**, 189-227.

Cattabriga, L., 1961, *Su un Problema al Contorno Relativo al Sistema di Equazioni di Stokes,* Rend. Sem. Mat. Padova **31**, 308-340.

Chang, I.D. & Finn, R., 1961, *On the Solutions of a Class of Equations Occurring in Continnum Mechanics with Application to the Stokes Paradox,* Arch. Ratl Mech. Anal. **7**, 388-401.

Farwig, R., 1991a, *A Variational Approach in Weighted Sobolev Spaces to the Operator* $-\Delta+\partial/\partial x_1$ *in Exterior Domains of* $\mathbb{R}^3$, Preprint, Universität Paderborn.

Farwig, R., 1991b, *The Stationary Exterior 3D-Problem of Oseen and Navier-Stokes Equations in Anisotropically Weighted Sobolev Spaces,* Preprint, Universität Paderborn.

Finn, R., 1965, *On the Exterior Stationary Problem for the Navier-Stokes Equations and Associated Perturbation Problems,* Arch. Ratl Mech. Anal **19**, 363-406

Finn, R. & Smith, D.R., 1967, *On the Linearized Hydrodynamical Equations in Two Dimensions,* Arch. Ratl Mech. Anal. **25**, 1-25.

Gagliardo, E., 1957, *Caratterizzazioni delle Tracce sulla Frontiera Relative ad Alcune Classi di Funzioni in* n *Variabili,* Rend. Sem. Mat. Padova **27**, 284-305.

Galdi, G.P., 1991, *The* L^q *Approach to the Two-Dimensional Stationary Navier-Stokes Problem in Exterior Domains with Nonzero Velocity at Infinity,* Preprint, University of Ferrara.

Galdi, G.P., 1992, *An Introduction to the Mathematical Theory of the Navier-Stokes Equations,* Springer Tracts in Natural Phylosophy, in press.

Galdi, G.P. & Simader, C.G., 1990, *Existence, Uniqueness and* L^q *Estimates for the Stokes Problem in Exterior Domain,* Arch. Ratl Mech. Anal. **112**, 291-318.

Galdi, G.P., Simader, C.G. & Sohr, H., 1992, *The Stokes Problem in Lipschitz Domains*, Preprint Università di Ferrara.

Kozono, H. & Sohr, H., 1991, *New a priori Estimates for the Stokes Equations in Exterior Domains*, Indiana Univ. Math. J. 40 (1), 1-27.

Lizorkin, P.I., 1963, *(L^p-L^q)-Multipliers of Fourier Integrals*, Dokl. Akad. Nauk SSSR 152 (4), 808-812 (*in Russian*).

Maremonti, P. & Solonnikov, V.A., 1985, *Estimates for the Laplace Operator in Exterior Domains*, Zap. Nauk. Sem. LOMI 146, 92-102 (*in Russian*).

Maremonti, P. & Solonnikov, V.A., 1986, *Su una Diseguaglianza per le Soluzioni del Problema di Stokes in Domini Esterni*, Preprint, Università di Napoli.

Maslennikova, V.N., 1990, *On the Stokes Problem in Exterior Domains*, Lecture delivered at the Department of Mathematics, Università di Ferrara, October 22.

Nečas, J., 1967, *Les Méthodes Directes en Théorie des Équations Elliptiques*, Masson et Cie.

Specovius-Neugebauer, M., 1986, *Exterior Stokes Problems and Decay at Infinity*, Math. Meth. in the Appl. Sci. 8, 351-367.

THE EXTERIOR PROBLEM FOR THE STATIONARY NAVIER-STOKES EQUATIONS : ON THE EXISTENCE AND REGULARITY

Rodolfo Salvi

Dipartimento di Matematica
Politecnico di Milano - Milano (ITALY)

0. Introduction

The present paper is connected with the boundary value problem posed by 3D-steady flow problem of a viscous incompressible fluid past one or several bodies. Let $\Omega \subset R^3$ be an unbounded domain with bounded Ω^c, the complementary set of Ω, and S be the boundary of Ω^c.
The flow velocity $u = (u_1, u_2, u_3)$ and the pressure p of the fluid which occupies the domain Ω, and subjected to the external force f, satisfy the Navier-Stokes system

$$\begin{aligned} - \Delta u + u\cdot\nabla u + \nabla p &= f \\ \nabla\cdot u &= 0 \end{aligned} \qquad 0.1$$

For simplicity, we assume the viscosity and the density of the fluid equal 1. We complete (0.1) with the following boundary conditions on S and at infinity

$$u_{|S} = \beta \quad \text{and} \quad u \to u_\infty \quad \text{as } |x| \to \infty \qquad 0.2$$

where β, and u_∞ (constant) are given vectors.
The boundary value problem (0.1), (0.2) is called the exterior Navier-Stokes boundary value problem or simply the exterior problem for Navier-Stokes equations.
After a suitable choice of the coordinate system, scale transformations, and linearizing (0.1) in u with respect to u_∞, we have that the function $v = u - u_\infty$ satisfies the Oseen's system

$$\begin{aligned} &\frac{\partial v}{\partial x_1} - \Delta v + \nabla p = f \\ &\nabla\cdot v = 0 \\ &v_{|S} = \beta \ ; \ v \to 0 \quad \text{as } |x| \to \infty \end{aligned} \qquad (0.3)$$

Many existence and regularity results about systems (0.1), (0.3) are well known (see [2], [3], [5], [6], [7], [8], [9], [10], [11], [13]).
These results show that a detailed analysis of Oseen's system is quite useful in the analysis of system (0.1). Furthermore Oseen's system is interesting on one's own mainly for the presence of the skew-symmetric differential operator $\partial/\partial x_1$.
Now the fundamental solution or the first fundamental solution of (0.3) was obtained by Oseen (see [13]). Observing that (0.3) is an elliptic system in the sense of Agmon - Douglis - Nirenberg (see [1]) one expects to obtain regularity results along the line of the theory developed in [1].
To apply this theory to Oseen's system, we need Poisson kernel or the second fundamental solution for (0.3) in an half-space.
This paper concerns mainly with the construction of the second fundamental solution of (0.3). This construction shall be performed along the line of the construction of the second fundamental solution of the linear nonstationary Navier-Stokes system (see [14]). Using the first and the second fundamental solutions of the Oseen's system, we study the differential properties of the solutions in the spaces W^1_p (Sobolev spaces) B^1_p (Besov spaces).

Section 1 is devoted to preliminaries. In section 2 we construct the first fundamental solution of (0.3) and we prove some estimates.
In section 3 we consider the problem (0.3) in an half-space and we construct the second fundamental solution, and we prove differential properties of the solutions of (0,3).

1. Preliminaries

Throughout this paper we shall deal with an exterior domain Ω in R^3 with bounded complementary Ω^c and S denotes the boundary of Ω^c . All functions in this paper are R-valued or R^3-valued. The letter c denotes a constant.
The points in R^3 are denoted by $x = (x_1, x_2, x_3)$. By R^3_+ we denote the domain $x_3 \geq 0$. We shall denote the point in R^2 by x' .

Here $x - y' = (x_1 - y_1, x_2 - y_2, x_3)$, $|x| = \sqrt{x_1^2 + x_1^2 + x_3^2}$, $|(x'.y_3)| = \sqrt{|x'|^2 + y_3^2}$, and $s(x) = 1 + |x| - x_1$; $x \cdot y = x_1y_1 + x_2y_2 + x_3y_3$

The convolution of two functions in R^2 or in R^3 will be denoted by *

$$f*g = \int_{R^3} f(x-y)g(y)dy \qquad f\tilde{*}g = \int_{R^2} f(x-y)g(y)dy$$

The Fourier transform of the function f is the function

$$F(f)= \tilde{f}(\alpha) = \frac{1}{(2\pi)^{3/2}} \int_{R^2} f(x)e^{-ix\cdot\alpha}dx \qquad (\alpha =(\alpha_1,\alpha_2,\alpha_3))$$

We have

$$F(f*g) = (2\pi)^{3/2}F(f)F(g) \ ; \quad F^{-1}(\tilde{f}\cdot\tilde{g}) = \frac{1}{(2\pi)^{3/2}} F^{-1}(\tilde{f})*F^{-1}(\tilde{g})$$

and

$$F^{-1}\left(\frac{e^{-x_3\alpha'}}{\alpha}\right) = \frac{1}{|x|} \ ; \quad F^{-1}(e^{-x_3\alpha'}) = \frac{x_3}{|x|^3} = Q(x)$$

(here $\alpha = |\alpha|$, and $\alpha' = |\alpha'|$).

It is well known that $\lim_{x_3 \to 0^+} Q(x)*\phi(x') = \phi(x')$

Now we introduce the functions

$$\Theta(x) = \frac{e^{x_1-|x|}}{4\pi|x|} \ ; \qquad \Gamma(x,t) = \frac{e^{-(|x_1-t|^2+ |x'|^2)/4t}}{t^{3/2}}$$

The function Γ is the first fundamental solution of the time dependent Oseen's system (see [12]). Moreover we have

$$\Theta(x) = \frac{1}{(4\pi)^{3/2}} \int_0^\infty \Gamma(x,t)dt \ ; \quad F(\Theta) = \frac{1}{(2\pi)^{3/2}(i\alpha_1+ \alpha^2)} \qquad 1.1$$

and $\lim_{x_3 \to 0^+} \Theta(x)\tilde{*}\phi(x') = \phi(x')$

Now we use the following notations:

$$D_i^m f = \frac{\partial^m f}{\partial_x{}^m} = \partial_{x_i}^m f \ ; \qquad \|f\|_{L_p(R^3)} = \left(\int_{R^2} |f(x)|^p dx \right)^{1/p}$$

About the definition of W_p^1, B_p^1, see [14]. L_p^1 and $\tilde{L}_p$ are the space of type W_p^1, and B_p^1 respectively in which we consider the L_p-norm of the derivatives of order 1 only.

2. Oseen's tensor

Let u(x) and p(x) be smooth functions defined in the whole space R^3 which vanish sufficiently fast at infinity, and satisfy

$$\begin{aligned} -\Delta u + \partial u/\partial x_1 + \nabla p &= f \\ \nabla\cdot u &= 0 \end{aligned} \qquad 2.1$$

Now we will express the functions u_i and p in term of f_i.
Let $\tilde{u}_i$, $\tilde{p}$, $\tilde{f}_i$ be the Fourier transform of the functions u_i, p, f_i with respect to all three variables; then system (2.1) becomes

$$\begin{aligned} i\alpha_1\tilde{u}_i + \alpha^2\,\tilde{u}_i + i\alpha_i\tilde{p} &= \tilde{f}_i \\ i\alpha_1\tilde{u}_1 + i\alpha_2\tilde{u}_2 + i\alpha_3\tilde{u}_3 &= 0 \end{aligned} \qquad 2.2$$

Here $\alpha = \sqrt{\alpha_1^2 + \alpha_2^2 + \alpha_3^2}$. Introducing the vectors $\tilde{v} = (\tilde{u}_1, \tilde{u}_2, \tilde{u}_3, \tilde{p})$ and $\tilde{F} = (\tilde{f}_1, \tilde{f}_2, \tilde{f}_3, 0)$, the system (2.2) becomes

$$Av = F$$

where A is the matrix

$$A = \begin{pmatrix} i\alpha_1^2 + \alpha & 0 & 0 & i\alpha_1 \\ 0 & i\alpha_1 + \alpha^2 & 0 & i\alpha_2 \\ 0 & 0 & i\alpha_1 + \alpha^2 & i\alpha_3 \\ i\alpha_1 & i\alpha_2 & i\alpha_3 & 0 \end{pmatrix}$$

The determinant of this matrix is $D(A) = \alpha^2(i\alpha_1 + \alpha^2)^2$.
Now $D(A) \neq 0$ for $\alpha \neq 0$.
We have

$$v = A^{-1}F$$

where

$$A^{-1} = \frac{1}{\alpha^2(i\alpha_1 + \alpha^2)} \times$$

$$\begin{pmatrix} \alpha_2^2 + \alpha_3^2 & -\alpha_1\alpha_2 & -\alpha_1\alpha_3 & -i\alpha_1(i\alpha_1 + \alpha^2) \\ -\alpha_1\alpha_2 & \alpha_1^2 + \alpha_3^2 & -\alpha_2\alpha_3 & -i\alpha_2(i\alpha_1 + \alpha^2) \\ -\alpha_1\alpha_3 & -\alpha_2\alpha_3 & \alpha_1^2 + \alpha_2^2 & -i\alpha_3(i\alpha_1 + \alpha^2) \\ -i\alpha_1(i\alpha_1 + \alpha^2 & -i\alpha_2(i\alpha_1 + \alpha^2) & -\alpha_3(i\alpha_1 + \alpha^2) & (i\alpha_1 + \alpha^2)^2 \end{pmatrix}$$

Hence

$$v_i = \frac{\tilde{f}}{i\alpha_1 + \alpha^2} - \sum_1^3{}_j \frac{\alpha_i\alpha_j}{\alpha^2(i\alpha_1 + \alpha^2)} \tilde{f} \quad ; \quad p = -\sum_1^3{}_j \frac{i\alpha_j}{\alpha^2} \tilde{f}_j .$$

Returning to the coordinate representation, we have

$$u_i(x) = \sum_1^3{}_j E_{i,j} * f_j \quad ; \quad p(x) = \sum_1^3{}_j e_j * f_j \qquad 2.3$$

where

$$\tilde{E}_{i,j}(\alpha) = \frac{1}{(2\pi)^{3/2}} \left[\frac{\delta_{ij}}{i\alpha_1 + \alpha^2} - \frac{\alpha_i\alpha_j}{\alpha^2(i\alpha_1 + \alpha^2)} \right]$$

$$\tilde{e}_j(\alpha) = -\frac{1}{(2\pi)^{3/2}} \frac{i\alpha_j}{\alpha^2}$$

Thanks to (1.1), we have

$$E_{i,j}(x) = \Theta + \partial^2/\partial x_i \partial x_j 1/|x| * \Theta \quad ; \quad e_j = -\frac{1}{4\pi} \partial_{x_j} \frac{1}{|x|}$$

The functions $E_{i,j}$ and e_j, constructed by Oseen (see [13]), are the fundamental solution of system (2.1), since

$$\partial E_{i,j}/\partial x_1 - \Delta E_{i,j} + \partial e_j/\partial x_i = \delta_{i,j}\ \delta(x_1)\partial(x_2)\delta(x_3),$$

$$\sum_{1}^{3}{}_i\ \partial E_{i,j}/\partial x_i = 0$$

Now we shall obtain estimates for the Oseen's tensor.

By direct computations we have

THEOREM 1 - The derivatives of the function $\Theta(x)$ satisfy the inequality

$$|D^m\Theta| \le c\ \frac{1}{|x|^{m/2}}\left(1 + \frac{1}{|x|^{m/2}}\right) e^{\mu(x_1-|x|)} \qquad (2.4)$$

where $\mu < 1/2$.

THEOREM 2 - The derivatives of the functions

$$W_{i,j}(x) = \Theta * \partial^2_{x_i x_j}(1/|x|)$$

satisfy the inequality $(|x| \neq 0)$

$$|D^m W_{i,j}(x)| \le \begin{cases} \dfrac{c}{|x|^{m+1}} & \text{for } |x| \le 1 \\ \dfrac{c}{(s(x)|x|)^{1+m/2}} & \text{for } |x| > 1 \end{cases} \qquad 2.5$$

In order to show (2,5), we use for $\Gamma(x,t)$ the well-known estimates of the heat-kernel (see [14]). Now

$$|D^m W_{i,j}(x)| \le \int_0^\infty |D^m\Gamma(x,t) * \partial_{x_i x_j}(1/|x|)dt \le$$

$$\le \int_0^\infty \frac{dt}{(|x_1 - t|^2 + t + |x'|^2)^{(3+m)/2}} = I$$

Now we estimate I.

In the case $|x| \le 1$ or $2x_1 \le |x'|$, we have

$$|t-x_1|^2 + t + |x'|^2 \ge c(\ t^2 + t + |x|^2)$$

consequently $I \le c/|x|^{m+1}$.

In the case $|x| > 1$ and $2x_1 > |x'|$, we set $\tau = |x'|^2 + x_1 - 1/4$.

Note $\tau > 3/4$.

We put $\vartheta(t) = (t - x_1 + 1/2)/\sqrt{\tau}$,

changing variables, and bearing in mind that $\tau \ge c\ (|x| s(x))$, we have

$$I \le c\, \tau^{-(m+2)/2} \int_{\vartheta(0)}^{\infty} \frac{d\vartheta}{(\vartheta^2 + 1)^{(3+m)/2}} \le c(|x| s(x))^{-1-m/2}.$$

Hence (2.5) is proved.

We note that the behaviour of fundamental solution of the Oseen's system near the origin is of the type of the fundamental solution of the Stokes system.

3. Boundary value problem for system (2.1) Construction of the second fundamental solution

In this section we investigate tha boundary value problem

$$\begin{aligned} \partial_{x_1} u - \Delta u + \nabla u &= 0 \\ \nabla \cdot u &= 0 \qquad\qquad 3.1 \\ u_{|x_3} &= a(x_1, x_2) \end{aligned}$$

The solution of this problem is sought in the half-space R^3_+.

We assume that a is smooth and vanishes sufficiently fast at infinity.

The solution should also vanish at infinity.

To find the solution we shall use a method developed by Solonnikov in [14] and [15] to study linear time dependent Navier-Stokes equations.

This method permits to express the unknown functions in terms of the Fourier transforms of the functions $a_i(x_1, x_2)$. Then returning to coordinate repre- representation, we have that the functions obtained are solutions of problem (3.1).

Now we introduce the four-dimensional vector V with the component

$v_1 = \tilde{u}_1$; $v_2 = \tilde{u}_2$; $v_3 = \tilde{u}_3$; $v_4 = \tilde{p}$ and the Fourier transform of

(3.1) with respect to the variables x_1, x_2.

Then we obtain a boundary value problem for a system of ordinary differential equations

$$(i\alpha_1 +^2\alpha)v_1 - \frac{\partial^2_1 v}{\partial x_3^2} + i\alpha_1 v_4 = 0$$

$$(i\alpha_1 + \alpha^2)v_2 - \frac{\partial^2 v_2}{\partial x_3^2} + i\alpha_1 v_4 = 0$$

$$(i\alpha_1 + \alpha^2)v_3 - \frac{\partial^2 v_3}{\partial x_3^2} + i\alpha_1 v_4 = 0 \qquad 3.2$$

$$i\alpha_1 v_1 + i\alpha_2 v_2 + i\alpha_3 \frac{\partial v_3}{\partial x_3} = 0$$

$$v_{i|x_3=0} = \tilde{a}_i(\alpha_1,\alpha_2), \ (i = 1,2,3); \quad v \to \infty \quad \text{as} \quad x_3 \to \infty .$$

We shall find solutions of system (3.2) in a three dimensional space Q with vector basis

$$e^1 = (1, 0, i\alpha_1)e^{-x_3 r} \ ; \qquad p^1 = 0$$

$$e^2 = (0, 1, i\alpha_2)e^{-x_3 r} \ ; \qquad p^2 = 0$$

$$e^3 = (i\alpha_1, i\alpha_2, -\alpha)e^{-x_3 r} \ ; \qquad p^3 = i\alpha_1 e^{-x_3 \alpha}$$

where $r = \sqrt{\alpha^2 + i\alpha_1}$, and $\alpha^2 = |\alpha|^2$.
Then $\tilde{u}$ is a linear combination of e_1, e_2, e_3 i.e.

$$\tilde{u} = \phi e^{-rx_3} + \psi(i\alpha_1/r, i\alpha_2/r, -\alpha)e^{-\alpha x_3}$$

with

$$\phi = (\phi_1, \phi_2, (i\alpha_1\phi_1 + i\alpha_2\phi_2)/r) \quad \text{and } \psi \text{ a function.}$$

We determine ϕ and ψ by satisfying the boundary conditions

$$\tilde{u}_{|x=0} = \tilde{a}(\alpha_1,\alpha_2)$$

We introduce the matrix T;

$$T = \begin{pmatrix} 1 & 0 & i\alpha_1 \\ 0 & 1 & i\alpha_2 \\ \frac{i\alpha_1}{r} & \frac{i\alpha_2}{r} & -\alpha \end{pmatrix}$$

and the vector $H = (\phi_1, \phi_2, \psi)$; ϕ and ψ satisfy the equation: $TH = \tilde{a}$

Now $$\det T = \frac{\alpha(\alpha - r)}{r}$$

Consequently

$$T^{-1} = \begin{pmatrix} -\alpha + \frac{\alpha_2^2}{r} & \frac{-\alpha_1\alpha_2}{r} & -i\alpha_1 \\ -\alpha_1\alpha_2 & -\alpha + \frac{\alpha_1^2}{r} & -i\alpha_2 \\ -\frac{i\alpha_1}{r} & -\frac{i\alpha_2}{r} & 1 \end{pmatrix} \times \frac{r}{\alpha(\alpha - r)}$$

So

$$H = T^{-1}\,\tilde{a}$$

Now we can express the solution $\tilde{u}$ in matricial form $\tilde{u} = \tilde{G}\,\tilde{a}$ with

$$\tilde{G} = \frac{1}{2\pi}\, e^{-x_3\alpha} \begin{pmatrix} 1 & 0 & 0 \\ 0 & 1 & 0 \\ 0 & 0 & 1 \end{pmatrix} +$$

$$+ \frac{1}{2\pi}\,\frac{e^{-x_3\alpha} - e^{-x_3 r}}{\alpha - r} \begin{pmatrix} \frac{\alpha_1^2}{\alpha} & \frac{\alpha_1\alpha_2}{\alpha} & \frac{i\alpha_1}{\alpha} r \\ \frac{\alpha_1\alpha_2}{\alpha} & \frac{\alpha_2^2}{\alpha} & \frac{i\alpha_2}{\alpha} r \\ i\alpha_1 & i\alpha_2 & -r \end{pmatrix}$$

Then returning to the variable x, we have

$$u = G^{*}a \;; \quad p = G_{4j}{}^{*}a \tag{3.3}$$

By direct calculation, we have

$$(\tilde{G}_{4j}) = \frac{1}{2\pi} \frac{-x_3\alpha}{\alpha - r} \left(-\frac{\alpha_1^2}{\alpha}, -\frac{\alpha_2\alpha_1}{\alpha}, -\frac{i\alpha_1}{\alpha} r\right)$$

In (3.3), we put $\qquad G_{i,j}(x) = \frac{1}{(2\pi)} \int_{R^2} \tilde{G}_{i,j}(\alpha, x_3) e^{i\alpha' x'} d\alpha'$.

Let us evaluate the integrals.

We note that

$$\frac{e^{-x_3\alpha} - e^{-x_3\sqrt{\alpha^2 + i\alpha_1}}}{\alpha - \sqrt{\alpha^2 + i\alpha_1}} = - \int_0^{x_3} e^{-(x_3 - y_3)\sqrt{\alpha^2 + i\alpha_1}} e^{-y_3\alpha} dy_3 \qquad 3.4$$

hence

$$\frac{1}{(2\pi)^2} \int_{R^2} \frac{e^{-x_3\alpha} - e^{-x_3\sqrt{\alpha^2 + i\alpha_1}}}{\alpha(\alpha - \sqrt{\alpha^2 + i\alpha_1})} e^{i\alpha' x'} d\alpha' =$$

$$- \frac{1}{(2\pi)^2} \int_0^{x_3} \int_{R^2} e^{-(x_3 - y_3)\sqrt{\alpha^2 + i\alpha_1}} \frac{e^{-y_3\alpha}}{\alpha} e^{i\alpha' x'} d\alpha' =$$

$$- \frac{1}{(2\pi)^2} \frac{1}{|x|} * \left(\int_0^{x_3} \int_{R^2} e^{-(x_3 - y_3)\sqrt{\alpha^2 + i\alpha_1}} e^{i\alpha' x'} d\alpha'\right.$$

Now we note that

$$e^{-x_3 \sqrt{\alpha^2 + i\alpha_1}} = \frac{1}{2\sqrt{\pi}} \int_0^\infty \frac{3}{t^{3/2}} e^{\left(-\alpha^2 t - \frac{x_3^2}{4t} - i\alpha_1 t\right)} dt.$$

Consequently

$$\frac{1}{(2\pi)^2} \int_{R^2} \frac{e^{-x_3\alpha} - e^{-x_3\sqrt{\alpha^2 + i\alpha_1}}}{\alpha(\alpha - \sqrt{\alpha^2 + i\alpha_1})} e^{i\alpha' x'} d\alpha' = - \frac{1}{\pi} \partial_{x_3} \tilde{\Theta}^* \frac{1}{|x|} \qquad 3.5$$

Performing analogous calculations for the other terms of $G_{i,j}$, we have

$$G_{ij} = -2\partial_{x_3}\Theta\delta_{ij} - \frac{1}{\pi}\partial_{x_j}C_i - \frac{1}{2\pi}\partial_{x_i}\frac{1}{|x|}\delta_{j3}$$

$$P_j = \frac{1}{2\pi}\partial^2_{x_j x_3}\frac{1}{|x|} + \frac{1}{2\pi}(\partial_{x_1} - \Delta')A - \frac{1}{2\pi}\frac{1}{|x|}\delta_{j3}. \qquad 3.6$$

where $\Delta' = \partial^2_{x_2} + \partial^2_{x_1}$, and

$$A(x) = \frac{1}{|x|}*\Theta(x') \quad ; \quad C_i(x) = \int_0^{x_3} \partial_{x_3}\tilde{\Theta}*\frac{x_i}{|x|^3}\,dx_3$$

The functions G_{ij}, P_j satisfy

$$\partial_{x_1}G_{ij} - \Delta G_{ij} + \nabla P_j = 0$$

$$\sum_{1}^{3}{}_i\, \partial_{x_i}G_{ij} = 0 \qquad 3.7$$

$$G_{ij|x_3=0} = \delta_{ij}\delta(x_1)\delta(x_2)$$

As in Theorem 1 we can prove

THEOREM 3 - The following estimates hold

$$|D^m A(x)| \le \begin{cases} \dfrac{1}{|x|^m} & \text{for } |x| \le 1 \\[2ex] \dfrac{1}{|x|^{(m+1)/2}(s(x))^{m/2}} & \text{for } |x| > 1 \end{cases} \qquad 3.8$$

$$|D^m_{x'}D^k_{x_3}\, G_{ij}(x)| \lesssim \begin{cases} \dfrac{c}{|x|^{m+2}\, x_3^k} & \text{for } |x| \le 1 \\[2ex] \dfrac{c}{|x|^{(m+2)/2}\, s(x)^{(m+1)/2}\, x_3^k} & \text{for } |x| > 1 \end{cases} \qquad 3.9$$

Then the solution of problem (3.1) can expressed by the second fundamental solution:

$$v_i = \sum_1^3 {}_i\, G_{ij} * a_j \; ;$$ 3.10

$$p = \frac{1}{2\pi} \sum_1^3 {}_i\, \partial^2_{x_i x_3} \frac{1}{|x|} * a_i - \frac{1}{\pi} \sum_1^3 {}_i\, (\partial_{x_1} - \Delta')(\partial_{x_i} A * a_i) + \frac{1}{2\pi|x|} * a_3$$

Using the Oseen tensor and (3.10) we can give a representation of the solution of the boundary value problem for (2.1) with nonhomogeneous right side i.e. for the following problem

$$\partial_{x_1} v - \Delta v + \nabla p = f$$

$$\nabla \cdot v = 0 \qquad 3.11$$

$$v_{|x_3} = a$$

If we extend f from R^3_+ to R^3 and introduce the new functions

$$w_i = v_i - \sum_1^3 {}_i\, E_{ij} * f_j \quad ; \; q = p - \sum_1^3 {}_i\, e_i * f_i$$

where E_{ij}, e_i is the Oseen's tensor, the functions w_i and q are solutions of the problem

$$\partial_{x_1} w - \Delta w + \nabla \cdot q = 0$$

$$\nabla \cdot w = 0 \qquad 3.12$$

$$w_{i|x_3} = a_i - \sum_1^3 {}_i\, E_{ij} * f_{j|x_3 = 0}$$

Thus the solution of problem (3.11) can be represented in the form

$$v_i = w_i + \sum_1^3 {}_i\, E_{ij} * f_j \; ; \quad p = q + \sum_1^3 {}_i\, e_i * f_i \qquad 3.13$$

Using the representation (3.13), thanks to theorems 1,2,3, to the Calderon-Zygmund theorem and Mihlin-Lizorkin theorem about the multipliers $i\alpha_i/(i\alpha_1 + \alpha^2)$, we have

THEOREM 4 - If the data are smooth enough, the solution of problem (3.1), with right side f, satisfies the inequality

$$\|\partial_{x_1} v\|_{L_p} + \|v\|_{L_p^{2l+2}} + \|\nabla p\|_{L_p^{2l}} \le c(\|f\|_{L_p^{2l}} + \|a\|_{B_p^{2l+2-1/p}});$$

with $l \ge 0$.

Concerning the weak solution of sistem (0.1), (0.2) (see [2], [6], [10], [11]), Theorem 4 implies in particular (with $\beta = 0$)

THEOREM 5 - If (u,p) is a weak solution of (0.1) with $f \in L^2 \cap L^p$, then the following inequality hold

$$\|\partial_{x_1} u\|_{L_p} + \|D^2 u\|_{L_p} + \|\nabla p\|_{L_p} \le c\|f\|_{L_p}$$

with $1 < p \le 3/2$.

REFERENCES

[1] Agmom, S., Douglis, A., Niremberg, L. : Estimates near the boundary of elliptic differential equations satisfying general boundary conditions II. Comm. Pure Appl. Math. 17 (1964), 35-92.

[2] Babenko, K.I.: On stationary solutions of the problem of flow past a body of viscous incompressible fluid. Math. USSR Sbornik 20 (1973), 1-25.

[3] Bemelmans, J.: Eine Außenraumaufgabe fur die instationaren Navier-Stokes Gleichungen. Math. Z. 162 (1978), 145-173.

[4] Cattabriga, L.: Su un problema al contorno relativo al sistema di equazioni di Stokes. Rend. Sem. Mat. Univ. Padova 31 (1961), 308-340.

[5] Farwig, R.: The stationary exterior 3D-problem of Oseen and Navier-Stokes equations in anisotropically weighted Sobolev spaces, Preprint.

[6] Finn, R.: On the steady state solutions of the Navier-Stokes partial differential equations. Arch. Rat. Mech. Anal. 3 (1959), 381-396.

[7] Finn, R. : On the exterior stationary problem for the Navier-Stokes equations, and associated pertubation problems. Arch. Rat. Mech. Anal. 19 (1965), 363-406.

[8] Fujita, H. : On the existence and regularity of the steady-state solutions of the Navier-Stokes equations. J. Fac. Sci. Univ. Tokyo 9 (1961), 59-102.

[9] Galdi, P.G. : Existence, Uniqueness and L^q - estimates for the Stokes problem in an exterior domain. Preprint.

[10] Ladyzhenskaya, O. A. : The mathematical theory of viscous incompressible flow. Gordon & Breach, New York 1969.

[11] Leray, J. : Etude de diverses equations integrales non lineaires et de quelques problemes que pose l'hydrodynamique. J. Math. Pures Appl. 12 series 9 (1933), 1-82.

[12] Mizumachi, R. : On the asymptotic behavior of incompressible viscous fluid motions past bodies. J. Math. Soc. Japan. 36 (1984), 497-522.

[13] Oseen, C.W. : Neure Methoden und Ergebnisse in der Hydrodynamik. Akademische Verlagsgesellschaft, Leipzig 1927.

[14] Solonnikov, V.A. : Estimates of the solutions of a nonstationary linearized system of Navier-Stokes equations. Amer. Math. Soc. Transl. (2) 75 (1968) 1-116.

[15] Solonnikov, V.A. : Estimates of solutions of an initial-boundary value problem for the linear nonstationary Navier-Stokes system. J. Sov. Math. (1978) 336-389.

This paper is in final form and no similar paper has been or is being submitted elsewhere.

Some results on the asymptotic behaviour of solutions to the Navier-Stokes equations

Maria E. Schonbek*
Department of Mathematics
University of California
Santa Cruz, CA 95064

§1. Introduction

We consider the asymptotic behaviour of solutions to the Navier-Stokes equations in $n \geq 2$ spatial dimensions

$$u_t + u \cdot \nabla u + \nabla p = \Delta u$$

$$\operatorname{div} u = 0$$

In earlier papers we discussed the upper bounds of rates of decay in three space dimensions with data $u_0 \in L^2 \cap L^p$, $1 \leq p \leq 2$ [6, 7]. There have been several extensions and improvements on these results [2, 11]. Here we first present a survey of results on lower bound of the L^2 rates of decay in two- and three-dimensions [8, 9] and then extend these results to $n \geq 3$ dimensions.

The study of the lower bounds is a much more subtle problem than the one corresponding to the upper bounds. The solutions to Navier-Stokes, unlike the solutions to the heat equation, do not decay at arbitrarily large algebraic rates or even exponentially depending on how oscillatory the initial data is. More precisely, solutions to Navier-Stokes outside a set M of radially equidistributed data have an algebraic lower bound of decay rate which is independent of the oscillations of the data and depends only on the number of dimensions of the space. The algebraic lower bound is a consequence of the nonlinear structure of the equations. The inertial term $\operatorname{div}(u \otimes u)$ in the Navier-Stokes equations appears to convert short-waves into long-waves reducing the decay rate. For data in M an example suggested by A. Majda shows that there are exponentially decaying solutions.

There are two cases to consider. First case: the average of the initial data is nonzero, i.e., the initial data has long waves. Here the argument relies on a comparison argument with solutions to the heat equation with the same data. Second case: the average of the data is zero. The

---

* This paper was partially funded by NSF grant No. DMS-9020941.

approach now is to find conditions on the data so that the corresponding solutions to the heat equations decay at a specific algebraic rate. These conditions will be met by the solution to the Navier-Stokes at some time $t_0 \geq 0$. The lower bound in the solution to the heat equation will be an essential tool to establish the corresponding estimate for solutions to Navier-Stokes.

§2. Main Results

The following notation will be used.

$$V(\mathbb{R}^n) = C_0^\infty(\mathbb{R}^n) \cap \{u\colon \nabla\cdot u = 0\},$$

$$H(\mathbb{R}^n) = H = \text{closure of } V \text{ in } L^2,$$

$$W_1 = \left\{u\colon \int_{\mathbb{R}^n} |x|^2|u|\,dx < \infty\right\}, \quad W_2 = \left\{u\colon \int_{\mathbb{R}^n} |u|^2|x|\,dx < \infty\right\},$$

$$|u|_{W_1} = \int_{\mathbb{R}^n} |x|^2|u|\,dx, \quad |u|_{W_2} = \int_{\mathbb{R}^n} |x||u|^2 dx.$$

Let $u \in \mathbb{R}^n$, $m_{ij} = \int_{\mathbb{R}^n} u_i u_j\,dx$, define $M = \{u\colon u \in \mathbb{R}^n, \text{ matrix } (m_{ij}) \text{ is scalar}\}$,

$$\alpha_i^j(t_0, u) = \int_0^{t_0} m_{ii} - m_{jj} ds, \quad \beta_i^j(t_0, u) = \int_0^{t_0} m_{ij}\,ds, \; i \neq j.$$

We recall that if the average of the initial data u_0 is nonzero, i.e., the initial data has long waves, the corresponding solutions to the heat equation have a lower bound of rate of decay of $(t + 1)^{-n/2}$ and the difference between the solution to the heat equation and solutions to Navier-Stokes decay at most like $(t + 1)^{-n/2-1}$. Hence, a straightforward comparison argument shows that.

Theorem 2.1: Let $u_0 \in H \cap L^1$ and $\hat{u}(0, t) = \int_{\mathbb{R}^n} u(x, t)\,dt \neq 0$ then there exist constants C_0 and C_1 such that

$$C_o(t + 1)^{-n/2} \leq |u(\cdot, t)|_{L^2}^2 \leq C_1(t + 1)^{-n/2}$$

with C_0, C_1 depending only on the L^2 and L^1 norms of the data.

Proof: See [8]. □

The case when $\widehat{u}(0, t) = \int u(x, t)dx = 0$ is more subtle. The reason being that the mass $\int u(x, t)dx$ is invariant with time and hence stays equal to zero for all time. Hence comparison with the heat equation cannot be expected to work in a straightforward way. There are two preliminary steps to be carried out. First find conditions on the data so that the corresponding solution to the heat equation decays at a slow rate. Second show that solutions to the Navier-Stokes equation, with zero average data lying in M^c, satisfy these conditions at some time $t_0 \geq 0$. In other words long waves will develop eventually.

Once this has been achieved a comparison argument will be used. We note that in this case the lower bound of the rate of decay for solutions to the heat equation and the upper bound of the rate of decay of the difference between solutions to the heat equation and Navier-Stokes equations is the same. Hence the comparison between these two equations is much more difficult.

The data theorem for lower bounds for the solutions to the heat equation can be stated as follows.

Theorem 2.2: Let $v_0 \in L^2(\mathbb{R}^n)$. Let v be a solution to the heat equation with data v_0. Suppose there exists function ℓ and h such that the Fourier transform of δ_0 for $|\xi| \leq \delta$, $\delta > 0$ admits the representation

$$\widehat{v}_0(\xi) = \xi \cdot \ell(\xi) + h(\xi) \qquad \ell = (\ell_1, \dots, \ell_n)$$

where ℓ and h satisfy

i. $|h(\xi)| \leq M_0|\xi|^2$, since $M_0 > 0$,

ii. ℓ is homogeneous of degree zero,

iii. $\alpha_1 = \int_{|w|=1} |w \cdot \ell(w)|^2 dw > 0.$

Let $M_1 = \sup_{|y|=1} |\ell(y)|$, $M_2 = \sup|\nabla \ell(y)|$, $K = \max(M_0, M_1, M_2)$ then there exists constants C_0 and C_1 such that

$$C_0(t+1)^{-(n/2+1)} \leq |v(\cdot, t)|_{L^2}^2 \leq C_1(t+1)^{-(n/2+1)},$$

where C_0 and C_1 depend only on M_0, M_1, δ, $|v_0|_{L^2}$ and C_0 also depends on K and α_1.

Proof: We give only the idea of the proof. Note that by the form of the initial data $|\widehat{v_0}(\xi)|^2 = O(|\xi|^2)$ and hence by Parseval

$$\int_{\mathbb{R}^n} |v|^2 dx = \int_{\mathbb{R}^n} |\widehat{v_0}(\xi)|^2 e^{-|\xi|^2 t} d\xi = \int_{\mathbb{R}^n} O(|\xi|^2) e^{-|\xi|^2 t} d\xi$$

which when made rigorous and after change of variables implies that $\int_{\mathbb{R}^n} |v|^2\, dx$ is of order $(t+1)^{-(n/2+1)}$.

For a detailed proof see [8]. □

Corollary 2.3 The conclusion of theorem 2.2 holds if i and ii of theorem 2.2 hold and iii is replaced by the following conditions.

iii. $\omega_0 \cdot \ell\,(\omega_0) = \alpha \neq 0$, for some $\omega_0 \in S^{n-1}$,

iii'. $\xi \cdot \ell(\xi) \in C^1(\mathbb{R}^n \backslash 0)$.

Proof: See [8]. □

For the second preliminary step there are two cases. 1. The zero of the data is of order one. 2. The zero of the data is of order greater than zero. In the first case we use the following.

Theorem 2.4: Let $g \in H \cap W_1 \cap W_2(\mathbb{R}^n)$, $n = 2, 3$. If g has a zero of order one, then there exists $\delta > 0$ such that for $|\xi| \leq \delta$

$$\widehat{g}(\xi) = \xi \cdot \ell(\xi) + h(\xi)$$

where ℓ and h satisfy the hypothesis of theorem 2.2 with $M_0 = \sup_{|\xi|\leq\delta} |\nabla^2 g(\xi)|$ and α_1 depending only on $\widehat{\nabla g}(0)$.

Proof: See [7]. The proof presented in [8] is valid for all n. □

For the second case the data $u_0 = u(x, 0)$ has to be in M^c. Before treating this case we give several auxiliary lemmas.

Lemma 2.4: Let $u_0 \in H^1(\mathbb{R}^n) \cap H(\mathbb{R}^n)$. Let $u(x, t)$ be a Leray-Hopf solution of the Navier-Stokes equations with data u_0. Then there exists $t_0 > 0$ such that

$$\|\nabla u(\cdot, t)\|_{L^2}^2 \leq C, \text{ for } t \leq t_0.$$

Proof: The proof is standard. In [9] we give a proof for $n = 3$. The main idea is to first multiply the equation by Δu and integrate in space. Using Agmon's inequality one derives easily an ordinary differential inequality for $|\nabla u|_{L^2}^2$. The solution of this inequality exists for $t \leq t_0$ where $t_0 = t_0(|\nabla u_0|_{L^2}, |u_0|_{L^2}, n)$. □

Lemma 2.5: Let $u_0 \in H^1(\mathbb{R}^n) \cap H(\mathbb{R}^n)$. Let $u(x, t)$ be a Leray-Hopf solution to the Navier-Stokes equation with data u_0, then there exist $t_0 > 0$ such that for $t \leq t_0$

$$|u(\cdot, t)|_{L^\infty} \leq C.$$

where C depends only on the L^2 norms of the data u_0 and the gradient of the data ∇u_0.

Proof: Follows by Agmon's standard inequality and the last lemma. □

The proof of the next lemma is formal for $n \geq 3$. In order to make it rigorous it should be applied to approximating solutions as the ones constructed by Caffarelli, Kohn and Nirenberg [1] for $n = 3$ or for $n \geq 3$ by Kayikiya and Miyikawa [2] or by von Wahl [13].

Lemma 2.6: Let $u_0 \in H \cap H^s(\mathbb{R}^n) \cap W_1 \cap W_2$ where $s = \left[\frac{n}{2}\right] + 2$. Let u be a weak solution to the Navier-Stokes equations with data u_0. Then there exists $t_0 > 0$ such that for $t \leq t_0$

$$a_{ij} = a_{ij}^0 + \xi \nabla_\xi a_{ij}(\bar{\xi}),$$

where $a_{ij} = a_{ij}(\xi, t) = \widehat{u_i u_j}(\xi, t)$, $a_{ij}^0 = (\xi, 0)$.

Note for $n = 3$ it suffices to have $s = 1$ and the result is valid for almost all t.

Proof: For $n = 3$ the theorem is an immediate consequence of Lemma (8.1) in [1]. Actually this lemma establishes the following bound for almost all t:

$$\frac{1}{2}\int_{\mathbb{R}^3} |u(x, t)|^2 |x| \, dx \leq A(t)$$

with $A(t)$ depending only on the L^2 and W_2 norms of the data. Hence letting

$$A = \left\{t : |\partial_\xi \widehat{u_i u_j}(\xi, t)| \leq A(t)\right\},$$

hence for all $t \in A^0$ the conclusion of the theorem follows in the case $n = 3$. for higher dimensions we use the well-known fact that solutions are regular for a short period of time (this argument can also be used for three dimensions). There are several ways of establishing short time regularity in particular for 3 dimensions, see Kato [4]. For higher dimensions, see Temam [10]. A simple way of obtaining short time regularity is to bound the Dirichlet norm for a short time. This can be done formally by multiplying the Navier-Stokes equations by Laplacian and integrating in space. Agmon's inequality for the L^∞ norm will yield an ODE for the Dirichlet norm from where the bound for short time follows. From here Temam's methods will give rigorous short time regularity.

To obtain the conclusion of the theorem for $n \geq 4$, let t_0 be such that the solutions are regular for $t \leq t_0$. It will be necessary to show that $\nabla_{\xi_k} a_{ij}$ is well defined or equivalently that for $t \leq t_0$

$$\int_{\mathbb{R}^n} |u(x, t)|^2 |x| \, dx \leq \infty.$$

Since this is an auxiliary result we will give a proof in an appendix at the end of the paper. □

We will use the notation for $i \neq j$

(2.1) $$\alpha_{ij} = \alpha_{ij}(t) = \int_{\mathbb{R}^n} u_i u_j \, dx, \qquad \alpha_{ij}^0 = \alpha_{ij}(0), \qquad i, j = 1,\ldots,n$$

(2.2) $$\beta_{ij} = \beta_{ij}(t) = \int_{\mathbb{R}^n} u_i u_j \, dx, \qquad \beta_{ij}^0 = \beta_{ij}(0), \qquad i, j = 1,\ldots,n$$

Lemma 2.7: Let $u_0 \in H \cap H^s(\mathbb{R}^n) \cap M^c$. Let $u(x, t)$ be a weak solution to the NS with data u_0. Then

i. If $\alpha_{ij}^0 \neq 0$ for some i, j then there exists t_0 such that

(2.3) $$\left| \mathrm{A}_{ij}(t) \right| = \left| \int_0^t \alpha_{ij}(x, s) \, ds \right| \geq \frac{t}{2} \alpha_{ij}^0,$$

for all $t \leq t_0$, t_0 depending only on the H^s norm of the data.

ii. If $\beta_{ij}^0 \neq 0$ for some i, j then there exists t_0 such that

(2.4) $$\left| \mathrm{B}_{ij}(t) \right| = \left| \int_0^t \beta_{ij}(x, s) \, ds \right| \geq \frac{t}{2} \beta_{ij}^0,$$

for all $t \leq t_0$, t_0 depending only on the H^s norm of the data.

Proof: Let t_0 be such that the solution is regular for $t < t_0$. The proof follows the same lines as in [8, 9]. The proof in [8] can be used for $n \geq 3$ since we do have short time regularity. Note that ii is a consequence of i since β_{ij} is a rotation of α_{ij} by an angle of $\pi/4$. □

Theorem 2.8: Let $u_0 \in H \cap H^m \cap W_2 \cap M^c(\mathbb{R}^n)$, $n \geq 2$, $m \geq \left[\frac{n}{2}\right] + 2$ (if $n = 2$ it suffices if $m = 1$). Let $u(x, t)$ be a solution to the NS equations with data u_0. If $\hat{u}_0$ has a zero greater than one at the origin there exists $t_0 > 0$ and $\delta > 0$ such that for $|\xi| \leq \delta$

$$\hat{u}_k(\xi, t_0) = \xi \cdot \ell_k(\xi. t_0) + h_k(\xi. t_0),$$

where t_0 depends on the H^m and W_2 norms of the data, ℓ_k and h_k satisfy the conditions of corollary 2.3.

Proof: For the proof we work in Fourier space. We will give the main outline. More details can be found in [8]. The proof follows the general lines of [8].

Note that the solution satisfies

$$\widehat{u}_t + |\xi|^2\widehat{u} = -\widehat{H}, \quad \widehat{u}_0(\xi) = \widehat{u}_0(\xi, 0),$$

where $\widehat{H} = \widehat{u \cdot \nabla u} + \widehat{\nabla p}$. Arguments of Wiegner [12] show that

$$\widehat{u}_k(\xi, t_0) = \sum_{j=1}^{n} (\delta_{jk} - \xi_k\xi_j|\xi|^{-2})\left[\widehat{u}_{0,j}e^{-|\xi|^2 t_0} - \int_0^{t_0} \widehat{u \cdot \nabla u_j}(\xi, s)e^{-|\xi|^2(t_0-s)}ds\right], \tag{2.5}$$

where $\widehat{u}_{0,j}$ is the j-th component of $\widehat{u}_0$ and t_0 is given by lemma 2.6. By hypothesis, $\widehat{u}_{0j}$ has a zero of order greater than one, hence we only have to consider the terms in

$$\sum_{j=1}^{n} (\delta_{kj} - \xi_k\xi_j|\xi|^{-2})\left[\int_0^{t_0} \widehat{u \cdot \nabla u_j}(\xi, s)e^{-|\xi|^2(t_0-s)}ds\right]. \tag{2.6}$$

Since $\widehat{u \cdot \nabla u_j} = \sum_{i=1}^{3} \xi_i\widehat{u_i u_j}$ and $e^{-|\xi|^2(t_0-s)} = 1 + 0(|\xi|^2)$ it follows by Lemma 2.5 that (2.6) can be rewritten as

$$-i\sum_{j,i=1}^{n} (\delta_{kj} - \xi_k\xi_j|\xi|^{-2})\left[\int_0^{t_0} \xi_i a_{ij}^0(s)\, ds\right] + K_j^k(\xi), \tag{2.7}$$

where $|K_j^k(\xi)| \le M|\xi|^2$, $M = M(\|u_0\|_{L^2}, \|u_0\|_{W^2}, t_0)$. Without loss of generality let $k = 1$. The sum in (2.7) will be rewritten as

$$-i\xi \cdot \ell_1(\xi, t_0) = -i\sum_{j=1}^{n} \xi_j\ell_1^j(\xi, t_0), \qquad \ell_i = (\ell_i^1, \cdots, \ell_i^n) \tag{2.8}$$

For this we first subdivide the sum in (2.7) into three parts.

a. When $i = j = 1$ the corresponding terms of the sum are

$$\xi_1\left[1 - \frac{|\xi_1|^2}{|\xi|^2}\right]\int_0^{t_0} a_{11}^0 ds = \xi_1\sum_{i=2}^{n}\frac{|\xi_i|^2}{|\xi|^2}\int_0^{t_0} a_{11}^0 ds .$$

b. When $j = 1$ and $i \geq 2$ or $i = 1, j \geq 2$, the terms of the sum are

$$\sum_{r=1}^{n} \xi_r \left[1 - 2\frac{|\xi_1|^2}{|\xi|^2}\right] \int_0^{t_0} a_{r1}^0 ds.$$

c. When $i \geq 1$ and $j \geq 1$, the terms of the sum are

$$-\sum_{i \neq j > 1}^{n} \frac{\xi_1 \xi_i \xi_j}{|\xi|^2} \int_0^{t_0} a_{ij}^0 ds - \sum_{i=2}^{n} \frac{\xi_1 |\xi_i|^2}{|\xi|^2} \int_0^{t_0} a_{ii}^0 ds .$$

Hence, $\ell_i(\xi, t) = (\ell_1^1, \cdots, \ell_1^n)$ can be defined as follows

$$\ell_1^i = \ell_r(\xi, t_0) = \sum_{i=2}^{n} \frac{|\xi_i|^2}{|\xi|^2} \int_0^{t_0} a_{11}^0 - a_{ii}^0\, ds.$$

$$\ell_1^r = \ell_1^r(\xi, t_0) = \left[1 - 2\frac{|\xi_1|^2}{|\xi|^2}\right] \int_0^{t_0} a_{r1}^0 ds - \sum_{r \neq j > 1}^{n} \frac{\xi_1 \xi_j}{|\xi|^2} \int_0^{t_0} a_{jr}^0 ds,$$

for $r \geq 2$. Hence the sum (2.7) can be expressed as described by (2.6). Combining (2.5), (2.6) and (2.8) yields

$$\widehat{u}_k(\xi, t_0) = \xi \cdot \ell_k(\xi, t_0) + h_k(\xi, t_0),$$

where ℓ_k and $h_k = \sum_{j=1}^{n} k_j^k(\xi)$ satisfy trivially conditions i, ii and iii' of corollary (2.3). To establish iii', let e_j be the canonical basis of $\mathbb{R}^n$. Let α_{ij}^0 and β_{ij}^0 be defined by (2.1) and (2.2), then by hypothesis either $\alpha_{ij}^0 \neq 0$ or $\beta_{ij}^0 \neq 0$. There are various cases to analyze.

i. For some i, j, $\alpha_{ij}^0 \neq 0$. Without loss of generality, let $i = 1$. Choose $\omega_0 = \frac{e_1 + e_j}{\sqrt{2}}$, then $\omega_0 \cdot \ell(\omega_0) = \mathcal{A}_{ij}(t_0) \neq 0$, where $\mathcal{A}_{ij}(t_0)$ was defined by (2.3) (see lemma (2.6)).

ii. For all i, j, $\alpha_{ij}^0 = 0$. Suppose $\beta_{ij}^0 \neq 0$ for some i, j. Suppose either i or j is one. Without loss of generality, let $i = 1$. Choose $\omega_0 = e_j$, then $\omega_0 \cdot \ell(\omega_0) = \mathrm{B}_{ij}(t_0) \neq 0$, where $\mathbf{B}_{ij}(t_0)$ was defined by (2.4) (see lemma (2.6)).

iii. For all i, j $\alpha_{ij}^0 = 0$ and $\beta_{ij}^0 = 0$. Suppose $\beta_{ij}^0 \neq 0$ for some i, j. Suppose i and j are not one. Choose $\omega_0 = \frac{1}{\sqrt{3}}(e_1 + e_j - e_j)$. Hence $\omega_0 \cdot \ell(\omega_0) \neq 0$. Note that multiplying by appropriate signs of $\mathcal{A}_{ij}$ or $\mathcal{B}_{ij}$, $\omega_0 \cdot \ell(\omega_0) > 0$. □

The next theorem was the essential step in establishing the lower bound of rate of decay for the L^2 norm in [8].

Theorem 2.9: Let $u_0 \in L^2 \cap W_2 \cap H(\mathbb{R}^n)$. Let v be a solution to the heat equation with data u_0. Suppose

$$C_0(1 + t)^{-(n/2+1)} \leq \|v(\cdot, t)\|_{L^2}^2 \; C_1(1 + t)^{-(n/2+1)}.$$

Let $u(x, t)$ be a solution to the NS equations with data u_0, then there exist constants M_0 and M_1 such that

$$M_0(1 + t)^{-(n/2+1)} \leq \|u(\cdot, t)\|_{L^2}^2 \; M_1(1 + t)^{-(n/2+1)}.$$

where M_0 and M_1 depend on C_1, n, the L^1 and the L^2 norm of u_0 and M_- also depends on the W_2 norm of u_0.

Note 1: The proof is based on the proof presented in [8], where the 2-dimensional case was established and the n-dimensional was outlined. We give only the changes necessary to complete the proof in [8].

Note 2: The outline of the proof in [8] is formal. To make it rigorous apply it to approximating sequences and pass to the limit.

Proof: There are two cases to be considered. Let $i \neq j$.

Case 1: Given t there exists $T > t$, such that for all pairs i, j

$$|A_{ij}(T)| < \beta\sqrt{C_0} \quad \text{or} \quad |B_{ij}(T)| < \beta\sqrt{C_0}.$$

Case 2: There exists T_0 such that for all $t \geq T_0$ and for at least one pair $1 \leq i, j \leq n$

$$|A_{ij}(t)| \geq \beta\sqrt{C_0} \quad \text{and} \quad |B_{ij}(t)| \geq \beta\sqrt{C_0}.$$

Here β is such that $16\beta^2 A_n = 1/16$ where $A_n = 2\sigma_i(2n)^{n/2+1}$ and σ_i = measure of the $n-1$ sphere of radius one. See [8] for reason of this choice.

The proof of case 1 is the same as the one presented in [9] if we replace $\widehat{H}$ by $\widehat{\mathcal{H}}$ where

$$\widehat{H}(\xi, t) = i\xi \cdot \left(\frac{|\xi_2|^2}{|\xi|^2} [a_{11}^0 - a_{22}^0], \left[1 - 2\frac{|\xi_1|^2}{|\xi|^2} \right] a_{12}^0 \right) + O(|\xi|^2),$$

$$\widehat{\mathcal{H}}(\xi, t) = i\xi \cdot \Gamma,\ \Gamma = (\Gamma_1, \ldots, \Gamma_n), \text{ where}$$

$$\Gamma_1 = \Gamma_1(\xi) = \sum_{i=2}^{n} \frac{|\xi_i|^2}{|\xi|^2} (a_{11}^0 - a_{ii}^0),$$

$$\Gamma_r = \Gamma_r(\xi) = \left[1 - 2\frac{|\xi_1|^2}{|\xi|^2} \right] a_{r1}^0 - \sum_{r \neq j > 1}^{n} \frac{\xi_1 \xi_j}{|\xi|^2} a_{jr}^0\, ds,\ r \geq 2.$$

The proof of case 2 is the same as the one given in [8] if we replace $\widehat{H}$ by $\widehat{\mathcal{H}}$ and note that the auxiliary lower bound for $\alpha_i = \int_{|\omega|=i} |\omega \cdot \ell_i(\omega)|^2 d\sigma$ can be given easily using that we can find a $\omega_0 \cdot \ell(\omega_0) = \alpha > 0$, as shown above and hence the smoothness of $\ell(\omega)$ ensures that in some neighborhood of ω_0, $\omega \cdot \ell(\omega) > \alpha/2$.

Combining theorems 2.2, 2.4, 2.7, 2.8, lemma 2.5, 2.6 and corollary 2.3 yields

Theorem 2.9: Let $u_0 \in L^1 \cap H(\mathbb{R}^n)$ be such that $\widehat{u}(0, t) = \int_{\mathbb{R}^n} u(x, t)dx = 0$.

i. If $\widehat{u}_0(\xi)$ has a zero of order one at the origin then also let $u_0 \in W_1 \cap W_2$.

ii. If $\widehat{u}_0(\xi)$ has a zero of order greater than one at the origin then also let

$u_0 \in H^m \cap M^c \cap W_2$, $m = \left[\frac{n}{2}\right] + 2$ (if $n = 3$ then $m = \left[\frac{n}{2}\right] + 1$ suffices). If u is a solution to the NS equations then there exist constants C_0 and C_1 such that

$$C_0(t+1)^{-\left(\frac{n}{2}+1\right)} \leq \|u(\cdot, t)\|_{L^2(\mathbb{R}^n)} \leq C_1(t+1)^{-\left(\frac{n}{2}+1\right)},$$

where in case i C_0 and C_1 depend only on the L^2, W_1 and W_2 norms of u_0 and in case ii C_0 and C_1 depend on the L^1 and L^2 norms of u_0 and C_0 also depends on the H^m norms of the data.

Proof: For upper bounds see [2, 7, 11]. The lower bounds are immediate consequences of theorems 2.2, 2.4, 2.8, 2.9, corollary 2.3, lemma 2.4 and 2.5. To obtain a rigorous proof it will be necessary to apply the above theorems and lemmas to approximating sequences and pass to the limit. See [1, 2, 3, 5, 12] for construction of such sequences. Note that the lower bounds will be first shown to be valid only a.e. in t. To obtain them for all t we use the following lemma.

Lemma 2.10: Let $A \subset \mathbb{R}$ be a set such that the Lebesgue measure $\mu(A^c) = 0$. If $u(x, t)$ is a solution to the Navier-Stokes equations and for $t \in A$

$$C_0(t+1)^{-\alpha} \leq \|u(\cdot, t)\|_2^2 \leq C_1(t+1)^{-\alpha}$$

then for all $t \in \mathbb{R}$

$$\left(\frac{1}{2}\right)^{\alpha} C_0(t+1)^{-\alpha} \leq \|u(\cdot, t)\|_2^2 \leq C_1(t+1)^{-\alpha} 2^{\alpha}.$$

Proof: Let $t_0 \in A^c$, $t_1, t_2 \in A$ such that $t_0 \in [t, t_2]$, $(t_0+1)(t_1+1)^{-1} \leq 2$. and $(t_0+1)(t_2+1)^{-1} \geq 1/2$. Then since $\|u(\cdot, t)\|$ is a decreasing function

$$\left(\frac{1}{2}\right)^{\alpha}(t_0+1)^{-\alpha} \leq \|u(\cdot, t)\|_2^2 \leq 2^{\alpha} C_1(t_0+1)^{-\alpha}.$$ □

Having data in M^c is essential since the example presented in [8, 9] of exponentially decaying vorticity can be easily extended to n dimensions (see [9]). This example was suggested by A. Majda for solutions in two spatial dimensions.

3. Appendix

Here we establish lemma 2.7 for $n \geq 3$.

Lemma A: Let $u_0 \in H \cap H^s(\mathbb{R}^n) \cap W_1 \cap W_2$ where $s = \left[\frac{n}{2}\right] + 2$. Let $u = (u_1, u_2, \ldots, u_n)$ be a Leray-Hopf solution to the Navier-Stokes solutions. Then there exists $t_0 > 0$, such that for $t \leq t_0$

$$a_{ij} = a_{ij}^0 + \xi \nabla_\xi \overline{a_{ij}(\xi)},$$

where $a_{ij} = a_{ij}(\xi, t) = \widehat{u_i u_j}(\xi, t)$, $a_{ij}^0 = a_{ij}(0, t)$.

Proof: Let t_0 be given by Lemma 2.4. It is only necessary to show that for $t \leq t_0$, $\nabla_{\xi_k} a_{ij}$ is well-defined or equivalently that

$$\int_{\mathbb{R}^n} |x|\, |u|^2\, dx < \infty.$$

Multiply the Navier-Stokes equations by $|x| u_i$ and integrating in space

$$\frac{d}{dt} \int_{\mathbb{R}^n} |x| \frac{|u_i|^2}{2}\, dx = -\sum_{j=1}^{n} \int_{\mathbb{R}^n} |x|\, u_i \partial_j u_j u_i\, dx - \int_{\mathbb{R}^n} |x| u_i\, \partial_i p\, dx + \int_{\mathbb{R}^n} |x| u_i \Delta u_i dx.$$

Note that the first term on the right can be bounded as follows.

$$-\sum_{j=1}^{n} \int |x|\, u_i u_j \partial_j u_i\, dx \leq C \int_{\mathbb{R}^n} |x|\, |u|^2 dx + C \int |x|\, |u|^2 |\nabla u|^2 dx \leq (C + |\nabla u|) \left(\int |x|\, |u|^2 dx \right) L^\infty(\mathbb{R}^n \times [0, T_0]$$

$$\leq C_1 \int |x|\, |u|^2 dx.$$

Further integration by parts in (2.1) yields after summation over the i index

$$\sum_{i=1}^{n} \frac{d}{dt} \int_{\mathbb{R}^n} |x| \frac{|u_i|^2}{2}\, dx = C_1 \int |x|\, |u|^2 dx.$$

$$+ \sum_{i=1}^{n} \int_{\mathbb{R}^n} u_i p dx - \sum_{i,j}^{n} \int_{\mathbb{R}^n} u_i \partial_j u_i dx - \int_{\mathbb{R}^n} |x|\, |\nabla u_i|^2 dx$$

Here we supposed that the integrated terms tend to zero.

$$\le C_1 \int_{\mathbb{R}^n} |x|\,|u|^2 dx + n\left(\int_{\mathbb{R}^n} |u|^2 dx \int_{\mathbb{R}^n} |p|^2 dx\right)^{1/2} + n^2\left(\int_{\mathbb{R}^n} |u|^2 dx \int_{\mathbb{R}^n} |\nabla u|^2 dx\right)^{1/2} -$$

$$- \int_{\mathbb{R}^n} |x|\,|\nabla u_i|^2 dx \le C_1 \int_{\mathbb{R}^n} |x|\,|u|^2 dx + C_2(t_0, n, |u_0|_{L^2}, |\nabla u_0|_{L^2}) - \int_{\mathbb{R}^n} |x|\,|\nabla u_i|^2 dx.$$

Hence by Gronwall's inequality for $t \le t_0$

$$\int |x|\,|u(x,t)|^2 dx \le k \exp C_1 t_0,$$

where $k = k(t_0, n, |u_0|_{L^2}, |\nabla u_0|_{L^2})$. And the lemma is proven. □

References

[1] L. Caffarelli, R. Kohn and L. Nirenberg. Partial regularity of suitable weak solutions of the Navier-Stokes equations, Comm. Pure Appl. Math XXXV (1982) 771-831.

[2] R. Kayikiya and T. Miyakawa. On the L^2 decay of weak solutions of the Navier-Stokes equations in $\mathbb{R}^n$. Math. Z. 192 (1986) 135-148.

[3] P. Galdi, P. Maremonti. Navier-Stokes equations in exterior domains. Archive for Rational Mechanics 94 (1986) 253-266.

[4] T. Kato. Non-stationary flows of viscous and ideal fluids in $\mathbb{R}^3$. Journal of Functional Analysis 9 (1972) 295-305.

[5] J. Leray. Sur le mouvement d'une liquide visqueux complissent l'espace. Acta Math 63, 1934, 193-248.

[6] M. E. Schonbek. L^2 decay for weak solutions of the Navier-Stokes equations, Archive for Rational Mechanics and Analysis, Vol. 88, 3, 1985, pp. 209-222.

[7] M. E. Schonbek. Large time behavior of solutions to the Navier-Stokes equations, Comm. in P.D.E. 11(7) (1986) 753-763.

[8] M. E. Schonbek. Lower bounds of rates of decay for solutions to the Navier-Stokes equations. Journal of American Mathematical Society, July 1991.

[9] M. E. Schonbek. Asymptotic behaviour of solutions to the three-dimensional Navier-Stokes equation. Preprint.

[10] Temam. Navier-Stokes Equations, Theory and Numerical Analysis. North Holland, Amsterdam and NY. 1972.

[11] M. Wiegner. Decay results for weak solutions of the Navier-Stokes equations in $\mathbb{R}^n$. J. London Math Soc. 35 (1987) 303-313.

[12] M. Wiegner. Some remarks concerning the approximation of weak solutions in exterior domains for $n = 3,4$. Private communication.

[13] M. von Wahl. The equations of Navier-Stokes and abstract parabolic equations. Aspects of Mathematics (Viewig-Verlag 1985).

Approximation of weak solutions of the Navier-Stokes equations in unbounded domains

Michael Wiegner
Mathematisches Institut der Universität
Postfach 101251, 8580 Bayreuth, Germany

Introduction

During the last years there has been considerable interest in the construction of weak solutions to the Navier-Stokes equations, which have some useful "additional" properties. Among these properties are local energy estimates, the generalized energy inequality and decay estimates for large times, cf. [G-M], [V], [M-S], [S-vW-W], [B-M], [W]. Due to uniqueness in two dimensions, we may restrict our attention to $n \geq 3$.

As it is unclear, whether Hopf's solutions inherit these additional properties, different methods of construction have been developped. These consist of solving strongly some modified equation (either by adding $\varepsilon\Delta^2 u_\varepsilon$ [V] or by regularizing the nonlinearity to $J_\varepsilon(u_\varepsilon) \cdot \nabla u_\varepsilon$ in various ways [S-vW-W]) and showing that the approximations u_ε converge weakly to some weak solution u with the claimed additional properties - e.g. an energy decay of the type $\|u(t)\|_2 \leq c(1+t)^{-\alpha}$.

Curiously enough it has turned out that it would be useful to have the additional information, that a subsequence $u_{\varepsilon_k}(t)$ converges also strongly in $L_2(\Omega)$, at least for almost all t. Namely, in order to characterize the long-time behaviour precisely, one needs also energy-estimates from below. This was done for the Cauchy-problem in a recent paper by M. E. Schonbek [Sch]; the lower estimates for $n \geq 3$ hold a.e., provided strong convergence of u_ε in $L_2(\mathbb{R}^n)$ is known (In fact, she addresses only the cases $n = 2$ and 3, but the same proof should work also in higher dimensions). Now for $n = 3$, this fact was implicitely contained in [G-M] by proving a uniform estimate of the type

$$\int_{|x|\geq 2R} |u_\varepsilon(x,t)|^2\, dx \leq c \int_{|x|\geq R} |u(x,0)|^2\, dx + \frac{C}{R}$$

which together with the strong convergence of a subsequence on compact subdomains implies the claim.

It is the aim of this note to verify this claim also for $n = 4$.

The approximation

Let $T > 0$, Ω an exterior domain in $\mathbb{R}^n$ (or $\Omega = \mathbb{R}^n$), P the projection from $L_2(\Omega)^n$ onto the divergence-free fields $L_{2,\sigma}$ and $A = -P\Delta$ the Stokes-operator with $D(A) = W_2^2 \cap \mathring{W}_2^1 \cap L_{2,\sigma}$. Define $\|f\|_{p,s} := (\int_0^T (\int_\Omega |f(t,x)|^p\, dx)^{\frac{s}{p}}\, dt)^{\frac{1}{s}}$ with the obvious interpretation for $s = \infty$.

From [M-S] we have the following:

If $a \in L_{2,\sigma}$ is given, there are strong solutions u_ε with

$$\begin{aligned} u_\varepsilon &\in L_2((0,T), D(A)) \cap C([0,T], D(A^{\frac{1}{2}})) \\ u_\varepsilon' &\in L_2((0,T), L_{2,\sigma}) \end{aligned}$$

of

$$(1)\quad \begin{aligned} &u_\varepsilon' - \Delta u_\varepsilon + J_\varepsilon(u_\varepsilon) \cdot \nabla u_\varepsilon + \nabla \pi_\varepsilon = 0 \\ &\operatorname{div} u_\varepsilon = 0 \\ &u_\varepsilon = 0 \text{ on } \partial\Omega \times (0,T) \\ &u_\varepsilon(x,0) = J_\varepsilon(a) := a_\varepsilon. \end{aligned}$$

Here $J_\varepsilon = (I + \varepsilon A)^{-1-[n/4]}$ is the Yoshida-approximation with $\|J_\varepsilon(u)\|_\infty \leq c(\varepsilon)\|u\|_2$, $\|J_\varepsilon(u)\|_2 \leq \|u\|_2$ and $\|J_\varepsilon(u) - u\|_2 \to 0$ for $u \in L_{2,\sigma}$.

The following energy-inequality holds uniformly

$$(2)\quad \|u_\varepsilon\|_{2,\infty}^2 + \|\nabla u_\varepsilon\|_{2,2}^2 \leq \|a\|_2^2$$

and the solution is given by the integral equation

$$(3)\quad u_\varepsilon(t) = e^{-tA} a_\varepsilon - \int_0^t e^{-(t-s)A} P(J_\varepsilon(u_\varepsilon) \cdot \nabla u_\varepsilon)\, ds.$$

Letting $v_\varepsilon(t) := e^{-tA} a_\varepsilon$, $w_\varepsilon(t) := u_\varepsilon(t) - v_\varepsilon(t)$, they show that there exists a pressure $\pi_\varepsilon := q_\varepsilon + \tilde{q}_\varepsilon$, with

$$(4)\quad \Delta v_\varepsilon - v_\varepsilon' = \nabla q_\varepsilon, \quad \Delta w_\varepsilon - w_\varepsilon' - J_\varepsilon(u_\varepsilon) \cdot \nabla u_\varepsilon = \nabla \tilde{q}_\varepsilon$$

and $\|\tilde{q}_\varepsilon\|_{r^*,r} \leq C_T$ uniformly for $r = \frac{n+2}{n+1}$, $\frac{1}{r^*} = \frac{1}{r} - \frac{1}{n}$.

Note, that in general one can not expect uniform estimates for the pressure q_ε up to $t = 0$ without further assumptions on the initial value a.

The result

Theorem: Let Ω denote an exterior domain in $\mathbb{R}^n$ or the whole of $\mathbb{R}^n$ with $n = 3$ or 4, and let u_ε denote the solutions of the approximated Navier-Stokes equations as described above. If $\Omega \neq \mathbb{R}^n$, assume additionally $a \in L_{p,\sigma} \cap D(A^\rho)$ for some $p < 2$, $\rho > 0$. Then a subsequence u_{ε_k} converges strongly in $L_2((0,T) \times \Omega)$ to a weak solution u of the Navier-Stokes equations.

Proof: We restrict to the case $n = 4$, as the estimates for $n = 3$ are essentially known. Assume first $\Omega = \mathbb{R}^4$. Then we have $q_\varepsilon = 0$ due to $a \in L_{2,\sigma}$ and $P\Delta = \Delta P$ for the Cauchy-problem.

Multiply (1) by $u_\varepsilon \cdot \phi$ and integrate over $\Omega \times (0,t)$, $t \leq T$. Here $\phi \in C_\infty(\mathbb{R}^4)$ with $0 \leq \phi \leq 1$,

$$\phi = \left\{ \begin{array}{l} 1 \text{ for } |x| \geq R \\ 0 \text{ for } |x| \leq R/2 \end{array} \right\} \text{ and } |\nabla \phi| \leq cR^{-1}.$$

Then

$$\int_{\mathbb{R}^4} |u_\varepsilon(x,t)|^2 \phi(x)\, dx \leq \int_{\mathbb{R}^4} |a_\varepsilon(x)|^2 \phi(x)\, dx +$$

$$+ c \int_0^T \int_{\mathbb{R}^4} |\nabla \phi| (|J_\varepsilon(u_\varepsilon)||u_\varepsilon|^2 + |\tilde{q}_\varepsilon||u_\varepsilon| + |\nabla u_\varepsilon||u_\varepsilon|)\, dx\, dt.$$

With $\delta(R) := 2 \int_{\mathbb{R}^4} |a(x)|^2 \phi(x)\, dx \searrow 0$ for $R \to \infty$, the first integral is estimated by $2\|a_\varepsilon - a\|_2^2 + \delta(R)$.

The second integral consists of three terms. We have, using (2),

$$\int_0^T \int_{\mathbb{R}^4} |u_\varepsilon||\nabla u_\varepsilon||\nabla \phi|\, dx\, dt \leq \frac{C}{R} \|u_\varepsilon\|_{2,\infty} \|\nabla u_\varepsilon\|_{2,2} \leq \frac{C}{R}.$$

Let us recall the wellknown interpolation inequality

$$\|v\|_p \leq c\|v\|_2^{1-a} \|\nabla v\|_2^a$$

with $a = n(\frac{1}{2} - \frac{1}{p}) \in (0,1)$.

With $a = \frac{2}{3}$ we get

$$\int_0^T \int_{\mathbb{R}^4} |\nabla \phi||J_\varepsilon(u_\varepsilon)||u_\varepsilon|^2\, dx\, dt \leq \frac{C}{R} \int_0^T \int_{\mathbb{R}^4} |u_\varepsilon|^3\, dx\, dt$$

$$\leq \frac{C}{R} \int_0^T \|u_\varepsilon\|_2 \|\nabla u_\varepsilon\|_2^2\, dt \leq \frac{C}{R}.$$

Last, $\|\tilde{q}_\varepsilon\|_{r^*,r} \le C_T$ with $r^* = \frac{12}{7}$, $r = \frac{6}{5}$ implies with $a = \frac{1}{3}$, that

$$\int_0^T \int_{\mathbb{R}^4} |\tilde{q}_\varepsilon||u_\varepsilon||\nabla\phi|\,dx\,dt \le \frac{C}{R}\int_0^T \|\tilde{q}_\varepsilon\|_{r^*}\|u_\varepsilon\|_{12/5}\,dt$$

$$\le \frac{C}{R}\int_0^T \|\tilde{q}_\varepsilon\|_{r^*}\|u_\varepsilon\|_2^{\frac{2}{3}}\|\nabla u_\varepsilon\|_2^{\frac{1}{3}}\,dt$$

$$\le \frac{C}{R}\|\tilde{q}_\varepsilon\|_{r^*,r}\|\nabla u_\varepsilon\|_{2,2}^{\frac{1}{3}}\|u_\varepsilon\|_{2,\infty}^{\frac{2}{3}} \le C_T R^{-1}.$$

We end with

$$(5)\qquad \sup_{t\le T}\int_{|x|\ge R} |u_\varepsilon(x,t)|^2\,dx \le 2\|a_\varepsilon - a\|_2^2 + \delta_T(R) := \gamma(\varepsilon, R)$$

with $\delta_T(R) \to 0$ for $R \to \infty$, independent of ε.

From [M-S] we know that a subsequence u_{ε_k} converges strongly in $L_2((0,T)\times B_R)$ for each R. Choosing first R, then ε_k, (5) implies that u_{ε_k} is also a Cauchy-sequence in $L_2((0,T)\times \mathbb{R}^4)$. As a consequence, $\lim_{k\to\infty}\|u_{\varepsilon_k}(t)\|_2 = \|u(t)\|_2$ for almost all t.

In the case that Ω is an exterior domain, we have to estimate the additional term

$$(6)\qquad \int_0^t\int_\Omega |q_\varepsilon||u_\varepsilon\phi||\nabla\phi|\,dx\,ds$$

where we used ϕ^2 instead of ϕ when testing the equation. The fact that $\|\nabla\phi\|_n$ does not decay with R causes an additional problem. From the definition of v_ε, we have

$$\int_s^T \|\nabla q_\varepsilon\|_p^p\,dt \le c\int_s^T \|Ae^{-tA}a_\varepsilon\|_p^p\,dt \le cs^{1-p}$$

as $a \in L_{p,\sigma}$ with some $p < 2$ (we may assume $p \ge \frac{4}{3}$). Let $0 < \alpha < 2 - p$, then

$$\int_0^T s^{-\alpha}\Big(\int_s^T \|\nabla q_\varepsilon\|_p^p\,dt\Big)\,ds \le cT^{2-p-\alpha}$$

and integration by parts gives

$$\int_0^T s^{1-\alpha}\|\nabla q_\varepsilon\|_p^p\,ds \le cT^{2-p-\alpha}.$$

By Lemma 3.2 of [M-S]

$$(7)\qquad \int_0^T s^{1-\alpha}\|q_\varepsilon\|_{p^*}^p\,ds \le C_T \text{ with } \frac{1}{p^*} = \frac{1}{p} - \frac{1}{4}.$$

Due to $a \in D(A^\rho)$, a similar argument gives

$$\int_s^T \|\nabla q_\varepsilon\|_2^2\, dt \le cs^{-1+2\rho},$$

hence

$$\int_0^T s^{1-\rho}\|q_\varepsilon\|_4^2\, ds \le cT^\rho. \tag{8}$$

Now we shall use (8) to estimate (6) for small t; we get with $\|\nabla\phi\|_4 \le c$:

$$\int_0^t \int_\Omega |q_\varepsilon||u_\varepsilon\phi||\nabla\phi|\, dx\, ds \le ct^{\frac{\rho}{2}}\Big(\int_0^t \|u_\varepsilon\phi\|_2^2 s^{\rho-1}\, ds\Big)^{\frac{1}{2}}.$$

Let $t_0 \le 1$ and $Y := \max_{t\le t_0} \|u_\varepsilon(t)\phi\|_2^2$; together with (5) we have

$$Y \le \gamma(\varepsilon, R) + ct_0^\rho Y^{\frac{1}{2}}$$

hence $Y \le 2\gamma(\varepsilon, R) + ct_0^{2\rho}$. Now

$$\begin{aligned}&\int_0^T \int_\Omega |q_\varepsilon||u_\varepsilon\phi||\nabla\phi|\, dx\, dt\\ &\le c\gamma(\varepsilon, R) + ct_0^{2\rho} + c\Big(\int_{t_0}^T \|q_\varepsilon\|_{p^*}\, dt\Big)\|\nabla\phi\|_m\|u_\varepsilon\|_{2,\infty}\\ &\le c\gamma(\varepsilon, R) + ct_0^{2\rho} + c(t_0^{-1})\cdot\|\nabla\phi\|_m\end{aligned}$$

with $m = \frac{4p}{3p-4} > 4$. Therefore

$$\int_\Omega |u_\varepsilon(x,t)|^2\phi^2\, dx \le c\gamma(\varepsilon, R) + ct_0^{2\rho} + c(t_0^{-1})R^{-2(\frac{2}{p}-1)}$$

and the same reasoning applies - just choose t_0 first, then R and then ε_k. This finishes the proof.

References

[B-M] W. Borchers - T. Miyakawa: Algebraic L_2 decay for Navier-Stokes flows in exterior domains, Acta Math. 165, 189-227(1990)

[G-M] G. Galdi - P. Maremonti: Monotonic decreasing and asymptotic behaviour of the kinetic energy for weak solutions of the Navier-Stokes equations in exterior domains, Archive Rat. Mech. & Anal. 94, 253-266(1986)

[M-S] T. Miyakawa - H. Sohr: On energy inequality, smoothness, and large time behaviour in L_2 for weak solutions of the Navier-Stokes equations, Math. Zeit. 199, 455-478(1988)

[Sch] M. E. Schonbek: Lower bounds of rates of decay for solutions to the Navier-Stokes equations, Journal of the AMS, to appear

[S-vW-W] H. Sohr - W. von Wahl - M. Wiegner: Zur Asymptotik der Gleichungen von Navier-Stokes, Nachr. der Akad. Wiss. Göttingen, Math. Phys. Kl. II, 3, 45-59(1986)

[V] B. da Veiga: On the suitable weak solutions to the Navier-Stokes equations in the whole space, J. Math. pures et appl. 64, 77-86(1985)

[W] M. Wiegner: Decay results for weak solutions of the Navier-Stokes equations in $\mathbb{R}^n$, J. London Math. Soc. 35, 303-313(1987).

This paper is in final form and no similar paper has been or is being submitted elsewhere.

On Chorin's Projection Method for the Incompressible Navier-Stokes Equations

Rolf Rannacher

Universität Heidelberg
Institut für Angewandte Mathematik
Im Neuenheimer Feld 293
D-6900 Heidelberg, Germany

Summary. Pseudo-compressibility methods are frequently used in computational fluid dynamics in order to cope with the algebraic difficulties caused by the incompressibility constraint. A popular example is the pressure stabilization (Petrov-Galerkin) method of T.J.R. Hughes, et al., which can be applied to the stationary as well as to the nonstationary Navier-Stokes problem. Also the classical projection method of A.J. Chorin can be interpreted as a variant of this method. This observation sheds some new light on the approximation properties of the projection method, particularly for the pressure.

1. Introduction

Let $\Omega \subset \mathbf{R}^n$ $(n = 2,3)$ be a bounded domain with a sufficiently regular boundary $\partial\Omega$. On the cylinder $\Omega\times[0,T]$, we consider the nonstationary Navier-Stokes problem

$$u_t - \nu\Delta u + (u\cdot\nabla)u + \nabla p = f\,, \quad \nabla\cdot u = 0\,, \quad \text{in } \Omega\times(0,T]\,, \qquad (1.1)$$
$$u_{|\partial\Omega} = 0\,, \quad u_{|t=0} = u^0\,.$$

Below, we will use the standard notation $L^2(\Omega)$, $H^k(\Omega)$, and $H_0^k(\Omega)$, for the Lebesgue and Sobolev spaces over Ω, and $(\cdot,\cdot)_\Omega$, $\|\cdot\|_\Omega$ and $\|\cdot\|_{k;\Omega}$ for the corresponding inner products and norms. $L_0^2(\Omega)$ is the subspace of all L^2-functions having mean-value zero. The subscript Ω is usually suppressed. Spaces of vector-valued functions are indicated by bold face type, e.g., $\mathbf{H}_0^1 = [H_0^1]^n$, while no distinction is made in the notation of inner products and norms. Further, the subspaces of solenoidal functions $\mathbf{J}_1 = \{v\in \mathbf{H}_0^1,\ \nabla\cdot v = 0\}$ and $\mathbf{J}_0 = \{v\in \mathbf{L}^2,\ \nabla\cdot v = 0\,,$ and $v\cdot n_{|\partial\Omega} = 0\,,$ weakly$\}$ are used. The $\mathbf{L}^2$-projection onto $\mathbf{J}_0$ is denoted by P.

This work has been supported by the Deutsche Forschungsgemeinschaft, SFB 123, Universität Heidelberg. This paper is in final form and no similar paper has been or is being submitted elsewhere.

One major computational difficulty in problem (1.1) arises from the incompressibility constraint which gives it a saddle point character. In the earlier days of computational fluid dynamics this has caused severe problems, particularly when a nonstationary flow had to be computed by an implicit time stepping scheme. One of the first approaches to overcome this problem was the so-called "projection method" of Chorin [2], [3] (see also Temam [14]). Here, for $m \geq 1$, one successively computes approximations

$$\text{i)}\quad \tilde{u}^m \in \mathbf{H}_0^1 : \quad \frac{1}{k}(\tilde{u}^m - u^{m-1}) - \nu\Delta\tilde{u}^m + (\tilde{u}^m \cdot \nabla)\tilde{u}^m = f^m \quad (\text{"Burgers step"}), \tag{1.2}$$

$$\text{ii)}\quad u^m \in \mathbf{J}_0 : \quad u^m = P\tilde{u}^m \quad (\text{"projection step"}), \tag{1.3}$$

where k is the time step and $u^0 \in \mathbf{J}_1$ is a suitable initial value. The intermediate approximations $\tilde{u}^m$ satisfy the correct boundary condition but violate the incompressibility constraint, while the u^m are divergence free but may have a non-zero tangential component. In view of the orthogonal splitting $\mathbf{L}^2 = \mathbf{J}_0 \oplus \{\nabla\psi, \psi \in H^1\}$ (see, e.g., [5]), the projection step (ii) can equivalently be written in the form $u^m = \tilde{u}^m - k\nabla p^m$, with some $p^m \in L_0^2 \cap H^1$, which is determined through the Neumann problem

$$\text{ii')}\quad \Delta p^m = \frac{1}{k}\nabla \cdot \tilde{u}^m, \text{ in } \Omega, \quad \partial_n p^m{}_{|\partial\Omega} = 0 . \tag{1.4}$$

The formulation (1.4) expresses that the pressure p^m in the projection method automatically satisfy the non-physical boundary condition $\partial_n p^m{}_{|\partial\Omega} = 0$. This has caused a lot of controversal discussion about the quality of these pressure approximations, which has even led to the believe that these are mere fictitious quantities without any physical relevancy (see, e.g., [15]). The main purpose of this paper is to show that this point of view is wrong. The quantities p^m are indeed proper approximations to the true pressure $p(t_m)$, even in the pointwise sense, in some distance to the boundary $\partial\Omega$. This view is also strongly supported in a recent paper of Gresho, et al., [6]. In fact, the practical success of the projection method has frequently been reported in the literature even for nonstationary flows with complicated vortex structures; see, e.g., Shen [12]. The best known theoretical result is the following error estimate due to Shen [11],

$$\|u^m - u(t_m)\| + \left(k\sum_{\mu=0}^{m} \|p^\mu - p(t_\mu)\|^2\right)^{1/2} = O(\sqrt{k}), \quad 0 < t_m \leq T, \tag{1.5}$$

which is proved for the classical projection method including a first order linearization in the step (i). Furthermore, for a slightly modified scheme Shen [13] obtained the optimal order $O(k)$ in (1.5). Whether this is true also for Chorin's original method was an open question. We will approach this problem through a re-interpretation of the projection method as a certain pressure stabilization method for which a well developped convergence analysis is available. To this end, we combine the relations (1.2) and (1.4) to obtain that

$$\frac{1}{k}(\tilde{u}^m - \tilde{u}^{m-1}) - \nu\Delta\tilde{u}^m + (\tilde{u}^m\cdot\nabla)\tilde{u}^m + \nabla p^{m-1} = f^m , \quad \tilde{u}^m{}_{|\partial\Omega} = 0 , \tag{1.6}$$

$$\nabla\cdot\tilde{u}^m - k\Delta p^m = 0 , \quad \partial_n p^m{}_{|\partial\Omega} = 0 , \tag{1.7}$$

where $p^0 = 0$. This scheme reminds of a simple pressure stabilization method, with a stabilization parameter $\varepsilon = k$ and an explicit coupling of the pressure in the momentum equation. In this method the quantities $\tilde{u}^m \in \mathbf{H}_0^1$ are viewed as the primary approximations to $u(t_m)$. Pseudo-compressibility methods have been frequently used in order to overcome the algebraic difficulties caused by the incompressibility constraint. In this approach additional terms are introduced in the continuity equation in order to stabilize the pressure, e.g.,

$$\nabla\cdot u + \varepsilon p_t = 0 , \quad p_{|t=0} = p^0 , \qquad \text{(artificial compressibility method)} \tag{1.8}$$

$$\nabla\cdot u + \varepsilon p = 0 , \qquad \text{(penalty method)} \tag{1.9}$$

$$\nabla\cdot u - \varepsilon\Delta p = 0 , \quad \partial_n p_{|\partial\Omega} = 0 . \qquad \text{(Petrov-Galerkin method)} \tag{1.10}$$

Obviously, Chorin's projection method can be interpreted as a pressure stabilization (Petrov-Galerkin) method, in which the pressure term in the momentum equation is treated explicitly and the quantities $\tilde{u}^m \in \mathbf{H}_0^1$ are viewed as the primary approximations to $u(t_m)$. As the main result we obtain that Chorin's original projection method is indeed of full first order accurate.

Theorem 1. *The classical projection method* (1.2), (1.3) *admits the error estimates*

$$\|u^m - u(t_m)\|_1 + \|p^m - p(t_m)\| = O(\sqrt{k}) , \tag{1.11}$$

$$\|u^m - u(t_m)\| + \|p^m - p(t_m)\|_{-1} = O(k) , \tag{1.12}$$

for $0 < t_m \le T$, *where* $\|\cdot\|_{-1}$ *denotes the norm of the dual space of* $L_0^2 \cap H^1$.

The proof of this result will be given below on the basis of a detailed error analysis for the pressure stabilization method. This analysis also indicates that the pressure in the projection method is indeed a reasonable approximation to the true pressure, at least in the interior of the domain Ω . In fact, the effect of the non-physical Neumann boundary condition decays exponentially with rate $\mathrm{dist}(x,\partial\Omega)/\sqrt{\nu k}$, and hence lives only in a narrow boundary layer of width $O(\sqrt{\nu k})$. A simple model example presented in [6] shows an analogous behavior.

2. Pressure Stabilization in the Stationary Stokes Problem

The pressure stabilization (Petrov-Galerkin) method was originally introduced by Hughes, et. al., [9], in the context of finite element discretizations of the steady Stokes and Navier-Stokes problem, in order to enhance the numerical stability properties of these schemes. For the stationary Stokes problem this method reads as follows,

$$\nu(\nabla u_h,\nabla\varphi) - (p_h,\nabla\cdot\varphi) = (f,\varphi)\,, \quad \forall\ \varphi\in \mathbf{H}_h\,, \tag{2.1}$$

$$(\nabla\cdot u_h,\chi) + h^2 \sum_{T\in \mathbf{T}_h} \alpha_T(\nabla p_h,\nabla\chi)_T = h^2 \sum_{T\in \mathbf{T}_h} \beta_T(f+\nu\Delta u_h,\nabla\chi)_T\,, \quad \forall\ \chi\in \mathbf{L}_h\,, \tag{2.2}$$

where $\alpha_T > 0$ and $\beta_T \geq 0$ are appropriate constants. This constitutes a stable discretization for most pairs of finite element trial spaces $\mathbf{H}_h\times\mathbf{L}_h \subset \mathbf{H}_0^1\times\{L_0^2\cap H^1\}$. In the case $\alpha_T = \beta_T$, the stabilization is fully consistent, since the exact solution satisfies $\nabla p = \nu\Delta u + f$ in Ω. Although, this method also implicitly imposes a Neumann boundary condition for the pressure in the discrete sense, "$\partial_n p_{h|\partial\Omega} = 0$", there was not much concern about this inconsistency. In fact, the method works well in practice and allows one to use very efficient solution procedures (e.g., pcg- or multi-grid methods). Furthermore, optimal order error estimates are available (see, e.g., [9], [16], and [7]),

$$\|u_h-u\|_1 + \|p_h-p\| = O(h)\,, \qquad \|u_h-u\| + \|p_h-p\|_{-1} = O(h^2)\,. \tag{2.3}$$

Corresponding estimates have also been proven in the pointwise sense (see [4], for $n = 2$, and [7], for $n = 3$). Furthermore, recent numerical results (see [1] and [7]) show that on almost uniform meshes an interior "super convergence" estimate holds true for the pressure,

$$\max_{x\in\Omega'} |(p_h-p)(x)| = O(h^2)\,, \quad \forall\ \Omega' \subset\subset \Omega\,. \tag{2.4}$$

On such meshes a refined error analysis leads to asymptotic error expansions of the form (see Blum [1]),

$$u_h(x) = u(x) + h^2 e^u(x) + h^2 E_h^u(x) + o(h^2)\,, \tag{2.5}$$

$$p_h(x) = p(x) + h^2 e^p(x) + h^2 E_h^p(x) + o(h^2)\,, \tag{2.6}$$

where e^u, e^p are functions independent of h, but E_h^u, E_h^p depend on the mesh like

$$\max_{x\in\Omega}|E_h^u(x)| \approx |\log(h)|^{3/2}\,, \quad \max_{x\in\Omega} |E_h^p(x)| \approx h^{-1}|\log(h)|\,. \tag{2.7}$$

The coefficients E_h^u, E_h^p are defined as the finite element approximations to certain Greens tensors satisfying $-\Delta E^u+\nabla E^p = 0$, $\nabla\cdot E^u = g_{\partial\Omega}(\cdot)$, in the distributional sense, where the functional $g_{\partial\Omega}(\cdot)$ is given in the form, for $\alpha_T \equiv \beta_T \equiv \alpha$,

$$g_{\partial\Omega}(\chi) = \int_{\partial\Omega} n\cdot\left(\tfrac{1}{12}D^2u - \alpha\Delta u\right)\chi\,ds\,, \quad \chi\in H^1(\Omega)\,. \tag{2.8}$$

Hence, E^p represents a Dirac measure concentrated along the boundary $\partial\Omega$. In view of this construction, the error coefficient E_h^p is suspected to decay exponentially like

$$E_h^p(x) \approx e^{-d(x)/(\sqrt{\nu}h)}\,, \quad d(x) = \mathrm{dist}(x,\partial\Omega)\,. \tag{2.9}$$

Such a behavior is well known, e.g., for the finite element L^2-projection of the Dirac point functional. Numerical tests indicate that indeed the error in the pressure is concentrated along the boundary $\partial\Omega$ and decays very rapidly (and smoothly) into the interior of Ω . Furthermore, it was observed in [1] that $D^2 = \Delta$, if bilinear elements are used on a uniform rectangular mesh. Therefore the particular choice $\alpha = 1/12$ implies $g_{\partial\Omega}(\cdot) \equiv 0$ and cancels out the "bad" pressure term $h^2E_h^p(x)$ in the expansion (2.6). This phenomenon was also observed in numerical tests. It particularly shows that, in contrast to the common believe, adding further terms in the continuity equation for enhancing the consistency in general cannot cure the problem of the bad pressure behavior along the boundary. The following plots (from [1]) show the error behavior of the pressure for a model Stokes problem with a polynomial solution. A globally smooth pressure approximation can be obtained simply by re-calculation of the "bad" boundary values by extrapolation from interior parts of the mesh. The resulting error is smooth enough to allow even for global h^2-defect correction.

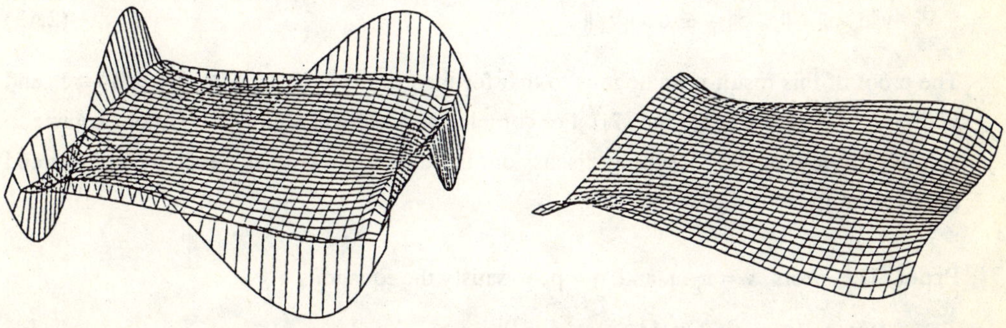

Figure 1. Pressure error plots for $\alpha = 1/10$ and $\alpha = 1/12$

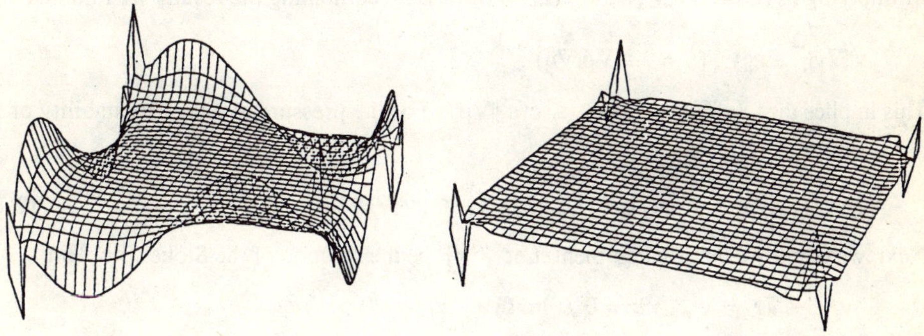

Figure 2. Pressure error plots (a) before, and (b) after h^2-defect correction (scaled by 5)

The above mentioned results for the pressure stabilization method also shed some new light on the projection method, particularly with regard to the pressure approximation. The following analysis is elaborated for a continuous situation in order to abstract from the technical complication of a particular discretization scheme,

$$-\nu\Delta u_\varepsilon + \nabla p_\varepsilon = f, \text{ in } \Omega, \quad u_{\varepsilon|\partial\Omega} = 0, \tag{2.10}$$

$$\nabla\cdot u_\varepsilon - \varepsilon\Delta p_\varepsilon = 0, \text{ in } \Omega, \quad \partial_n p_{\varepsilon|\partial\Omega} = 0. \tag{2.11}$$

However, all the results carry over to the spatially discrete case with minor modifications. Below, we will use the letter c for a positive generic constant which may vary with the context but only depends on Ω. The dependence on the viscosity ν will always be made explicit. We begin with the following basic result.

Theorem 2. *For the singularly perturbed problem* (2.10), (2.11), *there holds*

$$\nu\|u_\varepsilon - u\|_1 + \|p_\varepsilon - p\| + \sqrt{\nu\varepsilon}\|\nabla(p_\varepsilon - p)\| \le c\sqrt{\nu\varepsilon}\,\|\nabla p\|, \tag{2.12}$$

$$\nu\|u_\varepsilon - u\| + \|p_\varepsilon - p\|_{-1} \le c\nu\varepsilon\|\nabla p\|. \tag{2.13}$$

The proof of this result is analogous to that for the estimates (2.3), with $\alpha_T \equiv \beta_T \equiv 1$ and with $\varepsilon = h^2$ (see, e.g., [1], and [7]). For completeness we will supply the simple argument, below. In view of the foregoing discussion, the orders in ε in the estimates (2.12) and (2.13) seem to be best possible.

Proof. The errors $v = u_\varepsilon - u$ and $q = p_\varepsilon - p$ satisfy the equations

$$-\nu\Delta v + \nabla q = 0, \text{ in } \Omega, \quad v_{|\partial\Omega} = 0, \tag{2.14}$$

$$\nabla\cdot v - \varepsilon\Delta q = \varepsilon\Delta p, \text{ in } \Omega, \quad \partial_n q_{|\partial\Omega} = -\partial_n p_{|\partial\Omega}. \tag{2.15}$$

Multiplying in (2.14) by v, and in (2.15) by q, and combining the results we find that

$$\nu\|\nabla v\|^2 + \varepsilon\|\nabla q\|^2 = -\varepsilon(\nabla p, \nabla q).$$

This implies that $\sqrt{\nu}\|\nabla v\| + \sqrt{\varepsilon}\|\nabla q\| \le c\sqrt{\varepsilon}\,\|\nabla p\|$. For the pressure, we use the stability of the div-operator as follows,

$$\|q\| \le c\sup\left\{(q, \nabla\cdot\varphi),\ \varphi \in \mathbf{H}_0^1,\ \|\varphi\|_1 = 1\right\} \le c\nu\|\nabla v\|. \tag{2.16}$$

Next, we employ a duality argument. Let $\{z,r\}$ be the solution of the Stokes problem

$$-\nu\Delta z + \nabla r = v, \quad \nabla\cdot z = 0, \text{ in } \Omega, \quad z_{|\partial\Omega} = 0,$$

satisfying $\|\nabla r\| \le c\|v\|$. Then, there holds

$$\|v\|^2 = \nu(\nabla v, \nabla z) - (\nabla\cdot v, r) = \varepsilon(\nabla(q+p), \nabla r) \le \varepsilon\|\nabla p_\varepsilon\|\,\|\nabla r\|,$$

and, hence, $\|v\| \le c\varepsilon\|\nabla p\|$. Finally, using again the stability of the div-operator, we obtain

$$\|q\|_{-1} \le c \sup\left\{(q,\nabla\cdot\varphi),\ \varphi\in \mathbf{H}_0^1\cap\mathbf{H}^2,\ \|\varphi\|_2=1\right\} \le c\nu\|v\| , \tag{2.17}$$

which completes the proof. #

Next, we will prove a result which indicates that, indeed, the pressure error in the pressure stabilization method is significantly smaller in the interior of the domain Ω than along the boundary. To this end, let Δ_D and Δ_N denote the Laplacian operator corresponding to zero Dirichlet boundary conditions and to zero Neumann boundary conditions, respectively. Then, the error equations (2.14), (2.15) can be converted into the form

$$\left(\frac{1}{\nu}\nabla\cdot\Delta_D^{-1}\nabla - \varepsilon\Delta_N\right)q = \varepsilon\Delta p\,, \quad \text{in } \Omega\,, \quad \partial_n q_{|\partial\Omega} = -\partial_n p_{|\partial\Omega}\,, \tag{2.18}$$

in which the velocity error has been eliminated. The zero-order operator $\nabla\cdot\Delta_D^{-1}\nabla$ is defined on all of L^2 and positive definite. If we replace it in (2.18) by the identity operator, we arrive at the singularly perturbed Neumann problem

$$\left(\frac{1}{\nu}I - \varepsilon\Delta_N\right)\tilde{q} = \varepsilon\Delta p\,, \quad \text{in } \Omega\,, \quad \partial_n\tilde{q}_{|\partial\Omega} = -\partial_n p_{|\partial\Omega}\,. \tag{2.19}$$

For this simplified situation we have the following result.

Lemma 1. *For* $0<\delta<\mathrm{diam}(\Omega)/2$, *let* $\Omega_\delta = \{x\in\Omega,\ \mathrm{dist}(x,\partial\Omega)>\delta\}$. *Then, for the solution* $\tilde{q}\in L_0^2\cap H^1$ *of the singularly perturbed problem* (2.19), *there holds*

$$\|\tilde{q}\|_{\Omega_\delta} \le c e^{-\delta/(2\sqrt{\nu\varepsilon})}\sqrt{\nu\varepsilon}\|\nabla p\| + c\nu\varepsilon\|\Delta p\| , \tag{2.20}$$

where c *is a numerical constant independent of* Ω.

Proof. Let $d(x) = \mathrm{dist}\{x,\partial\Omega\}$, $\varepsilon' = \nu\varepsilon$, and $\sigma(x) = \min\{e^{d(x)/\sqrt{\varepsilon'}}, e^{\delta/\sqrt{\varepsilon'}}\}$. Then,

$$\|\tilde{q}\|_{\Omega_\delta}^2 + \varepsilon'\|\nabla\tilde{q}\|_{\Omega_\delta}^2 \le e^{-\delta/\sqrt{\varepsilon'}}\left\{(\sigma\tilde{q},\tilde{q}) + \varepsilon'(\sigma\nabla\tilde{q},\nabla\tilde{q})\right\} \equiv e^{-\delta/\sqrt{\varepsilon'}}A\,.$$

Further,

$$A = (\sigma\tilde{q},\tilde{q}) + \varepsilon'(\nabla(\sigma\tilde{q}),\nabla\tilde{q}) - \varepsilon'(\tilde{q}\nabla\sigma,\nabla\tilde{q}) ,$$

and, since $|\nabla\sigma| \le \sigma/\sqrt{\varepsilon'}$,

$$\varepsilon'|(\tilde{q}\nabla\sigma,\nabla\tilde{q})| \le \frac{1}{2}\left\{(\sigma\tilde{q},\tilde{q}) + \varepsilon'(\sigma\nabla\tilde{q},\nabla\tilde{q})\right\}.$$

Hence, absorbing terms and observing equation (2.19), we obtain

$$A \le 2\left\{(\sigma\tilde{q},\tilde{q}) + \varepsilon'(\nabla(\sigma\tilde{q}),\nabla\tilde{q})\right\} = 2\varepsilon'(\Delta p,\sigma\tilde{q}) + 2\varepsilon'(\sigma\tilde{q},\partial_n p)_{\partial\Omega}\,.$$

Since $\sigma = 1$ along $\partial\Omega$, there holds

$$\varepsilon'(\sigma\tilde{q},\partial_n p)_{\partial\Omega} = \varepsilon'(\nabla\tilde{q},\nabla p) + \varepsilon'(\tilde{q},\Delta p) \;=\; \varepsilon'(\nabla\tilde{q},\nabla p) + \varepsilon'(\tilde{q},\Delta p) \,.$$

From this we conclude that

$$(\sigma\tilde{q},\tilde{q}) + \varepsilon'(\nabla(\sigma\tilde{q}),\nabla\tilde{q}) \;\le\; 2A^{1/2}\Big\{\varepsilon'^2 e^{\delta/\sqrt{\varepsilon'}}\|\Delta p\|^2 + \varepsilon'\|\nabla p\|^2\Big\}^{1/2} ,$$

and, consequently,

$$A \le c\Big\{\varepsilon'^2 e^{\delta/\sqrt{\varepsilon'}}\|\Delta p\|^2 + \varepsilon'\|\nabla p\|^2\Big\} ,$$

which completes the proof. #

Notice that $\Delta p = \nabla\cdot f \equiv 0$, in the case of a conservative force. We believe that the estimate (2.20) also holds true for the pressure error $q = p_\varepsilon - p$, satisfying the original equation (2.18). This would imply that the pressure approximation p_ε has full first order accuracy in L^2 , and possibly even in L^∞ , at least outside a boundary layer of width

$$\delta = \sqrt{\nu\varepsilon}\,|\log(\nu\varepsilon)| \,. \tag{2.21}$$

Since we did not succeed proving this rigorously, we will supply the following somewhat weaker result concerning the interior super-approximation property of the pressure stabilization method.

Theorem 3. *For any subdomain* $\Omega' \subset\subset \Omega$, *there holds*

$$\|\nabla(u_\varepsilon - u)\|_{\Omega'} + \|p_\varepsilon - p\|_{\Omega'} \;\le\; c(\Omega')\nu\varepsilon\big\{\|\nabla p\| + \|\Delta p\|\big\} \,. \tag{2.22}$$

Proof. We set again $v = u_\varepsilon - u$ und $q = p_\varepsilon - p$, and choose some (non-negative) cut-off function $\sigma \subset C_0^\infty(\Omega)$ satisfying $\sigma_{|\Omega'} \equiv 1$. Multiplying in (2.14) by $\sigma^2 v$, and in (2.15) by $\sigma^2 q$, we obtain by a simple calculation that

$$\nu\|\sigma\nabla v\|^2 + \varepsilon\|\sigma\nabla q\|^2 \;=\; -\,\nu(v\nabla\sigma^2,\nabla v) + (q,v\nabla\sigma^2) - \varepsilon(\nabla q,q\nabla\sigma^2) + \varepsilon(\Delta p,\sigma^2 q) \,.$$

The four terms on the right are bounded as follows,

$$\nu|(v\nabla\sigma^2,\nabla v)| \;\le\; c\nu\|v\|^2 ,$$

$$|(q,v\nabla\sigma^2)| \;\le\; \frac{\alpha}{2\nu}\|\sigma q\|^2 + c_\alpha\nu\|v\|^2 \;,\quad \alpha > 0 ,$$

$$\varepsilon|(\nabla q,q\nabla\sigma)| \;\le\; c\varepsilon\|q\|^2 ,$$

$$\varepsilon|(\Delta p,\sigma q)| \;\le\; \frac{\alpha}{2\nu}\|\sigma q\|^2 + c_\alpha\nu\varepsilon^2\|\Delta p\|^2 \;,\quad \alpha > 0 \,.$$

Collecting these estimates, we obtain

$$\nu\|\sigma\nabla v\| \;\le\; \alpha\|\sigma q\| + c_\alpha\varepsilon\big\{\nu\|v\| + \sqrt{\nu\varepsilon}\|q\|\big\} + c_\alpha\nu\varepsilon\|\Delta p\| \,. \tag{2.23}$$

Now, we use again the stability of the div-operator to estimate,

$$\|\sigma q-\overline{\sigma q}\| \le c \sup\left\{(\sigma q-\overline{\sigma q},\nabla\cdot\varphi),\ \varphi\in \mathbf{H}_0^1(\Omega),\ \|\varphi\|_1=1\right\},$$

where $\overline{\sigma q}$ denotes the mean-value of σq over Ω. Since $|\overline{\sigma q}| \le c|\Omega|\,\|q\|_{-1}$, it follows that

$$\|\sigma q\| \le c \sup\left\{(\sigma q,\nabla\cdot\varphi),\ \varphi\in \mathbf{H}_0^1(\Omega),\ \|\varphi\|_1=1\right\} + c\|q\|_{-1}.$$

In view of (2.14), there holds

$$(\sigma q,\nabla\cdot\varphi) = (q,\nabla\cdot(\sigma\varphi)) - (q,\varphi\nabla\sigma) = \nu(\sigma\nabla v,\nabla\varphi) - \nu(\nabla v,\varphi\nabla\sigma) - (q,\varphi\nabla\sigma).$$

From this we conclude that

$$\|\sigma q\| \le c\left\{\nu\|\sigma\nabla v\| + \nu\|v\| + \|q\|_{-1}\right\}. \tag{2.24}$$

Combining (2.24) with (2.23) and choosing α sufficiently small, eventually yields

$$\nu\|\sigma\nabla v\| + \|\sigma q\| \le c\left\{\nu\|v\| + \sqrt{\nu\varepsilon}\|q\| + \|q\|_{-1}\right\} + c\nu\varepsilon\|\Delta p\|,$$

and, by the estimates of Theorem 1,

$$\nu\|\sigma\nabla v\| + \|\sigma q\| \le c\nu\varepsilon\left\{\|\nabla p\| + \|\Delta p\|\right\}. \tag{2.25}$$

Clearly, this completes the proof. #

We think that most of the foregoing results for the Stokes case in an appropriate sense carry over to the full nonstationary Navier-Stokes problem. These results for the pressure stabilization method could then be re-interpreted to apply also for the projection method. However, this program has not been completed yet. As a first step in the subsequent section we will establish the proof of Theorem 1 by showing that an optimal order (global) error estimate for the pressure stabilization method can be carried over to Chorin's projection method. The analysis of the "local" approximation properties of the projection method in the light of Theorems 2 and 3 will be the subject of further research.

3. The Projection Method for the Navier-Stokes Problem

The following analysis bases on optimal order error estimates for the pressure stabilization method applied to the full nonstationary Navier-Stokes problem,

$$u_{\varepsilon,t} - \nu\Delta u_\varepsilon + (u_\varepsilon\cdot\nabla)u_\varepsilon + \nabla p_\varepsilon = f, \ \text{ in } \Omega, \quad u_{\varepsilon|\partial\Omega} = 0, \quad u_{\varepsilon|t=0} = u^0, \tag{3.1}$$

$$\nabla\cdot u_\varepsilon - \varepsilon\Delta p_\varepsilon = 0, \ \text{ in } \Omega, \quad \partial_n p_{\varepsilon|\partial\Omega} = 0. \tag{3.2}$$

Complete proofs of these results will be presented in [10]. We note that alternatively to (3.1), one could approximate the convective term in a conservative form, $(u_\varepsilon\cdot\nabla)u_\varepsilon - (\nabla\cdot u_\varepsilon)u_\varepsilon/2$, without changing the approximation properties of the method. Below, we use the symbol C_T for a generic constant which depends on the solution $\{u,p\}$ of the Navier-Stokes equations only through the quantities $\|u^0\|_2$ and $\sup_{[0,T]}\{\|f\|+\|f_t\|+\|f_{tt}\|\}$, and a pre-assumed bound for $\sup_{[0,T]}\|\nabla u\|$. Through the use of the Gronwall inequality C_T generally grows exponentially with T and $1/\nu$. Further, we use the time weights $\tau = \min\{t,1\}$ and $\tau_m = \min\{t_m,1\}$, in order to describe a possible blow up of the error constants as $t \to 0$.

Theorem 4. *There holds, for* $0 < t \le T$,

$$\nu\|(u_\varepsilon-u)(t)\|_1 + \sqrt{\tau}\|(p_\varepsilon-p)(t)\| \le C_T\sqrt{\varepsilon}\,, \tag{3.3}$$

$$\nu\|(u_\varepsilon-u)(t)\| + \tau\|(p_\varepsilon-p)(t)\|_{-1} \le C_T\varepsilon\,. \tag{3.4}$$

The τ-factors in front of the pressure errors are due to the fact, that in the proof of Theorem 4 the negative Sobolev norm $\|u_{\varepsilon,t}-u_t\|_{-1}$ is estimated crudely by the L^2-norm $\|u_{\varepsilon,t}-u_t\|$. Whether this is necessary must be left as an open question; see also [8, I,II,IV] for a related problem.

Next, we consider the time discretization of the perturbed problem (3.1), (3.2), by the backward Euler scheme. Starting with $u_\varepsilon^0 = u^0$, let approximations u_ε^m to $u(t_m)$, $m \ge 1$, be successively defined through

$$d_t u_\varepsilon^m - \nu\Delta u_\varepsilon^m + (u_\varepsilon^m\cdot\nabla)u_\varepsilon^m + \nabla p_\varepsilon^m = f^m\,, \text{ in } \Omega\,, \quad u_{\varepsilon\,|\partial\Omega}^m = 0\,, \tag{3.5}$$

$$\nabla\cdot u_\varepsilon^m - \varepsilon\Delta p_\varepsilon^m = 0\,, \text{ in } \Omega\,, \quad \partial_n p_{\varepsilon\,|\partial\Omega}^m = 0\,. \tag{3.6}$$

where $d_t u_\varepsilon^m = (u_\varepsilon^m - u_\varepsilon^{m-1})/k$. For this scheme we have the following result.

Theorem 5. *There holds, for* $0 < t_m \le T$,

$$\nu\|u_\varepsilon^m-u(t_m)\|_1 + \sqrt{\tau_m}\,\|p_\varepsilon^m-p(t_m)\| \le C_T\{\sqrt{k}+\sqrt{\varepsilon}\}\,, \tag{3.7}$$

$$\nu\|u_\varepsilon^m-u(t_m)\| + \tau_m\,\|p_\varepsilon^m-p(t_m)\|_{-1} \le C_T\{k+\varepsilon\}\,. \tag{3.8}$$

As it was discussed in the introduction, the linearized projection method can be viewed as a perturbation of the time stepping scheme (3.5), (3.6), with the special choice $\varepsilon = k$. In the sequel, we will analyse the effect of these perturbations in some detail. There are several possibilities of handling the nonlinear convective term. The standard linearization within the framework of the pressure stabilization approach is

$$(u^m\cdot\nabla)u^m \approx (u^{m-1}\cdot\nabla)u^m\,, \tag{3.9}$$

or, in a conservative form,

$$(u^m\cdot\nabla)u^m - \frac{1}{2}(\nabla\cdot u^m)u^m \approx (u^{m-1}\cdot\nabla)u^m - \frac{1}{2}(\nabla\cdot u^{m-1})u^m . \tag{3.10}$$

The latter will not be pursued further since this analysis is quite similar to that for the simpler version (3.1) and leads to the same results. From the view point of the projection method the linearization

$$(u^m\cdot\nabla)u^m \approx (Pu^{m-1}\cdot\nabla)u^m \tag{3.11}$$

is more natural, since the projection $Pu^{m-1}\in J_0$ is known from the preceding time step.

Lemma 3. *If the convective term in the scheme* (3.5), (3.6) *is linearized accordingly either to* (3.9) *or to* (3.10), *the estimates* (3.7) *and* (3.8) *remain valid.*

This may be proved by the same type of argument as it is commonly used in the analysis of first order linearization processes for the Navier-Stokes problem; see, e.g., [8,IV] . For the sake of brevity, we suppress the proof and refer to [10] for further details. We will rather concentrate on the most significant difference between the projection method and the scheme (3.5), (3.6), which consists in the explicit coupling of the pressure term in the momentum equation. In view of the preceding discussion the following lemma establishes the proof of our main result stated in Theorem 1.

Lemma 4. *Let* $\{u^m,p^m\}$ *denote the solutions of the fully implicit scheme* (3.5), (3.6), *with* $\varepsilon = k$, *and* $\{\tilde{u}^m,\tilde{p}^m\}$ *the solutions of the projection method* (1.6), (1.7). *Then, for* $0 < t_m \le T$, *there holds*

$$\nu\|u^m-\tilde{u}^m\|_1 + \sqrt{\tau_m}\|p^m-\tilde{p}^m\| \le C_T\sqrt{k} , \tag{3.12}$$

$$\nu\|u^m-\tilde{u}^m\| + \tau_m\|p^m-\tilde{p}^m\|_{-1} \le C_T k . \tag{3.13}$$

Proof. We give the proof only for the case of the linear Stokes problem. The additional nonlinear convective term in the Navier-Stokes problem can be treated in the standard way by using certain Sobolev inequalities and absorbing the resulting lower order terms into the dominating dissipative terms. Then, the use of the Gronwall inequality results in an exponential growth of the error constant C_T with respect to T and to $1/\nu$; see [8,IV] for a similar analysis for a standard time discretization of the Navier-Stokes problem. Analogous arguments will also be used below in the proof of Lemma 5.

We begin with deriving an estimate for the velocity error. The error terms $v^m = \tilde{u}^m - u^m$ and $q^m = \tilde{p}^m - p^m$ satisfy $v^0 = 0$, $q^0 = 0$, and

$$d_t v^m - \nu\Delta v^m + \nabla q^{m-1} = - k\nabla d_t p^m , \quad v^m{}_{|\partial\Omega} = 0 , \tag{3.14}$$

$$\nabla\cdot v^m - k\Delta q^m = 0 , \quad \partial_n q^m{}_{|\partial\Omega} = 0 . \tag{3.15}$$

Multiplying in (3.14) by v^m, and in (3.15) by q^{m-1}, and combining the results yields

$$d_t\|v^m\|^2 + k\|d_t v^m\|^2 + 2\nu\|\nabla v^m\|^2 + 2k(\nabla q^{m-1},\nabla q^m) = - 2k(d_t p^m,\nabla\cdot v^m) .$$

By rearranging terms we transform this into

$$d_t\|v^m\|^2 + k\|d_t v^m\|^2 + 2\nu\|\nabla v^m\|^2 + 2k\|\nabla q^m\|^2 = 2k^2(\nabla q^m,\nabla d_t q^m) - 2k(d_t p^m,\nabla\cdot v^m) .$$

For the two terms on the right we use the relation (3.15),

$$2k^2(\nabla q^m,\nabla d_t q^m) - 2k(d_t p^m,\nabla\cdot v^m) = 2k(\nabla q^m,d_t v^m) + 2k^2(\nabla d_t p^m,\nabla q^m) ,$$

and, by absorbing terms, obtain

$$d_t\|v^m\|^2 + \nu\|\nabla v^m\|^2 + k\|\nabla q^m\|^2 \le ck^3\|\nabla d_t p^m\|^2 . \tag{3.16}$$

Then, summing this over m, yields

$$\|v^m\|^2 + k\sum_{\mu=1}^{m}\left(\nu\|\nabla v^\mu\|^2 + k\|\nabla q^\mu\|^2\right) \le ck^4\sum_{\mu=1}^{m}\|\nabla d_t p^\mu\|^2 . \tag{3.17}$$

Next, we multiply in (3.14) by $d_t v^m$, and in (3.15) by q^{m-1}, and combine the results to

$$2\|d_t v^m\|^2 + \nu d_t\|\nabla v^m\|^2 + \nu k^2\|\nabla d_t v^m\|^2 + 2k(\nabla q^{m-1},\nabla d_t q^m) = - 2k(\nabla d_t p^m,d_t v^m) .$$

Then, using the relation (3.15) and

$$(\nabla q^{m-1},\nabla d_t q^m) = \frac{1}{2}d_t\|\nabla q^m\|^2 - \frac{k}{2}\|\nabla d_t q^m\|^2 ,$$

we conclude that

$$d_t\left(\nu\|\nabla v^m\|^2 + k\|\nabla q^m\|^2\right) + \|d_t v^m\|^2 \le ck^2\left(\|\nabla d_t q^m\|^2 + (\nabla d_t p^m,\nabla d_t q^m)\right) . \tag{3.18}$$

Summing this over m, particularly yields

$$\nu\|\nabla v^m\|^2 + k\|\nabla q^m\|^2 + k\sum_{\mu=1}^{m}\|d_t v^\mu\|^2 \le ck^3\sum_{\mu=1}^{m}\left(\|\nabla d_t q^\mu\|^2 + \|\nabla d_t p^\mu\|^2\right) .$$

From this relation and observing additionally (3.17), we obtain

$$\nu\|\nabla v^m\|^2 + k\sum_{\mu=1}^{m}\|d_t v^\mu\|^2 \le ck^3\sum_{\mu=1}^{m}\|\nabla d_t p^\mu\|^2 . \tag{3.19}$$

In virtue of the a priori estimate of Lemma 5, below, it follows from (3.17) and (3.19) that

$$\|v^m\|^2 + \nu k\|\nabla v^m\|^2 + k\sum_{\mu=1}^{m}\left(\nu\|\nabla v^\mu\|^2 + k\|\nabla q^\mu\|^2 + k\|d_t v^\mu\|^2\right) \le C_T k^2 . \tag{3.20}$$

For the pressure error we use again the stability properties of the div-operator to obtain

$$\|q^{m-1}\|_{-1} \le c\{\|d_t v^m\|_{-2} + \nu\|v^m\| + k\|d_t p^m\|_{-1}\}, \tag{3.21}$$

$$\|q^{m-1}\| \le c\{\|d_t v^m\|_{-1} + \nu\|\nabla v^m\| + k\|d_t p^m\|\}. \tag{3.22}$$

For the last terms on the right in (3.21) and (3.22) we have, again in view of Lemma 5 and since $\tau_m \ge k$,

$$k\|d_t p^m\|_{-1} \le C_T \tau_m^{-1} k, \qquad k\|d_t p^m\| \le C_T \tau_m^{-1/2}\sqrt{k}. \tag{3.23}$$

The second terms on the right we have already shown to be of the order $O(k)$ and $O(\sqrt{k})$, respectively. On the first terms on the right, however, we use the crude estimate $\|d_t v^m\|_{-2} + \|d_t v^m\|_{-1} \le c\|d_t v^m\|$, since we do not see how to make use of the "negative" Sobolev norm. This will result in the blow up of the error constant of the pressure as $t \to 0$.

The estimate of the L^2-norm $\|d_t v^m\|$ requires a series of steps. We begin with taking difference quotients in (3.14) and (3.15),

$$d_t^2 v^m - \nu\Delta d_t v^m + \nabla d_t q^{m-1} = k d_t^2 \nabla p^m, \tag{3.24}$$

$$\nabla \cdot d_t v^m - k\Delta d_t q^m = 0. \tag{3.25}$$

Then, we multiply in (3.24) by $d_t v^m$, and in (3.25) by $d_t q^{m-1}$, and combine the results to

$$d_t\|d_t v^m\|^2 + k\|d_t^2 v^m\|^2 + 2\nu\|\nabla d_t v^m\|^2 + 2k(\nabla d_t q^m, \nabla d_t q^{m-1}) = -2k(d_t^2 p^m, d_t \nabla\cdot v^m).$$

Rearranging terms leads us to

$$d_t\|d_t v^m\|^2 + k\|d_t^2 v^m\|^2 + 2\nu\|\nabla d_t v^m\|^2 + 2k\|\nabla d_t q^m\|^2 =$$

$$= 2k(\nabla d_t q^m, \nabla d_t^2 q^m) - 2k(d_t^2 p^m, d_t \nabla\cdot v^m).$$

In view of (3.25), we have

$$2k(\nabla d_t q^m, \nabla d_t^2 q^m) - 2k(d_t^2 p^m, d_t \nabla\cdot v^m) = 2k(d_t^2 v^m, \nabla d_t q^m) + 2k^2(\nabla d_t^2 p^m, \nabla d_t q^m).$$

Then, absorbing terms yields

$$d_t\|d_t v^m\|^2 + \nu\|\nabla d_t v^m\|^2 + k\|\nabla d_t q^m\|^2 \le ck^3\|\nabla d_t^2 p^m\|^2. \tag{3.26}$$

We multiply this by τ_m^r, for $r = 1,2$,

$$d_t\{\tau_m^r\|d_t v^m\|^2\} + \tau_m^r\{\nu\|\nabla d_t v^m\|^2 + k\|\nabla d_t q^m\|^2\} \le c\tau_m^r k^3\|\nabla d_t^2 p^m\|^2 + r\tau_m^{r-1} k\|d_t v^m\|^2,$$

and sum over m, to obtain

$$\tau_m^r\|d_tv^m\|^2 + k\sum_{\mu=2}^{m} \tau_\mu^r\left\{\nu\|\nabla d_tv^\mu\|^2 + k\|\nabla d_tq^\mu\|^2\right\} \le \tag{3.27}$$

$$\le k^{r-2}\|v^1\|^2 + ck^4\sum_{\mu=2}^{m} \tau_\mu^r\|\nabla d_t^2p^\mu\|^2 + rk\sum_{\mu=2}^{m} \tau_\mu^{r-1}\|d_tv^\mu\|^2 .$$

In virtue of (3.20), the first term on the right is bounded by C_Tk^r. From (3.27), with $r = 2$, and (3.19), with $m = 1$, we conclude that

$$\tau_m^2\|d_tv^m\|^2 + k^2\sum_{\mu=2}^{m} \tau_\mu^2\|\nabla d_tq^\mu\|^2 \le ck^4\sum_{\mu=2}^{m} \tau_\mu^2\|\nabla d_t^2p^\mu\|^2 + 2k\sum_{\mu=2}^{m} \tau_\mu\|d_tv^\mu\|^2 . \tag{3.28}$$

To estimate the last term on the rigt hand side, we multiply in (3.18) by τ_m, to obtain

$$d_t\left(\tau_m\left\{\nu\|\nabla v^m\|^2 + k\|\nabla q^m\|^2\right\}\right) + \tau_m\|d_tv^m\|^2 \le$$

$$\le ck^2\tau_m\left\{\|\nabla d_tq^m\|^2 + (\nabla d_tp^m,\nabla d_tq^m)\right\} + \nu\|\nabla v^m\|^2 + k\|\nabla q^m\|^2 .$$

Summing this over m, and observing (3.17), with $m = 1$, and (3.20), we conclude that

$$k\sum_{\mu=2}^{m} \tau_\mu\|d_tv^\mu\|^2 \le C_Tk^2 + k\sum_{\mu=2}^{m}\left\{(\alpha k\tau_\mu^2 + ck^2\tau_\mu)\|\nabla d_tq^\mu\|^2 + \frac{1}{\alpha}k^3\|\nabla d_tp^\mu\|^2\right\}, \tag{3.29}$$

for some $\alpha > 0$, which will be appropriately chosen below. Again from (3.27), this time with $r = 1$, we obtain

$$k^2\sum_{\mu=2}^{m} \tau_\mu\|\nabla d_tq^\mu\|^2 \le k^{-1}\|v^1\|^2 + ck^4\sum_{\mu=2}^{m} \tau_\mu\|\nabla d_t^2p^\mu\|^2 + k\sum_{\mu=2}^{m} \|d_tv^\mu\|^2 ,$$

and, consequently, in view of (3.20),

$$k^3\sum_{\mu=2}^{m} \tau_\mu\|\nabla d_tq^\mu\|^2 \le C_Tk^2 + ck^5\sum_{\mu=2}^{m} \tau_\mu\|\nabla d_t^2p^\mu\|^2 . \tag{3.30}$$

Combining this with (3.29) and trading powers of k for powers of τ_m, yields

$$k\sum_{\mu=2}^{m} \tau_\mu\|d_tv^\mu\|^2 \le C_Tk^2 + \alpha k^2\sum_{\mu=2}^{m} \tau_\mu^2\|\nabla d_tq^\mu\|^2 + ck^4\sum_{\mu=2}^{m}\left\{\tau_\mu^2\|\nabla d_t^2p^\mu\|^2 + \frac{1}{\alpha}\|\nabla d_tp^\mu\|^2\right\}.$$

Now we insert this into (3.28) and choose α sufficiently small to obtain

$$\tau_m^2\|d_tv^m\|^2 \le C_Tk^2 + ck^4\sum_{\mu=2}^{m}\left\{\tau_\mu^2\|\nabla d_t^2p^\mu\|^2 + \|\nabla d_tp^\mu\|^2\right\}. \tag{3.31}$$

Hence, in view of Lemma 5, it follows that

$$\|d_tv^m\| \le C_T\tau_m^{-1}k . \tag{3.32}$$

Using this in (3.21) and (3.22) completes the proof. #

Finally, we supply the a priori estimates for the solution $\{u^m,p^m\}$ of the time stepping scheme (3.5), (3.6), with $\varepsilon = k$, which have been used in the proof of Lemma 4.

Lemma 5. *For the solution* $\{u^m,p^m\}$ *of the scheme* (3.5), (3.6), *with* $\varepsilon = k$, *there holds the a priori estimate*

$$\tau_m\|d_t p^m\|^2 + k^2\sum_{\mu=2}^{m}\Big\{\|\nabla d_t p^\mu\|^2 + \tau_\mu^2\|\nabla d_t^2 p^\mu\|^2\Big\} \le C_T . \tag{3.33}$$

Proof. Since the argument for proving these estimates is rather standard, we only sketch the main steps. For the convective term we make frequent use of the estimates

$$|(u\cdot\nabla v,w)| \le c\|u\|_\alpha\|v\|_\beta\,\|w\|_\gamma, \quad u,v,w \in \mathbf{H}_0^1, \tag{3.34}$$

where $\alpha,\beta,\gamma \in \{0,1,2\}$, with $\alpha+\beta+\gamma = 3$. These are are easily proved for $n \in \{2,3\}$, by using various Hölder-, Poincaré- and Sobolev-inequalities. Norms involving the force term, e.g., $\|d_t f^m\|$, are always bounded by C_T.

In the first step we provide some lower order a priori estimates. The error estimates of Theorems 2 and 3 imply that

$$\max_{0\le t_m\le T}\{\|\nabla u^m\| + \|d_t u^m\|\} \le C_T, \tag{3.35}$$

and, furthermore,

$$\max_{0\le t_m\le T}\{\|u^m\|_2\} \le C_T . \tag{3.36}$$

For (3.36), we use (3.5) and (3.35) to get a bound for $\|\Delta u^m\|$ in terms of $\|\nabla p^m\|$. Through (3.6) the latter is related to $\|u^m-u_k^m\|/k$, and this, in turn, can be bounded by using the error estimate (3.8) for the pressure stabilization method. From the identities (3.5) and (3.6), with $m = 1$, we conclude the initial estimates

$$\|d_t u^\mu\| + \sqrt{\nu k}\|\nabla d_t u^\mu\| + k\|\nabla d_t p^\mu\| \le C_T, \quad \mu = 1,2 . \tag{3.37}$$

Below, the bounds (3.35), (3.36) and (3.37) will frequently be used without mentioning.

The proof of the estimate (3.33) is now accomplished in a series of steps. Taking difference quotients in (3.5) and (3.6) yields

$$d_t^{r+1}u^m - \nu\Delta d_t^r u^m + \nabla d_t^r p^m = d_t^r f^m - d_t^r[(u^m\cdot\nabla)u^m], \quad r \ge 0, \tag{3.38}$$

$$\nabla\cdot d_t^s u^m - k\Delta d_t^s p^m = 0, \quad s \ge 0 . \tag{3.39}$$

First, we take $r = 1$, $s = 1$, and multiply in (3.35) by $d_t u^m$, and in (3.36) by $d_t p^m$, to obtain

$$d_t\|d_t u^m\|^2 + \nu\|\nabla d_t u^m\|^2 + k\|\nabla d_t p^m\|^2 \le C_T .$$

Summing this over m, and observing (3.34), yields the first result

$$k\sum_{\mu=1}^{m}\left\{\nu\|\nabla d_t u^m\|^2 + k\|\nabla d_t p^m\|^2\right\} \le C_T . \tag{3.38}$$

Now, we take $r = 1$, $s = 2$, and multiply in (3.35) by $d_t^2 u^m$, and in (3.36) by $d_t^2 p^m$, and combine the results to obtain

$$\|d_t^2 u^m\|^2 + d_t\left(\nu\|\nabla d_t u^m\|^2 + k\|\nabla d_t p^m\|^2\right) \le C_T + \nu\|\nabla d_t u^m\|^2 .$$

We multiply this by τ_m and obtain

$$\tau_m\|d_t^2 u^m\|^2 + d_t\left(\nu\tau_m\|\nabla d_t u^m\|^2 + k\tau_m\|\nabla d_t p^m\|^2\right) \le C_T + 2\nu\|\nabla d_t u^m\|^2 + k\|\nabla d_t p^m\|^2 .$$

Summing this over m, and observing (3.38), yields

$$\nu\tau_m\|\nabla d_t u^m\|^2 + k\sum_{\mu=2}^{m}\tau_\mu\|d_t^2 u^\mu\|^2 \le C_T . \tag{3.39}$$

Finally, we take $r = 2$, $s = 2$, and multiply in (3.35) by $d_t^2 u^m$, and in (3.36) by $d_t^2 p^m$, and combine the results to

$$d_t\|d_t^2 u^m\|^2 + \nu\|\nabla d_t^2 u^m\|^2 + k\|\nabla d_t^2 p^m\|^2 \le C_T\left\{1 + \|d_t^2 u^m\|^2 + \|\nabla d_t u^m\|^2\right\} .$$

We multiply this by τ_m^2 and obtain

$$d_t\left(\tau_m^2\|d_t^2 u^m\|^2\right) + \tau_m^2\left\{\nu\|\nabla d_t^2 u^m\|^2 + k\|\nabla d_t^2 p^m\|^2\right\} \le C_T\tau_m\left\{1 + \|d_t^2 u^m\|^2 + \|\nabla d_t u^m\|^2\right\} .$$

Summing this over m, and observing (3.38) and (3.39), then yields the second result

$$\tau_m^2\|d_t^2 u^m\|^2 + k^2\sum_{\mu=3}^{m}\tau_\mu^2\|\nabla d_t^2 p^\mu\|^2 \le C_T . \tag{3.40}$$

Finally, we use the stability of the div-operator to obtain

$$\|d_t p^m\| \le c\left\{\|d_t^2 u^m\| + \nu\|\nabla d_t u^m\| + \|d_t[(u^m\cdot\nabla)u^m]\| + \|d_t f^m\|\right\} .$$

Consequently, in virtue of the foregoing estimates, we obtain the last result

$$\tau_m\|d_t p^m\| \le C_T . \tag{3.41}$$

This completes the proof. #

References

1. Blum,H.: Asymptotic error expansion and defect correction in the finite element method. Habilitationsschrift, Universität Heidelberg, 1991.
2. Chorin,A.J.: Numerical solution of the Navier-Stokes equations. Math.Comp. 22, 745-762 (1968).
3. Chorin,A.J.: On the convergence of discrete approximations of the Navier-Stokes equations. Math.Comp. 23, 341-353 (1969).
4. Durán,R., Nochetto,R.H.: Pointwise accuracy of a stable Petrov-Galerkin approximation to the Stokes problem. SIAM J.Numer.Anal. 26, 1395-1406 (1989).
5. Girault,V., Raviart, P.A.: Finite Element Methods for Navier-Stokes Equations. Springer, Berlin-Heidelberg 1986.
6. Gresho,P.M.: On the theory of semi-implicit projection methods for viscous incompressible flow and its implementation via a finite element method that also introduces a nearly consistent mass matrix. Part 1: Theory. Int.J.Numer.Meth.Fluids 11, 621-659 (1990). Part 2: Implementation. Int.J.Numer.Meth.Fluids 11, 587-620 (1990).
7. Harig,J.: Eine robuste und effiziente Finite-Elemente-Methode zur Lösung der inkompressiblen 3-D-Navier-Stokes-Gleichungen auf Vektorrechnern. Dissertation, Universität Heidelberg, 1991.
8. Heywood.J.G., Rannacher,R.: Finite element approximation of the nonstationary Navier-Stokes problem. I. Regularity of solutions and second order error estimates for spatial discretization. SIAM J.Numer.Anal. 19, 275-311 (1982). II. Stability of solutions and error estimates uniform in time. SIAM J.Numer.Anal. 23, 750-777 (1986). III. Smoothing property and higher order estimates for spatial discretization. SIAM J.Numer.Anal. 25, 489-512 (1988). IV. Error analysis for second-order time discretization. SIAM J.Numer. Anal. 27, 353-384 (1990).
9. Hughes,T.J.R., Franca,L.P., Balestra,M.: A new finite element formulation for computational fluid mechanics: V. Circumventing the Babuska-Brezzi condition: A stable Petrov-Galerkin formulation of the Stokes problem accommodating equal order interpolation. Comp.Meth.Appl.Mech.Eng. 59, 85-99 (1986).
10. Rannacher,R.: On pseudo-compressibility methods for the incompressible Navier-Stokes equations. In preparation.
11. Shen,J.: On error estimates of projection methods for the Navier-Stokes equations: First order schemes. To appear in SIAM J.Numer.Anal. 1991.
12. Shen,J.: Hopf bifurcation of the unsteady regularized driven cavity flows. Preprint Dept. of Mathematics, Indiana University-Bloomington, to appear in J.Comp.Phys. 1991.
13. Shen,J.: On error estimates of higher order projection and penalty-projection methods for Navier-Stokes equations. Indiana University-Bloomington, to appear in J.Comp. Phys. 1991.
14. Temam,R.: Sur l'approximation de la solution des equations de Navier-Stokes par la méthode des pas fractionaires II. Arch.Rati.Mech.Anal. 33, 377-385 (1969).
15. Temam,R.: Remark on the pressure boundary condition for the projection method. Technical Note 1991, to appear in Theoretical and Computational Fluid Dynamics.
16. Verfürth,R.: On the stability of Petrov-Galerkin formulations of the Stokes equations. Preprint, Universität Zürich, 1990.

Analysis of the spectral Lagrange-Galerkin method for the Navier-Stokes equations [1]

Endre Süli and Antony F. Ware
Oxford University Computing Laboratory
11 Keble Road, Oxford OX1 3QD

1 Introduction

The spectral Lagrange-Galerkin method for advection-dominated diffusion problems, introduced in [9], is a direct descendant of the finite element Lagrange-Galerkin method (or transport-diffusion algorithm), which has been the subject of extensive research over the past decade [1], [5], [6], [7]. The underlying motivation for the introduction of this method has been the observation that advection-dominated diffusion problems exhibit a nearly hyperbolic behaviour. Thus the advection term is treated in a manner related to the method of characteristics for hyperbolic problems, whereby the Lagrangian material derivative appearing in the advection term is discretised *along particle trajectories*. The resulting schemes possess remarkable stability properties, and exhibit lower time truncation errors than standard time stepping procedures.

The *finite element* Lagrange-Galerkin approximation of the Navier-Stokes equations has been analysed by Pironneau [5] and subsequently by Süli in [7], where optimal-order error estimates have been obtained in both the L^2 and H^1 norms and the non-linear stability of the scheme has been demonstrated subject to certain constraints on the size of the time step in terms of the mesh spacing. Unfortunately, quadrature is necessary in the practical implementation of the method, and it has been shown that the resulting scheme is unstable for linear advection and advection-diffusion problems [4], [8]. On the other hand, for linear problems, the *spectral* Lagrange-Galerkin method remains unconditionally stable even with the use of quadrature, [9]. Our aim in this paper is to describe the spectral Lagrange-Galerkin approximation of the Navier-Stokes equations, and to derive L^2 and H^1 error estimates that correspond to those in [7]. The stability and convergence results obtained are subject to weaker constraints on the time step than for the finite element version of the method. In particular, the spectral Lagrange-Galerkin method is shown to be unconditionally nonlinearly stable. The effect of quadrature in the nonlinear case will be the subject of future investigation. A related technique, combining a high order Lagrangian time stepping procedure with a spectral element method, has been proposed by Ho *et al.* [2]. The spectral Lagrange-Galerkin method has been successfully implemented on two dimensional problems. These results, together with further details of the implementation, will be published elsewhere [12].

1.1. Notation. We set $\Omega = (0, 2\pi)^n$; $H^s_{\#}(\Omega)$ is the space of restrictions to Ω of real-valued 2π-periodic distributions which are in the Sobolev-Slobodetskii space $H^s(\Theta)$ for

[1]This paper is in final form and no similar paper has been or is being submitted elsewhere.

every open bounded set Θ. On these spaces we can use the Fourier series representation

$$u(x) = \sum_{p \in \mathbf{Z}^n} \hat{u}(p)\phi_p(x),$$

where $\phi_p(x) = e^{ip \cdot x}$, $x \in \Omega$, and $\overline{\hat{u}(p)} = \hat{u}(-p)$ since u is real. Then $u \in L^2(\Omega)$ if and only if $\sum_p |\hat{u}(p)|^2 < \infty$, and $u \in H^s_{\#}(\Omega)$ if and only if $\sum_p (1 + |p|^2)^s |\hat{u}(p)|^2 < \infty$. The square root of the last expression defines a norm $\|\cdot\|_s$ on $H^s_{\#}(\Omega)$ equivalent to that induced by $H^s(\Omega)$. We denote by Π_N the orthogonal projector in $L^2(\Omega)$ onto the space

$$S_N = \{u = \sum_{|p|_\infty \le N} \hat{u}(p)\phi_p\}.$$

Plainly $S_N \subset H^s_{\#}(\Omega)$ for all real s, and, when $s \ge 0$, Π_N is also the orthogonal projector in $H^s_{\#}(\Omega)$ onto S_N.

We define the space $\dot{L}^2(\Omega)$ (resp. $\dot{H}^s_{\#}(\Omega)$, $s \ge 0$) as the set of all functions u in $L^2(\Omega)$ (resp. $H^s_{\#}(\Omega)$) such that $\int_\Omega u\,dx = 0$.

Let $\nu = (\nu_1, \ldots, \nu_n)$ denote the unit outward normal to $\partial\Omega$ and let Γ_i (resp. Γ_{i+n}) be the face of Ω with $\nu_i = 1$ (resp. $\nu_i = -1$). Two spaces central to our discussion will be

$$H := \{u \in \dot{L}^2(\Omega)^n | \nabla \cdot u = 0, u_{i|\Gamma_i} = -u_{i|\Gamma_{i+n}}, i = 1, \ldots, n\},$$

where the latter condition is that $u \cdot \nu$, with ν the outward normal, is periodic (the trace of $u \cdot \nu$ on $\partial\Omega$ exists when $u \in L^2(\Omega)^n$ and $\nabla \cdot u \in L^2(\Omega)$, [11]); and

$$V := \{u \in \dot{H}^1_{\#}(\Omega)^n | \nabla \cdot u = 0\}.$$

For $s \ge 1$ we set $V_s = H^s_{\#}(\Omega)^n \cap V$, and define a norm on V_s equivalent to that induced by $H^s_{\#}(\Omega)^n$ by

$$\|u\|_{V_s} = \left(\sum_p |p|^{2s} |\hat{u}(p)|^2\right)^{1/2}.$$

Let $T > 0$; for a real Banach space X, $L^p(X)$, $H^s(X)$ and $C(X)$ will denote the function spaces $L^p(0,T;X)$, $H^s(0,T;X)$ and $C([0,T];X)$, respectively. Let M be a positive integer; for $m = 0, \ldots, M$, let $t^m := m\Delta t$, where $\Delta t = T/M$, and $\{u^m\}_{m=0}^M \subset X$. Then

$$\|u\|_{l^\infty(X)} := \max_{0 \le m \le M} \|u^m\|_X,$$

and

$$\|u\|_{l^p(X)} := (\Delta t \sum_{0 \le m \le M} \|u^m\|_X^p)^{1/p}, \quad 1 \le p < \infty.$$

We denote by $C^{0,1} = C^{0,1}(\overline{\Omega})$ the space of Lipschitz-continuous functions on the closure of Ω.

2 The equations and their approximation

2.1. Statement of the problem. The Navier-Stokes equations for a viscous incompressible fluid in two or three space dimensions take the form

$$u_t + (u \cdot \nabla)u - \nu\Delta u + \nabla p = f \quad \text{on } \Omega \times (0,T), \tag{1}$$

$$\nabla \cdot u = 0 \quad \text{on } \Omega \times (0,T), \tag{2}$$

$$u(x,0) = u_0(x) \quad \text{on } \Omega, \tag{3}$$

where $u(x,t)$ is the velocity of the fluid, $p(x,t)$ the kinematic pressure, ν the kinematic viscosity, $f(x,t)$ the density of body force per unit mass, and u_0 the initial velocity. The problem is completed in this case by periodic boundary conditions

$$u(x+2\pi e_i,t) = u(x,t) \quad \forall x \in \mathbf{R}^n,\ \forall t \in [0,T], \tag{4}$$

where $e_1,\dots,e_n$ is the canonical basis for $\mathbf{R}^n$. We assume that the mean flow is zero.

Let us recall the following functional-analytic formulation of the initial boundary value problem (1)–(4) (see [11]):

For f and u_0 given,

$$f \in L^2(H), \tag{5}$$

$$u_0 \in V, \tag{6}$$

find the *strong solution* u satisfying

$$u \in L^2(H^2_\#(\Omega)^n) \cap L^\infty(V), \tag{7}$$

$$\frac{d}{dt}(u,v) + ((u\cdot\nabla)u,v) + \nu(\nabla u,\nabla v) = (f,v) \quad \forall v \in V, \tag{8}$$

and

$$u(0) = u_0. \tag{9}$$

For $n=2$, there exists a unique solution to problem (5)–(9) satisfying

$$u \in L^2(H^2_\#(\Omega)^n) \cap C(V) \cap H^1(H). \tag{10}$$

For $n=3$, given $f \in L^\infty(H)$ instead of (5), the same result holds for T small enough. Under the same assumptions, $D_t u = \partial u/\partial t + (u\cdot\nabla)u$, the material derivative of u, belongs to $L^2(L^2(\Omega)^n)$, so that (8) may be rewritten as

$$(D_t u,v) + \nu(\nabla u,\nabla v) = (f,v) \quad \forall v \in V. \tag{11}$$

The crucial aspect of the Lagrange-Galerkin method is the discretisation of the material derivative along particle trajectories. To this end we seek to cast the equations into a Lagrangian form.

2.2. Lagrangian form. We assume here that the solution u of (7)–(9) satisfies, in addition to (10),

$$u \in C(C^{0,1}(\overline{\Omega})^n). \tag{12}$$

Using von Neumann's measurable selection theorem, it is possible to give a description of a Lagrangian representation of the flow under weaker assumptions on u (see for example [11]). Since, however, we shall need u to satisfy (12) for most of the results in this paper, we are happy to assume it here.

For $x \in \mathbf{R}^n$, $t \in [0,T]$, let $X_u(x,t;\cdot)$ denote the trajectory of the particle of fluid whose motion is governed by the velocity field u and which is at position x at time t. Then $X_u(x,t;\cdot)$ is the solution of the initial value problem

$$\frac{d}{ds}X_u(x,t;s) = u(X_u(x,t;s),s), \quad s \in [0,T]\backslash\{t\}, \tag{13}$$

$$X_u(x,t;t) = x. \tag{14}$$

The map $x \rightarrow X_u(x,t;s)$ is, for each $s,t \in [0,T]$, a quasi-isometric homeomorphism from $\mathbf{R}^n$ onto itself, and by virtue of the Rademacher-Stepanov theorem it is differentiable almost everywhere. Moreover, since u is divergence-free, the map has the volume-preserving property, i.e. its Jacobian is equal to 1 almost everywhere.

Associated with $X_u(x,t;s)$, for each $t,s \in [0,T]$, we can define the linear operator $E_u(t;s)$ on $\dot{H}^0_\#(\Omega)^n$ (which coincides exactly with $\dot{L}^2(\Omega)^n$) as follows. Each $v \in \dot{H}^0_\#(\Omega)^n$ may be periodically extended to the whole of $\mathbf{R}^n$. We then define $(E_u(t;s)v)(x) := v(X_u(x,t;s))$, $x \in \mathbf{R}^n$. The restriction of $E_u(t;s)v$ to Ω is in $\dot{H}^0_\#(\Omega)^n$. Moreover $E_u(t;s)$ is invertible (with inverse $E_u(s;t)$) and, because of the volume preserving property,

$$\|E_u(t;s)v\| = \|v\|.$$

Under the assumptions (5), (6), together with the additional assumption (12), we may replace the term $D_t u$ in (11) by $\left(\frac{d}{ds}u(X_u(\cdot,t;s),s)\right)\Big|_{s=t}$. The first step in obtaining the discrete equations described in the next subsection is to replace this time derivative by a first-order backward difference formula. Higher order backward difference formulae may also be used; their properties have been analysed in [12].

2.1 Spectral Lagrange-Galerkin approximation

Having described how we will discretise (11) in time, we form our fully discrete equations by employing a spectral Galerkin method for our spacial discretisation. Thus we seek $\mathcal{U} = (U^0,\ldots,U^M)^T \in (\Pi_N V)^{M+1}$ satisfying

$$\begin{aligned}\frac{1}{\Delta t}(U^k,v) + \nu(\nabla U^k,\nabla v) &= \frac{1}{\Delta t}(E_{U^{k-1}}(t^k;t^{k-1})U^{k-1},v) + (f^k,v) \\ &\quad \forall v \in V_N = \Pi_N V \subset V, \quad k = 1,\ldots,M, \end{aligned} \qquad (15)$$

with $U^0 = \Pi_N u_0$, where, for $x \in \Omega$ and $t \in [0,T]$, we define $X_{U^{k-1}}(x,t;s)$ by

$$\frac{d}{ds}X_{U^{k-1}}(x,t;s) = U^{k-1}(X_{U^{k-1}}(x,t;s)), \quad s \in [0,T]\backslash\{t\}, \qquad (16)$$

$$X_{U^{k-1}}(x,t;t) = x, \qquad (17)$$

and then $E_{U^{k-1}}(t;s)$ is defined by

$$\left(E_{U^{k-1}}(t;s)w\right)(x) = w(X_{U^{k-1}}(x,t;s)), \quad x \in \Omega.$$

We note that, similarly to $E_u(t;s)$, $E_{U^{k-1}}(t;s)$ is volume-preserving and invertible (with inverse $E_{U^{k-1}}(s;t)$). These properties will be used repeatedly in the subsequent analysis.

Given the solution U^{k-1} at time t^{k-1}, $X_{U^{k-1}}(x,t^k;t^{k-1})$ is found by solving (16), (17) to whatever accuracy is required. The right-hand side of (15) is then well defined, giving a diagonal system to solve for U^k. In practice, the solution of (16), (17) is carried out by using a forward Euler discretisation to reduce it to a functional equation, which is solved at equally spaced values of x by Newton iteration. These values of x are the quadrature points (for the trapezium rule) which are then used in the evaluation of the integrals implicit in the right hand side of (15). An additional approximation is employed, taking the form of a piecewise Chebyshev interpolant of U^{k-1}, in order to speed up the evaluation of $E_{U^{k-1}}(t^k;t^{k-1})U^{k-1}$. The effect of these approximations is analysed in the linear case in [9] and shown to be minimal. Their effect in the nonlinear case will be the subject of future investigation.

3 Error analysis

For simplicity of presentation, we shall sometimes write $u(\cdot,t)$ as $u(t)$, and we define without ambiguity, for $k=1,\dots,M$,

$$u^k := u(t^k),\ \ E_u := E_u(t^k;t^{k-1}),\ \ E_{U^{k-1}} := E_{U^{k-1}}(t^k;t^{k-1}),$$
$$X_u := X_u(x,t^k;t^{k-1}),\ \ X_{U^{k-1}} := X_{U^{k-1}}(x,t^k;t^{k-1}).$$

We set $\xi^k := \Pi_N u^k - U^k$, $\eta^k := (I-\Pi_N)u^k$ and $\zeta^k := \xi^k + \eta^k$, and derive, from (11) and (15),

$$\frac{1}{\Delta t}(\xi^k,v)+\nu(\nabla\xi^k,\nabla v)=\frac{1}{\Delta t}(E_{U^{k-1}}\xi^{k-1},v)+\frac{1}{\Delta t}((E_u-E_{U^{k-1}})u^{k-1},v)$$
$$+\frac{1}{\Delta t}(E_{U^{k-1}}\eta^{k-1},v)+(\frac{1}{\Delta t}(u^k-E_u u^{k-1})-D_t u^k,v)\quad\forall v\in V_N. \tag{18}$$

Our first error estimate will be obtained from (18), taking $v=\xi^k$. First, however, we present some preliminary results.

Lemma 1 *Let $v\in\dot{S}_N=\Pi_N\dot{L}^2(\Omega)^n$, $n=2,3$. Then*

$$\|v\|_\infty \le D_n(N)\|v\|_1,$$

where

$$\begin{aligned} D_2(N) &\le [\pi(2+4\ln(1+N^2))]^{1/2},\\ D_3(N) &\le [\pi(9+12\ln(1+N^2)+32\sqrt{3}N)]^{1/2}. \end{aligned}$$

Lemma 2 *For $k=1,\dots,M$, let $c_k=|u^k|_{C^{0,1}}$: then*

$$\|X_u - X_{U^{k-1}}\| \le \frac{1}{c_{k-1}}(e^{c_{k-1}\Delta t}-1)(\|\zeta^{k-1}\|+\|\frac{du}{dt}\|_{L^1(t^{k-1},t^k;L^2(\Omega)^n)}). \tag{19}$$

Proof. Integrating (13) and the corresponding equation for $X_{U^{k-1}}$ between t^k-t and t^k we obtain

$$(X_u-X_{U^{k-1}})(\cdot,t^k,t^k-t)=-\int_0^t\{E_{U^{k-1}}(t^k;t^k-s)(u^{k-1}-U^{k-1})$$
$$+(E_u-E_{U^{k-1}})(t^k;t^k-s)u^{k-1}+E_u(t^k;t^k-s)(u(t^k-s)-u^{k-1})\}ds,$$

so that

$$\|(X_u-X_{U^{k-1}})(\cdot,t^k,t^k-t)\|\le t(\|\zeta^{k-1}\|+\|\tfrac{du}{dt}\|_{L^1(t^{k-1},t^k;L^2(\Omega)^n)})$$
$$+c_{k-1}\textstyle\int_0^t\|(X_u-X_{U^{k-1}})(\cdot,t^k,t^k-s)\|ds.$$

A straightforward application of Gronwall's lemma yields the required result.

Lemma 3 *For $k = 1, \ldots, M$, $v \in V$, and $t, s \in [0, T]$,*

$$\|(E_{U^{k-1}}(t;s) - I)v\| \le |t-s| \|U^{k-1}\|_\infty \|\nabla v\|. \tag{20}$$

Proof. We have

$$\begin{aligned}(E_{U^{k-1}}(t;s) - I)v &= \int_t^s \frac{d}{d\tau}(E_{U^{k-1}}(t;\tau)v)\, d\tau \\ &= \int_t^s E_{U^{k-1}}(t;\tau)\left(U^{k-1}\cdot\nabla v\right) d\tau.\end{aligned}$$

The result then follows by taking the norm of both sides.

The final lemma deals with the time truncation error in the backward Euler approximation of the material derivative along particle trajectories.

Lemma 4 *For $k = 1, \ldots, M$,*

$$\|(D_t u)^k - \frac{u^k - E_u u^{k-1}}{\Delta t}\| \le \|D_t^2 u\|_{L^1(t^{k-1}, t^k; L^2(\Omega)^n)}. \tag{21}$$

3.1. L^2 error estimates. We assume that

$$f \in C(H) \quad \text{and} \quad u_0 \in C^{0,1}(\overline{\Omega})^n \cap V_s, \quad s > n/2, \tag{22}$$

and that the corresponding solution u of (2) satisfies

$$\begin{aligned} u &\in C(C^{0,1}(\overline{\Omega})^n \cap V_s), \quad s > n/2, &(23)\\ du/dt &\in L^2(H) \quad \text{and} &(24)\\ D_t^2 u &\in L^2(H). &(25)\end{aligned}$$

Then we have the following

Theorem 1 *Suppose that u satisfies (8), (9), $\mathcal{U} = (U^0, \ldots, U^M)^T$ satisfies (15)–(17), and that (22)–(25) hold. Then there exist two positive constants Δt_0 and N_0 such that, for all $\Delta t \le \Delta t_0$ and all $N \ge N_0$,*

$$\|u - \mathcal{U}\|_{l^\infty(H)} \le C_3(C_4 N^{-s} + C_5 \Delta t) \tag{26}$$

where

$$C_3 = \exp\{2|u|^2_{l^2(C^{0,1})} + |u|_{l^1(C^{0,1})}\},$$

$$C_4 = \|u\|_{L^\infty(V_s)} + \left(\frac{4C_1^2\|u\|^2_{L^\infty(V_s)}}{\nu} + 1\right)^{1/2} \|u\|_{l^2(V_s)},$$

$$C_5 = \left(\|\frac{du}{dt}\|_{L^2(H)} + \left(\frac{2}{\nu}\right)^{1/2} \|D_t^2 u\|_{L^2(H)}\right),$$

and C_1 is a positive constant which depends only on s.

The remainder of this subsection will be taken up with the proof of this theorem.
Proof. We choose $v = \xi^k$ in (18) to obtain

$$\begin{aligned}
&\frac{1}{\Delta t}\|\xi^k\|^2 + \nu\|\nabla\xi^k\|^2 \\
\leq\ & \frac{1}{\Delta t}\|\xi^k\|\|\xi^{k-1}\| + \frac{1}{\Delta t}\|(E_u - E_{U^{k-1}})u^{k-1}\|\|\xi^k\| \\
&+\frac{1}{\Delta t}|(E_{U^{k-1}}\eta^{k-1}, \xi^k)| + \|\frac{u^k - E_u u^{k-1}}{\Delta t} - D_t u^k\|\|\xi^k\| \\
=\ & A_1 + \ldots + A_4. \qquad (27)
\end{aligned}$$

We shall bound the terms A_2 to A_4 in turn. We have, for A_2,

$$\|(E_u - E_{U^{k-1}})u^{k-1}\| \leq c_{k-1}\|X_u - X_{U^{k-1}}\|,$$

and so, from Lemma 2,

$$\begin{aligned}
A_2 \leq\ & \frac{1}{\Delta t}(e^{c_{k-1}\Delta t} - 1)\left(\|\zeta^{k-1}\| + \|\frac{du}{dt}\|_{L^1(t^{k-1},t^k;L^2(\Omega)^n)}\right)\|\xi^k\| \\
\leq\ & \|\xi^k\|\left\{\frac{1}{\Delta t}(e^{c_{k-1}\Delta t} - 1)\ \|\xi^{k-1}\|\right. \\
&\left. +c_{k-1}e^{c_{k-1}\Delta t}\left(\|\eta^{k-1}\| + \|\frac{du}{dt}\|_{L^1(t^{k-1},t^k;L^2(\Omega)^n)}\right)\right\},
\end{aligned}$$

where we have written ζ^{k-1} as $\xi^{k-1} + \eta^{k-1}$.

In order to bound A_3 we make use of the fact that E_u and $E_{U^{k-1}}$ are invertible and volume preserving. Thus we have, applying Lemma 1 and Lemma 3,

$$\begin{aligned}
A_3 \leq\ & \frac{1}{\Delta t}|(\eta^{k-1}, (E_{U^{k-1}}^{-1} - I)\xi^k)| \\
\leq\ & \|\eta^{k-1}\|\|U^{k-1}\|_\infty\|\nabla\xi^k\| \\
\leq\ & \|\eta^{k-1}\|\|\nabla\xi^k\|\left(\|\Pi_N u^{k-1}\|_\infty + D_n(N)\|\nabla\xi^{k-1}\|\right).
\end{aligned}$$

Now, by Sobolev's imbedding theorem and the contractivity of Π_N in V_r for any real r, it follows that there is a constant C_1, independent of N but dependent on $s > n/2$, such that $\|\Pi_N u\|_{l^\infty(L^\infty)} \leq C_1\|u\|_{L^\infty(V_s)} =: C_2$. Including the bound on A_4 which follows from Lemma 4 we find, repeatedly making use of the inequality $ab \leq \frac{\epsilon}{2}a^2 + \frac{1}{2\epsilon}b^2$ (valid for all real a, b and for $\epsilon > 0$) and gathering like terms, that

$$\begin{aligned}
&\frac{1}{\Delta t}\|\xi^k\|^2 + \nu\|\nabla\xi^k\|^2 \\
\leq\ & \frac{e^{c_{k-1}\Delta t}}{\Delta t}\|\xi^k\|\|\xi^{k-1}\| + \|\nabla\xi^k\|\|\eta^{k-1}\|\left(C_2 + D_n(N)\|\nabla\xi^{k-1}\|\right) \\
&+\|\xi^k\|\left\{c_{k-1}e^{c_{k-1}\Delta t}\left(\|\frac{du}{dt}\|_{L^1(t^{k-1},t^k;L^2(\Omega)^n)} + \|\eta^{k-1}\|\right)\right. \\
&\left. + \|D_t^2 u\|_{L^1(t^{k-1},t^k;L^2(\Omega)^n)}\right\} \\
\leq\ & \frac{1}{2\Delta t}\left(\|\xi^k\|^2 + e^{2c_{k-1}\Delta t}\|\xi^{k-1}\|^2\right) + c_{k-1}^2\|\xi^k\|^2 + \frac{\nu}{2}\|\nabla\xi^k\|^2
\end{aligned}$$

$$+\frac{e^{2c_{k-1}\Delta t}}{2}\left(\|\frac{du}{dt}\|^2_{L^1(t^{k-1},t^k;L^2(\Omega)^n)}+\|\eta^{k-1}\|^2\right)$$
$$+\frac{2(\|\eta^{k-1}\|D_n(N))^2}{\nu}\|\nabla\xi^{k-1}\|^2$$
$$+\frac{1}{\nu}\|D_t^2u\|^2_{L^1(t^{k-1},t^k;L^2(\Omega)^n)}+\frac{2C_2^2}{\nu}\|\eta^{k-1}\|^2,$$

where we have also made use of the fact that $\|v\|\le\|\nabla v\|$ for $v\in V$.

Because of the smoothness of u, there is an N_0 such that for $N\ge N_0$,

$$2\left(\|\eta\|_{C(H)}D_n(N)\right)^2\le\nu^2/4.$$

We assume that we are dealing with such an N, so that multiplying through by $2\Delta t$ we obtain

$$\begin{aligned}(1-2\Delta tc_{k-1}^2)\|\xi^k\|^2+\nu\Delta t\|\nabla\xi^k\|^2&\le e^{2c_{k-1}\Delta t}\|\xi^{k-1}\|^2\\&+\Delta t\left(e^{2c_{k-1}\Delta t}+\frac{4C_2^2}{\nu}\right)\|\eta^{k-1}\|^2+\frac{\nu\Delta t}{2}\|\nabla\xi^{k-1}\|^2\\&+\Delta t^2\left(e^{2c_{k-1}\Delta t}\|\frac{du}{dt}\|^2_{L^2(t^{k-1},t^k;L^2(\Omega)^n)}+\frac{2}{\nu}\|D_t^2u\|^2_{L^2(t^{k-1},t^k;L^2(\Omega)^n)}\right).\end{aligned}\tag{28}$$

We proceed under the assumption that $\Delta t\le\Delta t_0=\min\left(T,1/(4|u|^2_{l^\infty(C^{0,1})})\right)$. For such Δt, we can bound $(1-2\Delta tc_{k-1}^2)$ from below by $e^{-4\Delta tc_{k-1}^2}$. Then we deduce from (28) that

$$\begin{aligned}&\|\xi^k\|^2+e^{4\Delta tc_{k-1}^2}\nu\Delta t\|\nabla\xi^k\|^2\\&\le\ e^{4\Delta tc_{k-1}^2+2\Delta tc_{k-1}}\left\{\|\xi^{k-1}\|^2+\frac{\nu\Delta t}{2}e^{4\Delta tc_{k-2}^2}\|\nabla\xi^{k-1}\|^2+\Delta t\left(1+\frac{4C_2^2}{\nu}\right)\|\eta^{k-1}\|^2\right.\\&\quad\left.+\Delta t^2\|\frac{du}{dt}\|^2_{L^2(t^{k-1},t^k;L^2(\Omega)^n)}+\frac{2\Delta t^2}{\nu}\|D_t^2u\|^2_{L^2(t^{k-1},t^k;L^2(\Omega)^n)}\right\}.\end{aligned}\tag{29}$$

This inequality is valid for all $k\ge1$ if we set $c_{-1}=0$. It is slightly less sharp than (28), since we have multiplied some of the terms on the right hand side of (28) by terms greater than unity. Our reason for doing this is to enable us to make use of the following lemma, whose proof (by induction) is straightforward.

Lemma 5 *Let $\{a_k\}$, $\{b_k\}$, $\{d_k\}$ and $\{\lambda_k\}$ be sequences of non-negative real numbers satisfying, for $k=1,\ldots,M$,*

$$a_k+b_k\le e^{\lambda_{k-1}}\left(a_{k-1}+\frac{1}{2}b_{k-1}+d_{k-1}\right).\tag{30}$$

Then the following inequality holds:

$$a_M+\frac{1}{2}\sum_{k=0}^{M}b_k\le\left(\prod_{k=0}^{M-1}e^{\lambda_k}\right)\left(a_0+\frac{b_0}{2}+\sum_{k=0}^{M-1}d_k\right).\tag{31}$$

If we denote $2\Delta t\left(\sum_{k=0}^{M-1} 2c_k^2 + c_k\right)$ by α, then from (29)–(31) we deduce that

$$\|\xi^M\|^2 \;+\; \frac{\nu\Delta t}{2}\sum_{k=1}^{M} e^{4\Delta t c_{k-1}^2}\|\nabla\xi^k\|^2 \le e^{\alpha}\sum_{k=1}^{M}\Delta t\left(1+\frac{4C_2^2}{\nu}\right)\|\eta^{k-1}\|^2$$
$$+\;\Delta t^2 e^{\alpha}\left(\|\frac{du}{dt}\|^2_{L^2(L^2(\Omega)^n)} + \frac{2}{\nu}\|D_t^2 u\|^2_{L^2(L^2(\Omega)^n)}\right). \tag{32}$$

Since the same bound holds for $\|\xi^k\|^2$, for any $k \le M$, the conclusions of the theorem follow.

We note that the result of this theorem is such that, given sufficient smoothness and decay of u, it is valid for a semi-infinite time interval.

3.2. Error estimate in V

Theorem 2 *Suppose that the conditions of Theorem 1 hold. Suppose also that N and Δt are related by*

$$\Delta t = \begin{cases} O((\ln N)^{-1/2}), & n = 2 \\ O(N^{-1/2}), & n = 3, \end{cases} \tag{33}$$

and that Δt_0, N_0, C_1, C_3, C_4 and C_5 are as in Theorem 1. Then, for $\Delta t \le \Delta t_0$ and $N \ge N_0$,

$$\|u - U\|_{l^\infty(V)} \le C_6(C_7 N^{1-s} + C_8\Delta t) \tag{34}$$

where C_6, C_7 and C_8 depend on $C_1, \ldots, C_5$.

The proof of this result is similar to that in Theorem 1 and, for the sake of brevity, it is ommitted here: a detailed proof is given in [10].

4 Stability

Although in the previous section we have demonstrated the convergence of the spectral Lagrange-Galerkin method for the Navier-Stokes equations, we have not yet addressed the pertinent question of the stability properties of the scheme. These properties become important, for example, in the presence of rounding errors. Here we carry out a stability analysis within the framework introduced by López-Marcos and Sanz-Serna [3], and we begin with a brief discussion of this framework.

4.1. Definition of stability. Let u be a solution of the equation $\Phi(u) = 0$, where Φ is a mapping from a Banach space $\mathcal{X}$ into a Banach space $\mathcal{Y}$. (The Navier-Stokes equations as described earlier may be set in this form). Let $\mathcal{H}$ be a set of positive numbers (or vectors in $\mathbf{R}^2$ with positive entries) with zero infinum. For each $h \in \mathcal{H}$, let $\mathcal{U}_h$ be a numerical approximation to u, obtained by solving

$$\Phi_h(\mathcal{U}_h) = 0, \tag{35}$$

where Φ_h is a fixed mapping with domain $\mathcal{D}_h \in \mathcal{X}_h$ and taking values in $\mathcal{Y}_h$, with both $\mathcal{X}_h$ and $\mathcal{Y}_h$ being finite-dimensional linear spaces. We choose a norm $\|\cdot\|_{\mathcal{X}_h}$ in $\mathcal{X}_h$, a norm $\|\cdot\|_{\mathcal{Y}_h}$ in $\mathcal{Y}_h$ and a 'target element' u_h in the interior of $\mathcal{D}_h$, which should be some discrete representation of the analytical solution u in $\mathcal{X}_h$. Then we have the following definition of stability for (35) (c.f. [3]).

Definition 1 *For $h \in \mathcal{H}$, let $R_h \in (0,\infty)$. The discretisation (35) is said to be stable, restricted to the thresholds R_h, if there exist positive constants h_0 and S such that for $h \in \mathcal{H}$, $|h| \le h_0$, the open ball $B(u_h, R_h)$ is contained in $\mathcal{D}_h$ and for any $\mathcal{V}_h, \mathcal{W}_h$ in that ball,*

$$\|\mathcal{V}_h - \mathcal{W}_h\|_{\mathcal{X}_h} \le S\|\Phi_h(\mathcal{V}_h) - \Phi_h(\mathcal{W}_h)\|_{\mathcal{Y}_h}. \tag{36}$$

In order to make use of this definition we have to cast the spectral Lagrange-Galerkin method into an appropriate form. This is the aim of the next subsection.

4.2. Reformulation of the problem and stability analysis. In the previous section the discrete equations were parameterised by small positive numbers h. In our case, however, there are two parameters, Δt and N, associated with the discrete equations. We thus denote by h the pair $(\Delta t, 1/N) \in (0,\infty)^2$. In fact we invoke the limitations on Δt and N involved in Theorem 1, so that $h \in (0, \Delta t_0] \times (0, 1/N_0]$.

The spaces $\mathcal{X}_h$ and $\mathcal{Y}_h$ will be distinguished only by their norms. They will both consist of vectors

$$\mathcal{U} = (U^0, \dots, U^M)^T \in \left(\dot{S}_N\right)^{M+1}.$$

For $\mathcal{U} \in \mathcal{X}_h$ we define

$$\|\mathcal{U}\|_{\mathcal{X}_h} := \|\mathcal{U}\|_{l^\infty(H)} + \nu^{1/2}\|\mathcal{U}\|_{l^2(V)},$$

and for $\mathcal{U} \in \mathcal{Y}_h$ we define

$$\|\mathcal{U}\|_{\mathcal{Y}_h} := \|U^0\| + \left(\Delta t \sum_{k=1}^{M} \|U^k\|^2\right)^{1/2}.$$

We suppose that u satisfies (8)–(9) and (22)–(25), for some $s > (n+2)/2$. Then we set our target element $u_h \in \mathcal{X}_h$ to be $\Pi_N u$. We are now in a position to define the operator $\Phi_h : \mathcal{X}_h \to \mathcal{Y}_h$. Let $\mathcal{V} = (V^0, \dots, V^M)^T \in \mathcal{X}_h$. For each $k = 1, \dots, M$, we have $V^{k-1}, V^k \in \dot{S}_N$, and we construct $F^k \in \dot{S}_N$ as follows. For $(x,t) \in \Omega \times [0,T]$ we define $X_{V^{k-1}}(x,t;\cdot)$ to be the solution of the initial value problem

$$\begin{aligned} \frac{d}{ds} X_{V^{k-1}}(x,t;s) &= V^{k-1}(X_{V^{k-1}}(x,t;s)), \quad s \in [0,T]\backslash\{t\}, \\ X_{V^{k-1}}(x,t;t) &= x. \end{aligned}$$

We then define, for $s,t \in [0,T]$, $E_{V^{k-1}}(t;s) : H \to H$ analogously to $E_{U^{k-1}}(t;s)$ in the discussion of convergence, and similarly denote $E_{V^{k-1}}(t^k;t^{k-1})$ by $E_{V^{k-1}}$. Furthermore we define $F^k \in \dot{S}_N$ to be the solution of

$$(F^k, v) = \frac{1}{\Delta t}((V^k - E_{V^{k-1}} V^{k-1}), v) + \nu(\nabla V^k, \nabla v) - (f^k, v) \quad \forall v \in V_N,$$

and F^0 to be the solution of

$$(F^0, v) = (V^0 - \Pi_N u_0, v) + \frac{1}{2}\nu\Delta t\left(\nabla(V^0 - \Pi_N u_0), \nabla v\right) \quad \forall v \in V_N.$$

In this way we construct $\mathcal{F} = (F^0, \dots, F^M)^T$ from $\mathcal{V}$, and we define $\Phi_h : \mathcal{X}_h \to \mathcal{Y}_h$ by

$$\Phi_h(\mathcal{V}) = \mathcal{F}.$$

The spectral Lagrange-Galerkin method for the Navier-Stokes equations may then be succinctly described as follows: find $\mathcal{U} \in \mathcal{X}_h$ such that

$$\Phi_h(\mathcal{U}) = 0. \tag{37}$$

Theorem 3 *There is a pair of positive numbers $(\Delta t_0, 1/N_0)$ such that for $N \geq N_0$ and $\Delta t \leq \Delta t_0$ the spectral Lagrange-Galerkin method (37) is non-linearly stable with stability threshold $R_h = (ND_n(N))^{-1}$.*

Proof. Choose $\mathcal{V}, \mathcal{W} \in \mathcal{X}_h \cap B(u_h, R_h)$, and write $\mathcal{F} = \Phi_h(\mathcal{V})$ and $\mathcal{G} = \Phi_h(\mathcal{W})$. For $k = 0, \ldots, M$, we write $\zeta^k = V^k - W^k$. Thus, for $k = 1, \ldots, M$ and for all $v \in V_N$, we have

$$\frac{1}{\Delta t}\left(\zeta^k - (E_{V^{k-1}}V^{k-1} - E_{W^{k-1}}W^{k-1}), v\right) + \nu(\nabla\zeta^k, \nabla v) = (F^k - G^k, v),$$

so that, taking $v = \zeta^k$, we find

$$\frac{1}{\Delta t}\|\zeta^k\|^2 + \nu\|\nabla\zeta^k\|^2 \quad \leq \quad \|\zeta^k\| \left\{\frac{1}{\Delta t}\|E_{V^{k-1}}V^{k-1} - E_{W^{k-1}}W^{k-1}\| + \|F^k - G^k\|\right\}. \quad (38)$$

The key part of the proof is the bound on $\|E_{V^{k-1}}V^{k-1} - E_{W^{k-1}}W^{k-1}\|$. We write

$$\|E_{V^{k-1}}V^{k-1} - E_{W^{k-1}}W^{k-1}\| \leq \|\zeta^{k-1}\| + \|(E_{V^{k-1}} - E_{W^{k-1}})W^{k-1}\|,$$

where we have bounded $\|E_{V^{k-1}}\zeta^{k-1}\|$ by $\|\zeta^{k-1}\|$. The second term may be bounded by $\|X_{V^{k-1}} - X_{W^{k-1}}\||W^{k-1}|_{C^{0,1}}$, where $|\cdot|_{C^{0,1}}$ is the Lipschitz seminorm, and it is relatively straightforward to show, by use of Gronwall's lemma, that

$$\|X_{V^{k-1}} - X_{W^{k-1}}\| \leq \Delta t\|\zeta^{k-1}\|e^{\Delta t|W^{k-1}|_{C^{0,1}}}.$$

Under the assumptions that we have on Δt and $\mathcal{W}$, and making use of Sobolev's embedding theorem, we have

$$\begin{aligned} \Delta t|W^{k-1}|_{C^{0,1}} &\leq \Delta t D_n(N)\|W^{k-1} - \Pi_N u^{k-1}\|_{V_2} + \Delta t|\Pi_N u^{k-1}|_{C^{0,1}} \\ &\leq \Delta t^{1/2} N D_n(N)\nu^{-1/2}R_h + \Delta t C_{10}\|u\|_{l^\infty(V_s)}, \end{aligned} \quad (39)$$

and so according to the assumptions of the theorem $e^{\Delta t|W^{k-1}|_{C^{0,1}}}$ may be bounded by a constant C_{11}. Thus we deduce from (38), making use of the relation $\|\zeta^k\| \leq \|\nabla\zeta^k\|$, that

$$\begin{aligned} \frac{1}{\Delta t}\|\zeta^k\|^2 \;&+\; \nu\|\nabla\zeta^k\|^2 \leq \frac{1}{2\Delta t}\left(\|\zeta^k\|^2 + \|\zeta^{k-1}\|^2\right) + \frac{\nu}{2}\|\nabla\zeta^k\|^2 \\ &+\; \frac{1}{\nu}\left(C_{11}^2|W^{k-1}|_{C^{0,1}}^2\|\zeta^{k-1}\|^2 + \|F^k - G^k\|^2\right). \end{aligned}$$

Multiplying through by $2\Delta t$ we obtain

$$\|\zeta^k\|^2 + \nu\Delta t\|\nabla\zeta^k\|^2 \quad \leq \quad \|\zeta^{k-1}\|^2\left(1 + \frac{C_{11}^2\Delta t}{\nu}|W^{k-1}|_{C^{0,1}}^2\right) + \frac{2\Delta t}{\nu}\|F^k - G^k\|^2.$$

Bounding $1 + \frac{C_{11}^2\Delta t}{\nu}|W^{k-1}|_{C^{0,1}}^2$ by $\exp\{\frac{C_{11}^2\Delta t}{\nu}|W^{k-1}|_{C^{0,1}}^2\}$ and applying Lemma 5 gives that

$$\|\mathcal{V} - \mathcal{W}\|_{\mathcal{X}_h} \leq 2\left(\frac{1+\nu}{\nu}\right)^{1/2} e^{\frac{C_{11}^2}{2\nu}|\mathcal{W}|_{l^2(C^{0,1})}^2}\|\Phi_h(\mathcal{V}) - \Phi_h(\mathcal{W})\|_{\mathcal{Y}_h}.$$

We bound $|\mathcal{W}|_{l^2(C^{0,1})}$ by employing an argument identical to (39), and the (unconditional) nonlinear stability of the spectral Lagrange-Galerkin method follows.

References

[1] Douglas, J., Jr. and Russell, T. F. (1982). Numerical methods for convection-dominated diffusion problems based on combining the method of characteristics with finite element or finite difference procedures. *SIAM J. Numer. Anal.*, 19:871–885.

[2] Ho, L.-W., Maday, Y., Patera, A., and Ronquist, E. M. (1989). A high-order Lagrangian-decoupling method for the incompressible Navier-Stokes equations. *ICASE Report*, 89-57.

[3] López-Marcos, J. and Sanz-Serna, J. (1987). Stability and convergence in numerical analysis III: Linear investigation of nonlinear stability. *IMA J. Numer. Anal.*, 8: 71–84.

[4] Morton, K. W., Priestley, A., and Süli, E. (1988). Stability of the Lagrange-Galerkin method with non-exact integration. *RAIRO M²AN*, 22:123–151.

[5] Pironneau, O. (1982). On the transport-diffusion algorithm and its applications to the Navier-Stokes equations. *Numer. Math.*, 38:309–332.

[6] Süli, E. (1985). Lagrange-Galerkin mixed finite element approximation of the Navier-Stokes equations. In *Numerical Methods for Fluid Dynamics*, Morton, K. W. and Baines, M. J., Eds., pp. 439–448. Oxford University Press.

[7] Süli, E. (1988). Convergence and nonlinear stability of the Lagrange-Galerkin method for the Navier-Stokes equations. *Numer. Math.*, 53:459–483.

[8] Süli, E. (1988). Stability and convergence of the Lagrange-Galerkin method with non-exact integration. In *The Mathematics of Finite Elements and Applications VI*, Whiteman, J. R., Ed., pp. 435–442. Academic Press.

[9] Süli, E. and Ware, A. (1991). A spectral method of characteristics for first-order hyperbolic equations. SIAM J. Numer. Anal., 28:423–445.

[10] Süli, E. and Ware, A. (1989). A spectral Lagrange-Galerkin method for the Navier-Stokes equations: convergence and non-linear stability. Oxford University Computing Laboratory Report, 89/10.

[11] Temam, R. (1983). Navier-Stokes Equations and Nonlinear Functional Analysis. CBMS-NSF Regional Conference Series in Applied Mathematics, SIAM, Philadelphia.

[12] Ware, A.F. (1991). A spectral Lagrange-Galerkin method for convection-diffusion problems. D.Phil. Thesis, University of Oxford.

A Fractional Step Method for Regularized Navier–Stokes Equations

Werner Varnhorn
Fachbereich Mathematik, Technische Hochschule Darmstadt
Schloßgartenstr. 7, D–6100 Darmstadt

1. Introduction

For $\varepsilon \geq 0$ we consider the following modified initial boundary value problem, which describes the motion of viscous incompressible fluid flow:

$$\begin{aligned} &\partial_t v - \nu\Delta v + \nabla p + (v_\varepsilon \cdot \nabla)v = F \quad \text{in} \quad \Omega_T \\ &\nabla \cdot v = 0 \quad \text{in} \quad \Omega_T, \qquad v_\varepsilon(t) = v(t-\varepsilon), \\ &v(s,x) = v_0(s,x) \quad \text{for} \quad -\varepsilon \leq t \leq 0, \quad x \in \Omega, \\ &v(t,x) = 0 \quad \text{for} \quad t > 0, \qquad x \in \partial\Omega. \end{aligned} \tag{N_ε}$$

Here the case $\varepsilon = 0$ corresponds to the nonlinear Navier–Stokes equations: v is the velocity and p the pressure of the fluid. The function F is the given external force density, the constant $\nu > 0$ is the kinematic viscosity, and the condition div $v = 0$ expresses the incompressibility of the fluid. In addition, we have a given initial velocity distribution v_0 (depending also on $t \in [-\varepsilon, 0]$ if $\varepsilon > 0$), and we require the no–slip condition $v = 0$ on the boundary $\partial\Omega$. Throughout this paper we consider (N_ε) in $\Omega_T = (0,T) \times \Omega$, where $\Omega \subset \mathbb{R}^3$ is a bounded domain with a smooth boundary $\partial\Omega$ and $T > 0$ is given.

If the above mentioned parameter ε is positive, then the nonlinearity vanishes, and the Navier–Stokes equations reduce to some regularized equations with a time delay in the convective term. It can be shown [12] that the solution $v = v_\varepsilon$ of (N_ε) is unique and exists globally in time. On the other hand, if ε tends to zero, then there always exists a subsequence of $(v_\varepsilon)_\varepsilon$ with limit v such that v is a weak solution (Hopf [5]) of the Navier–Stokes equations (N_0). Up to now, such a solution is the only one, whose existence globally in time has been proved.

In the present paper we assume a fixed positive time delay $\varepsilon = T/N > 0$ $(N \in \mathbb{N})$ and consider the following time stepping procedure of Crank–Nicholson type for (N_ε):

Let

$$h = \varepsilon/n > 0 \quad (n \in \mathbb{N}), \qquad t_k = k \cdot h \quad (k = -n, -n+1, \dots, -1, 0, 1, \dots, nN),$$

be a time grid on $[-\varepsilon, T]$. Then we can replace the first equation of (N_ε) by both

$$\frac{(v^k - v^{k-1/2})}{h/2} - \nu\Delta v^k + \nabla p^k = \frac{2}{h} \int_{(k-1/2)h}^{kh} F(t)\,dt - (v^{k-n-1/2} \cdot \nabla)v^{k-1/2} \tag{1.1}$$

or

$$\frac{(v^{k-1/2} - v^{k-1})}{h/2} - \nu\Delta v^{k-1} + \nabla p^{k-1} = \frac{2}{h} \int_{(k-1)h}^{(k-1/2)h} F(t)\,dt - (v^{k-n-1/2} \cdot \nabla)v^{k-1/2}. \tag{1.2}$$

Here (1.1) treats the viscosity term $\nu\Delta v$ implicitly and the convective term $(v_\varepsilon \cdot \nabla)v$ explicitly, whereas (1.2) does vice versa. Summing up these equations and replacing $v^{j-1/2}$ by $\frac{1}{2}(v^j + v^{j-1})$, we can approximate the solution v, p of (N_ε) at time t_k by the solution v^k, p^k $(k = 1, 2, \dots, nN)$ of

$$\begin{aligned} &\frac{(v^k - v^{k-1})}{h} - \tfrac{\nu}{2}\Delta(v^k + v^{k-1}) + \tfrac{1}{2}\nabla(p^k + p^{k-1}) = \\ &\qquad \tfrac{1}{h} \int_{(k-1)h}^{kh} F(t)\,dt - \tfrac{1}{4}\big((v^{k-n} + v^{k-n-1}) \cdot \nabla\big)(v^k + v^{k-1}) \quad \text{in } \Omega, \\ &\operatorname{div} v^k = 0 \quad \text{in } \Omega, \qquad v^k = 0 \quad \text{on } \partial\Omega \quad (k = 1, 2, \dots, nN), \\ &v^i = v_0(t_i) \quad \text{in } \Omega \quad (i = -n, -n+1, \dots, 0). \end{aligned} \tag{N_ε^h}$$

This fractional step method is implicit for the sum $(v^k + v^{k-1})$. Similar time stepping procedures for Stokes– and Navier–Stokes equations have been studied in [3, 4, 6, 8, 9, 10, 11, 12].

In the present paper we assume $\boldsymbol{H}^2$–regularity of the initial data (see the notations below) such that the corresponding solution v of (N_ε) is strongly H^2–continuous uniformly in $t \in [0, T]$. Under these assumptions we study the stability of the solution v^k of (N_ε^h) as $h = \varepsilon/n \longrightarrow 0$ ($\varepsilon = T/N > 0$ remains fixed) and show, in particular, that $(v^k + v^{k-1})$ is bounded uniformly in $k = 1, 2, \dots, nN$ with respect to the $\boldsymbol{H}^2$–norm (Theorem 2.2). We consider the discretization error $e^k = v^k - v(t_k)$ and prove that its $\boldsymbol{L}^2$–norm vanishes with a linear order of convergence uniformly for $k = 1, 2, \dots, nN$. Moreover, the above mentioned stability property even implies convergence of the sum $e^k + e^{k-1}$ with respect to the $\boldsymbol{H}^2$–norm uniformly in k (Theorem 2.3).

2. Notations and Results

Throughout this paper, $G \subset \mathbb{R}^3$ is a bounded domain having a compact boundary ∂G of class C^2. In the following, all functions are real valued. As usual, $C_0^\infty(G)$ denotes the spaces of smooth functions defined in G with compact support, and $L^p(G)$ is the Lebesgue (Banach) space equipped with the norm $\|f\|_{0,p}$ ($1 \le p \le \infty$; if $p = \infty$ we use $\|f\|_\infty$ instead of $\|f\|_{0,\infty}$). The space $L^2(G)$ is a Hilbert space with scalar product and norm

$$\langle f, g\rangle = \int_G f(x)\, g(x)\, dx, \quad \|f\| = \|f\|_{0,2} = \langle f, f\rangle^{1/2},$$

respectively.

The Sobolev (Hilbert) space $H^m(G)$ ($m \in \mathbb{N} = \{0, 1, 2, \ldots,\}$) is the space of functions f such that $\partial^\alpha f \in L^2(G)$ for all $\alpha = (\alpha_1, \alpha_2, \alpha_3) \in \mathbb{N}^3$ with $|\alpha| = \alpha_1 + \alpha_2 + \alpha_3 \le m$. Its norm is denoted by

$$\|f\|_m = \|f\|_{H^m(G)} = \Big(\sum_{|\alpha|\le m} \|\partial^\alpha f\|^2\Big)^{1/2}, \qquad \partial^\alpha = \partial_1^{\alpha_1}\partial_2^{\alpha_2}\partial_3^{\alpha_3},$$

where $\partial_k = \partial/\partial x_k$ ($k = 1, 2, 3$) is the distributional derivative.

The completion of $C_0^\infty(G)$ with respect to the norm $\|.\|_m$ is denoted by $H_0^m(G)$ $\big(H_0^0(G) = H^0(G) = L^2(G)\big)$.

The spaces $\boldsymbol{C}_0^\infty(G)$, $\boldsymbol{L}^2(G)$, $\boldsymbol{H}^m(G), \ldots$ with fat initials are the corresponding spaces of vector fields $u = (u_1, u_2, u_3)$. Here norm and scalar product are denoted as in the scalar case, hence, for example,

$$\langle u, v\rangle = \textstyle\sum_{k=1}^3 \langle u_k, v_k\rangle, \quad \|u\| = \langle u, v\rangle^{1/2} = \int_G |u(x)|^2\, dx^{1/2},$$

where $|u(x)| = \big(u_1(x)^2 + u_2(x)^2 + u_3(x)^2\big)^{1/2}$ is the Euclidian norm of $u(x) \in \mathbb{R}^3$.

The following relations are well known ($a, b \in L^2(G)$, $c \in H_0^1(G)$):

$$\begin{aligned} \langle a - b, a + b\rangle &= \|a\|^2 - \|b\|^2, && (2.1)\\ \langle a - b, 2a\rangle &= \|a\|^2 - \|b\|^2 + \|a - b\|^2, && (2.2)\\ 2\langle a, b\rangle &= 2\|a\|\,\|b\| \le \|a\|^2 + \|b\|^2, && (2.3)\\ \|c\|^2 &\le C\|\nabla c\|^2. && (2.4) \end{aligned}$$

In Poincaré's inequality (2.4) the constant C depends on the size of the domain G. We will also use the imbeddings

$$\|a\|_\infty \le C_1\|a\|_2, \quad \|b\|_{0,p} \le C_2\|\nabla b\| \quad (2 \le p \le 6), \tag{2.5}$$

valid for functions $a \in H^2(G)$ and $b \in H_0^1(G)$, respectively [1]. Of course, the relations (2.1),. . .,(2.5) are also valid for the corresponding vector functions. The completion of

$$\boldsymbol{C}_{0,\sigma}^\infty(G) = \{u \in \boldsymbol{C}_0^\infty(G) \mid \operatorname{div} u = 0\}$$

with respect to the norm $\|.\|$ and $\|.\|_1$ are basic spaces for the treatment of Stokes and Navier–Stokes equations [9]. We denote them by $\boldsymbol{H}(G)$ and $\boldsymbol{V}(G)$, respectively. In $\boldsymbol{H}_0^1(G)$ and $\boldsymbol{V}(G)$ we also use

$$\langle \nabla u, \nabla v \rangle = \sum_{k,j=1}^{3} \langle \partial_k u_j, \partial_k v_j \rangle, \quad \|\nabla u\| = \langle \nabla u, \nabla u \rangle^{1/2}$$

as scalar product and norm. Let

$$P: \boldsymbol{L}^2(G) \longrightarrow \boldsymbol{H}(G) \tag{2.6}$$

denote the orthogonal projection such that

$$\boldsymbol{L}^2(G) = \boldsymbol{H}(G) \oplus \{v \in \boldsymbol{L}^2(G) \mid v = \nabla p \quad \text{for some} \quad p \in H^1(G)\}$$

(see [9]). Then the stationary linear Stokes equations

$$-\Delta u + \nabla p = f \quad \text{in } G, \quad \nabla \cdot u = 0 \quad \text{in } G, \quad u_{|\partial G} = 0, \tag{2.7}$$

transform into

$$Au = Pf \quad \text{in } G \tag{2.8}$$

with the Stokes operator $A = -P\Delta$ in $\boldsymbol{H}(G)$, having domain of definition $D(A) = \boldsymbol{H}^2(G) \cap \boldsymbol{V}(G)$. It is well known [2] that, given $f = Pf \in \boldsymbol{H}(G)$, (2.8) is uniquely solvable with $u \in \boldsymbol{H}^2(G) \cap \boldsymbol{V}(G)$ satisfying

$$\|u\|_2 \leq C\|P\Delta u\| = C\|Au\| = C\|f\|. \tag{2.9}$$

Finally, we use the B–valued spaces $C^m(J, B)$ and $H^m(a, b, B)$, $m \in \mathbb{N}$, where $J \subset \mathbb{R}$ is a compact interval, where $a, b \in \mathbb{R}$ $(a < b)$, and where B is any of the spaces above. In case of $C^0(,)$ we simply write $C(,)$, and we use $\boldsymbol{H}, \boldsymbol{V}, H^m, \ldots$ instead of $\boldsymbol{H}(G), \boldsymbol{V}(G), H^m(G), \ldots$, if the domain of definition is clear from the context.

Concerning the asymptotic behaviour of functions $g : (0, \infty) \to \mathbb{R}$, we sometimes use the notation

$$g(h) = o(h^\alpha) \quad \text{as} \quad h \to 0$$

to express that $h^{-\alpha} g(h) \to 0$ as $h \to 0$.

All constants appearing in this note are generic, i.e. its value may be different in different estimates. Throughout the paper we deal with two kinds of constants. Constants, which may depend on the domain G (on its size or on the regularity of its boundary ∂G) and on the viscosity ν, but not on the data v_0 and F, are always denoted by $C, C_1, C_2, \ldots$. Constants, which, in addition, may also depend on the data v_0 and F, on T, and on the delay ε, are always denoted by $K, K_1, K_2, \ldots$. No constant appearing in this paper depends on the stepsize h.

Throughout this paper, we keep $\varepsilon = T/N > 0$ fixed and consider the equations (N_ε) under the following regularity assumptions on the data F and v_0:

$$\begin{aligned} F \in H^1(0,T,\boldsymbol{H}), \quad & v_0 \in C([-\varepsilon,0],\boldsymbol{H}^2 \cap \boldsymbol{V}) \\ \text{with} \quad & \partial_t v_0 \in C([-\varepsilon,0],\boldsymbol{H}) \cap \boldsymbol{L}^2(-\varepsilon,0,\boldsymbol{V}). \end{aligned} \tag{2.10}$$

The next proposition has been proved in [10] (compare also [12]).

Proposition 2.1: *In case of (2.10) there is an unique solution $v, \nabla p$ of the equations (N_ε) such that $\nabla p \in C([0,T],\boldsymbol{L}^2)$ and that v has on $[0,T]$ the same regularity properties as v_0 on $[-\varepsilon,0]$. Moreover, there is a constant $K = K(G,T,\varepsilon,\nu,F,v_0)$ such that*

$$\int_0^T \|\nabla\partial_t v(t)\|^2 dt \le K, \quad \|v(t)\|_2^2 \le K, \quad \|\partial_t v(t)\|^2 \le K \quad (t \in [0,T]). \tag{2.11}$$

There is a corresponding result for the discrete system (N_ε^h), $h = \varepsilon/n > 0$ $(n \in \mathbb{N})$:

Theorem 2.2: *(Stability)*

In case of (2.10) there is an unique solution $v^k \in \boldsymbol{H}^2 \cap \boldsymbol{V}$, $\nabla p^k \in \boldsymbol{L}^2$ of (N_ε^h) for all $k = 1,2,\ldots,nN$ and a constant $K = K(G,T,\varepsilon,\nu,F,v_0)$ independent of h and k such that for all $k = 1,2,\ldots,nN$

$$\|v^k\|_1^2 \le K, \quad \|P\Delta(v^k + v^{k-1})\|^2 \le K, \quad \left\|\frac{(v^k - v^{k-1})}{h}\right\|^2 \le K. \tag{2.12}$$

The estimates (2.11) and (2.12) are essential for the investigation of the discretization error (defect) $e^k := v^k - v(t_k)$. Here v^k is the solution of (N_ε^h) and $v(t_k)$ the solution of (N_ε) at time $t_k = k \cdot h$. The following theorem will be proved in the next section.

Theorem 2.3: *(Convergence)*

In case of (2.10) there is a constant $K = K(G,T,\varepsilon,\nu,F,v_0)$ such that for all $k = 1,2,\ldots,nN$ the discretization error $e^k = v^k - v(t_k)$ has the following properties:

$$\|e^k\|^2 + h\tfrac{\nu}{4}\sum_{j=1}^{k}\|\nabla(e^j + e^{j-1})\|^2 \le Kh^2, \tag{2.13}$$

$$\|e^k + e^{k-1}\|_2 = o(1) \quad \text{as} \quad h \to 0, \tag{2.14}$$

$$\|e^k - e^{k-1}\|^2 + h\tfrac{\nu}{4}\sum_{j=2}^{k}\|\nabla(e^j - e^{j-2})\|^2 = o(h^2) \quad \text{as} \quad h \to 0. \tag{2.15}$$

Proof of Theorem 2.2: Because existence and uniqueness is known (see [9]), it remains to prove the estimates. In the following, for abbreviation we set

$$V_+^k := v^k + v^{k-1}, \qquad V_-^k := v^k - v^{k-1},$$

and we firstly prove all estimates for $k = 1, 2, \ldots, n$. Using P from (2.6), (N_e^h) implies

$$V_-^k - \tfrac{h\nu}{2} P\Delta V_+^k = \int_{(k-1)h}^{kh} F(t)\,dt - \tfrac{h}{4} P(V_+^{k-n} \cdot \nabla)V_+^k. \tag{2.16}$$

Scalar multiplication in $\boldsymbol{L}^2$ with V_+^k yields

$$\begin{aligned} \|v^k\|^2 - \|v^{k-1}\|^2 + \tfrac{h\nu}{2}\|\nabla V_+^k\|^2 &\le 2\sqrt{\tfrac{h\nu}{4}}\,\|\nabla V_+^k\| \cdot \sqrt{\tfrac{1}{h\nu}}\,C\|\int_{(k-1)h}^{kh} F(t)\,dt\| \\ &\le \tfrac{h\nu}{4}\|\nabla V_+^k\|^2 + C_1 \int_{(k-1)h}^{kh} \|F(t)\|^2 dt. \end{aligned}$$

Here we used (2.1), (2.3), (2.4), and the orthogonality relation (see [9])

$$\langle (u \cdot \nabla)w, w\rangle = 0 \quad (u \in \boldsymbol{V}, w \in \boldsymbol{H}_0^1). \tag{2.17}$$

It follows for all $k = 1, 2, \ldots, n$ (even for $k = 1, \ldots, nN$)

$$\|v^k\|^2 + \tfrac{h\nu}{4}\sum_{j=1}^{k}\|\nabla V_+^j\|^2 \le \|v^0\|^2 + K_1 \tag{2.18}$$

with some constant $K_1 = K_1(\Omega, F)$.
Multiplying (2.16) with V_-^k instead, we find

$$\begin{aligned} &\|V_-^k\|^2 + \tfrac{h\nu}{2}\{\|\nabla v^k\|^2 - \|\nabla v^{k-1}\|^2\} \\ &\le 2 \cdot \tfrac{1}{2}\|V_-^k\|\,\|\int_{(k-1)h}^{kh} F(t)\,dt\| + 2 \cdot \tfrac{h}{4}\|V_+^{k-n}\|_\infty \|\nabla V_+^k\| \cdot \tfrac{1}{2}\|V_-^k\| \\ &\le \tfrac{1}{2}\|V_-^k\|^2 + Kh^2\|\nabla V_+^k\|^2 + h\int_{(k-1)h}^{kh} \|F(t)\|^2 dt, \end{aligned}$$

because $\|V_+^{k-n}\|_\infty^2 \le K$ for $k = 1, 2, \ldots, n$ by (2.5), (2.9), and (2.11). Using (2.18) again, we obtain for all $k = 1, 2, \ldots, n$

$$\|\nabla v^k\|^2 + \tfrac{1}{h\nu}\sum_{j=1}^{k}\|V_-^j\|^2 \le \|\nabla v^0\|^2 + K_2 \tag{2.19}$$

with some constant $K_2 = K_2(\Omega, F, v_0)$. Now (2.18) and (2.19) imply the first estimate of (2.12) for $k = 1, 2, \ldots, n$.

We need another estimate, which follows from (2.16) by scalar multiplication in L^2 with $-P\Delta V_+^k$:

$$\begin{aligned}
\|\nabla v^k\|^2 - \|\nabla v^{k-1}\|^2 + \tfrac{h\nu}{2}\|P\Delta V_+^k\|^2 &\le 2\sqrt{\tfrac{h\nu}{8}}\,\|P\Delta V_+^k\|\cdot\sqrt{\tfrac{2}{h\nu}}\,\|\int_{(k-1)h}^{kh} F(t)\,dt\| \\
&\quad +2\sqrt{\tfrac{h\nu}{8}}\,\|P\Delta V_+^k\|\cdot\sqrt{\tfrac{h}{2\nu}}\|V_+^{k-n}\|_\infty\,\|\nabla V_+^k\| \\
&\le \tfrac{h\nu}{4}\|P\Delta V_+^k\|^2 + C_2\int_{(k-1)h}^{kh}\|F(t)\|^2 dt + Kh\|\nabla V_+^k\|^2.
\end{aligned}$$

Hence, for all $k = 1, 2, \ldots, n$:

$$\|\nabla v^k\|^2 + h\tfrac{\nu}{4}\sum_{j=1}^{k}\|P\Delta V_+^j\|^2 \le \|\nabla v_0\|^2 + K_3 \tag{2.20}$$

with some constant $K_3 = K_3(\Omega, F, v_0)$.

Because (2.16) is equivalent to

$$AV_+^k := -P\Delta V_+^k = \tfrac{2}{\nu}\left(\tfrac{1}{h}\int_{(k-1)h}^{kh} F(t)\,dt - \tfrac{1}{h}V_-^k - \tfrac{1}{4}P(V_+^{k-n}\cdot\nabla)V_+^k\right), \tag{2.21}$$

we see from (2.8) and (2.9) that the second estimate of (2.12) hold for $k = 1, \ldots, n$, if the L^2–norm of the right hand side in (2.21) is bounded independent of h and k for $k = 1, \ldots, n$. Because

$$\begin{aligned}
\|\tfrac{1}{h}\int_{(k-1)h}^{kh} F(t)\,dt\| &\le \max_{t\in[0,T]}\|F(t)\| \le K, \\
\|(V_+^{k-n}\cdot\nabla)V_+^k\| &\le 2\|V_+^{k-n}\|_\infty\cdot\max_{0\le k\le n}\|\nabla v^k\| \le K, \qquad \text{(see (2.19))}
\end{aligned}$$

our next estimate to prove is

$$\|V_-^k\| \le Kh \tag{2.22}$$

for $k = 1, 2, \ldots, n$.

Because

$$V_+^k - V_+^{k-1} = V_-^k + V_-^{k-1}\,, \tag{2.23}$$

from (2.16) we obtain ($k \ge 2$)

$$\begin{aligned}
V_-^k - V_-^{k-1} - \tfrac{h\nu}{2}\,P\Delta(V_-^k + V_-^{k-1}) &= \int_{(k-1)h}^{kh}(F(t) - F(t-h))\,dt \\
-\tfrac{h}{4}\,P\Big[\big((V_-^{k-n} + V_-^{k-n-1})\cdot\nabla\big)V_+^k &+ (V_+^{k-n-1}\cdot\nabla)(V_-^k + V_-^{k-1})\Big].
\end{aligned} \tag{2.24}$$

Scalar multiplication in L^2 with $(V_-^k + V_-^{k-1})$ gives

$$\|V_-^k\|^2 - \|V_-^{k-1}\|^2 + \tfrac{h\nu}{2}\,\|\nabla(V_-^k + V_-^{k-1})\|^2$$

$$\leq 2\sqrt{\tfrac{h\nu}{8}}\,\|\nabla(V_-^k + V_-^{k-1})\|\,\sqrt{\tfrac{2}{h\nu}}\cdot C\,\|\int_{(k-1)h}^{kh}(F(t) - F(t-h))\,dt\|$$

$$+2\sqrt{\tfrac{h\nu}{8}}\,\|\nabla(V_-^k + V_-^{k-1})\|\cdot\sqrt{\tfrac{h}{8\nu}}\,\|V_-^{k-n} + V_-^{k-n-1}\|\,\|V_+^k\|_\infty$$

$$\leq \tfrac{h\nu}{4}\,\|\nabla(V_-^k + V_-^{k-1})\|^2 + C_3h^2\int_{(k-2)h}^{kh}\|\partial_t F(t)\|^2dt + Kh^3\|P\Delta V_+^k\|^2.$$

Here for the last estimate we used that for $k = 2,\ldots,n$

$$\|V_-^{k-n} + V_-^{k-n-1})\|^2 \leq 4\max_{1\leq j\leq n}\|V_-^{j-n}\|^2 \leq Kh^2$$

and the first inequality of (2.5).

Hence by (2.20) we obtain for all $k = 2,\ldots,n$:

$$\|V_-^k\|^2 + \tfrac{h\nu}{4}\sum_{j=2}^{k}\|\nabla(V_-^j + V_-^{j-1})\|^2 \leq \|V_-^1\|^2 + K_4h^2, \tag{2.25}$$

and (2.21) holds for $k = 1,\ldots,n$ if only $\|V_-^1\|^2 = \|v^1 - v^0\|^2 \leq K_5h^2$.

To see this, consider (2.16) for $k = 1$: Because

$$V_-^1 - \tfrac{h\nu}{2}\,P\Delta V_-^1 = h\nu P\Delta v^0 + \int_0^h F(t)\,dt - \tfrac{h}{4}\,P(V_+^{1-n}\cdot\nabla)V_+^1\,,$$

we obtain by scalar multiplication in L^2 with V_-^1:

$$\begin{aligned}\|V_-^1\|^2 + \tfrac{h\nu}{2}\,\|\nabla V_-^1\|^2 \;&\leq\; 2\cdot\tfrac{1}{\sqrt{6}}\,\|V_-^1\|\cdot C_4h\,\|P\Delta v^0\| + 2\cdot\tfrac{1}{\sqrt{6}}\,\|V_-^1\|\cdot C_4\|\int_0^h F(t)\,dt\|\\ &\quad +2\cdot\tfrac{1}{\sqrt{6}}\,\|V_-^1\|\cdot C_4h\|V_+^{1-n}\|_\infty\|\nabla V_+^1\|\\ &\leq\; \tfrac{1}{2}\|V_-^1\|^2 + Kh^2.\end{aligned}$$

It follows

$$\|V_-^1\|^2 + h\nu\|\nabla V_-^1\|^2 \leq K_5h^2 \tag{2.26}$$

and thus (2.22). Hence all estimates asserted in (2.12) for $k = 1,\ldots,nN$ are valid for $k = 1,\ldots,n$, due to (2.18), (2.19), (2.25) and (2.26). Repeating this procedure a finite number of times, we obtain (2.12) for all $k = 1,\ldots,nN$ successively, i.e. for $k = n+1,\ldots,2n$, for $k = 2n+1,\ldots,3n,\ldots,$ for $k = (N-1)n+1,\ldots,Nn$.

This proves the theorem.

3. Proof of Convergence

To show the asserted convergence properties of the solution of (N_ε^h) if h tends to zero ($\varepsilon > 0$ remains fixed), we use the Lax–Richtmyer approach "stability + consistency $\longrightarrow$ convergence". Let us set

$$
\begin{aligned}
V_+^k &= v^k + v^{k-1}, & V_-^k &= v^k - v^{k-1},\\
U_+^k &= v(t_k) + v(t_{k-1}), & U_-^k &= v(t_k) - v(t_{k-1}),\\
U_+^{\varepsilon,k} &= v(t_{k-n}) + v(t_{k-n-1}), & U_-^{\varepsilon,k} &= v(t_{k-n}) - v(t_{k-n-1}),
\end{aligned}
$$

for abbreviation, and define

$$
\begin{aligned}
\big(\Pi_h\{v^j\}\big)(t_k) &= V_-^k - \tfrac{h\nu}{2} P\Delta V_+^k + \tfrac{h}{4} P(V_+^{k-n}\cdot\nabla)V_+^k,\\
\big(\Pi_h v\big)(t_k) &= U_-^k - \tfrac{h\nu}{2} P\Delta U_+^k + \tfrac{h}{4} P(U_+^{k-n}\cdot\nabla)U_+^k.
\end{aligned}
$$

For the defect $e^k = v^k - v(t_k)$ we set, analogously,

$$
E_+^k = e^k + e^{k-1}, \qquad E_-^k = e^k - e^{k-1}.
$$

This leads to the identity

$$
E_-^k - \tfrac{h\nu}{2} P\Delta E_+^k + \tfrac{h}{4} P\big((E_+^{k-n}.\nabla)V_+^k + (U_+^{k-n}.\nabla)E_+^k\big) = (\Pi_h\{v^j\})(t_k) - (\Pi_h v)(t_k) = F^k, \quad (3.1)
$$

which is used to obtain estimates of e^k in terms of F^k ($\approx$ Stability). Then the behaviour of

$$
\begin{aligned}
F^k = &\int\limits_{(k-1)h}^{kh} \big\{\partial_t v(t) - \nu P\Delta v(t) + P(v_\varepsilon(t)\cdot\nabla)v(t)\big\}\, dt\\
&-\big\{U_-^k - \tfrac{h\nu}{2} P\Delta U_+^k + \tfrac{h}{4} P\big(U_+^{\varepsilon,k}\cdot\nabla\big)U_+^k\big\}\\
= &\int\limits_{(k-1)h}^{kh} \Big\{ -\nu P\Delta\big(v(t) - \tfrac{U_+^k}{2}\big)\\
&+P\Big(\big(v_\varepsilon(t) - \tfrac{U_+^{\varepsilon,k}}{2}\big)\cdot\nabla\Big)v(t) + P\big(\tfrac{U_+^{\varepsilon,k}}{2}\cdot\nabla\big)\big(v(t) - \tfrac{U_+^k}{2}\big)\Big\}\, dt
\end{aligned}
\quad (3.2)
$$

as h tends to zero ($\approx$ Consistency) follows from the regularity properties of the exact solution of the equations (N_ε).

Proof of Theorem 2.3: Multiplying (3.1) scalar in L^2 by E_+^k gives

$$
\begin{aligned}
\|e^k\|^2 - \|e^{k-1}\|^2 + h\tfrac{\nu}{2}\|\nabla E_+^k\|^2 &\le 2\sqrt{\tfrac{h\nu}{16}}\,\|\nabla E_+^k\|\cdot\sqrt{\tfrac{4h}{\nu}}\,\|V_+^k\|_\infty\|E_+^{k-n}\| + \langle F^k, E_+^k\rangle\\
&\le \tfrac{h\nu}{16}\|\nabla E_+^k\|^2 + hK\|E_+^{k-n}\|^2 + |\langle F^k, E_+^k\rangle|,
\end{aligned}
$$

where we used (2.3), (2.5), and (2.12) to estimate $\|V_+^k\|_\infty$, and the orthogonality relation (2.17). The term $|\langle F^k, E_+^k\rangle|$ can be estimated as follows:

From (3.2) we find, setting $\alpha := \|\nabla E_+^k\|$ for abbreviation,

$$\begin{aligned}
|\langle F^k, E_+^k\rangle| \;\le\;& \nu\alpha \cdot \Big\| \int_{(k-1)h}^{kh} \nabla\big(v(t) - \tfrac{U_+^k}{2}\big)\, dt \Big\| + \alpha \cdot \Big\| \int_{(k-1)h}^{kh} \big(v_\varepsilon(t) - \tfrac{U_+^{\varepsilon,k}}{2}\big)\, dt \Big\| \cdot \max_{t\in[0,T]} \|v(t)\|_\infty \\
&+ \alpha \cdot \Big\| \int_{(k-1)h}^{kh} \big(v(t) - \tfrac{U_+^k}{2}\big)\, dt \Big\| \cdot \max_k \|U_+^{\varepsilon,k}\|_\infty \\
\le\;& 3\tfrac{h\nu}{16}\alpha^2 + C_1 \int_{(k-1)h}^{kh} \Big\| \int_{(k-1)h}^{kh} |\nabla \partial_\sigma v(\sigma)|\, d\sigma \Big\|^2 dt \\
&+ K \int_{(k-1)h}^{kh} \Big[\Big\| \int_{(k-1)h}^{kh} |\partial_\sigma v_\varepsilon(\sigma)|\, d\sigma \Big\|^2 + \Big\| \int_{(k-1)h}^{kh} |\partial_\sigma v(\sigma)|\, d\sigma \Big\|^2 \Big]\, dt\,.
\end{aligned}$$

Here we used the Taylor expansion

$$2v(t) - U_+^k = 2v(t) - (v(t_k) + v(t_{k-1})) = \int_{(k-1)h}^{t} \partial_\sigma v(\sigma)\, d\sigma - \int_{t}^{kh} \partial_\sigma v(\sigma)\, d\sigma,$$

which implies

$$|2v(t) - U_+^k| \le \int_{(k-1)h}^{kh} |\partial_\sigma v(\sigma)|\, d\sigma.$$

It follows

$$|\langle F^k, E_+^k\rangle| \;\le\; 3\tfrac{h\nu}{16}\, \|\nabla E_+^k\|^2 + Kh^2 \int_{(k-1)h}^{kh} \big(\|\nabla \partial_t v(t)\|^2 + \|\partial_t v_\varepsilon(t)\|^2\big)\, dt,$$

and we obtain for all $k = 1, \ldots, n$

$$\|e^k\|^2 + h\tfrac{\nu}{4} \sum_{j=1}^{k} \|\nabla E_+^k\|^2 \le K_1 h^2, \tag{3.3}$$

with some constant K_1 independent of h and k.

Here we used, in particular, $\|e^0\| = \|v^0 - v(0)\| = 0$, and

$$\|E_+^{k-n}\| \le \|e^{k-n}\| + \|e^{k-n-1}\| \le Kh. \tag{3.4}$$

Note that on the present level, i.e. for $k = 1, \ldots, n$, we even have $\|E_+^{k-n}\| = 0$, but we only use (3.4) in order to keep this technique of estimation working on subsequent levels, too.

Because of (3.1) we have

$$-P\Delta E_+^k = \tfrac{2}{h\nu}(F^k - E_-^k) - \tfrac{1}{2\nu}P\Big((E_+^{k-n}\cdot\nabla)V_+^k + (U_+^{k-n}\cdot\nabla)E_+^k\Big),$$

where

$$\begin{aligned}\|\tfrac{1}{h}F^k\|^2 &\le \tfrac{K}{h}\int_{(k-1)h}^{kh}\Big\{\|P\Delta(2v(t)-U_+^k)\|^2 + \|P\Delta(2v_\epsilon(t)-U_+^{\epsilon,k})\|^2\Big\}\,dt\\ &\le K\max_{\substack{\sigma,\tau\in[-\epsilon,T]\\ |\sigma-\tau|\le h}}\|P\Delta(v(\sigma)-v(\tau))\|^2 = o(1)\end{aligned}$$

as $h\to 0$ due to the strong H^2–continuity of v, where

$$\|(E_+^{k-n}\cdot\nabla)V_+^k\| \le C_1\|E_+^{k-n}\|_{0,3}\,\|\nabla V_+^k\|_{0,6} \le C_2\|\nabla E_+^{k-n}\|\|P\Delta V_+^k\| \le K_1 h$$

due to (2.12) and (3.3), and where

$$\|(U_+^{k-n}\cdot\nabla)E_+^k\| \le \|U_+^{k-n}\|_\infty\|\nabla E_+^k\| \le K_2 h.$$

Thus it remains to prove

$$\|E_-^k\|^2 = o(h^2) \quad \text{as} \quad h\longrightarrow 0, \tag{3.5}$$

and we obtain (2.14) with help of Cattabriga's estimate (2.9).

To prove (3.5), we consider (3.1) for k and $k-1$, obtaining for $2\le k\le n$

$$\begin{aligned}E_-^k - E_-^{k-1} - h\tfrac{\nu}{2}P\Delta(E_-^k + E_-^{k-1}) = -\tfrac{h}{4}P\Big(\big((E_-^{k-n} + E_-^{k-n-1})\cdot\nabla\big)V_+^k\\ +(E_+^{k-n-1}\cdot\nabla)(V_-^k + V_-^{k-1}) + \big((U_-^{k-n} + U_-^{k-n-1})\cdot\nabla\big)E_+^k\\ +(U_+^{k-n-1}\cdot\nabla)(E_-^k + E_-^{k-1})\Big) + F^k - F^{k-1}.\end{aligned}$$

Now scalar multiplication in $\boldsymbol{L}^2$ by $(E_-^k + E_-^{k-1})$ gives

$$\|E_-^k\|^2 - \|E_-^{k-1}\|^2 + h\tfrac{\nu}{2}\|\nabla(E_-^k + E_-^{k-1})\|^2 \le \sum_{j=1}^{5} s_j, \tag{3.6}$$

and, setting $\alpha := \|\nabla(E_-^k + E_-^{k-1})\|$, we estimate as follows:

$$\begin{aligned}s_1 &\le \tfrac{h}{4}\alpha\|V_+^k\|_{0,3}\,\|E_-^{k-n} + E_-^{k-n-1}\|_{0,6}\\ &\le K_1 h\alpha\|\nabla(E_-^{k-n} + E_-^{k-n-1})\|\\ &\le h\tfrac{\nu}{32}\alpha^2 + hK_2\|\nabla(E_-^{k-n} + E_-^{k-n-1})\|^2,\end{aligned}$$

$$\begin{aligned}s_2 &\le \tfrac{h}{4}\alpha\|E_+^{k-1-n}\|_\infty\|V_-^k + V_-^{k-1}\|\\ &\le h\tfrac{\nu}{32}\alpha^2 + Kh^3\|P\Delta E_+^{k-1-n}\|^2,\end{aligned}$$

$$
\begin{aligned}
s_3 &\leq \tfrac{h}{4}\alpha\|E_+^k\|\|U_-^{k-n}+U_-^{k-n-1}\|_\infty \\
&\leq h\tfrac{\nu}{32}\alpha^2 + Kh^3 \max_{\substack{|\sigma-\tau|\leq h\\ \sigma,\tau\in[-\epsilon,T]}}, \|P\Delta\big(v(\sigma)-v(\tau)\big)\|^2,
\end{aligned}
$$

$$
s_4 = 0 \qquad (\text{see } (2.17)),
$$

$$
\begin{aligned}
s_5 &= \langle F^k - F^{k-1}, E_-^k + E_-^{k-1}\rangle \\
&\leq \alpha\cdot\Big\|\int_{(k-1)h}^{kh} \nabla\Big(v(t) - \tfrac{U_+^k}{2} - v(t-h) + \tfrac{U_+^{k-1}}{2}\Big)\,dt\Big\| \\
&\quad +\alpha\int_{(k-1)h}^{kh} \|v_\epsilon(t) - \tfrac{U_+^{\epsilon,k}}{2} - v_\epsilon(t-h) + \tfrac{U_+^{\epsilon,k-1}}{2}\|\cdot\|v(t)\|_\infty\,dt \\
&\quad +\alpha\int_{(k-1)h}^{kh} \|v_\epsilon(t-h) - \tfrac{U_+^{\epsilon,k-1}}{2}\|\cdot\|v(t)-v(t-h)\|_\infty\,dt \\
&\quad +\alpha\int_{(k-1)h}^{kh} \Big\|\frac{U_+^{\epsilon,k}-U_+^{\epsilon,k-1}}{2}\Big\|\cdot\Big\|v(t)-\frac{U_+^k}{2}\Big\|_\infty dt \\
&\quad +\alpha\int_{(k-1)h}^{kh} \Big\|\frac{U_+^{\epsilon,k-1}}{2}\Big\|_\infty\cdot\Big\|v(t)-\frac{U_+^k}{2}-v(t-h)+\frac{U_+^{k-1}}{2}\Big\|\,dt \\
&=: \sum_{j=1}^{5}\sigma_j,
\end{aligned}
$$

where

$$
\begin{aligned}
\sigma_1 &\leq h\tfrac{\nu}{32}\alpha^2 + c_1\int_{(k-1)h}^{kh}\|\int_{(k-1)h}^{kh}|\partial_\sigma\nabla\big(v(\sigma)-v(\sigma-h)\big)|\,d\sigma\|^2 dt \\
&\leq h\tfrac{\nu}{32}\alpha^2 + c_1h^2\int_{(k-1)h}^{hk}\|\partial_t\nabla\big(v(t)-v(t-h)\big)\|^2 dt,
\end{aligned}
$$

$$
\begin{aligned}
\sigma_2 &\leq h\tfrac{\nu}{32}\alpha^2 + K\int_{(k-1)h}^{kh}\|\int_{(k-1)h}^{kh}|\partial_\sigma\big(v_\epsilon(\sigma)-v_\epsilon(\sigma-h)\big)|\,d\sigma\|^2 dt \\
&\leq h\tfrac{\nu}{32}\alpha^2 + Kh^3\max_{\substack{|\sigma-\tau|\leq h\\ \sigma,\tau\in[-\epsilon,T]}}\|\partial_t\big(v(\tau)-v(\sigma)\big)\|^2 \\
&= h\tfrac{\nu}{32}\alpha^2 + Kh^3\cdot o(1) \quad \text{as} \quad h\to 0,
\end{aligned}
$$

$$\begin{aligned}
\sigma_3 &\le h\tfrac{\nu}{32}\alpha^2 + c_1 \max_{\substack{|\sigma-\tau|\le h\\ \sigma,\tau\in[-\epsilon,T]}} \|P\Delta\big(v(\tau)-v(\sigma)\big)\|^2 \cdot \int\limits_{(k-1)h}^{kh} \| \int\limits_{(k-2)h}^{(k-1)h} |\partial_\sigma v_\epsilon(\sigma)|\, d\sigma\|^2 dt \\
&\le h\tfrac{\nu}{32}\alpha^2 + Kh^3 \cdot \max_{\substack{|\sigma-\tau|\le h\\ \sigma,\tau\in[-\epsilon,T]}} \|P\Delta\big(v(\tau)-v(\sigma)\big)\|^2 \\
&= h\tfrac{\nu}{32}\alpha^2 + Kh^3 o(1) \quad \text{as} \quad h\to 0,
\end{aligned}$$

$$\begin{aligned}
\sigma_4 &\le h\tfrac{\nu}{32}\alpha^2 + Kh^3 \cdot \max_{\substack{|\sigma-\tau|\le h\\ \sigma,\tau\in[-\epsilon,T]}} \|P\Delta\big(v(\tau)-v(\sigma)\big)\|^2 \\
&= h\tfrac{\nu}{32}\alpha^2 + Kh^3 o(1) \quad \text{as} \quad h\to 0,
\end{aligned}$$

$$\begin{aligned}
\sigma_5 &\le h\tfrac{\nu}{32}\alpha^2 + K \int\limits_{(k-1)h}^{kh} \|v(t) - \tfrac{U_+^k}{2} - v(t-h) + \tfrac{U_+^{k-1}}{2}\|^2 dt \\
&\le h\tfrac{\nu}{32}\alpha^2 + Kh^3 \max_{\substack{|\sigma-\tau|\le h\\ \sigma,\tau\in[-\epsilon,T]}} \|\partial_t\big(v(\tau)-v(\sigma)\big)\|^2 \\
&= h\tfrac{\nu}{32}\alpha^2 + Kh^3 o(1) \quad \text{as} \quad h\to 0.
\end{aligned}$$

Now collecting all estimates for s_1, s_2, s_3, and $\sigma_1, \ldots, \sigma_5$, we obtain from (3.6):

$$\begin{aligned}
\|E_-^k\|^2 - \|E_-^{k-1}\|^2 + h\tfrac{\nu}{4}\|\nabla(E_-^k + E_-^{k-1})\|^2 &\le K\Big(h\|\nabla(E_-^{k-n} + E_-^{k-n-1})\|^2 \\
&+ h^3\|P\Delta E_+^{k-1-n}\|^2 + h^3 o(1) + h^2 \int\limits_{(k-1)h}^{kh} \|\partial_t\nabla\big(v(t)-v(t-h)\big)\|^2 dt\Big).
\end{aligned}$$

This implies for all $k = 2, \ldots, n$:

$$\begin{aligned}
&\|E_-^k\|^2 + h\tfrac{\nu}{4}\sum_{j=2}^{k}\|\nabla(E_-^j + E_-^{j-1})\|^2 \le \\
&\le \|E_-^1\|^2 + K\Big(h\sum_{j=2}^{k}\|\nabla(E_-^{j-n} + E_-^{j-n-1})\|^2 + h^2 T\big(\max_{j=1,\ldots,n}\|P\Delta E_+^{j-n}\|^2\big) \\
&\quad + h^2 T o(1) + h^2 \int\limits_h^T \|\partial_t\nabla\big(v(t)-v(t-h)\big)\|^2 dt\Big).
\end{aligned} \tag{3.7}$$

Because $e^0 = E_+^{1-n} = 0$, from (3.1) we find

$$E_-^1 - h\tfrac{\nu}{2}P\Delta E_-^1 = F^1 - \tfrac{h}{4}P(U_+^{1-n}\cdot\nabla)E_+^1\ ,$$

and scalar multiplication in $\boldsymbol{L}^2$ by E_-^1 gives

$$\begin{aligned}
\|E_-^1\|^2 + h\tfrac{\nu}{2}\|\nabla E_-^1\|^2 &\le |\langle F^1, E_-^1\rangle| + 2\sqrt{\tfrac{1}{8}}\,\|E_-^1\| \cdot \tfrac{h}{\sqrt{8}}\|U_+^{1-n}\|_\infty\|\nabla E_+^1\| \\
&\le \tfrac{1}{2}\|E_-^1\|^2 + o(h^2),
\end{aligned}$$

hence

$$\|E_-^1\|^2 + h\nu\|\nabla E_-^1\|^2 \le o(h^2). \tag{3.8}$$

Now (3.7) and (3.8) imply (3.5) and thus (2.14) and (2.15) for all $k = 1, \dots, n$. Repeating this procedure a finite number of times as pointed out at the end of Section 2, the theorem is proved.

References

[1] Adams, R.A.: Sobolev Spaces. New York et al., Academic Press 1975

[2] Cattabriga, L.: Su un Problema al Contorno Relativo al Sistema di Equazioni di Stokes. Sem. Mat. Univ. Padova 31 (1964) 308–340

[3] Girault, V., Raviart, P.A.: Finite Element Approximation of the Navier–Stokes Equations. Berlin et al., Springer 1979

[4] Heywood, J.G., Rannacher, R.: Finite Element Approximation of the Nonstationary Navier–Stokes Problem. II. Stability of Solutions and Error Estimates Uniform in Time. Siam J. Numer. Anal. 23 (1986) 750-777

[5] Hopf, E.: Über die Anfangswertaufgabe für die hydrodynamischen Grundgleichungen. Math. Nachr. 4 (1951) 213–231

[6] Kaniel, S., Shinbrot, M.: The Initial Value Problem for the Navier–Stokes Equations. Arch. Rat. Anal. 21 (1966) 270–285

[7] Lax, P.D., Richtmeyer, R.D.: Survey of the Stability of Linear Finite Difference Equations. Com. Pure Appl. Math. 9 (1956) 267–293

[8] Rautmann, R.: Zur Konvergenz des Rothe–Verfahrens für instationäre Stokes–Probleme in dreidimensionalen Gebieten. Z. Angew. Math. Mech. (ZAMM) 64 (1984) T387–388

[9] Temam, R.: Navier–Stokes Equations. Amsterdam et al., North Holland 1977

[10] Varnhorn, W.: Zur Numerik der Gleichungen von Navier–Stokes. Universität Paderborn, Dissertation 1985

[11] Varnhorn, W.: Time Stepping Procedures for the Nonstationary Stokes Equations. Preprint 1353 Technische Hochschule Darmstadt (1991) 1–20 (to appear in Math. Meth. Appl. Sci)

[12] Varnhorn, W.: Time Delay and Finite Differences for the Nonstationary Nonlinear Navier–Stokes Equations. Preprint 1359 Technische Hochschule Darmstadt (1991) 1–27 (to appear in Math. Meth. Appl. Sci)

Finite Difference Vorticity Methods

Brian T. R. Wetton
Department of Mathematics. University of British Columbia.
Vancouver, BC, Canada V6T 1Y4

Abstract

Computations are presented for finite difference vorticity and primitive methods for the time dependent incompressible Navier Stokes equations in bounded domains. To directly compare solutions of the methods, modifications must be made to the primitive method and these are described. Both methods are based on semi-discrete methods that have been proven to be second order accurate. The computational results show second order uniform convergence in the velocities and vorticity using arbitrary (not compatible at time 0) initial data showing the validity of the time discrete methods for which only partial convergence results are known. An increase in the expected convergence order of the boundary vorticity using Thom's first order boundary condition is observed. Comparisons between the methods and between different boundary conditions in the vorticity method are given.

1 Introduction

The purpose of this paper is to examine computationally the convergence properties of two finite difference methods for the incompressible Navier Stokes equations in bounded domains. One of the two methods is based on the vorticity - stream function form of the equations and the other on the primitive (velocity and pressure) form of the equations. The primitive method is considered in this paper on vorticity methods for comparative purposes and also because it will furnish approximations to the vorticity by differencing the velocity values. Both methods considered are based on semi-discrete methods that have been proven second order accurate [7, 8]. The time discretization used here handles the convective terms in a leap frog manner and are initialized with a single forward euler step. The diffusive terms are handled in a Crank Nicholson fashion. In the primitive method, the pressure is solved using an iteration scheme. The computational convergence studies described here were designed to close some gaps in the convergence theory of these methods: the use of the initial forward euler step, the issue of smoothing of incompatible initial data, and the use of a single iteration for the pressure in the primitive method instead of iteration to convergence. A final question of interest is which method performs better in practice and so the primitive and vorticity methods (using several different numerical vorticity boundary conditions) are compared.

The computations model the time evolution of a perturbed Poiseuille flow in a periodic channel at reynolds number 180. This can be considered an easy computation since the reynolds number is well below the critical value for Poiseuille flow. Even though the finite difference methods are implicit, they can be solved very cheaply: after a discrete fourier transform in the periodic direction, the equations reduce to a tri- or penta-diagonal system for each component which can be solved efficiently. There are also other factors that make this example a good one. For instance, unlike the case of flow past a cylinder, there is no need to consider boundary conditions on an artificial computational surface. Other flows, such as that in a driven cavity or over a step, have boundaries with corners where singularities in the flow can be expected to develop. The modeling of these singularities adds another source of error and may confuse the errors coming from the underlying methods.

In the periodic channel, there is a hidden boundary condition which is represented by the net flow through the channel or the net pressure drop through the channel over one period. When the vorticity method is used the natural condition to apply is that the net flow stays constant. In the primitive method, the natural condition is that the net pressure drop is zero. To directly compare the methods, the pressure algorithm is modified in a stable way that approximates the constant net flow condition to second order accuracy.

In the next section, the two finite difference methods are presented. In section 3, the issues to be resolved by the computations are summarized. In section 4, the computations and results of the convergence and comparison studies are described. This is followed by a brief discussion.

2 Numerical Methods

Finite difference methods based on the primitive (velocity and pressure) formulation and the vorticity-stream function formulation of the equations are presented below. They are formulated in the two-dimensional periodic channel Ω: a unit square periodic in the horizontal x direction with two walls at $y = 0$ and $y = 1$. Modifications to the primitive algorithm to make it model the correct upstream boundary condition are also discussed.

2.1 Primitive Method

The two dimensional incompressible Navier-Stokes equations are given below

$$\mathbf{u}_t = -\mathbf{u} \cdot \nabla \mathbf{u} + \nu \Delta \mathbf{u} - \nabla p \qquad (1)$$

$$\nabla \cdot \mathbf{u} = 0 \qquad (2)$$

where $\mathbf{u} = (u, v)$ are the velocities and p is the pressure. Homogeneous boundary values for u and v are given on the upper and lower walls and it is assumed that both $\mathbf{u}$ and p are 1-periodic in the x-direction. The condition

$$\int p = 0 \qquad (3)$$

fixes the choice of the arbitrary constant in the pressure. Initial data $\mathbf{u}_0 = (u_0, v_0)$ is given and it is assumed that $\nabla \cdot \mathbf{u}_0 = 0$.

We turn now to the specification of the finite difference algorithm to approximate these equations. The MAC staggered grid [3] with grid spacing k in time and h in space is used in the approximation of the velocities and the pressure. That is, we consider approximations

$$\mathbf{U}^n_{i,j}(t) = (U^n_{i,j}, V^n_{i,j}) \text{ and } P^n_{i,j} \tag{4}$$

to

$$u((ih, (j-1/2)h; nk), \; v((i-1/2)h, jh; nk) \text{ and } p((i-1/2)h, (j-1/2)h; nk). \tag{5}$$

The discrete divergence of a vector field $\mathbf{U} = (U, V)$ is defined as

$$\mathbf{D} \cdot \mathbf{U}_{i,j} = D^x_- U_{i,j} + D^y_- V_{i,j} \tag{6}$$

and the discrete gradient of a scalar field P is given by

$$\mathbf{G}P_{i,j} = (D^x_+ P_{i,j}, D^y_+ P_{i,j}) \tag{7}$$

where D_+ and D_- are forward and backward differences, respectively. Note that $\mathbf{D} \cdot \mathbf{U}$ is given by differences centered on the scalar pressure points and the differences in $\mathbf{G}P$ are centered at the vector velocity points. Following Anderson [1], we define the reduced divergence operator $\mathbf{D}^* \cdot$ as follows: it equals $\mathbf{D} \cdot$ at scalar points with $2 \leq j \leq N-1$ and at $j = 1$ it is given by

$$\mathbf{D}^* \cdot \mathbf{U}_{i,1} = D^x_- U_{i,1} + V_{i,1}/h \tag{8}$$

and at $j = N$ by

$$\mathbf{D}^* \cdot \mathbf{U}_{i,N} = D^x_- U_{i,N} - V_{i,N-1}/h. \tag{9}$$

Note that $\mathbf{D}^* \cdot$ involves only interior variables and that $\mathbf{D}^* \cdot \mathbf{U}$ equals $\mathbf{D} \cdot \mathbf{U}$ when $\mathbf{U}$ has homogeneous normal values at the boundary as in our case.

Assume that $\mathbf{U}^0$ and $\mathbf{U}^1$ are given approximations to the velocity at time zero and time k that satisfy $\mathbf{D} \cdot \mathbf{U}^0 = \mathbf{D} \cdot \mathbf{U}^1 = 0$. The discretization for the momentum equations is:

$$\frac{\mathbf{U}^{n+1} - \mathbf{U}^{n-1}}{2k} = -N(\mathbf{U}^n) - \mathbf{G}P^n + \frac{1}{2}\nu\Delta_h(\mathbf{U}^{n+1} + \mathbf{U}^{n-1}) \tag{10}$$

where $N(\mathbf{U}^n)$ or N^n are the nonlinear convection terms evaluated using centered differences and Δ_h is the standard 5-point discrete Laplacian using homogeneous boundary conditions for V and homogeneous reflection boundary conditions for U (since the values of U do not lie on the boundary). If we solve formally for $\mathbf{U}^{n+1}$ we get

$$\mathbf{U}^{n+1} = (I - k\nu\Delta_h)^{-1}(-2kN^n - 2k\mathbf{G}P^n + (I + k\nu\Delta_h)\mathbf{U}^{n-1}) \tag{11}$$

where I is the identity matrix on vector fields. Simply by applying $\mathbf{D}^* \cdot$ to (11) and setting the resulting expression to zero (to ensure that the velocity at the next step is divergence free) we get the equation for the pressure:

$$2k\mathbf{D}^* \cdot (I - k\nu\Delta_h)^{-1}GP^n = \mathbf{D}^* \cdot (-2kN^n + (I + k\nu\Delta_h)\mathbf{U}^{n-1}). \tag{12}$$

Using the symmetry of the vector operator Δ_h and the fact that it is negative definite, we can show that this equation has a unique solution for P^n with the side condition

$$\sum_{i,j=1}^{N} P^n_{i,j} = 0. \tag{13}$$

The vorticity at position (ih, jh) can be computed from the velocities as follows:

$$D_+^x V_{i,j} - D_+^y U_{i,j}. \tag{14}$$

To get the initial data, proceed as follows: let $\tilde{\mathbf{U}}_{i,j} = (u_0(ih,(j-1/2)h), v_0((i-1/2)h, jh))$. Now take $\mathbf{U}^0 = \mathcal{P}\tilde{\mathbf{U}}$, where $\mathcal{P}$ represents the projection onto discrete divergence free fields (see [8]), and get $\mathbf{U}^1$ by using a single forward euler projection step. The semi-discrete version of this scheme converges in max norm for both velocities and pressure with second order accuracy [8]. The analysis can be easily extended to show that the vorticity given by (14) converges with first order on the boundary and second order in the interior. The techniques of the proof are careful energy estimates and an asymptotic characterization of the error.

Suppose that k is related to h by the fixed ratio $k = h/\lambda$ where $\lambda = \alpha \sup |\mathbf{u}|$ with $\alpha > 1$. Then fully discrete version (11)-(12) can be shown to converge with second order in space and time under the additional assumption that $\mathbf{U}^1$ match the asymptotic error of the rest of the scheme to a high order of accuracy [15]. Since the first step is calculated using another method, this is clearly not the case. Some results by Michelson [9] for pure hyperbolic problems suggest that this difficulty in the proof may be handled using additional, initial layer error terms.

Unfortunately, we cannot compute the solution to (12) directly in a computationally efficient way. It is possible to solve (10) and $\mathbf{D}^* \cdot \mathbf{U}^{n+1} = 0$ as a coupled system for $\mathbf{U}^{n+1}$ and P^n. However, we reduce the size of the implicit system by using an iterative scheme. A second superscript is used to denote the iteration level. We take P^{n-1} as the initial guess for the value of P^n, i.e. $P^{n,0} = P^{n-1}$ (the initial forward euler step gives a value for P^0). Then we compute:

$$\begin{aligned} \mathbf{U}^{n+1,j+1} &= (I - k\nu\Delta_h)^{-1}(-2kN^n - 2k\mathbf{G}P^{n,j} + (I + k\nu\Delta_h)\mathbf{U}^{n-1}) && (15) \\ 2k\mathbf{D}^* \cdot \mathbf{G}P^{n,j+1} &= \mathbf{D}^* \cdot (-2kN^n + (I + k\nu\Delta_h)\mathbf{U}^{n-1} + k\nu\Delta_h \mathbf{U}^{n,j+1}). && (16) \end{aligned}$$

This scheme is similar to that proposed by Bell, Colella and Glaz [2] although they use a more complicated expression for the nonlinear convection terms. The above iteration has a fixed point the equations (10), (12) as desired. It can be shown that the above iteration, when considered as a map for the discrete pressure, is a contraction. However, there is a certain difficulty since the contraction parameter for the pressure map tends to 1 as h tends to zero as in the following formula:

$$\frac{\nu}{ch^2/k + \nu} \tag{17}$$

for some $c > 0$ depending on the domain only. In the computational example described below this difficulty does not manifest itself: we observe convergence of the scheme using a single iteration after which the velocities are updated using (15).

In the periodic channel geometry, the iterations corresponding to (15)-(16) can be performed quickly by taking the discrete fourier transform in the horizontal direction. In this case, the implicit equations become tri-diagonal in each fourier component.

2.2 Vorticity Methods

The two dimensional Navier-Stokes in terms of the vorticity ω and stream function ψ are given below:

$$\omega_t + u\omega_x + v\omega_y = \nu\Delta\omega \tag{18}$$

$$\Delta\psi = -\omega \tag{19}$$

$$u = \psi_y \tag{20}$$

$$v = -\psi_x \tag{21}$$

The boundary conditions can be written in terms of the stream function as

$$\psi(x,0;t) = 0 \quad , \quad \psi(x,1;t) = F \tag{22}$$

$$\frac{\partial\psi}{\partial y}(x,0;t) = 0 \quad , \quad \frac{\partial\psi}{\partial y}(x,1;t) = 0 \tag{23}$$

where F represents the net horizontal flow through the channel

$$F = \int_\Omega u. \tag{24}$$

An $N \times N$ grid (with spacing $h = 1/N$) is laid on the periodic channel Ω and we consider approximations $\Psi^n_{i,j}$ to $\psi(ih, jh; nk)$. Approximations $\Omega^n_{i,j}$, $U^n_{i,j}$, and $V^n_{i,j}$ are defined similarly.

In formulating the implicit equations approximating (18)-(21), the discrete equations are best written directly in terms of the stream function. Assume that Ψ^0 and Ψ^1 are given approximations to the stream function at time zero and time k. The discretization for the vorticity transport equation is:

$$\frac{\Delta_h(\Psi^{n+1} - \Psi^{n-1})}{2k} = N^n + \frac{1}{2}\nu\Delta_h\Delta_h(\Psi^{n+1} + \Psi^{n-1}) \tag{25}$$

where N^n represents the non-linear convection terms at the n-th time step, calculated in a straight forward way using centered differences. Thom's vorticity boundary condition

$$\Delta_h\Psi^n_{i,0} = -\Omega^n_{i,0} = \frac{2}{h^2}\Psi^n_{i,1} \tag{26}$$

is used to eliminate the terms $\Delta_h\Psi$ on the boundary from the diffusion term in (25) at every time level n. A similar formula is used on the upper boundary, modified to handle the fact that stream function has a non-zero value F there. This type of boundary condition is called a vorticity - stream function boundary condition because it relates the vorticity on the boundary to the stream function in the interior. Thom's condition (26) is one of the simplest: other conditions will be discussed below.

Using (26), (25) is now an implicit equation for the interior values of the stream function at time step $n+1$. In the case of the periodic channel, one can take the discrete fourier transform of (25) in the x-direction, reducing it to a five-diagonal system for each fourier component which can be solved efficiently. As in the primitive method, Ψ^1 is obtained using a single forward euler step.

The second order max-norm convergence of interior vorticity and velocities in the semi-discrete analog of (25) is shown in [7]. The convergence of boundary vorticity is limited to first order since (26) is only first order accurate. The author expects that the fully discrete method can be proved convergent assuming a CFL condition on the time steps. The proof should proceed in a similar way to the fully discrete analysis for the projection method but the details are more complicated due to the presence of higher derivatives in the convection terms. It is relatively easy, though, to prove convergence for a discrete method for the Stokes equations (in which the nonlinear convection terms are not present) and this is presented below. In this case, we consider the two-level scheme:

$$\frac{\Delta_h(\Psi^{n+1}-\Psi^n)}{k} = \frac{1}{2}\nu\Delta_h\Delta_h(\Psi^{n+1}+\Psi^n) \tag{27}$$

using the vorticity boundary conditions (26).

Theorem 1 *The approximate solutions Ψ of (27) and corresponding vorticity values $\Omega = -\Delta_h\Psi$ converge uniformly with second order accuracy in space and time in the interior to the exact solution of the Stokes equation. The vorticity boundary values converge uniformly with first order accuracy.*

Proof: First, we will address the question of stability. To begin with, we rewrite two results obtained by summation by parts due to Meth [10]:

$$\begin{aligned} -h^2\sum_{i=1}^{N}\sum_{j=1}^{N-1}\Psi_{i,j}\Delta_h\Psi_{i,j} &= h^2\sum_{i=1}^{N}\sum_{j=1}^{N}[(D_-^x\Psi_{i,j})^2+(D_+^y\Psi_{i,j})^2] \\ &:= \|\Psi\|_{1,2}^2 \end{aligned} \tag{28}$$

$$\begin{aligned} h^2\sum_{i=1}^{N}\sum_{j=1}^{N-1}\Psi_{i,j}\Delta_h\Delta_h\Psi_{i,j} &= h^2\sum_{i=1}^{N}\sum_{j=1}^{N-1}(\Delta_h\Psi_{i,j})^2+\frac{2}{h^2}\sum_{i=1}^{N}[(\Psi_{i,1})^2+(\Psi_{i,N-1})^2] \\ &:= \|\Psi\|_{2,2}^2 \end{aligned} \tag{29}$$

where these operators have the no-flow $\Psi = 0$ and vorticity boundary conditions (26). Equations (28)-(29) essentially describe the definiteness of the operators Δ_h and $\Delta_h\Delta_h$ with the boundary conditions we have given. Note that these operators are also symmetric. If we take the inner product of (27) with $\Psi^{n+1}+\Psi^n$, *i.e.* multiply the equation by $h^2(\Psi_{i,j}^{n+1}+\Psi_{i,j}^n)$ and sum, and use the symmetry of Δ_h and equations (28)-(29), we obtain

$$\|\Psi^{n+1}\|_{1,2}^2 = \|\Psi^n\|_{1,2}^2 - k\nu\|\Psi^{n+1}+\Psi^n\|_{2,2}^2.$$

This proves the stability of the method in the $\|\cdot\|_{1,2}$-norm which is the l_2 norm in velocity, the natural energy norm. To show convergence, we must use the asymptotic error analysis developed in [7]. This analysis allows us to handle the truncation errors at the boundary, to pass the convergence to the vorticity and to sharpen the estimates from l_2 to maximum norm.

Let us now consider some other vorticity boundary conditions that can be used. We consider two boundary conditions presented in Orszag and Israeli's review article [12]:

$$\Omega_{i,0} = \frac{-8\Psi_{i,1}+\Psi_{i,2}}{2h^2} \tag{30}$$

$$\Omega_{i,0} = -\frac{3\Psi_{i,1}}{h^2}-\frac{1}{2}\Omega_{i,1}. \tag{31}$$

Similar conditions are defined on the upper boundary, again taking into account that $\Psi = F$ there. Condition (30) was proposed by Wilkes and Pearce and (31) by Woods. Both (30) and (31) are second order accurate. Recall that (26) was first order in vorticity but only introduced a second order error in the interior. Using the other two methods, convergence will be second order in the interior and at the boundary. Thus the advantage of using Wilkes' or Wood's formulas is that second order convergence for the boundary vorticity is obtained.

We now turn to the stability of (30)-(31) in the diffusive term. All can be shown stable following the same plan as laid out in the proof of theorem 1: showing that the quadratic form using the matrix $\Delta_h \Delta_h$ is positive definite using these boundary conditions (the matrix is no longer symmetric). In general, taking the inner product of Ψ with $\Delta_h \Delta_h \Psi$ we obtain

$$h^2 \sum_{i=1}^{N} \sum_{j=1}^{N-1} (\Delta_h \Psi_{i,j})^2 + \sum_{i=1}^{N} [\Psi_{i,1} \Delta_h \Psi_{i,0} + \Psi_{i,N-1} \Delta_h \Psi_{i,N-1}] \tag{32}$$

after summation by parts. The interior sum has the right sign and the lower boundary term together with the term $j = 1$ in the interior sum is considered using formula (30) for $\Omega_{i,0} = -\Delta_h \Psi_{i,0}$:

$$\frac{1}{h^2}[4\Psi_{i,1}^2 - \frac{\Psi_{i,1}\Psi_{i,2}}{2} + (\Psi_{i+1,1} + \Psi_{i-1,1} + \Psi_{i,2} - 4\Psi_{i,1})^2]. \tag{33}$$

Suppose for now that Ψ does not depend on i. Then (33) becomes:

$$\begin{aligned} & \frac{1}{h^2}[4\Psi_{\cdot,1}^2 - \frac{\Psi_{\cdot,1}\Psi_{\cdot,2}}{2} + (\Psi_{\cdot,2} - 2\Psi_{\cdot,1})^2] \qquad (34) \\ = & \frac{1}{h^2}[8\Psi_{\cdot,1}^2 - \frac{9\Psi_{\cdot,1}\Psi_{\cdot,2}}{2} + \Psi_{\cdot,2}^2] \geq 0 \end{aligned}$$

using the inequality $2|ab| \leq \gamma a^2 + b^2/\gamma$ that holds for any a and b and positive γ. The case where Ψ does depend on i can be handled by taking the discrete fourier transform in i and analyzing each component. The bound in (34) is the bound for the zeroth order fourier component.

We now consider the last boundary condition (31). Using the fact that $\Omega_{i,1} = -\Delta_h \Psi_{i,1}$, (31) can be written as

$$\Delta_h \Psi_{i,0} = -\Omega_{i,0} = \frac{8\Psi_{i,1} - \Psi_{i,2}}{2h^2} - \frac{\Psi_{i+1,1} + \Psi_{i-1,1} - 2\Psi_{i,1}}{2h^2}. \tag{35}$$

This reveals two things. First, it shows that in the one dimensional setting, (30) is the same as (31) which explains the computational observations of Orszag and Israeli [12]. Secondly, it shows the stability of (31) since the first term in (35) is handled as in (30) and the second term (when multiplied by $\Psi_{i,1}$ and summed over i) is positive.

The above discussion shows the definiteness of $\Delta_h \Delta_h$ using the boundary conditions (30)-(31). Convergence of method (27) for the Stokes equations with these boundary conditions follows as in theorem 1 but with second order convergence in the boundary vorticity. Second order convergence for these boundary conditions in a semi-discrete scheme for the full nonlinear equations can also be shown in a manner similar to that presented in [7].

An interesting question is whether the use of higher order formulas (30)-(31) will give more accurate answers even for the interior vorticity and the velocities. The methods are compared on the computations described in section 4 to answer this question.

2.3 Modifications to the Primitive Method

We introduce the quantity a, the net pressure gradient through the channel:

$$a = \int \frac{\partial p}{\partial x}. \tag{36}$$

If we integrate the first component of the Navier-Stokes equations (1) over the channel and cancel those terms that integrate to zero, we obtain

$$\dot{F} = -a + \nu \int u_{yy} \tag{37}$$

where the dot denotes time derivative. The integral in (37) can be reduced to an integral on the boundary. At this point it is appropriate to say that the choice of $\dot{F}(t)$ or $a(t)$ is arbitrary. They correspond to an "upstream" boundary condition and are related through the flow by relationship (37). In the vorticity algorithm, we set F equal to a constant so $\dot{F} = 0$. In the original formulation of the primitive algorithm, the pressure is periodic so $a = 0$ and the flows computed using the two methods cannot be compared directly as they stand. The primitive method is modified below to allow direct comparison.

We set the discrete pressure equal to $(A^n, 0) + GP^n$ where P^n is a periodic function and compute A^n so that the net flow in the channel remains constant. One might be tempted to discretize (37) directly (setting $\dot{F} = 0$) but this is the wrong approach for stability reasons, just as a direct implementation of the equations and boundary conditions for the pressure is ill-advised. Instead, we impose the discrete version of the F constant condition directly, *i.e.*

$$h^2 \sum_{i,j} U^n_{i,j} = h^2 \sum_{i,j} U^0_{i,j} \tag{38}$$

for all n and let this condition implicitly dictate the choice of A^n. Just as in the case of the pressure, this determines A^n uniquely. The stability of the term $(A^n, 0)$ in the pressure can be shown since the error in velocities will have zero net horizontal flow and so will be orthogonal to the term $(A^n, 0)$. Also, the physical condition of F =constant is satisfied up to second order (since (38) is a mid-point rule approximation of this) and so second order convergence can be shown. In terms of the iterative algorithm, we define iterates $A^{n,j}$ and calculate as follows:

$$\begin{aligned}
\mathbf{U}^{n+1,j+1} &= (I - k\nu\Delta_h)^{-1}[-2kN^n - 2k\mathbf{G}P^{n,j} - 2k(A^{n,j}, 0) + \\
&\quad (I + k\nu\Delta_h)\mathbf{U}^{n-1}] && (39)\\
2k\mathbf{D}^* \cdot \mathbf{G}P^{n,j+1} &= \mathbf{D}^* \cdot (-2kN^n + (I + k\nu\Delta_h)\mathbf{U}^{n-1} + k\nu\Delta_h \mathbf{U}^{n,j+1}) && (40)\\
A^{n,j+1} &= h^2 \sum_{i,j} [-N^n_{(1),i,j} + \frac{\nu}{2}\Delta_h(U^{n+1,j+1} + U^{n-1})] && (41)
\end{aligned}$$

where the subscript (1) on the nonlinear terms in the last equation means first component. The above iteration has a fixed point with velocities that satisfy

$$h^2 \sum_{i,j} U^{n+1}_{i,j} = h^2 \sum_{i,j} U^{n-1}_{i,j} \tag{42}$$

as desired. The iterations can be viewed as disjoint maps for P^n and A^n, both of which are contractions. Therefore the iteration scheme will converge.

3 Issues

In the discussion of the previous section, we have already seen some of the issues that need to be resolved. The first is the use of a single forward euler step to start off the leap-frog schemes. In addition, we would like to know if we can still observe second order accuracy in the primitive method if we use only one iteration in the scheme to solve for the pressure. Also, we are interested in comparing the performance of the primitive method and the vorticity method using different boundary conditions.

There are also two other issues. First of all, the convergence results in [7, 8] for the semi-discrete versions of the algorithms described in section 2 are all of the form: error $\leq K(t)h^2$ where $K(t)$ is growing exponentially in time at a rate depending on high order derivatives of the solution. For unstable flows, this may be the best one can do, but when computing stable flows, one hopes to do better and obtain the error estimates for K uniform in time. Results along these lines are proved in [5]. In this paper, this will be tested computationally.

The final issue concerns incompatibility of initial data and this problem is discussed below: Consider the Navier Stokes equations written in vorticity-stream function formulation (18)-(21) in the periodic channel Ω with initial data $\psi(x, y, 0) = \Psi(x, y)$. There is a hierarchy of compatibility conditions that Ψ must satisfy if the solution ψ is to be smooth at $t = 0$:

first Clearly, $\Psi(x, 0) = 0$, $\Psi(x, 1) = F$, $\Psi_y(x, 0) = 0$ and $\Psi_y(x, 1) = 0$ to have a continuous solution ψ.

second Let $g^{(1)}(x, y) = \psi_t(x, y, 0)$. By differentiating the boundary conditions, one obtains $g^{(1)} = g_y^{(1)} = 0$ on the boundary. From equations (18)-(21), we have

$$\Delta g^{(1)} = -\Psi_y \Delta \Psi_x + \Psi_x \Delta \Psi_y + \nu \Delta\Delta\Psi := \gamma^{(1)}. \tag{43}$$

It appears that $g^{(1)}$ is overdetermined (both Dirichlet and Neumann boundary conditions for $g^{(1)}$ are given) so $\gamma^{(1)}$ must be in the class of functions $\mathcal{C}$ for which (43) can be solved. In this case, the class $\mathcal{C}$ consists of all functions γ such that

$$\int_\Omega \gamma u = 0 \tag{44}$$

for all harmonic functions u in Ω [14]. Assuming that the solvability condition is met, $g^{(1)}$ can be determined in terms of Ψ.

Higher order compatibility conditions can also be derived. In principle, one would expect that smooth solutions will exist up to $t = 0$ if these compatibility conditions are satisfied. See the paper by Temam [13] for a proof of the sufficiency of nonlocal compatibility conditions for smoothness in the case of the Navier Stokes equations written in primitive variables. The semi-discrete convergence of the methods we are considering in this paper were proved under the assumption that the compatibility conditions are satisfied. Heywood and Rannacher point out that this is not a reasonable assumption since the conditions are nonlinear and nonlocal and can not be verified in practice for anything but trivial flows. In [4, 6] they describe what behaviour the solutions have as t goes to zero when these compatibility conditions are not met and show convergence of finite element approximations for the primitive equations in this case. Naughton [11] shows convergence of a vorticity-stream function method in a linear one-dimensional model problem with incompatible initial data. Thus there is some evidence that we can expect the finite difference methods we are using to have some smoothing properties that will allow us to specify incompatible initial data and still obtain the predicted convergence rates away from $t = 0$. This is investigated in the computations described below.

4 Computations

In this section, calculations using the primitive and vorticity methods discussed in section 2 are presented and discussed. The calculation is performed in the periodic channel Ω with homogeneous boundary conditions on the velocities. The initial velocities are given below:

$$u(x,y) = 6y(1-y) + 16(2y - 6y^2 + 4y^3)\sin(2\pi x)/2\pi \qquad (45)$$

$$v(x,y) = -16(y^2 - 2y^3 + y^4)\cos(2\pi x) \qquad (46)$$

where $u = 6y(1-y)$ is the underlying Poiseuille flow and the other terms are a perturbation. The calculations for the primitive variable method described in section 2 are initialized with the velocities above and the vorticity-stream function method is initialized with the corresponding stream function. It is appropriate here to discuss the use of solutions for the Navier-Stokes problem made exact using forcing terms. Using such solutions certainly makes the convergence calculations easier but there are two disadvantages to this approach. Firstly, the initial conditions will be automatically compatible with the equations. Secondly, the solution chosen is often very simple and may have give zero error at the predicted order. The author is guilty of this in [15], where a solution was chosen for which Thom's vorticity boundary condition is actually second order accurate! (invalidating the computations which were designed to verify the predicted result that Thom's boundary condition gave second order convergence in the interior even when the boundary values were only first order accurate). Meth [10] was also guilty of this mistake. Computing the solution with the initial data above without forcing terms will allow us to see the effect of higher order incompatibility. Also, the basic Poiseuille flow does give a non-zero first order error in Thom's boundary condition.

The viscosity ν is taken to be 0.01 and the maximum velocities observed in the computation are 1.8 giving a reynolds number of 180. Rather conservatively, k is taken to be $h/4$. The computed vorticity using the vorticity method with Thom's boundary conditions and $N = 128$ is shown in figure 1 at time intervals 1/2 up to time 2. The vorticity in the problem has the following symmetry

$$\omega(x + 1/2, 1 - y, t) = -\omega(x, y, t) \,. \qquad (47)$$

which can be observed in the initial vorticity in figure 1. For later times only the negative vorticity is shown. The damping of the perturbation and decay to Poiseuille flow $\omega = 6(2y - 1)$ is clearly observed.

The computations were performed using $N = 16, 32, 64, 128, 256$. Estimated convergence rates of the velocities, interior vorticity and boundary vorticity for the vorticity

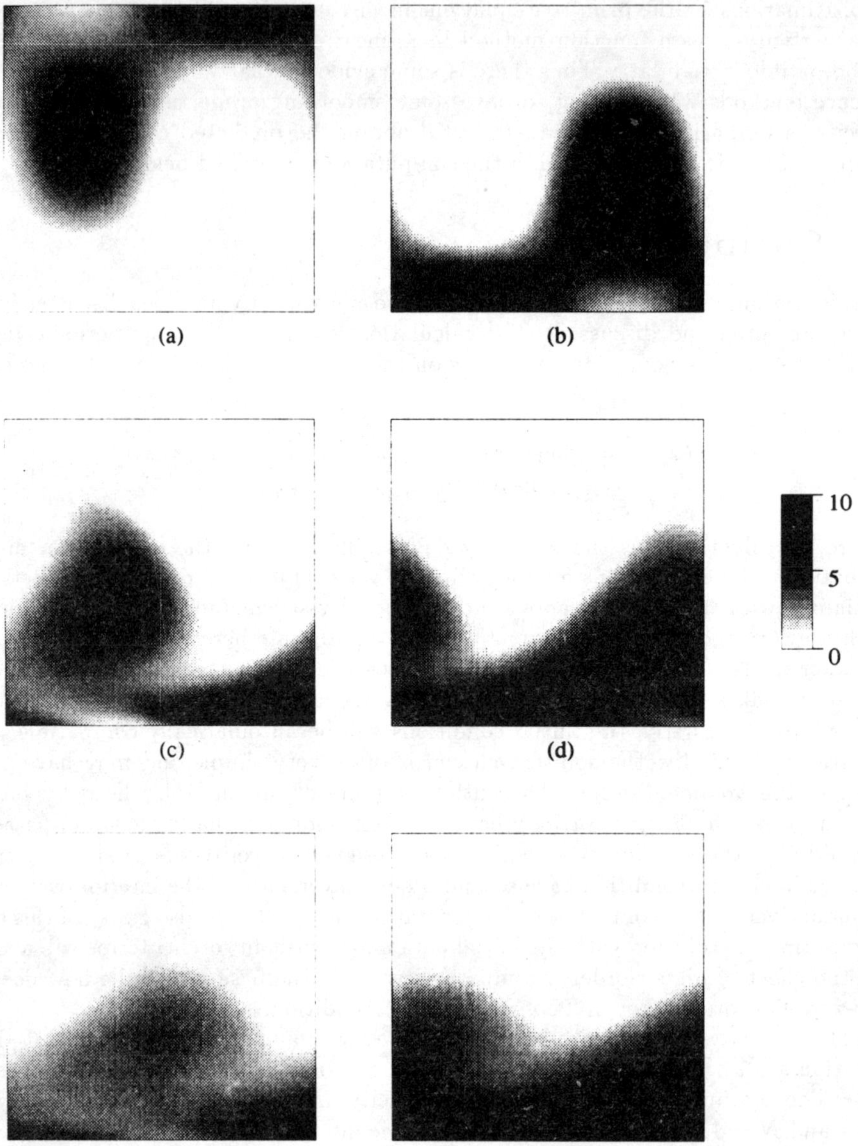

Figure 1: Initial vorticity (a) positive and (b) negative. Negative vorticity at times (c) 0.5 (d) 1.0 (e) 1.5 and (f) 2.0.

method with Thom's boundary condition are shown in figure 2. The estimated convergence rate $\rho(N)$ in a quantity f centered at the computations on an N by N grid is

$$\rho(N) = \log_2 \frac{\|f(N) - f(N/2)\|}{\|f(2N) - f(N)\|}. \tag{48}$$

This formula is based on the asymptotic character of the error. The estimated convergence rates in figure 2 are for the maximum norm. Second order convergence in velocities and interior vorticity are clearly observed. It is clear from the graphs that the velocities are better resolved than the vorticity which is natural. The convergence rate calculations for the boundary vorticity show a surprising result: The convergence rate is first order as predicted at $t = 0$ but increases to second order convergence for $t > 0$. This is definitely not intuitive, since Poiseuille flow, which the computed flow is decaying to, gives a nonzero error contribution in Thom's boundary condition as noted previously.

Computational convergence analysis for the primitive method using a single iteration for the pressure gives similar results, including the increase in expected accuracy at in the boundary vorticity. The results using the vorticity method and Wood's and Wilkes' formulas show second order convergence in all quantities including time 0 as expected. A comparison between the methods shows that they are all converging to the same solution. Similar results for all methods are obtained when fourth order Runge-Kutta time integration is used.

A comparison between the methods is given in figure 3. The maximum norm errors between the $N = 64$ and 128 calculations for the primitive and vorticity methods using Thom's and Wood's boundary conditions are shown for velocities, the interior vorticity and the boundary vorticity (these errors can be related to the errors to the exact solution by multiplying by 4/3 for quantities converging with second order and 2 for those converging with first order). The errors from Wilkes' boundary conditions are very similar to those of Wood's. The errors are normalized by the maximum values of the quantities. The errors from the methods are comparable except in the boundary values for the vorticity where the second order formulas are clearly superior especially at $t = 0$. In general, the results from Wood's formula are the best except for velocities where the primitive method performs as well. The vorticity methods are about half as expensive to run, however: 1470 seconds vs 3024 seconds for the $N = 128$ calculation. These times are from a Silicon Graphics 4D/320S machine. The algorithms used are $O(N^3 \log N)$ in operation count for the whole calculation.

5 Discussion

The results of the computations described in the previous section show that the convergence rate of the method is not affected by the use of incompatible initial data or by the use of a single forward euler step to initialize the scheme. In addition, the error constants do not grow exponentially as time increases for this stable flow. These are all results we expected to obtain based on previous analysis for related problems.

An interesting question, whether the use of higher order accurate boundary conditions will give better results for the interior vorticity, has been investigated for this flow and the answer is yes.

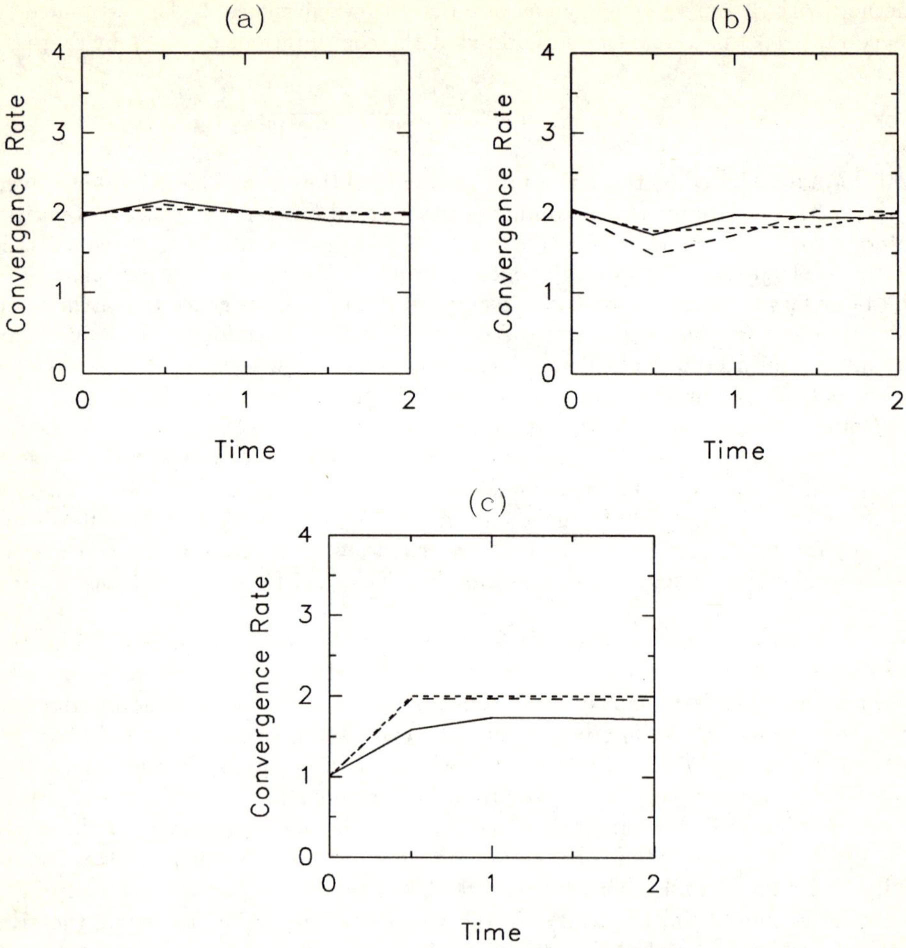

Figure 2: Approximate max norm convergence rates for the vorticity method using Thom's boundary conditions (a) velocities (b) interior vorticity (c) boundary vorticity. The solid lines are the estimates for $N = 32$, the dashed for $N = 64$ and the dotted for $N = 128$.

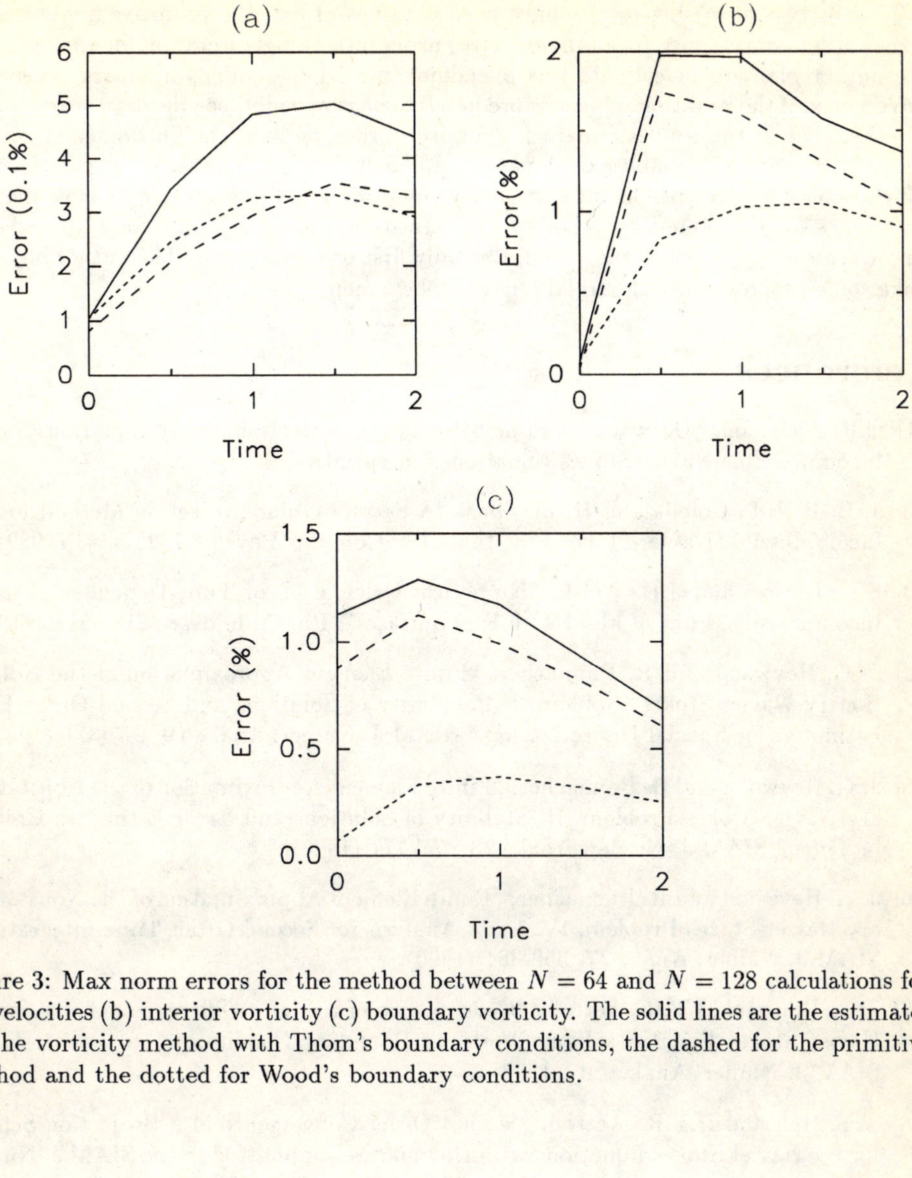

Figure 3: Max norm errors for the method between $N = 64$ and $N = 128$ calculations for (a) velocities (b) interior vorticity (c) boundary vorticity. The solid lines are the estimates for the vorticity method with Thom's boundary conditions, the dashed for the primitive method and the dotted for Wood's boundary conditions.

Finally, two surprising results have been obtained. First, the primitive method gives second order convergence (even in vorticity) using just a single iteration for the pressure. The author plans to investigate this phenomenon for low reynolds numbers, where the convergence of the iteration scheme is predicted to behave badly: see the discussion around equation (17). If the results remain second order, then perhaps the phenomenon is more generally true and the analytic explanation can be discovered. The second surprise is that the vorticity method with Thom's boundary condition and the primitive variable method give values for the boundary vorticity that are second order accurate for $t > 0$ when a simple error analysis shows they should be only first order accurate. The author hopes to make some progress towards explaining this phenomenon.

References

[1] C.R. Anderson, "Derivation and Solution of the Discrete Pressure Equations for the Incompressible Navier-Stokes Equations," preprint.

[2] J. B. Bell, P. Colella and H. M. Glaz, "A Second-Order Projection Method for the Incompressible Navier-Stokes Equations," J. Comput. Phys., **85**, 257-283 (1989).

[3] F.H. Harlow and J.E. Welsh, "Numerical Calculation of Time-Dependent Viscous Incompressible Flow of Fluid with Free Surfaces," Phys. Fluids, **8**, 2181-2189 (1965).

[4] J. G. Heywood and R. Rannacher, "Finite Element Approximation of the Nonstationary Navier-Stokes Problem. I. Regularity of Solutions and Second-Order Error Estimates for Spacial Discretization," SIAM J. Numer. Anal., **19**, 275-311 (1982).

[5] J. G. Heywood and R. Rannacher, "Finite Element Approximation of the Nonstationary Navier-Stokes Problem. II. Stability of Solutions and Error Estimates Uniform in Time." SIAM J. Numer. Anal., **23**, 750-777 (1986).

[6] J. G. Heywood and R. Rannacher, "Finite Element Approximation of the Nonstationary Navier-Stokes Problem. IV. Error Analysis for Second Order Time Integration." SIAM J. Numer. Anal., **27**, 353-384 (1990).

[7] T.Y. Hou and B.T.R. Wetton, "Convergence of a Finite Difference Scheme for the Navier Stokes Equations Using Vorticity Boundary Conditions," to appear in the SIAM J. Numer. Anal., July, 1992.

[8] T.Y. Hou and B.T.R. Wetton, "Second Order Convergence of a Projection Scheme for the Navier Stokes Equations with Boundaries," submitted to the SIAM J. Numer. Anal.

[9] D. Michelson, "Convergence Theorems for Finite Difference Approximations for Hyperbolic Quasi-Linear Initial-Boundary Value Problems," Math. Comp., **49**, 445-459 (1987).

[10] K. Meth, *A Vortex and Finite Difference Hybrid Method to Compute the Flow of an Incompressible, Inviscid Fluid Past a Semi-infinite Plate*, Thesis, New York University (1988).

[11] M.J. Naughton, *On Numerical Boundary Conditions for the Navier-Stokes Equations*, Thesis, California Institute of Technology (1986).

[12] S. A. Orszag and M. Israeli, "Numerical Simulation of Viscous Incompressible Flows," Ann. Rev. of Fluid Mech., **6**, 281-318 (1974).

[13] R. Temam, "Behaviour at Time $t = 0$ of the Solutions of Semi-linear Evolution Equations," J. Diff. Eq., **43**, 73-92 (1982).

[14] F. V. Valz-Gris and L. Quartapelle, "Projection Conditions on the Vorticity in Viscous Incompressible Flows," Int. J. Numer. Math. Fluids, **1**, 453 (1981).

[15] B.T.R. Wetton, *Convergence of Numerical Approximations for the Navier-Stokes Equations with Boundaries: Vorticity and Primitive Formulations*, Thesis, Courant Institute (1991).

THE CLOSURE PROBLEM FOR THE CHAIN OF THE FRIEDMAN-KELLER MOMENT EQUATIONS IN THE CASE OF LARGE REYNOLDS NUMBERS

A.V. Fursikov
Department of Mechanics and Mathematics, Moscow University, Lenin Hills, 119899, Moscow, USSR

§ 1. Introduction

This paper is devoted to investigation of the chain of moment equations corresponding to differential equation

$$\partial_t u(t) + Au + B(u) = 0, \qquad (1.1)$$

where $\partial_t u = \partial u/\partial t$, A is linear, B is quadratic operators. The main example of equation (1.1) is the Navier-Stokes system, which describes the flow of viscous incompressible fluid. The case of turbulent flows is the most interesting for us. For their study one idea of O.Reynolds [1] is quite important. He offered to decompose a turbulent flow u(t) on the sum of mean flow $M_1(t)$ and random fluctuation $w(t)$: $U(t) = M_1(t) + w(t)$ and to study firstly mean flows. The flow $M_1(t)$ together with means of higher order $M_k(t)$ satisfy the chain of moment equations, deduced by Friedman and Keller in [2] (see also [3]). This chain has the form

$$\partial_t M_k(t) + A_k M_k + B_k M_{k+1}(t) = 0, \quad k = 1,2,\dots, \qquad (1.2)$$

where A_k, B_k - are linear operators, which are constructed by A, B from (1.1).

As is shown below, in subsection 2.4 of this paper, the time means

$$M_k(t) = \frac{1}{\tau_0}\int_0^{\tau_0} \overset{k}{\otimes} u(t + \tau)dt$$

satisfy (1.2), for example, where $\tau_0 > 0$ is fixed, $u(t)$ is a solution of (1.1), $\overset{k}{\otimes} u(s) = u(s)\otimes \dots \otimes u(s)$ (k times) is a tensor product of k copies of vector u(s). More generally, the moments of arbitrary statistical solution of (1.1) satisfy (1.2) (see [3]).

The characteristic peculiarity of (1.2) resulting from non-linearity of (1.1) is the presence of k+1-th moment M_{k+1} in the equation for k-th moment M_k. Just because of it, system (1.2) is unclosed or, more precisely, is infinite.

The topic on closure of (1.2) is important for applications, i.e., the approximation the solution of infinite chain (1.2) by solutions of certain finite systems. A number of applied papers is devoted to this problem (see review [4]), where certain methods of closure are offered, but without mathematical substantiation of the fact that solutions of approximated and approximating systems are near. The closure problem with substantiation of nearness was studied in [3], where the case of small Reynolds numbers was considered. This problem in the case of large Reynolds numbers for two-dimensional Navier-Stokes system is investigated in [5].

In the present paper, as in [5], the solution M of (1.2) is approximated by solutions $M^N = \{M_j^N\}$ of the same chain, such that $M_j^N \equiv 0$ for $j > N$. The main difficulty is to find the initial conditions for approximating problem. The method of determination of this initial conditions utilises the conditional well-posed formulation of Cauchy problem (see [6], [7]) and is reduced to solution of certain extremal problem. We suppose, that there exists the solution $M = \{M_k\}$ of Cauchy problem for (1.2), such that $\|M\|_{Y_R^\alpha} < \infty$, $\alpha > n-1$, (the definition of $\|\cdot\|_{Y_R^\alpha}$ see below in (2.16)). Here, α is a smoothness index, which shows on the smoothness of functions M_k, R is Reynolds number, n is the dimension of Navier-Stokes system. If this assumption and some other conditions is fulfilled, then it was proved in [5], that

$$\|M - M^N\|_{Y_{R_1}^{\alpha_1}} \longrightarrow 0 \quad \text{as} \quad N \longrightarrow \infty, \qquad (1.3)$$

where $\alpha < \alpha_1$, R_1 is large enough, M^N is a solution of approximating problem. All this assumptions are fulfilled, when n=2.

By the methods of present paper the result (1.3) of [5] is improved. Here, (1.3) is proved, when $\alpha_1 = \alpha$, and R_1 satisfies certain strict conditions. Thus, the result of the present paper isnear to the unimproved one.

It is necessary to point out, that (1.3) is proved, when certain condition on operator S(t) is fulfilled, where S(t) is the operator assigning to an initial condition u_0 the solution u(t) of Cauchy problem of (1.1). It follows from the existence of solution $M(t) \in Y_R^\alpha$ of (1.2) that the support of the measure μ_0, whose moments equal M(o), is contained in the domain Ω_T^α of operator $S(t), t \in [0,T]$. We assume that supp μ_0 is contained in the connected component of Ω_T^α, which contains zero. This condition is fulfilled for two-dimensional Navier-Stokes system. In this paper we offer the sufficient condition, which garantees that condition on connected component takes place in three-dimensional case. It is sufficient only to suppose that the solution $M(t) = \{M_k(t)\}$ of (1.2) exists for $t \in (0, \infty)$ and belongs to the space Y_R^{α} of functions defined for $t \in (0, \infty)$ (see below (2.16)). Then (1.3) can be proved for $n = 3$, when $\alpha = \alpha_1$, $R_1 > R$.

Note, that results connected with three-dimensional case are announceed in [8].

§ 2. Preliminaries

2.1. We issue from the following ordinary differential equation in Banach space

$$\partial_t u(t) + Au(t) + B(u(t)) = 0, \tag{2.1}$$

$$\gamma_0 u = u_0, \tag{2.2}$$

where $\partial_t u(t) = du(t)/dt$, γ_0 is the operator of restriction of function $u(t)$, at $t = 0$: $\gamma_0 u = u(0)$, A is a linear, B is a quadratic operators. Describe this operators. Let H be real Hilbert space, $A: H \longrightarrow H$ be a positive self-adjoint operator, which has a compact inverse operator $A^{-1}: H \longrightarrow H$. Denote the eigenvectors of A by $e_1, e_2, \ldots$ and the corresponding eigenvalues by $0 < \lambda_1 \leq \lambda_2 \leq \ldots$, i.e., $Ae_j = \lambda_j e_j$. We suppose that $\lambda_j \sim j^{2/n}$ as $j \longrightarrow \infty$. Usually in applications A will be the Laplace operator defined on functions of n variables. Let $E_\infty = \{\sum u_j e_j\}$ be a set of finite linear combinations of e_j,

$$\|u\|_\alpha = \Big(\sum_{j=1}^{\infty} \lambda_j^{\alpha} |u_j|^2\Big)^{1/2} \quad \text{where } \alpha \in R,\ u = \sum_{j=1}^{\infty} u_j e_j . \tag{2.3}$$

Let H^α be a completion of E_∞ with respect to the norm (2.3). We denote by $(\cdot,\cdot)_\alpha$ the scalar prodict in H^α and the duality engendered by it.

Let $B(u,v)$ be bilinear operator for which estimate

$$\|B(u,v)\|_{-\alpha_3} \leq c\|u\|_{\alpha_1} \|v\|_{\alpha_2+1} \tag{2.4}$$

holds, where $\alpha_j \in R$, $j = 1,2,3$ satisfy conditions

$$\alpha_i + \alpha_j \geq 0 \quad \forall\, i \neq j,\ \alpha_1 + \alpha_2 + \alpha_3 > n/2 . \tag{2.5}$$

We define the quadratic operator $B(u)$ in (2.1) by the formula

$$B(u) = B(u,u). \tag{2.6}$$

Note that equation (2.1) contains, as a particular case, the Burgers equation

$$\partial u(t,x)/\partial t - \partial^2 u/\partial x^2 + u\, \partial u/\partial x = 0$$

with periodic or zero boundary conditions, when number n in (2.5) equals 1. If $n = 2$ or $n = 3$, then two-dimensional or, correspondingly, three-dimensional Navier-Stokes system

$$\frac{\partial u(t,x)}{\partial t} + \sum_{j=1}^{n} u_j \frac{\partial u}{\partial x_j} = \Delta u - \nabla p(t,x), \ \operatorname{div} u = 0, \ u = (u_1, \dots, u_n)$$

with periodic boundary conditions can be included to equation (2.1). This inclusion has done in [3, 9], for example.

2.2. The main object, which is studied in this paper, is the chain of moment equations for (2.1). This chain has deduced from (2.1) in [3], for instance. We would remined certain definitions from [3, 10] connected with it. Set

$$\overline{j^k} = (j_1, \dots, j_k) \in N^k, \ \lambda^{\alpha}_{\overline{j^k}} = \lambda^{\alpha}_{j_1} \cdot \dots \cdot \lambda^{\alpha}_{j_k}, \ |\lambda_{\overline{j^k}}| = \lambda_{j_1} + \dots \lambda_{j_k},$$

where N is the set of natural numbers, $\alpha \in R$, λ_j are eigenvalues of operator A. Let

$$H^{\alpha}(k) = H^{\alpha} \otimes \dots \otimes H^{\alpha} \quad (k \text{ times}) \tag{2.7}$$

be the tensor product of spaces H^{α} with norm (2.3). In accordance with the definition of tensor product of Hilbert spaces, the vectors

$$e(\overline{j^k}) = e_{j_1} \otimes \dots \otimes e_{j_k} \qquad \overline{j^k} \in N^k$$

forms a basis in $H^{\alpha}(k)$, because the eigenvectors e_j of operator A forms a basis in H^{α}. Set

$$H^{\alpha}_q(k) = \Big\{ w = \sum_{\overline{j^k} \in N^k} w(\overline{j^k}) e(\overline{j^k}) : \|w\|^2_{H^{\alpha}_q(k)} = \sum_{\overline{j^k} \in N^k} \lambda^{\alpha}_{\overline{j^k}} |\lambda_{\overline{j^k}}|^q |w(\overline{j^k})|^2 < \infty \Big\} \tag{2.8}$$

where $\alpha \in R$, $q \in R$. Evidently, $H^{\alpha}_0(k) = H^{\alpha}(k)$. We denote by $SH^{\alpha}_q(k)$ the subspace of space $H^{\alpha}(k)$, which consists of elements $w = \sum w(\overline{j^k})$ where $w(\overline{j^k})$ are symmetric functions on variables $j_1, \dots, j_k$.

Set

$$A_k = \sum_{j=1}^{k} I(j-1) \otimes A \otimes I(k-j), \tag{2.9}$$

where A is operator from (2.1), $I(l) = I \otimes \dots \otimes I$ (l-times), and I is identity operator. The following assertion is deduced from the definition of $SH^{\alpha}_q(k)$.

Lemma 2.1. Operator A_k: $SH^{\alpha}_1(k)$ —— $SH^{\alpha}_{-1}(k)$ is continuous and its norm equals to unity.

Let

$$G_q^\alpha(k) = L_2(0,T;\ SH_q^\alpha(k)),\quad \|u\|^2_{G_q^\alpha(k)} = \int_0^T \|u(t)\|^2_{H_q^\alpha(k)}\, dt$$

$$Y^\alpha(k) = \left\{u \in G_1^\alpha(k):\ \partial_t u \in G_{-1}^\alpha(k)\right\};\quad \|u\|^2_{Y^\alpha(k)} = \|u\|^2_{G_1^\alpha(k)} + \|\partial_t u\|^2_{G_{-1}^\alpha(k)} \tag{2.10}$$

The following assertion can be proved, as in [3, p.26].

Lemma 2.2. The embedding $Y^\alpha(k) \subset C(0,T;\ SH^\alpha(k))$ is continuous and

$$\|\gamma_t u\|_{H^\alpha(k)} \leqslant \|u\|_{C(0,T;H^\alpha(k))} \leqslant c\, \|u\|_{Y_k^\alpha}$$

where γ_t is the operator of restriction of function $u(\cdot)$ at time moment t : $\gamma_t u = u(t)$, and c does not depend on u, α, k.

We will suppose that bilinear operator $B(u,v)$ from (2.6) satisfies the condition

Condition 2.1. The linear continuous operator

$$\hat{B} : H_1^\alpha(2) \longrightarrow H^{\alpha-1} \tag{2.11}$$

exists and satisfies the equality

$$\hat{B}(u \otimes v) = (B(u,v) + B(v,u))/2, \tag{2.12}$$

where $B(u,v)$ is operator from (2.6) and $\alpha > \frac{n}{2} - 1$, as $n \geqslant 2$, $\alpha \geqslant 0$, as $n = 1$.

As it was shown in [10], the quadratic operators of Burgers equation and of Navier-Stokes system satisfy condition 2.1.

Let $\alpha > \frac{n}{2} - 1$, as $n \geqslant 2$, and $\alpha \geqslant 0$, as $n = 1$. We set

$$B_k = \sum_{j=1}^{k} I(j-1) \otimes \hat{B} \otimes I(k-j), \tag{2.13}$$

where B is operator (2.11), (2.12). The following assertion can be proved, as lemma 2.4, from [10].

Lemma 2.3. Let $\alpha > \frac{n}{2} - 1$ and $0 \leqslant \varepsilon < \alpha - \frac{d}{2} + 1$, $\varepsilon \leqslant 1$

Then

$$\|B_k u\|^2_{H^\alpha_{1+\varepsilon}(k)} \leqslant c k^\varepsilon\, \|u\|^2_{H_1^\alpha(k+1)} \tag{2.13'}$$

where c does not depend on k, ε, u.

The infinity system of equations

$$\partial_t M_k(t) + A_k M_k(t) + B_k M_{k+1}(t) = 0,\quad k \in N \tag{2.14}$$

is called the Friedman-Keller chain of moment equations, where A_k, B_k are operators (2.9), (2.13).

The following functional spaces are connected naturally with chain of equations (2.14):

$$\mathcal{H}^{\alpha}_{R} = \prod_{k=1}^{\infty} SH^{\alpha}(k), \quad \|m\|^2_{\mathcal{H}^{\alpha}_{R}} = \sum_{k=1}^{\infty} R^{-2k}\|m_k\|^2_{H^{\alpha}(k)} \tag{2.15}$$

$$Y^{\alpha}_{R} = \prod_{k=1}^{\infty} Y^{\alpha}(k), \quad \|M\|^2_{Y^{\alpha}_{R}} = \sum_{k=1}^{\infty} R^{-2k}\|M_k\|^2_{Y^{\alpha}(k)} \tag{2.16}$$

$$L_2(0,T;\mathcal{H}^{\alpha}_{q,k}) = \prod_{k=1}^{\infty} L_2(0,T;H^{\alpha}_{q}(k)), \quad \|M\|^2_{L_2(0,T;\mathcal{H}^{\alpha}_{q,R})} = \tag{2.17}$$

$$= \sum_{k=1}^{\infty} R^{-2K}\|M_k\|^2_{L_2(0,T;H^{\alpha}_{q}(k))}$$

2.3. We remind some necessary information on statictical solutions. The Borel σ -algebra of sets of Banach space X is denoted by $\mathcal{B}(X)$.

Let $\mu_o(w)$, $w\in\mathcal{B}(\overset{\alpha}{H})$ be a probability measure on $\mathcal{B}(H^{\alpha})$ where $\alpha\geqslant 0$ is fixed. The closed set $w\in\mathcal{B}(\overset{\alpha}{H})$, such that $\mu(w) = 1$, is called the support of measure μ_o, if for arbitrary closed $w'\subset w$, $w' \neq w$ the inequality $\mu_o(w')<1$ holds. The support of μ_o is denoted by supp μ_o. Suppose that on supp μ_o the operator $S(t) : H^{\alpha} \longrightarrow H^{\alpha}$, $t\in[0,T]$ assigning to the initial value u_o the solution $u(t)$ of problem (2.1), (2.2), i.e., $S(t)\, u_o = u(t)$, is defined and continuous. Denote by

$$S(t)^{-1}w = \{u_o\in \overset{\alpha}{H} : S(t)u_o\in w\} \tag{2.18}$$

the pre-image of w

The family of measures

$$\mu(t,w) = \mu_o(S(t)^{-1}w) \qquad w\in\mathcal{B}(H^{\alpha}),t\in[0,T] \tag{2.19}$$

is called a statistical solution of equation (2.1).

The equality

$$\int f(u)\,\mu(t,du) = \int f(S(t)u)\,\mu_o(du) \tag{2.20}$$

follows from (2.19) and is true for such f(u) for which one of integrals from (2.20) is defined.

We will consider statistical solutions $\mu(t,w)$ contained in the ball

$$B^{\alpha}_{R} = \{ u\in H^{\alpha} : \|u\|_{\alpha} < R\} \tag{2.21}$$

for arbitrary $t\in[0,T]$. It means that for certain $R>0$

$$\operatorname{supp} \mu(t,w) \subset B_R \quad \forall t \in [0,T] \; . \tag{2.22}$$

We introduce the class of functionals

$$\Psi_0 = \left\{ \Psi(t,v) \in C^1([0,T] \times B_R^\alpha) : \| \partial_v \Psi(t,v) \|_{1-\alpha} \leq c \; \forall (t,v) \right\} \tag{2.23}$$

where $\partial_v \Psi(t,v)$ is the Frechet derivative along the functional variable v.

The following theorem is proved by methods of book [3].

Theorem 2.1. Let $\mu(t,w), t \in [0,T]$ be a statistical solution which satisfies (2.22) for certain $R > 0$. Then the energy inequality

$$\int \|u\|_\alpha^{2k} \mu(t,du) + \int_0^t \int \|u\|_{\alpha+1}^2 \|u\|_\alpha^{2k-2} \mu(\tau,du) d\tau \leq c^k \int \|u\|_\alpha^{2k} \mu_0(du) \tag{2.24}$$

holds, where k is arbitrary natural and c does not depend on k. Besides, the Hopf-Foias equation

$$\int q(T,u) \mu(T,du) - \int q(0,u) \mu_0(du) - \int_0^T \int \Big[\partial_t q(t,u) - \\ - (\partial_u q(t,u), Au + B(u))_0 \Big] \mu(\tau,du) d\tau = 0 \tag{2.25}$$

holds for any $q \in \Psi_0$.

Let $\mu(w), w \in \mathcal{B}(H)$ be a probability measure and $\operatorname{supp} \mu \subset B_R^\alpha$. The vector $m_k \in SH(k)$ satisfies condition

$$(m_k, \varphi)_{H^0(k)} = \int (\overset{k}{\otimes} u, \varphi)_{H^0(k)} \mu(du) \quad \forall \varphi \in H^{-\alpha}(k) \tag{2.26}$$

is called k-th moment of the measure μ.

The existence and uniqueness of moments $m_k \in H^\alpha(k)$ of measure μ follow from Riesz theorem (see [3, p.81]).

The system of moments $m_k \in SH^\alpha(k), k \in N$ is called positive-definite if such probability measure $\mu(w), w \in \mathcal{B}(H^\alpha)$ exists that for each k m_k is the k-th moment of measure μ.

Lemma 2.4. Let $\mu(w), w \in \mathcal{B}(H^\alpha)$ be a probability measure and $\operatorname{supp} \mu \subset B_R^\alpha$ for certain $R > 0$. Then the collection $m = \{m_k, k \in N\}$ of moments of μ belongs to the space H_R^α defined in (2.15). Inversely, if a vector $m = \{m_k\} \in H_R^\alpha$ is positive-definite, and μ is the probability measure corresponding to it, then $\operatorname{supp} \mu \subset \overline{B_R^\alpha}$, where the line means the operation of closure.

The proof of the first part of lemma is evident, and the second part is proved in [10, § 8].

Lemma 2.4 shows on connection between parameter R from the space

$\mathcal{H}_R^\alpha$ and Reynolds number of measure μ, which is defined by radius R of the ball B_R^α containing the support of the measure μ.

The following theorem ascertains the connection between a statistical solution and Friedman-Keller chain of equations.

Theorem 2.2. Suppose that $R>0$, $\alpha>n-1$. Then two assertions hold: a) Let $\mu(t,w)$, $w\in\mathcal{B}(H^\alpha)$ be a statistical solution, which satisfies (2.22). Then the collection of its moments $M(t)=\{M_k(t), k\in N\}$ belongs to Y_R^α, satisfies the chain of equations (2.14) and initial conditions $M_k(0)=m_k$, where $\{m_k\}$ are moments of initial measure μ_0; b) Conversselly, let a collection $M(t)=\{M_k\}\in Y_R^\alpha$ be positive-definite at $t=0$, $\{M_k(o)\}=\{m_k\}$, and satisfy (2.14). Then the collection $M_k(t)$ is positive-definite at arbitrary $t\in[0,T]$, and the measure $\mu(t,w)$ corresponding to it is a statistical solution and

$$\operatorname{supp}\mu(t,w)\subset\overline{B_R^\alpha}. \tag{2.27}$$

We note the following concerning the proof of this theorem a) The inclusion $M(t)\in Y_R^\alpha$ is deduced from (2.22), (2.24), and correctness of (2.14) is proved in [3]. b) This assertion is proved in [10].

2.4. Now we give an example of statistical solution connected with the time averages of solutions of (2.1), (2.2), i.e.,

$$\bar u(t)=\tau_0^{-1}\int_0^{\tau_0} S(t+\tau)u_0\,d\tau, \qquad t\in[0,T]. \tag{2.28}$$

Here S(t) is the inverse operator to (2.1), (2.2), $\tau_0>0$ is fixed. We assume that function $S(t)u_0\in H^\alpha$ is defined as $t\in(0,T+\tau_0)$.

We denote by Γ the following curve in H^α:

$$\Gamma=\{S(t)u,\ t\in[0,\tau_0]\}$$

Let $w\subset H^\alpha$ be an open set and $w\cap\Gamma\neq\emptyset$. Then

$$w\cap\Gamma=\bigcup_i\{S(t)u_0,\ t\in(t_i,\hat t_i)\} \tag{2.29}$$

where index i runs by finite or account set and $0<t_1<\hat t_1<\dots$ $\dots<t_k<\hat t_k<\dots T$. Define the measure $\mu_0(W)$ by the formula

$$\mu_0(w)=\mu_0(w\cap\Gamma)=\begin{cases}\sum_i(t_i-t_i)\tau_0^{-1} & \text{as } w\cap\Gamma\neq\emptyset\\ 0 & \text{as } w\cap\Gamma=\emptyset\end{cases} \tag{2.30}$$

The measure μ_0 is prolonged by natural way from open sets onto $\mathcal{B}(H^\alpha)$. By virtue of (2.30) the equality

$$\int f(u)\,\mu_0(du)=\frac{1}{\tau_0}\int_0^{\tau_0} f(S(\tau)u_0)\,d\tau \tag{2.31}$$

holds for arbitrary μ_o - integrable functional f(u). Let $\mu(t,w)$ be the statistical solution with initial measure μ_o, defined in (2.30). Then by (2.20), (2.31)

$$\int f(u)\,\mu(t,du) = \frac{1}{\tau_o}\int_0^{\tau_o} f(S(t+\tau_o)u_o)\,d\tau \qquad (2.32)$$

It follows from (2.26), (2.32) that the moments $M_k(t)$ of this statistical solution are defined by formula

$$M_k(t) = \frac{1}{\tau_o}\int_0^{\tau_o} \overset{k}{\otimes}(S(t+\tau)u_o)\,d\tau$$

from which we obtain, when k = 1, that $M_1(t) = \overline{u}(t)$, where $\overline{u}(t)$ is the time mean (2.28).

§ 3. The theorem on density

3.1. In this section the problem on possibility of approximation of solution $M = \{M_1, \dots M_k, \dots\}$ of (2.14) by solutions $M^N = \{M_k^N\}$ with $M_j^N \equiv 0$, when $j > N$, will be studied. Let $Q_N = \{M^N = \{M_k^N\} \in Y_R :$ $: M_k^N \equiv 0 \, \forall k > N,\ M^N$ satisfied (2.14)$\}$ (3.1)

We want to know is it possible to approximate an arbitrary positive-definite solution $M \in Y_R^\alpha$ of (2.14) by elements of sets Q_N in the space $Y_{R_1}^\alpha$ with certain $R_1 > R$.

We begin from the lemma.

Lemma 3.1. There exists such $C > 0$ that inequality

$$\|M\|_{Y_{R_1}^\alpha} \leq C\,\|M\|_{L_2(0,T;\mathcal{H}_{1,R_1}^\alpha)} \qquad (3.2)$$

holds for arbitrary $M \in Y_{R_1}^\alpha$ satisfying (2.14).

Proof. It follows from (2.14) and lemmas 2.1, 2.3 that

$$\|\partial_t M\|_{L_2(0,T;\mathcal{H}_{-1,R_1}^\alpha)} \leq (\|M\|_{L_2(0,T;\mathcal{H}_{1,R_1}^\alpha)} + b\|M\|_{L_2(0,T;\mathcal{H}_{1,R_1}^\alpha)}) \qquad (3.3)$$

We obtain (3.2) by (3.3) and definition (2.16), (2.10) of $Y_{R_1}^\alpha$-norm. ∎

By virtue of this lemma, it is enough to ascertain that a positive definite solution $M \in Y_R^\alpha$ is approximated by elements of sets Q_N in $L_2(0,T;\mathcal{H}_{1,R_1}^\alpha)$. We prove the Lemma.

Lemma 3.2. Let $p = \{p_k\} \in L_2(0,T;\mathcal{H}_{-1,R_1^{-1}}^{-\alpha})$, $\alpha > \frac{n}{2} - 1$ satisfies condition $p \perp Q_N \ \forall N>0$ with respect to scalar product in $L_2(0,T;\mathcal{H}^{o}_{o,1})$. Then the functions q_k defined by equations

$$-\partial_t q_1 + Aq_1 = p_1, \quad q_1(T) = 0 \tag{3.4}$$

$$-\partial_t q_k + A_k q_k + B^{*}_{k-1} q_{k-1} = p_k, \ q_k(T) = 0, \ k \geqslant 2 \tag{3.5}$$

satisfy equalities

$$q_k(0) = 0 \quad \forall k \geqslant 1 \tag{3.6}$$

Here $B^{*}_k : SH_1^{-\alpha}(k) \text{ —— } SH_{-1}^{-\alpha}(k+1)$ is the operator adjoint to operator (2.13).

Proof. One can easily deduce from (3.4), (3.5) that for each k $q_k \in Y^{-\alpha}(k)$. By condition $p \perp Q_N$, and (3.4), (3.5), we obtain for arbitrary $M \in Q_N$ that

$$0 = \langle M,p \rangle = \int_0^T (M_1, -\partial_t q_1 + Aq_1)_o dt +$$

$$+ \sum_{k=2}^{N} \int_0^T (M_k, -\partial_t q_k + A_k q_k + B^{*}_{k-1} q_{k-1})_{H^o(k)} dt =$$

$$= \sum_{k=1}^{N-1} \int_0^T (\partial_t M_k + A_k M_k + B_k M_{k+1}\ q_k)_{H^o(k)} dt + \int_0^T (\partial_t M_N + A_N M_N, q_N)_{H^o(N)} dt +$$

$$+ \sum_{k=1}^{N} (M_k(0), q_k(0))_{H^o(k)}$$

Equalities (3.6) follow from this equations. ■

3.2. In this subsection several technical assertions are collected, which will be used in the proof of theorem on density. Besides H^{α}, we need in analogous complex space, which we denote CH^{α}.

Let $\Omega_T^{\alpha} \subset CH^{\alpha}$ be the domain of definition of functional $u \longrightarrow \|S(\cdot)u\|_{L_2(0,T;CH^{\alpha+1})}$, where $S(t)$ is inverse operator to (2.1), (2.2).

Lemma 3.3. Let $\alpha > \frac{n}{2} - 1$, $v \in \Omega^{\alpha}_T$, $u(t) = S(t)v$. Then

$$\|u(t)\|_{\alpha+1} \leqslant \frac{\|v\|}{\sqrt{2e}\, t^{1/2}} + C\, \|u\|_{L_\infty(0,T;CH^{\alpha})} \|u\|_{L_2(0,T;CH^{\alpha})}$$

where C does not depend on v.

By lemma 3.3 the following assertion is proved.

Lemma 3.4. Let $\alpha > \frac{n}{2} - 1$, $R>0$ and $\mu(t,w)$, $w \in \mathcal{B}(H^{\alpha})$ be a statistical solution satisfying (2.22). Then

$$\int \|u\|^2_{\alpha+1} \mu(t,du) \leqslant K(t^{-1}+1)$$

$$\int_t^T \int \|u\|^2_{\alpha+2} \mu(\tau,du)d \leqslant K(t^{-1}+1)$$

where $t>0$, and constant K depends on supp μ_0.

We suppose that supp μ_0 satisfies condition;

Condition 3.1. The set supp μ_0 is contained in $\Omega^\alpha_T(0)$, where $\Omega^\alpha_T(0)$ is the connected component of Ω^α_T, which contains the zero of CH^α.

Set

$$R_1 > \max\left\{R, \inf_U \sup_{v\in U} \|S(\cdot)v\|_{C(0,T;CH^\alpha)}\right\} \tag{3.7}$$

where infinum is taken on all open connected sets $U\subset\Omega^\alpha_T$, which contain supp μ_0 and the point $\{0\}$ of CH^α.

Let $p(t) = \{p_k(t)\}\in L_2(0,T;\mathcal{H}^{-\alpha}_{-1,R_1^{-1}})$ where R_1 is the number from (3.7).

We construct the analytic function

$$p(t,u) = \sum_{k=0}^{\infty} (p_k(t), \overset{k}{\otimes} u)_{H^\alpha(k)} \tag{3.8}$$

by the collection $\{p_k(t)\}$. It is clear that serie (3.8) converges in the ball $CB^\alpha_{R_1} = \{u\in CH^\alpha : \|u\|_\alpha < R_1\}$. We want to solve the Cauchy problem

$$-\partial_t q(t,u) + (\partial_u q(t,u), Au + B(u))_0 = p(t,u), \tag{3.9}$$

$$q(t,u)|_{t=T} = 0. \tag{3.10}$$

We obtain, solving this problem formally by characteristic method, that

$$q(t,v) = \int_t^T p(\tau, S(\tau - t)v)d\tau \tag{3.11}$$

Let us show that (3.11) satisfies (3.9), (3.10). Set

$$D = \{(t,v)\in[0,T]\times(CB^\alpha_R\cap\Omega^\alpha_{T-t}) : \|S(\cdot)v\|_{C(0,T-t;CH^\alpha)} < R_1\}, \tag{3.12}$$

where CB^α_R is the complex ball(2.21).

Lemma 3.5. Let q is function (3.11) with p defined in (3.8). Then a) $q(t,v)\in C(D)$, b) If $(t,v)\in D\cap([0,T]\times CH^{\alpha+1})$ then $q(t,v)$ is differentiable on v and

$$\left|\int_t^T (\partial_v q(\tau,v), Av + B(v))_0 d\tau\right| < C(1 + \|v\|^2_{\alpha+2}$$

for $v\in CH^{\alpha+2}$. c) Let $\theta \in [0,T]$ be the Lebesque points set of the function $t \longrightarrow \{p_k(t)\}\in \mathcal{H}^{-\alpha}_{-1,R^{-1}}$. If $(t,v)\in \mathcal{D}\cap(\theta\times H^{\alpha+2})$, then $q(t,v)$ is differentiable on t, satisfies (3.9), (3.10), and estimate

$$\Big|\int_t^T \partial_\tau q(\tau,v)d\tau\Big| \leqslant C(1+\|v\|^2_{\alpha+2}) .$$

By means of lemmas 3.4, 3.5 the following important assertion is ascertained.

Lemma 3.6. Let $\mu(t,w)$ be a statistical solution, satisfying (2.22), $q(t,v)$ be function (3.11). Then the following analog of Hopf-Foias equation (2.25) holds:

$$-\int q(t,v)\,\mu(t,dv) = \int_t^T \Big(\Big\{\partial_\tau q(\tau,v)-(\partial_v q(\tau,v),Av+B(v))_o\Big\}\mu(\tau,dv)d\tau \quad (3.13)$$

where $0<t<T$.

3.3. Now let us prove the main result of this section

Theorem 3.1. Let $\alpha > n-1$, R_1 satisfy (3.7), $M(t) = \{M_k(t)\}\in Y^{\alpha}_R$ be a positive-definite solution of (2.14), and condition 3.1 holds. Then $M(t)$ is approximated in $Y^{\alpha}_{R_1}$ by elements of sets Q_N defined in (3.1).

Proof. By lemma 3.1, it is sufficiently to prove that the set $\bigcup_N Q_N$ approximates an arbitrary positive-definite solution $M(t)\in Y^{\alpha}_R$ in the space $L_2(0,T;\mathcal{H}^{\alpha}_{1,R_1})$.

Assume that such positive-definite solution $M = \{M_k\}\in Y^{\alpha}_R$ of (2.14) exists, which cannot be approximated by elements of $\bigcup_N Q_N$, i.e., $M\notin\overline{\bigcup_N Q_N}$, where the line means the closure in the topology of the space $L_2(0,T;\mathcal{H}^{\alpha}_{1,R_1})$. Then such $p = \{p_k\}\in (L(0,T;\mathcal{H}^{\alpha}_{1,R_1})^{*} = L_2(0,T;\mathcal{H}^{-\alpha}_{-1,R_1^{-1}})$ exists that

$$p\in Q_N^{\perp}\ \forall\, N>0 \quad\text{and}\quad (M,p)_{L_2(0,T;\mathcal{H}^o_{1,0})} \neq 0 \qquad (3.14)$$

Let $\mu(t,w)$ be a statistical solution, whose moments coincide with $M(t) = \{M_k(t)\}$. Then by (2.26)

$$(M,p)_{L_2(0,T;\mathcal{H}^o_{0,1})} = \sum_{k=1}^{\infty}(M_k,p_k)_{L_2(0,T;H^o(k))} = \int_0^T\!\!\int p(t,u)\mu(t,du)dt \quad (3.15)$$

where $p(t,u)$ is serie (3.8).

Let q be function (3.11). It satisfies equation (3.9) in virtue of lemma 3.5. We substitute the left hand side of (3.9) into the right hand side of (3.15) instead of p . Then we obtain by lemma 3.6 that

$$(M,p)_{L_2(0,T;\mathcal{H}^o_{o,1})} = \int q(t,v)\,\mu(t,dv) + \int_0^t p(\tau,v)\,\mu(\tau,dv)d\tau \qquad (3.16)$$

We pass to the limit in (3.16) as $t \longrightarrow 0$, using the section a) of lemma 3.5. Then we obtain that

$$(M,p)_{L_2(0,T;\mathcal{H}^o_{o,1})} = \int q(0,v)\,\mu_o(dv) \qquad (3.17)$$

One can easily ascertain just as in [11], [3] that the analyticness on u of right hand side of (3.9) involves the analyticness on u of solution q(t,u) of this equation, when u belongs to a neighbourhood of zero:

$$q(t,u) = \sum_{k=1}^{\infty} q_k(t,u,\ldots,u) \quad u\in B^{\alpha}_{\varepsilon}, \quad \varepsilon<0 \qquad (3.18)$$

where $q_k(t,u_1,\ldots,u_k) : H^{\alpha}\times\ldots\times H^{\alpha} \longrightarrow C[0,T]$ are k-linear symmetric operators. It is known [10, Lemma 5.2] that

$$q_k(t,u_1,\ldots,u_k) = (q_k(t), \bigotimes_{j=1}^{k} u_j)_{H^o(k)} \quad \forall u_1,\ldots,u_k\in H^{\beta} \qquad (3.19)$$

where $q_k(t)\in L_2(0,T;SH^{-\beta}(k))$ and $\beta>\alpha_1+\frac{n}{2}$.

Substituting series (3.18), (3.8) into (3.9), equating members which have identical homogeneity on u we obtain by (3.19) that $q_k(t)$ satisfy recurrent chain of equations (3.4), (3.5). Lemma 3.2 involves that $q_k(0) = 0$ for each k. Therefore, by (3.19), (3.18) the equation $q(0,u)\equiv 0$ holds for $u\in B^{\alpha}_{\varepsilon}$.

Set

$$D_o = \{v\in CH^{\alpha} : (0,v)\in \mathcal{D}\}$$

where $\mathcal{D}$ is set (3.12). It follows from (3.11) with t = 0 that the function q(0,v) is analytic for $v\in D_o \cap \Omega^{\alpha}_T(0)$, where, as we would remind, $\Omega^{\alpha}_T(0)$ is connected component of Ω^{α}_T contained zero of CH^{α}. We have supp $\mu_o\subset D_o\cap\Omega^{\alpha}_T(0)$ by condition 3.1.

Since $q(0,v)\equiv 0$ for v belonging to a neighbourhood of zero, then by analyticness of q the equality $q(0,v)\equiv 0$ holds for $v\in \text{supp}\,\mu_o\in D_o\cap \cap\Omega^{\alpha}_T(0)$. Hence (3.14), (3.17) contain a contradiction:

$$0 \neq (M,p)_{L_2(0,T;\mathcal{H}^o_{o,1})} = \int q(0,v)\,\mu_o(dv) = 0 \quad \blacksquare$$

Thus, the problem on approximation the positive-definite solution of (2.14) by elements of Q_N from (3.1) has been reduced to the verification of condition 3.1. We consider two cases when this condition is fulfilled.

Theorem 3.2. Let $n = \dim \Omega = 2$. Then condition 3.1 is fulfilled and hence theorem 3.1 holds.

Proof. If n = 2 then the theorems on existence, uniqueness, and smoothness of solutions hold for Navier-Stokes equations. This results involve the inclusion $H^{\alpha} \subset \Omega^{\alpha}_T(0)$. Since $\operatorname{supp}\mu_o \subset H^{\alpha}$ then condition 3.1 is true. ■

Theorem 3.3. Let n = 3, $\alpha > 2$, $R > 0$ and positive-definite solution M(t) of (2.14) belongs to the space Y^{α}_R of functions defined on $[0,T]$ with $T = \infty$. Then condition 3.1 fulfills, and theorem 3.1 holds with arbitrary $R_1 > R$.

Proof. It follows from [10] that the assumption on existence of positive-definite solution $M(t) \in Y^{\alpha}_R$ of (2.14) involves that operator S(t)v is defined for each $t \geq 0$ and $v \in \operatorname{supp}\mu_o$, where μ_o is initial measure with collection of moments M(0). It is clear that

$$S(t)v \longrightarrow 0 \quad \text{as } t \longrightarrow \infty \quad \forall v \in \operatorname{supp}\mu_o. \tag{3.20}$$

By means of (3.20) one can connect the sets $\{0\}$ and $\operatorname{supp}\mu_o$ by the curves $S(t)v, t \in R_+$ which, evidently, belongs to Ω^{α}_T. This proves that condition 3.1 is true. ■

Note that density theorem and, hence, the closure problem can be solved for n = 3 also in the case, when it is known that $M(t) \in Y^{\alpha}_R$ and Y^{α}_R is the space of functions defined on a finite segment $[0,T]$. But in this case the sets Q_N are defined by some more complicate way than in (3.1) (see [8]).

§ 4. A solution of the closure problem

4.1. To solve the closure problem we use a conditionally well-posed formulation of Cauchy problem for chain (2.14). More precisely, we consider chain of moment equations (2.14):

$$\partial_t M_k + A_k M_k + B_k M_{k+1} = 0, \quad k \in N \tag{4.1}$$

and assume that the solution $M = \{M_k\} \in Y^{\alpha}_R$ of (4.1) satisfies the condition

$$\|M\|^2_{L_2(0,T;\mathcal{H}^{\alpha}_{1,R})} \leq \gamma^2 \tag{4.2}$$

where γ is a known constant. Besides, we suppose that the collection of moments $m = \{m_k\} \in \mathcal{H}^{\alpha}_R$ is known, which is approximation of initial magnitude M(0) of solution of (4.1):

$$\|M(0) - m\|^2_{\mathcal{H}^{\alpha}_R} \leq \varepsilon^2 \tag{4.3}$$

where $\varepsilon > 0$ is a known small constant. The number ε is the accuracy of the measurement of initial moments $M(0) = \{M_k(0)\}$.

We suppose that problem (4.1) to (4.3) satisfies the assumption:

Condition 4.1. There exists a positive-definite solution $M(t) = \{M_k\} \in Y_R^\alpha$ of (4.1), which satisfies conditions (4.2), (4.3).

Evidently, condition 4.1 satisfies not for all collections (m, γ, ε). The aim of this section is to give the method of construction of solution $M^N = \{M_k^N\} \in Q_N$ of (4.1), which approximates the solution M(t) from condition 4.1.

Let $M^1 = \{M_k^1\}$, $M^2 = \{M_k^2\} \in Y_R$ satisfy conditions (4.1) to (4.3). Set $L = M^1 - M^2 \in Y_R^\alpha$. We prove one assertion on estimation of L which is similar to Tikhonov theorem [6].

Lemma 4.1. Let $\alpha > n - 1$, $\rho > R$ and condition 4.1 is true for arbitrary $\varepsilon > 0$ and fixed γ. Then there exists such continuous function $\varphi(t) > 0$, $t \in [0,1]$, $\varphi(0) = 0$ that

$$\|L\|^2_{L_2(0,T;\mathcal{H}^\alpha_{1,\rho})} \leq \varphi(\varepsilon) \tag{4.5}$$

Proof. Since $L = M^1 - M^2$, where M^1, M^2 satisfy (4.1) to (4.3), then

$$\partial_t L_k + A_k L_k + B_k L_{k+1} = 0, \quad k \in N, \tag{4.6}$$

$$\|L\|^2_{L_2(0,T;\mathcal{H}^\alpha_{1,R})} \leq (2\gamma)^2, \tag{4.7}$$

$$\|L(0)\|^2_{\mathcal{H}^\alpha_R} \leq (2\varepsilon)^2. \tag{4.8}$$

We write $L = \{L_k\}$ in the form

$$L_k(t) = L_k^1(t) + L_k^2(t), \tag{4.9}$$

where L_k^1 is a solution of the problem

$$\partial_t L_k^1 + A_k L_k^1 = 0, \quad L_k^1\big|_{t=0} = L_k(0), \quad k \geq 1, \tag{4.10}$$

and L_k^2 is a solution of the problem

$$\partial_t L_k^2 + A_k L_k^2 = -B_k L_{k+1}, \quad L_k^2\big|_{t=0} = 0, \quad k \geq 1. \tag{4.11}$$

It follows from (4.10), (4.8) that

$$\|L^1\|_{Y_R} \leq C\varepsilon. \tag{4.12}$$

By virtue of (4.9), (4.12) we see that to prove (4.5) it is enough to ascertain the estimate

$$\|L^2\|_{L_2(0,T;\mathcal{H}^\alpha_{1,\rho})} \leq \varphi(\varepsilon). \tag{4.13}$$

Let θ be the set of solutions $\{L_k^2\}$ of (4.11) where $L = \{L_k\}$ runs the set of vector-functions satisfying (4.7). We show that θ is compact in $L_2(0,T;\mathcal{H}^{\alpha}_{1,\rho})$ with $\rho > R$. Indeed, by (4.11), (4.7) and lemma 2.3 the inequalities hold:

$$\|L^2\|^2_{Y^{\alpha}_R} \leq c \sum_k R^{-2k} \|B_k L_{k+1}\|^2_{L_2(0,T;H^{\alpha}_{-1}(k))} \leq c\gamma^2. \qquad (4.14)$$

If $L = \{L_k, k>0\}$ then we set $\Pi_N L = \{L_k \; 0<k\leq N\}$. Let $\Theta_N = \Pi_N \Theta = \{\Pi_N L^2$, where $L^2 \in \Theta\}$. Since for $\rho > R$ we have

$$\sum_{k>N} R^{-2k} \|L^2_k\|_{Y^{\alpha}(k)} \leq c_1 \gamma^2 (\tfrac{R}{\rho})^{2N},$$

then to prove the compactness of Θ we have to prove the compactness of set Θ_N in the space $\prod_{1\leq k\leq N} L_2(0,T;H^{\alpha}_1(k))$ for arbitrary N.

By lemma 2.3

$$\|B_k L_{k+1}\|_{L_2(0,T;H^{\alpha}_{-1+\varepsilon}(k))} \leq ck^{\varepsilon} \|L_k\|_{L_2(0,T;H^{\alpha}_1(k))} \qquad (4.15)$$

where $0<\varepsilon<\alpha - \frac{n}{2} + 1$, $\varepsilon < 1$. It follows from (4.11), (4.15) that

$$\sum_{k=1}^{N} \rho^{-2k} \Big(\|L^2_k\|^2_{L_2(0,T;H^{\alpha}_{1+\varepsilon}(k))} + \|\partial_t L^2_k\|^2_{L_2(0,T;H^{\alpha}_{-1+\varepsilon}(k))} \Big) \leq c \sum_{k=1}^{N} \rho^{-2k} \|L_k\|^2_{L_2(0,T;H^{\alpha}_1(k))} \leq c\gamma^2 \qquad (4.16)$$

when $L^2 \in \Theta_N$. The embedding $\{y \in L_2(0,T;H^{\alpha}_{1+\varepsilon}(k)) : \partial_t y \in L_2(0,T;H^{\alpha}_{-1+\varepsilon}(k)\} \subset L_2(0,T;H^{\alpha}_1(k))$ is compact (see [3]). Hence, in virtue of (4.16) the embedding Θ_N into $\prod_{1\leq k\leq N} L_2(0,T;H^{\alpha}_1(k))$ is compact too. To prove (4.13) it is sufficient to ascertain that the left hand side of this equation tends to zero, as $\varepsilon \longrightarrow 0$. If we assume the inverse and take in mind that for arbitrary ε $L^{2,(\varepsilon)} \in \Theta$ and $\Theta \Subset L_2(0,T;\mathcal{H}^{\alpha}_{1,\rho})$, then we will be able to select such subsequence $\varepsilon_j \longrightarrow 0$ that

$$\lim_{j\to\infty} \|L^{2(\varepsilon_j)}\|_{L_2(0,T;\mathcal{H}^{\alpha}_{1,\rho})} = \|L^{2(0)}\|_{L_2(0,T;\mathcal{H}^{\alpha}_{1,\rho})} > 0 \qquad (4.17)$$

By (4.9), (4.12), (4.17) we obtain

$$L \equiv L^{(\varepsilon_j)} = L^{1(\varepsilon_j)} + L^{2(\varepsilon_j)} \longrightarrow L^{2(0)} \neq 0 \quad \text{as } \varepsilon_j \longrightarrow 0$$

Thus, $L^{2(0)}(t)$ is a solution of (4.1) and by (4.11) $L^{2(0)}(0)=0$. But this equation and (4.17) contradict to the uniqueness of solution of Cauchy problem for (4.1) which was proved in [10]. ■

4.2. We describe the method of construction of approximate solution of (4.1) to (4.3). Let us consider the extremal problem

$$\partial_t M_k^N + A_k M_k^N + B_k M_{k+1}^N = 0, \ k = 1, \dots, N-1; \ \partial_t M_N^N + A_N M_N^N = 0, \tag{4.18}$$

$$\sum_{k=1}^{N} R_1^{-2k} \|M_k^N\|^2_{L_2(0,T;H_1^{\alpha}(k))} \leq \gamma^2, \tag{4.19}$$

$$\sum_{k=1}^{N} R_1^{-2k} \|M_k^N(0) - m_k\|^2_{H^{\alpha}(k)} \longrightarrow \inf, \tag{4.20}$$

where $(m_1, \dots, m_N)$ are first N components of vector m from (4.3), R_1 is a number satisfying inequalities

$$R < R_1 < \rho \tag{4.21}$$

Lemma 4.2. There exists unique solution $(M_1^N, \dots, M_N^N) = M^N$ of (4.18) to (4.20).

Proof. The set

$\mathfrak{A} = \left\{ M^N = (M_1^N, \dots, M_N^N) \in \prod_{j=1}^{N} Y^{\alpha}(j) \ : \ M^N \text{ satisfy } (4.18), (4.19) \right\}$ is called the set of admissible elements. Evidently, $\mathfrak{A}$ is convex and closed in $\prod_{j=1}^{N} Y^{\alpha}(j)$ and therefore is closed with respect to weak convergence in this space. The set $\mathfrak{A}$ is compact with respect to weak convergence. Functional (4.20) is convex and continuous on $\prod Y^{\alpha}(j)$. Hence, it is semicontinuous from below with respect to weak convergence. Since a semicontinuous from below nonnegative functional reaches its minimum on a compact set, then the solution of (4.18) to (4.20) exists. The uniqueness of the solution of (4.18) to (4.20) follows from the strict convexity of functional from (4.20) on the initial conditions space $\prod_{j=1}^{N} H^{\alpha}(j)$, and from uniqueness theorem for Cauchy problem for system (4.18). ■

The solution M^N of (4.18) to (4.20) is called quasisolution of problem (4.1) to (4.3). Our aim is to prove that quasisolution M^N approximates the solution M of (4.1) to (4.3) from condition 4.1 if N is large enough.

Theorem 4.1. Let conditions 3.1, 4.1 hold for each $\varepsilon < 0$. Then for sufficiently large N the estimate*)

---

*) In (4.22) and in any similar expressions written below we think that $M^N = \{M_k^N\}$, where k runs all natural numbers, and $M_k^N \equiv 0$ for $k > N$.

$$\|M - M^N\|^2_{L_2(0,T;\mathcal{H}^\alpha_{1,\rho})} \leq \varphi\,(2\varepsilon) \tag{4.22}$$

takes place, where φ is the function, defined in lemma 4.1, $\alpha > n-1$, $\rho > R_1$, and R_1 satisfies (3.7).

Proof. Let $M(t) \in Y^\alpha_R$ be a solution of (4.1) to (4.3). By theorem 3.1 for each $\delta > 0$ there exist such $N = N_\delta$, and $\tilde{M}^N = \{\tilde{M}^N_k\} \in Y^\alpha_R$ that $\tilde{M}^N_k = 0$ for $k > N = N_\delta$, and

$$\|M - \tilde{M}^N\|_{Y^\alpha_{R_1}} < \delta \tag{4.23}$$

It follows from (4.2), (4.23) that

$$\|\tilde{M}^N\|_{L_2(0,T;\mathcal{H}^\alpha_{1,R_1})} \leq \|M-\tilde{M}^N\|_{Y^\alpha_{R_1}} + \|M\|_{L_2(0,T;\mathcal{H}^\alpha_{1,R_1})} \leq \gamma + \delta \tag{4.24}$$

Set

$$\overline{M^N} = \frac{\gamma}{\gamma + \delta}\tilde{M}^N \tag{4.25}$$

Then $\overline{M^N}$ satisfies (4.19) because of (4.24), i.e.,

$$\|\overline{M^N}\|_{L_2(0,T;\mathcal{H}^\alpha_{1,R_1})} \leq \gamma \tag{4.26}$$

We show that $\overline{M}^N$ approximates M too. Using (4.23) to (4.25) and lemma 3.1 we have

$$\|M-\overline{M^N}\|_{Y^\alpha_{R_1}} \leq \|M-\tilde{M}^N\|_{Y^\alpha_{R_1}} + \left(1 - \frac{\gamma}{\gamma+\delta}\right)\|\tilde{M}^N\|_{Y^\alpha_{R_1}} \leq$$

$$\leq \delta + \frac{\delta}{\gamma+\delta}\, c\|\tilde{M}^N\|_{L_2(0,T;\mathcal{H}^\alpha_{1,R_1})} \leq \delta(1 + c) \tag{4.27}$$

It follows from (4.27) and lemma 2.2 that

$$\|M(0) - \overline{M^N}(0)\|_{\mathcal{H}^\alpha_{R_1}} \leq c_1\|M - \overline{M^N}\|_{Y^\alpha_{R_1}} \leq c_1\delta\,(1 + C)$$

and therefore

$$\|m-\overline{M^N}(0)\|_{\mathcal{H}^\alpha_{R_1}} \leq \|m-M(0)\|_{\mathcal{H}^\alpha_{R_1}} + \|M(0)-M^N(0)\|_{\mathcal{H}^\alpha_{R_1}} \leq \varepsilon + c_1\delta(1 + c) \leq 2\varepsilon \tag{4.28}$$

if δ is small enough.

Let M^N be a quasisolution of (4.1) to (4.3), i.e., be a solution of extremal problem (4.18) to (4.20). By (4.26) $\overline{M^N}$ is admissible element of (4.18) to (4.20), and hence (4.28) involves inequality

$$\left\| m-M^N(0) \right\|_{\mathcal{H}^{\alpha}_{R_1}} \leq \left\| m-\overline{M^N}(0) \right\|_{\mathcal{H}^{\alpha}_{R_1}} \leq 2\mathcal{E}$$

Therefore, if we estimate $M-M^N$ by means of lemma 4.1, in which R is substituted by R_1, and $\mathcal{E}$ is substituted by $2\mathcal{E}$, we obtain estimate (4.22). ■

Thus, we have proved that quasisolution M^N of (4.1) to (4.3) approximates the solution M. At last, we would like to point out that the boundary value problem determining quasisolution M^N can be easily written by means of Lagrange principle.

Reference

1 Reynolds O. On the dynamical theory of incompressible viscous fluids and the determination of the criterion, Phil. Trans. Roy. Soc., London, 186(1894), 123-161.

2 Keller L.V., Friedman A.A. Differentialgleichung für die turbulent Bewegung einer kompressiblen Flüssigkeit, Proc. 1st Intern. Congr. Appl. Delft, (1924), 395-405.

3 Vishik M.I., Fursikov A.V. Mathematical problems of statistical hydromechanics. - Kluwer academic publishers, Dordrecht, Boston, London, 1988.

4 Monin A.S., Jaglom A.M. Statistical hydromechanics, v.1. Nauka Moscow 1965; v.2. Nauka, Moscow, 1967 (in Russian). English translation (revised). MIT Press. Cambridge, Mass, 1977.

5 Fursikov A.V. On the closure problem of the chain of moment equations in the case of large Reynolds numbers. - Nonclassical equations and equations of mixed type. Inst of Math. SOANSSSR Novosibirsk (1990), 231-250 (in Russian).

6 Tikhonov A.N., Arsenin V.Ia. Methods of solution of ill-posed problems, Nauka, Moscow, 1979 (in Russian).

7 Fursikov A.V. The Cauchy problem for a second-order elliptic equation in a conditionally well-posed formulation. - Trans Moscow Math. Soc. (1990), 139-176.

8 Fursikov A.V. The closure problem of chains of moment equations corresponding to three-dimensional Navier-Stokes system in the case of large Reynolds numbers. - Dokl. Akad. Nauk SSSR, v.319, N 1 (1991), 83-87 (in Russian).

9 Fursikov A.V. Properties of solutions of certain extremal problems connected with the Navier-Stokes equations. - Math USSR Sbornik, 46(3) (1983), 323-351.

10 Fursikov A.V. On uniqueness of the solution of the chain of moment equations corresponding to the three-dimensional Navier-Stokes system. - Math. USSR Sbornik, 62 N 2 (1989), 465-490.
11 Vishik M.I., Fursikov A.V. Analytical first integrals of non-linear parabolic, in the sense of I.G.Petrovsky, Systems of differential equations and their applications, Russian Math surveys, 29(2) (1974), 123-153 (in Russian).

A TINY STEP TOWARDS A THEORY OF FUNCTIONAL DERIVATIVE EQUATIONS —A STRONG SOLUTION OF THE SPACE–TIME HOPF EQUATION[0]

ATSUSHI INOUE

Department of Mathematics, Tokyo Institute of Technology

Dedicated to Prof.Nobuyuki IKEDA for his retirement

ABSTRACT. In this note, we construct a strong solution of the space-time Hopf equation,

$$(\frac{\partial}{\partial t} - \nu\Delta)\frac{\delta Z(\eta)}{\delta\eta_\ell(x,t)} = i\left[\tilde{T}^*\left\{\frac{\delta^2 Z(\eta)}{\delta\eta_j(x,t)\delta\eta_k(x,t)}\right\}\right]^\ell + if^\ell(x,t)Z(\eta)$$

with certain subsidary conditions, which is an example of functional derivative equations and is manageable using a tiny part of "analysis in functional spaces".

§1. PROBLEMS

One of my problems is to construct a theory of functional derivative equations (FDE). Such a theory should contain following equations, which are not yet treated in mathematical sense:

(I) Airy type FDE.

$$\text{(A)}\qquad \begin{cases} (\Box + M^2)\dfrac{\delta Z(\eta)}{\delta\eta(x,t)} = \dfrac{i}{\hbar}\eta(x,t)Z(\eta) - i\lambda\hbar\dfrac{\delta^2 Z(\eta)}{\delta\eta(x,t)^2}, \\ Z(0) = 1. \end{cases}$$

(II) Fokker-Planck type FDE.

$$\text{(FPNS)}\qquad \begin{cases} \dfrac{\partial}{\partial t}P(t,u) = \displaystyle\int_{\mathbb{R}^3} d^3x \frac{\delta}{\delta u^i(x)}\Bigg\{\Bigg(u^j(x)\nabla_j u^i(x) + \frac{1}{\rho}\nabla^i p(x) - \nu\Delta u^i(x) \\ \qquad\qquad + f^i(x)\Bigg)P(t,u)\Bigg\} + \dfrac{k_B T\nu}{\rho}\displaystyle\int_{\mathbb{R}^3} d^3x\left(\nabla_j\frac{\delta}{\delta u^i(x)}\right)^2 P(t,u), \\ P(0,u) = P_0(u). \end{cases}$$

[0]This paper is in final form and no similar paper has been or is being submitted elsewhere.

Typeset by $\mathcal{A}\mathcal{M}\mathcal{S}$-TEX

Here,

$$p(x) = \frac{\rho}{4\pi}\int_{\mathbb{R}^3} d^3x' \frac{(\nabla'_i u^j(x'))(\nabla'_j u^i(x')) + \nabla'_i f^i(x')}{|x - x'|}$$

and the functional derivatives are taken with respect to transverse velocity fields

$$u^j(x) = \int_{\mathbb{R}^3} \frac{d^3\xi}{(2\pi)^3}\left(\delta_{jk} - \frac{\xi_j\xi_k}{|\xi|^2}\right)u^k(x)e^{i\xi x}.$$

The equation (A) was introduced by Gelfand: "Some aspects of functional analysis and algebra" (the talk of I.C.M. at Amsterdam in 1954). That is, after his saying below, he proposed to consider the above functional derivative equation of Airy type as an example of his claim.

" ... Although the branches of functional analysis mentioned above are comparatively new and are continuing to develop rapidly, so to say, a "personal" character. This is not true for the group of questions covered by the last section of this report. We shall deal here with problems and procedures which are just beginning to appear; it is, however, possible that in the future they will occupy a central place in functional analysis as a whole. There are a number of physical problems in which, apart from difficulties of physical character, other difficulties arise, due to the absence of a sufficiently general, adequate mathematical apparatus. Some questions of quantum electrodynamics, the theory of turbulence, etc. are of this type. Lately such a mechanism has begun to take shape. It might be called analysis in functional spaces."

Remarks.

(0) See for example, Inoue [I1-2] for definition of functional derivatives. The meaning of traces of higher order functional derivatives is very vague in general. This is the main reason why even physicist hesitates to try to solve FDE (but see, Fried [Fr]).

(1) From author's knowledge, there is no mathematical paper treating this FDE (A). But, Fradkin [Fr1-2] used this equation as a model of QED and he claimed that he applied the so-called non-perturbation method to this, which is outside my scope at least for the time being.

(2) (FPNS) is derived by physicist as the Fokker-Planck equation corresponding to the Navier-Stokes equations (for example, Edwards [E], Edwards & McComb [EM], Graham [G]). On the other hand, Vishik-Komech-Fursikov [VKF] derived the backward Kolmogorov equation corresponding to the Navier-Stokes equations, which gives us something like a solution of (FPNS). More precisely speaking, if $k_B T = 0$ and $\mu(t, du)$ has a density $P(t, u)$ with respect to the notorious Feynman

measure $d_F u$, that is, $\mu(t,du) = P(t,u)d_F u$, then multiplying a test functional $\Phi(t,u)$ to (FPNS) and integrating by parts, we have the Hopf-Foiaş equation mentioned below, which was solved by Foiaş [F1].

For reader's sake, we derive heuristically Airy type FDE (A): Let $S = \mathbb{R}^{m+1} = \mathbb{R}^m \times \mathbb{R}$ be a configuration manifold, on which we consider a Lagrangian $L(u)$ given by

$$L(u) = \int_S \left(\frac{1}{2}(u_t^2 - |\nabla_x u|^2 - M^2 u^2) - \frac{\lambda}{3}u^3\right) dxdt \tag{A.1}$$

and we want to study the quantity

$$e^{\frac{i}{\hbar}L(u)} d_F u. \tag{A.2}$$

Taking the Fourier-Borel transformation of the above "measure" (A.2) on an unspecified function space $X = X(S)$, i.e.

$$\begin{gathered} Z(\eta) = C\int_X e^{\frac{i}{\hbar}L(u)} e^{-\frac{i}{\hbar}\langle u,\eta\rangle} d_F u, \\ C^{-1} = \int_X e^{\frac{i}{\hbar}L(u)} d_F u, \quad [u,\eta] = \int_S u(x,t)\eta(x,t)dxdt, \end{gathered} \tag{A.3}$$

we try to find the equation which "characterizes" this functional on X^*=the dual space of X.

Putting $\hat{Z}(u) = e^{\frac{i}{\hbar}L(u)}$, we get readly

$$\frac{\delta \hat{Z}(u)}{\delta u(x,t)} = -\frac{i}{\hbar}((\Box + M^2)u(x,t) + \lambda u(x,t)^2)\hat{Z}(u). \tag{A.4}$$

Formally calculating, we have also

$$\frac{\delta Z(\eta)}{\delta \eta(x,t)} = -\int_X \frac{i}{\hbar}u(x,t)\hat{Z}(u)e^{-\frac{i}{\hbar}\langle u,\eta\rangle} d_F u, \tag{A.5}$$

$$\frac{\delta^2 Z(\eta)}{\delta\eta(x,t)\delta\eta(y,s)} = \int_X \left(-\frac{i}{\hbar}\right)^2 u(x,t)u(y,s)\hat{Z}(u)e^{-\frac{i}{\hbar}\langle u,\eta\rangle} d_F u. \tag{A.6}$$

Supposing that the following limit (an example of the trace of higher functional derivatives) exists in suitable sense,

$$\lim_{(y,s)\to(x,t)} \frac{\delta^2 Z(\eta)}{\delta\eta(x,t)\delta\eta(y,s)} = \frac{\delta^2 Z(\eta)}{\delta\eta(x,t)^2} = \int_X \left(-\frac{i}{\hbar}\right)^2 u(x,t)^2 \hat{Z}(u)e^{-\frac{i}{\hbar}\langle u,\eta\rangle} d_F u, \tag{A.7}$$

and applying $\Box + M^2$ to (A.5) and inserting (A.7) into that, we have

$$(\Box + M^2)\frac{\delta Z(\eta)}{\delta\eta(x,t)} = \int_X \left(\frac{\delta \hat{Z}(u)}{\delta u(x,t)} + \frac{i}{\hbar}\lambda u(x,t)^2 \hat{Z}(u)\right) e^{-\frac{i}{\hbar}\langle u,\eta\rangle} d_F u. \tag{A.8}$$

Integrating by parts (the use of Lebesgue-like character of the Feynman measure $d_F u$) in the first term of the right-hand side of (A.8), i.e.,

$$\int_X \frac{\delta \hat{Z}(u)}{\delta u(x,t)} e^{-\frac{i}{\hbar}\langle u,\eta\rangle} d_F u = \frac{i}{\hbar}\eta(x,t)Z(\eta), \tag{A.9}$$

we get the desired equation. In other word, though there exists no Feynman measure $d_F u$ on X, if we could solve (A), then we might define the quantity (A.2).

§2. Mathematically formulated problems.

Let (M,g) be a compact Riemannian manifold of dimension d with boundary ∂M. We denote by $\overset{\circ}{\mathbf{X}}_\sigma(M)$ and $\overset{\circ}{\Lambda}{}^1_\sigma(M)$, the space of solenoidal vector fields on M which vanish near the boundary and that of divergence free 1-forms on M which vanish near the boundary, respectively. $\mathbf{H}$ (or $\tilde{\mathbf{H}}$) stands for the completion of the space $\overset{\circ}{\mathbf{X}}_\sigma(M)$ (or $\overset{\circ}{\Lambda}{}^1_\sigma(M)$) with respect to $\mathbf{L}^2$-norm (or $\tilde{\mathbf{L}}^2$-norm). Closures of $\overset{\circ}{\mathbf{X}}_\sigma(M)$ (or $\overset{\circ}{\Lambda}{}^1_\sigma(M)$) in the Sobolev space of order s are denoted by $\mathbf{V}^s$ (or $\tilde{\mathbf{V}}^s$) for $s \in \mathbb{R}$. We put $\mathbf{V}^1 = \mathbf{V}$ and $\tilde{\mathbf{V}}^1 = \tilde{\mathbf{V}}$. Here and what follows, using Einstein's convention, we put

$$(\nabla_u u)^j(x) = u^k(x)\frac{\partial}{\partial x^k}u^j(x) + \Gamma^j_{k\ell}(x)u^k(x)u^\ell(x) \quad \text{for} \quad u = u^j\frac{\partial}{\partial x^j},$$

$$(\nabla_k \eta)_j(x) = \frac{\partial}{\partial x^k}\eta_j(x) - \Gamma^\ell_{jk}(x)\eta_\ell(x) \quad \text{for} \quad \eta = \eta_j dx^j,$$

$$(\tilde{T}\eta)_{jk}(x) = \frac{1}{2}((\nabla_j\eta)_k(x) + (\nabla_k\eta)_j(x)).$$

As is well-known, the Navier-Stokes equation is given as follows:

$$\text{(NS)} \qquad \begin{cases} \dfrac{\partial}{\partial t}u^j(x,t) - \nu(\Delta u)^j(x,t) + (\nabla_u u)^j(x,t) + \dfrac{\partial}{\partial x^j}p(x,t) = f^j(x,t), \\ (\nabla_j u)^j(x,t) = 0, \\ u^j(x,t)|_{x\in\partial M} = 0 \quad \text{and} \quad u^j(x,0) = u_0^j(x). \end{cases}$$

(I) Hopf equations for NS.

(H) Find a functional $W(t,\eta)$ on $[0,\infty) \times \tilde{\mathbf{H}}$, satisfying

$$\frac{\partial}{\partial t}W(t,\eta) = \int_M \Big(-i(\nabla_k\eta)_j(x)\frac{\delta^2 W(t,\eta)}{\delta\eta_j(x)\delta\eta_k(x)} + \nu(\Delta\eta)_j(x)\frac{\delta W(t,\eta)}{\delta\eta_j(x)} + i\eta_j(x)f^j(x,t)W(t,\eta)\Big)\, d_g x, \tag{H.1}$$

$$\frac{1}{\sqrt{g(x)}}\frac{\partial}{\partial x^j}\left(\sqrt{g(x)}\frac{\delta W(t,\eta)}{\delta\eta_j(x)}\right) = 0, \tag{H.2}$$

$$W(t,0) = 1 \quad \text{and} \quad W(0,\eta) = W_0(\eta). \tag{H.3}$$

for $\eta = \eta(x) = \eta_j(x)dx^j \in \mathring{\Lambda}^1_\sigma(M)$ and $t \in (0,\infty)$. Here $f(x,t) = f^j(x,t)\partial/\partial x^j \in L^2(0,\infty : \mathbf{V}^{-1})$ and $W_0(\eta)$ is a given positive definite functional on $\tilde{\mathbf{H}}$ satisfying

$$\text{(H.4)} \qquad W_0(0) = 1 \quad \text{and} \quad \frac{1}{\sqrt{g(x)}}\frac{\partial}{\partial x^j}\left(\sqrt{g(x)}\frac{\delta W_0(\eta)}{\delta \eta_j(x)}\right) = 0.$$

(HF) Find a family of Borel measures $\{\mu(t,\cdot)\}_{0<t<\infty}$ on $\mathbf{H}$ satisfying

$$\text{(HF)} \quad -\int_0^\infty \int_{\mathbf{H}} \frac{\partial \Phi(t,u)}{\partial t}\mu(t,du)dt - \int_{\mathbf{H}} \Phi(0,u)\mu_0(du)$$
$$= \int_0^\infty \int_{\mathbf{H}} \int_M ((\nabla_u u)^j(x) - \nu\Delta u^j(x) - f^j(x,t))\,\frac{\delta\Phi(t,u)}{\delta u^j(x)}d_g x\,\mu(t,du)dt$$

for suitable 'test functionals' $\Phi(t,u)$. Here, $\mu_0(\cdot)$ is a given measure on $\mathbf{H}$.

Remarks.

(1) The equation (H.1) was introduced by Hopf [H] when M is a bounded domain in $\mathbb{R}^3$, which is the reason why (H.1) is called the Hopf equation. But there, he didn't mention the meaning of the second order functional derivatives (see also Hopf & Titt [H&T]).

(2) Foiaş [F1] is the first one who introduced (HF) when M is a bounded domain in $\mathbb{R}^d$ with $2 \le d \le 4$. As the idea of introducing this equation is same as Hopf, we call this equation as the Hopf-Foiaş equation.

(3) In order to treat the trace of second order derivatives in (H.1), it seems natural to consider the Navier-Stokes equation in a general Riemannian manifold (see Inoue [I2]).

(4) The definitions which are used without precision will be given below.

(II) Space-time Hopf equations for NS.

Let $T \le \infty$. To treat the space-time Hopf equations, we need to prepare function spaces on $M \times [0,T)$:

$$\mathcal{Y}_T = C_b(0,T : \mathbf{H}_w) \cap L^2(0,T : \mathbf{V}),$$
$$\mathcal{Z}_T = C_b(0,T : \mathbf{V}^{-s_1}) \cap L^2(0,T : \mathbf{H})$$

or

$$\tilde{\mathcal{Y}}_T = C_b(0,T : \tilde{\mathbf{H}}_w) \cap L^2(0,T : \tilde{\mathbf{V}}),$$
$$\tilde{\mathcal{Z}}_T = C_b(0,T : \tilde{\mathbf{V}}^{-s_1}) \cap L^2(0,T : \tilde{\mathbf{H}})$$

where $s_1 > (d/2) + 1$ arbitrary fixed, $\mathbf{H}_w$ (or $\tilde{\mathbf{H}}_w$) stands for the Hilbert space endowed with its weak topology and $C_b(I : X)$ denotes the set of continuous and bounded functions on I with values in X. As scalar products, we use

$$(u,\eta)=\int_M u^j(x)\eta_j(x)d_gx \quad \text{for} \quad u=u^j(x)\frac{\partial}{\partial x^j} \quad \eta=\eta_j(x)dx^j,$$

$$[u,\eta]=\int_0^T\int_M u^j(x,t)\eta_j(x,t)d_gxdt \quad \text{for} \quad u=u^j(x,t)\frac{\partial}{\partial x^j} \quad \eta=\eta_j(x,t)dx^j.$$

(Ar) Find a probability Radon measure P on $\mathcal{Y}_T$ satisfying the following.

(Ar.1) $$\int_{\mathcal{Y}_T} F(u)\frac{L(u,\eta)^2}{1+L(u,\eta)^2}P(du)=0,$$

(Ar.2) $$(\gamma_0^*P)(\omega_0)\equiv P(\gamma_0^{-1}\omega_0)=P(\{u(x,t)\in\mathcal{Y}_T:\gamma_0u=u(\cdot,0)\in\omega_0\})=\mu(\omega_0)$$

for any functional $F(u)\in C_b(\mathcal{Y}_T)$ and $\eta\in C_0^\infty([0,T))\otimes\overset{\circ}{\Lambda}{}^1_\sigma(M)$. Here, we define

$$L(u,\eta)=[\nabla_uu-\nu\Delta u-f,\eta]$$

and $\gamma_0:u\to u(\cdot,0)$ for $u\in\mathcal{Y}_T$ or $u\in\mathcal{Z}_T$.

Remark. This formulation is due to Arsen'ev [Ar] and the measure P is called a statistical solution of the Navier-Stokes equation with an initial measure μ_0.

(VKFbis) Let $T<\infty$. Find a probability measure $P(\omega)$, $\omega\in\mathcal{B}(\mathcal{Z}_T)$ with an initial measure $\mu_0(\cdot)$ on $\mathbf{H}$ and an external force $f\in L^2(0,T:\mathbf{V}^{-1})$, which satisfies the following properties:

(VKFbis.1) $$\int_{\mathcal{Z}_T}\left[\frac{\partial}{\partial t}u-\nu\Delta u+\nabla_u u-f,\phi\right]e^{i[u,\eta]}P(du)=0 \quad \text{for} \quad \phi,\eta\in C_0^\infty([0,T))\otimes\overset{\circ}{\Lambda}{}^1_\sigma(M).$$

(VKFbis.2) $$(\gamma_0^*P)(\omega_0)\equiv P(\gamma_0^{-1}\omega_0)=\mu_0(\omega_0) \quad \text{for} \quad \omega_0\in\mathcal{B}(\mathbf{H}).$$

Remark. Above problem (VKFbis) is a modification of a space-time statistical solution of the Navier-Stokes equation introduced by Vishik, Komech & Fursikov [VKF], which will be explained in the next section.

(STH) Find a functional $Z(\eta)$ on $\tilde{\mathcal{Z}}_T$, satisfying

(STH.1) $$\left(\frac{\partial}{\partial t}-\nu\Delta\right)\frac{\delta Z(\eta)}{\delta\eta_\ell(x,t)}=i\left[\tilde{T}^*\left\{\frac{\delta^2Z(\eta)}{\delta\eta_j(x,t)\delta\eta_k(x,t)}\right\}\right]^\ell+if^\ell(x,t)Z(\eta),$$

(STH.2) $$\frac{1}{\sqrt{g(x)}}\frac{\partial}{\partial x^j}\left\{\sqrt{g(x)}\frac{\delta Z(\eta)}{\delta\eta_j(x,t)}\right\}=0,$$

(STH.3) $$Z(0)=1 \quad \text{and} \quad Z(\delta(\cdot)\eta^0(\cdot))=W_0(\eta^0).$$

for $\eta=\eta(x,t)=\eta_j(x,t)dx^j\in C_0^\infty([0,T):\overset{\circ}{\Lambda}{}^1_\sigma(M))$ and $\eta^0\in\overset{\circ}{\Lambda}{}^1_\sigma(M)$. Here, $\tilde{T}$ is considered as a mapping from $\tilde{H}^1_{\sigma,N}(M)$ to $\tilde{\mathbf{L}}^2(ST_2(M))$ and $\tilde{T}^*$ is the formal adjoint of $\tilde{T}$. $ST_2(M)$ stands for a space of symmetric 2 tensors on M.

Remark. The equation (STH.1) is appeared as the first time in Inoue [I3].

§3. KNOWN RESULTS

Definition 3.1 [F1]. *A real functional $\Phi(\cdot,\cdot)$ defined on $[0,\infty)\times \mathbf{V}$ is called a test functional of type F (denoted by $\mathcal{TF}_F$) if it satisfies the following:*

(1) *$\Phi(\cdot,\cdot)$ is continuous on $[0,\infty)\times\mathbf{V}$.*

(2) *$\Phi(\cdot,\cdot)$ is Fréchet **H**-differentiable in the direction **V** for any fixed t and is differentiable in t for any fixed $u\in\mathbf{V}$.*

(3) *$\Phi_u(\cdot,\cdot)$ is continuous from $[0,\infty)\times\mathbf{V}$ to $\tilde{\mathbf{V}}^{s_1}$ with $s_1>(d/2)+1$ arbitrary fixed. Moreover, $\Phi_u(\cdot,\cdot)$ and $\Phi_t(\cdot,\cdot)$ satisfy the following: There exist constants c_1-c_3 depending on Φ such that*

$$\|\Phi_u(t,u)\|_{s_1}\le c_1 \quad \text{and} \quad |\Phi_t(t,u)|\le c_2+c_3|u|^2$$

for any $(t,u)\in[0,\infty)\times\mathbf{V}$.

(4) *A test functional $\Phi(\cdot,\cdot)$ is said to have a compact support in t if there exists a constant T depending on Φ such that*

$$\Phi(t,\cdot)=0 \quad \text{for} \quad t\ge T.$$

Definition 3.2 [F1]. *A family of Borel measures on **H** is called a strong solution for Problem (HF) on $(0,\infty)$ if it satisfies the following:*

(1) $\int_{\mathbf{H}}(1+|u|^2)\mu(\cdot,du)\in L^\infty_{loc}(0,\infty)$

(2) $\int_{\mathbf{H}}\|u\|^2\mu(\cdot,du)\in L^1_{loc}(0,\infty)$

(3) *$\int_{\mathbf{H}}\Phi(u)\mu(t,du)$ is measurable in t for any non-negative, weakly continuous funtions $\Phi(\cdot)$ on **H**.*

(4) *$\mu(t,du)$ satisfies (HF) for any $\Phi(\cdot,\cdot)\in\mathcal{TF}_F$ with compact support in t.*

Remark. In order to introduce a weak solution of (HF), author tagged the solution of Foiaş as strong. Weak solutions of (H) and also (HF) are constructed in Inoue [I6].

Theorem 3.3 [F1]. *For any Borel measure μ_0 on **H** satisfying*

$$\int_{\mathbf{H}}(1+|u|^2)\mu_0(du)<\infty,$$

and any $f(\cdot) \in L^2(0,\infty;\mathbf{V}^{-1})$, there exists a strong solution $\{\mu(t,\cdot)\}_{0<t<\infty}$ of Problem (HF).

Moreover, it satisfies the following energy inequality of strong form.

$$\text{(EIS)}\quad \frac{1}{2}\int_{\mathbf{H}} \psi(|u|^2)\mu(t,du) + \nu\int_0^t \left[\int_{\mathbf{H}} \psi'(|u|^2)|u|^2\mu(\tau,du)\right] d\tau$$
$$\leq \frac{1}{2}\int_{\mathbf{H}} \psi(|u|^2)\mu_0(du) + \int_0^t \left[\int_{\mathbf{H}} \psi'(|u|^2)(f(\tau),u)\mu(\tau,du)\right] d\tau$$

for $0 < t < \infty$ and $\psi \in C^1[0,\infty)$ satisfying

$$0 \leq \psi'(t) \leq \sup_{s\in[0,\infty)} \psi'(s) < \infty.$$

Definition 3.4 [I2]. *A functional defined on $[0,T)\times\tilde{\mathbf{H}}$, $(T \leq \infty)$ will be called a strong solution of Problem (H) on $(0,T)$ if there exists a set $\tilde{D} \supset \mathring{\Lambda}^1_\sigma(M)$, such that:*

(1) *For each $\eta \in \tilde{D}$, $W(t,\eta)$ belongs to $L^1_{loc}[0,T)$ and continuous in t at $t = 0$ and $W(t,\eta)$ is twice differentiable at $\eta \in \tilde{D}$ for a.e.t. Moreover,*

$$\frac{\delta^2 W(t,\eta)}{\delta\eta_j(x)\delta\eta_k(x)}\frac{\partial}{\partial x^j}\otimes\frac{\partial}{\partial x^k}$$

exists for almost every t on $(0,T)$ as a distributional element in $ST_2(M)$ for $\eta \in \tilde{D}$.

(2) *$W(t,\eta)$ satisfies (H.1)-(H.3) as distributions for each $\eta \in \tilde{D}$.*

Here, $ST_2(M)$ stands for the set of symmetric 2-tensors on M.

Remark. In the following, we put $\tilde{D} = \mathring{\Lambda}^1_\sigma(M)$, but in general, $\tilde{D}$ may be a set containing $\mathring{\Lambda}^1_\sigma(M)$ which is dense in $\tilde{\mathbf{V}}^s$, for some s.

Theorem 3.5 [I2]. *Suppose a positive definite functional $W_0(\cdot)$ on $\tilde{\mathbf{H}}$ satisfy*

$$\mathrm{trace}_{\tilde{\mathbf{H}}\to\mathbf{H}}[-W_{0\eta\eta}(0)] < \infty.$$

For any $f(\cdot) \in L^2(0,\infty;\mathbf{V}^{-1})$, there exists a strong solution $W(t,\eta)$ of Problem (I).

Definition 3.6 [VKF]. *Let $T < \infty$. A probability measure $P(\omega)$, $\omega \in \mathcal{B}(\mathcal{Z}_T)$, is called a space-time statistical solution of the Navier-Stokes equation corresponding to an initial measure $\mu_0(\cdot)$ on $\mathbf{H}$ and an external force $f \in L^2(0,\infty:\mathbf{V}^{-1})$, if it satisfies the following properties:*

(1) $P(\mathcal{U}_T) = 1$ where

$$\mathcal{U}_T = \left\{u \in L^2(0,T;\mathbf{V}) \cap L^\infty(0,T;\mathbf{H}) : \frac{\partial}{\partial t}u \in L^\infty(0,T;\mathbf{H}^{-s})\right\}.$$

(2) *There is a set $\mathcal{W}_T \subset \mathcal{U}_T$, closed in $\mathcal{U}_T$, such that (a) $\mathcal{W}_T \in \mathcal{B}(\mathcal{Z}_T)$, (b) $P(\mathcal{W}_T) = 1$, (c) $\mathcal{W}_T$ consists of weak solutions of the Navier-Stokes equation.*

(3) *The restriction of P to $t = 0$ coincides with μ_0:*

$$(\gamma_0^* P)(\omega_0) \equiv P(\gamma_0^{-1}\omega_0) \equiv P(\{u(t,x) \in \mathcal{Z}_T; u(0,\cdot) \in \omega_0\}) = \mu_0(\omega_0).$$

(4) *For any $\epsilon > 0$ and $t > 0$,*

$$\int_{\mathcal{Z}_T} \left(|u(\cdot,t)|^2 + (2\nu - \epsilon) \int_0^t \|u(\cdot,s)\|^2 ds \right) P(du) \le \bar{E}_0 + \frac{1}{\epsilon} \int_0^t \|f(\cdot,s)\|_{-1}^2 ds.$$

(5) *Moreover, we have*

$$\int_{\mathcal{Z}_T} \left(\|u\|^2_{L^\infty(0,T:\mathbf{H})} + \|\frac{\partial}{\partial t} u\|_{L^\infty(0,T:\mathbf{V}^{-s})} \right) P(du) \le C(1 + \bar{E}_0 + \|f\|^2_{L^\infty(0,T:\mathbf{V}^{-1})}) < \infty.$$

Theorem 3.7 [VKF]. *Let $T < \infty$. Assume that the initial measure μ_0 satisfies*

$$\bar{E}_0 = \int_{\mathbf{H}} |u_0|^2 \mu_0(du_0) < \infty. \tag{3.2}$$

Then, there exists a space-time statistical solution of the Navier-Stokes equation.

Remark. The solution constructed in [VKF] also satisfies (VKFbis). Therefore, from the point of view of solving FDE, it seems natural that they construct a solution of (VKFbis) which satifies also properties in Theorem 3.7.

§4. A strong solution of (STH)

Definition 4.1. *We put*

$$\tilde{\mathfrak{D}} = C_0^\infty(0,T) \otimes \mathring{\Lambda}^1_\sigma(M).$$

A positive definite functional Z on $\mathcal{Z}_T$ is called a strong solution of (STH) if it satisfies the following:

(1) *It is two times Fréchet differentiable in the direction $\tilde{\mathfrak{D}}$ and for any $\eta \in \tilde{\mathfrak{D}}$,*

$$Z_\eta(\eta)(x,t) \equiv \frac{\delta Z(\eta)}{\delta \eta_j(x,t)} \frac{\partial}{\partial x^j} \in \mathcal{D}'(M \times (0,T) : X_\sigma(M)),$$

$$Z_{\eta\eta}(\eta)(x,t) \equiv \frac{\delta^2 Z(\eta)}{\delta \eta_j(x,t) \delta \eta_k(x,t)} \frac{\partial}{\partial x^j} \otimes \frac{\partial}{\partial x^k} \in \mathcal{D}'(M \times (0,T) : ST_2(M)).$$

(2) *It satisfies also*

$$(\mathrm{STH})_S \quad \begin{cases} [Z_\eta(\eta), (\frac{\partial}{\partial t} - \nu\Delta)\phi] = i[Z_{\eta\eta}(\eta), \tilde{T}\phi] + i[f,\phi] Z(\eta) \quad \text{for} \quad \eta, \phi \in \tilde{\mathfrak{D}}, \\ Z(\delta(\cdot)\eta^0(\cdot)) = W_0(\eta^0) \quad \text{for} \quad \eta^0 \in \mathring{\Lambda}^1_\sigma(M), \\ Z(0) = 1. \end{cases}$$

Here, W_0 is a given positive definite functional on $\tilde{\mathbf{H}}$ and the value of Z at $\delta(\cdot)\eta^0(\cdot)$ is assumed to be well defined and

$$(\tilde{T}\phi)(x,t) = (\tilde{T}\phi)_{jk}(x,t)dx^j \otimes dx^k$$

with

$$(\tilde{T}\phi)_{jk}(x,t) = \frac{1}{2}((\nabla_k\phi)_j(x,t) + (\nabla_j\phi)_k(x,t)) \quad \text{for} \quad \phi \in \tilde{\mathfrak{D}}.$$

Theorem 4.2. *Let $T < \infty$. There exists a strong solution of (STH).*

To prove this theorem, we use arguments in §6 of Inoue [I2] and [I4].

Let $\hat{P}$ be a Fourier-Stieltjes transform of a bounded Borel measure P on $\mathcal{Z}_T$ given by

$$\hat{P}(\eta) = \int_{\mathcal{Z}_T} e^{i[u,\eta]} P(du) \quad \text{for} \quad \eta \in \mathcal{Z}_T^*,$$

which is called also as the characteristic functional of P.

Concerning spaces $\mathcal{Z}_T$ and $\mathcal{Z}_T^*$, we have

$$\mathcal{Z}_T \subset (L^\infty(0,T)\tilde{\otimes}\mathbf{V}^{-s_1}) \cap (L^2(0,T)\tilde{\otimes}\mathbf{H}),$$

$$\mathcal{Z}_T^* \supset (\mathcal{M}(0,T)\tilde{\otimes}\tilde{\mathbf{V}}^{s_1}) \cup (L^2(0,T)\tilde{\otimes}\tilde{\mathbf{H}}).$$

Here, $\mathcal{M}(0,T)$ is the space of signed Radon measures on $(0,T)$ with locally convex topology defined by weak convergence and $\mathcal{A}\tilde{\otimes}\mathcal{B}$ stands for the completion of the tensor product space $\mathcal{A}\otimes\mathcal{B}$ with the projective topology.

Notation. For $u, v \in L^2(0,T:\mathbf{H})$ and $\tilde{v} \in L^2(0,T:\tilde{\mathbf{H}})$ defined by $\tilde{v}_k(x,t) = g_{kj}(x)v^j(x,t)$, we put

$$[u,\tilde{v}] = \int_0^T \int_M g_{jk}(x)u^j(x,t)v^k(x,t)d_gx dt \quad \text{and} \quad \|u\|^2 = [u,\tilde{u}],$$

$$[\![u,\tilde{v}]\!] = \int_0^T \int_M g_{jk}(x)(\nabla u)^j(x,t)(\nabla v)^k(x,t)d_gx dt \quad \text{and} \quad |\!|\!|u|\!|\!|^2 = [\![u,\tilde{u}]\!].$$

Putting

$$H^s(M\times(0,T):X(M)) = \{u = u_j(x,t)\frac{\partial}{\partial x^j}; u_j(x,t) \in H^s(M\times(0,T))\},$$

$$\tilde{H}^s(M\times(0,T):\Lambda^1(M)) = \{\eta = \eta^j(x,t)dx^j; \eta^j(x,t) \in H^s(M\times(0,T))\},$$

we define, for $u, v \in H^s(M\times(0,T):X(M))$ and $v \in \tilde{H}^s(M\times(0,T):\Lambda^1(M))$,

$$[\![u,\tilde{v}]\!]_s = \int_0^T \int_M \sum_{\ell+|\alpha|\le s} g_{jk}(x)(\partial_t^\ell\nabla^\alpha u)^j(x,t)(\partial_t^\ell\nabla^\alpha u)^k(x,t)d_gx dt$$

and

$$|\!|\!|u|\!|\!|_s^2 = [\![u,\tilde{u}]\!]_s.$$

Lemma 4.3. *(1) If P is a Borel measure on $\mathcal{Z}_T$ satisfying*

$$\int_{\mathcal{Z}_T} \| u \|^2 P(du) < \infty, \tag{4.1}$$

then $\hat{P}$ is positive definite on $\mathcal{Z}_T^*$ and differentiable in the direction $\tilde{\mathfrak{D}}$, and its Fréchet differential $\hat{P}_\eta(\eta) \in L^2(0,T:\mathbf{H})$ satisfies

$$\frac{d}{d\epsilon}\hat{P}(\eta+\epsilon\phi)\Big|_{\epsilon=0} = [\hat{P}_\eta(\eta),\phi] = i\int_{\mathcal{Z}_T}[u,\phi]e^{i[u,\eta]}P(du) \quad \text{for} \quad \eta\in\tilde{\mathfrak{D}}, \phi\in L^2(0,T:\tilde{\mathbf{H}}). \tag{4.2}$$

Moreover, $\hat{P}_{\eta\eta}$ exists as an operator from $\tilde{\mathcal{Z}}_T$ to $\mathcal{Z}_T$ satisfying

$$\hat{P}_{\eta\eta}(\eta)\zeta = -\int_{\mathcal{Z}_T}[u,\zeta]ue^{i[u,\eta]}P(du) \quad \text{for} \quad \zeta\in\tilde{\mathcal{Z}}_T. \tag{4.3}$$

(2) In addition to (4.1), if we suppose that

$$\int_{\mathcal{Z}_T} |\!|\!| u |\!|\!|^2 P(du) < \infty, \tag{4.4}$$

then $\hat{P}_\eta(\eta) \in L^2(0,T:\mathbf{V})$ and

$$[\![\hat{P}_\eta(\eta),\phi]\!] = i\int_{\mathcal{Z}_T}[\![u,\phi]\!]e^{i[u,\eta]}P(du) \quad \text{for} \quad \eta\in\tilde{\mathfrak{D}}, \phi\in L^2(0,T:\tilde{\mathbf{V}}). \tag{4.5}$$

Proof. By Taylor's theorem, there exist constants $\theta_1,\theta_2\in(0,1)$ depending on λ such that

$$e^{i\lambda} = 1+i\lambda+\frac{(i\lambda)^2}{2}e^{i\lambda\theta_1} \quad \text{and} \quad e^{i\lambda} = 1+i\lambda e^{i\lambda\theta_2}.$$

Using the first equality, we have

$$\begin{aligned}\frac{1}{\epsilon}\Big|\hat{P}(\eta+\epsilon\phi)-\hat{P}(\eta)-i\epsilon\int_{\mathcal{Z}_T}[u,\phi]e^{i[u,\eta]}P(du)\Big| \\ = \frac{1}{\epsilon}\Big|\int_{\mathcal{Z}_T}(e^{i[u,\eta+\epsilon\phi]}-e^{i[u,\phi]}-i\epsilon[u,\phi]e^{i[u,\eta]})P(du)\Big| \le \frac{\epsilon}{2}\|\phi\|^2\int_{\mathcal{Z}_T}\|u\|^2P(du),\end{aligned}$$

which implies (4.2). Moreover,

$$\frac{1}{\sqrt{g}}\frac{\partial}{\partial x^j}\left(\sqrt{g}\frac{\delta\hat{P}(\eta)}{\delta\eta_j(x,t)}\right) = 0.$$

In fact, putting $(\widetilde{d_x p})^j = g^{jk}(x)\partial p/\partial x^k$, we have

$$[\hat{P}_\eta(\eta),\widetilde{d_x p}] = \int_{\mathcal{Z}_T}[u,\widetilde{d_x p}]e^{i[u,\eta]}P(du) = 0 \quad \text{for any} \quad p\in C_0^1((0,T)\times M).$$

Above means that $\hat{P}_\eta(\eta) \in L^2(0,T : \mathbf{H})$ and it is represented as

$$\hat{P}_\eta(\eta) = i \int_{Z_T} u e^{i[u,\eta]} P(du) \quad \text{for} \quad \eta \in \tilde{\mathfrak{D}}. \tag{4.6}$$

Using (4.6) and Taylor's theorem, we have

$$\begin{aligned} &\frac{1}{\epsilon}\left\| \hat{P}_\eta(\eta+\epsilon\zeta) - \hat{P}_\eta(\eta) + \epsilon \int_{Z_T} [u,\zeta] u e^{i[u,\eta]} P(du) \right\|_{L^2(0,T:\mathbf{H})} \\ &\leq \frac{1}{\epsilon} \int_{Z_T} \|u\| \left| e^{i[u,\epsilon\zeta]} - 1 - i\epsilon[u,\zeta] \right| P(du) \\ &\leq \frac{\epsilon \|\zeta\|^2}{2} \int_{\{u;u\in Z_T, \|u\| \leq r\}} \|u\|^3 P(du) + 2\|\zeta\| \int_{\{u;u\in Z_T, \|u\|\geq r\}} \|u\|^2 P(du) \end{aligned}$$

for any $r \in (0,\infty)$. Let $\delta > 0$. By (4.1), there exists an r_δ such that

$$2 \int_{\{u;u\in Z_T, \|u\| \geq r_\delta\}} \|u\|^2 P(du) < \frac{\delta}{2}.$$

This implies (4.3). (4.4) and (4.5) derived analogously. □

Proposition 4.4. *Let P be a Borel measure on Z_T satisfying*

$$\int_{Z_T} \|u\|_{s_2}^2 P(du) < \infty \tag{4.7}$$

with $s_2 > (d+1)/2$. Then, for any j,k and any $(x,t),(y,s) \in \bar{M} \times [0,T]$, we have

$$\langle \hat{P}_{\eta\eta}(\eta)\delta_{(x,t)}e^j, \delta_{(y,s)}e^k \rangle = \int_{Z_T} u^j(x,t) u^k(y,s) e^{i[u,\eta]} P(du). \tag{4.8}$$

Here $\delta_{(x,t)}e^j$ denotes the functional on $H^s(M \times (0,T) : X(M))$ defined by $\delta_{(x,t)}e^j(u) = u^j(x,t)$. Moreover, (LHS) of (4.8) is continuous in $(x,t),(y,s) \in \bar{M} \times [0,T]$.

Proof. Combining (4.3) with (4.7), we have

$$|[\hat{P}_{\eta\eta}(\eta)\zeta^{(1)}, \zeta^{(2)}]| \leq \int_{Z_T} \|u\|_s^2 P(du) \|\zeta^{(1)}\|_{-s} \|\zeta^{(2)}\|_{-s}$$

for $\zeta^{(1)}, \zeta^{(2)} \in \tilde{\mathfrak{D}}$. Therefore, the bilinear functional

$$\zeta^{(1)}, \zeta^{(2)} \to [\hat{P}_{\eta\eta}(\eta)\zeta^{(1)}, \zeta^{(2)}] \tag{4.9}$$

extends by continuity to a bilinear functional on the whole $\tilde{H}^{-s_2}(M \times (0,T) : \Lambda(M))$ whose value at $\phi^{(1)}, \phi^{(2)} \in \tilde{H}^{-s_2}(M \times (0,T) : \Lambda(M))$ is denoted by $\langle \hat{P}_{\eta\eta}(\eta)\phi^{(1)}, \phi^{(2)} \rangle$. On the other hand, $H^{s_2}(M \times (0,T) : X(M))$ is continuously imbedded in the set $C^0(M \times (0,T) : X(M))$. This implies that the Dirac functional $\delta_{(x,t)}e^j$ is well-defined and continuous on $H^{s_2}(M \times (0,T) : X(M))$ for all $(x,t) \in \bar{M} \times [0,T]$ and $j = 1,2,\cdots,d$. Thus by (4.9) and (4.3), we have

$$\begin{aligned} \langle \hat{P}_{\eta\eta}(\eta)\delta_{(x,t)}e^j, \delta_{(y,s)}e^k \rangle &= \int_{Z_T} \langle u, \delta_{(x,t)}e^j \rangle \langle u, \delta_{(y,s)}e^k \rangle e^{i[u,\eta]} P(du) \\ &= \int_{Z_T} u^j(x,t) u^k(y,s) e^{i[u,\eta]} P(du). \quad \square \end{aligned} \tag{4.10}$$

Definition 4.5. *Let a Borel measure P on $\mathcal{Z}_T$ satisfying (4.1) be given. We define an operator $\mathcal{L}$ by*

$$[(\mathcal{L}\hat{P})(\eta), \tilde{v}] = i\int_{\mathcal{Z}_T}\left(\int_0^T b(u,u,v)dt\right)e^{i[u,\eta]}P(du)$$

for $\eta \in \mathcal{Z}_T^*$ and $v \in \mathfrak{D}$.

Theorem 4.6. *Let P be a Borel measure on $\mathcal{Z}_T$ satisfying (4.7).*

(1) *$\delta^2\hat{P}(\eta)/\delta\eta_j(x,t)\delta\eta_k(x,t)$ exists and belongs to $C(\bar{M}\times[0,T])$ for each j,k.*

(2) For each $v \in \mathcal{D}(M\times(0,T) : X_\sigma(M))$,

$$[(\mathcal{L}\hat{P})(\eta), \tilde{v}] = \left\langle \frac{\delta^2\hat{P}(\eta)}{\delta\eta_j(x,t)\delta\eta_k(x,t)}, (\tilde{T}\tilde{v})_{jk}(x,t)\right\rangle \tag{4.11}$$

(3)

$$\frac{\delta^2\hat{P}(\eta)}{\delta\eta_j(x,t)\delta\eta_k(x,t)}\frac{\partial}{\partial x^j}\otimes\frac{\partial}{\partial x^k}$$

belongs to $\mathcal{D}'(M\times(0,T) : ST_2(M))$.

Proof. By the kernel theorem of Schwartz, (4.9) implies that there exist distribution kernels $K(\eta)^{jk}((x,t),(y,s))$ on $(M\times(0,T))\times(M\times(0,T))$ such that for $\zeta^{(1)},\zeta^{(2)}$

$$[\hat{P}_{\eta\eta}(\eta)\zeta^{(1)},\zeta^{(2)}] = \iint_{(M\times(0,T))\times(M\times(0,T))} K(\eta)^{jk}((x,t),(y,s))\zeta_j^{(1)}(x,t)\zeta_k^{(2)}(y,s)d_gxdtd_gyds$$

where the integral on the right hand side must be symbolically interpreted. Combining this with (4.8) and the definition of functional derivatives, we have, as distributions,

$$\frac{\delta^2\hat{P}(\eta)}{\delta\eta_j(x,t)\delta\eta_k(y,s)} = -\int_{\mathcal{Z}_T} u^j(x,t)u^k(y,s)e^{i[u,\eta]}P(du) = \langle\hat{P}_{\eta\eta}(\eta)\delta_{(x,t)}e^j, \delta_{(y,s)}e^k\rangle. \tag{4.12}$$

Since the above holds as continuous functions on $(\bar{M}\times[0,T])\times(\bar{M}\times[0,T])$, we get (1). Using the definition of $\mathcal{L}$ and Proposition 4.4, we have

$$\begin{aligned}
[(\mathcal{L}\hat{P})(\eta), \tilde{v}] &= i\int_{\mathcal{Z}_T}\left(\int_0^T b(u,u,v)dt\right)e^{i[u,\eta]}P(du)\\
&= -i\int_{\mathcal{Z}_T}\left(\int_0^T b(u,v,u)dt\right)e^{i[u,\eta]}P(du)\\
&= -i\int_{\mathcal{Z}_T}\left(\int_0^T\int_M(\nabla_k\tilde{v})_j(x,t)u^j(x,t)u^k(x,t)d_gxdt\right)e^{i[u,\eta]}P(du)\\
&= -i\int_0^T\int_M(\nabla_k\tilde{v})_j(x,t)\left[\int_{\mathcal{Z}_T}u^j(x,t)u^k(x,t)e^{i[u,\eta]}P(du)\right]d_gxdt
\end{aligned} \tag{4.13}$$

where we used (4.7) and Fubini's theorem. So we have

$$[(\mathcal{L}\hat{P})(\eta),\tilde{v}] = i\int_0^T\int_M (\nabla_k\tilde{v})_j(x,t)\langle \hat{P}_{\eta\eta}(\eta)\delta_{(x,t)}e^j,\delta_{(x,t)}e^k\rangle d_gx\,dt.$$

After replacing the term $(\nabla_k\tilde{v})_j(x,t)$ with $(\tilde{T}\tilde{v})_{jk}(x,t)$, we have an element $\langle \hat{P}_{\eta\eta}(\eta)\delta_{(x,t)}e^j,\delta_{(x,t)}e^k\rangle$ in $\mathcal{D}'(M\times(0,T):ST_2(M))$. This implies our result. □

Corollary 4.7. *Let P be a Borel measure on $\mathcal{Z}_T$ satisfying (4.1). There exists a distribution $K(\eta)^{jk}\partial/dx^j\otimes\partial/dx^k$ such that*

$$(4.14)\qquad [(\mathcal{L}\hat{P})(\eta),\tilde{v}] = i[K(\eta)^{jk}(x,t),(\tilde{T}\tilde{v})_{jk}(x,t)]$$

for $\tilde{v}\in\tilde{\mathfrak{D}}$.

Proof. Using the third equality in (4.13), which is valid when P satisfies (4.7), we have

$$[(\mathcal{L}\hat{P})(\eta),\tilde{v}] = -i\int_{\mathcal{Z}_T}\left(\int_0^T\int_M(\tilde{T}\tilde{v})_{jk}(x,t)u^j(x,t)u^k(x,t)d_gx\,dt\right)e^{i[u,\eta]}P(du).$$

This implies

$$|[(\mathcal{L}\hat{P})(\eta),\tilde{v}]| \le \|\tilde{T}\tilde{v}\|_{s-1}\int_{\mathcal{Z}_T}\|u\|\,\|u\|P(du).$$

Therefore, $K(\eta)^{jk}\partial/dx^j\otimes\partial/dx^k\in\mathcal{D}'(M\times(0,T):ST_2(M))$ which satisfies (4.14). □

Remark. Putting $P_\epsilon(du) = e^{-\epsilon\|u\|_s^2}P(du)$ for a measure P satisfying (4.1), we may put natural that

$$(4.15)\qquad \frac{\delta^2\hat{P}(\eta)}{\delta\eta_j(x,t)\delta\eta_k(x,t)}\frac{\partial}{\partial x^j}\otimes\frac{\partial}{\partial x^k} \equiv \lim_{\epsilon\to 0}\frac{\delta^2\hat{P}_\epsilon(\eta)}{\delta\eta_j(x,t)\delta\eta_k(x,t)}\frac{\partial}{\partial x^j}\otimes\frac{\partial}{\partial x^k}$$

in $\mathcal{D}'(M\times(0,T):ST_2(M))$.

Proof of Theorem 4.2. Let P be a strong solution of (VKFbis) with initial data μ_0 satisfying $\bar{E}_0<\infty$. Let $Z(\eta)$ denote the characteristic functional of $P(\cdot)$ and let $W_0(\cdot)$ be that of μ_0. We need to interprete Theorem 3.7, which gives us the following:

Lemma 4.8. *For any η, $\phi\in\tilde{\mathfrak{D}}$ and $\eta^0\in\mathring{\Lambda}^1_\sigma(M)$, we have*

$$(4.16)\qquad \begin{cases} [Z_\eta(\eta),(-\dfrac{\partial}{\partial t}-\nu\Delta)\phi] = i\displaystyle\int_{\mathcal{Z}_T}\left(\int_0^T b(u,u,\tilde{\phi})dt\right)e^{i[u,\eta]}P(du) + i[f,\phi]Z(\eta),\\ Z(\delta(\cdot)\eta^0(\cdot)) = W_0(\eta^0).\end{cases}$$

Proof. By the energy inequality, we have

$$\int_{\mathcal{Z}_T} \|u\|^2 P(du) < \infty.$$

The first equality in (4.16) is readily obtained by (VKFbis.1) and (4.2). The property $(\gamma_0^* P)(\omega_0) = \mu_0(\omega_0)$ for any $\omega_0 \in \mathcal{B}(\mathrm{H})$ implies that

$$\int_{\mathcal{Z}_T} \psi(\gamma_0 u) P(du) = \int_{\mathbf{H}} \psi(u_0)\mu_0(du_0)$$

for any Borel measurable function ψ. As P satisfies supp $P \subset \mathcal{Y}_T$, we replace $\mathcal{Z}_T$ in (LHS) of above equality with $\mathcal{Y}_T$ and we have

$$\int_{\mathcal{Y}_T} e^{i(\gamma_0 u, \eta^0)} P(du) = \int_{\mathbf{H}} e^{i(u_0, \eta^0)} \mu_0(du_0) = W_0(\eta^0)$$

As the dual Banach space of $\mathcal{Y}_T$ is contained by $(\mathcal{M}(0,T)\tilde{\otimes}\tilde{\mathbf{V}}^s) \cup L^2(0,T : \tilde{\mathbf{V}}^{-1})$, we have

$$(\gamma_0 u, \eta^0) = \int_M u^j(x,0)\eta_j^0(x) d_g x = [u, \delta(\cdot)\eta^0(\cdot)]$$

from which we get the desired result. □

Lastly, remarking (4.12), (4.14) and (4.15), we have proved Theorem 4.2. □

Remark. In order to treat the case $T = \infty$ in Theorem 4.2, we need to modify the arguments of VKF.

References

[E] S.F. Edwards, *The statistical dynamics of homogeneous turbulence*, J.Fluid Mech. **18** (1964), 239-273.

[E&M] S.F. Edwards & W.D. McComb, *Statistical mechanics far from equilibrium*, J.Phys.A **2** (1969), 157-171.

[F1] C. Foiaş, *Statistical study of Navier-Stokes equations I*, Rend. Sem. Mat. Padova **48** (1973), 219-349.

[F2] ———, *Statistical study of Navier-Stokes equations II*, Rend. Sem. Mat. Padova **49** (1973), 9-123.

[Fr] H.M. Fried, *Functional Methods and Models in Quantum Field Theory*, The MIT Press, Cambridge, Massachusetts and London, 1972.

[G] R. Graham, *Models of stochastic behavior in non-equilibrium steady state*, Scattering techniques applied to supramolecular and nonequilibrium systems (S.H. Chen, B. Chu & R. Nossal, ed.), vol. 73, Nato Advanced Study B, 1982?, pp. 559-612.

[H] E. Hopf, *Statistical hydrodynamics and functional calculus*, J.Rat.Mech.Anal. **1** (1952), 87-123.

[H&T] E. Hopf & E.W. Titt, *On certain special solution of the Φ-equation of statistical hydrodynamics*, J.Rat.Mech. Anal. **2** (1953), 587-591.

[I1] A. Inoue, *Some examples exhibiting the procedures of renormalization and gauge fixing–Schwinger-Dyson equations of first order*, Kodai Math.J. **9** (1986), 134-160.

[I2] ———, *Strong and classical solutions of the Hopf equation –an example of Functional Derivative Equation of second order*, Tôhoku Math.J. **39** (1987), 115-144.

[I3] ———, *On functional derivative equations I,II – An invitation to analysis in functional spaces*, Suugaku (in Japanese) **42** (1990), 170-177,261-265.

[I4] ———, *On Hopf type functional derivative equations for* $\Box u + cu + bu^2 + au^3 = 0$ *on* $\Omega \times \mathbf{R}$. *I. Existence of solutions*, J.Math.Anal.Appl. **152** (1990), 61-87.

[I5] ———, *A remark on Functional Derivative Equations*, (to appear).

[I6] ———, *Construction of a weak solution of the Hopf equation* (to appear).

[Sm-F] O.G. Smolyanov & S.V. Fomin, *Measures on linear topological spaces*, Russian Math.Surveys **31** (1976), 1-53.

[V-F] M.I. Vishik & A.V. Fursikov, *L'équation de Hopf, les solutions statistiques, les moments correspondants aux systèmes des équations paraboliques quasilinéaires*, J.Math.pures et appl. **56** (1977), 85-122.

[V-K] M.I. Vishik & A.I. Komech, *On the solvability of the Cauchy problem for the Hopf equation corresponding to a nonlinear hyperbolic equation*, Amer.Math.Soc.Transl.(2) **118** (1982), 161-184.

[V-K-F] M.I. Vishik, A.I. Komech & A.V. Fursikov, *Some mathematical problems of statistical hydromechanics*, Russian Math.Surveys **34** (1979), 149-234.

OH-OKAYAMA, MEGURO-KU, TOKYO, 152, JAPAN

Initial value problems for the Navier-Stokes equations with Neumann conditions

Gerd Grubb

Initial value problems for the Navier-Stokes equations with data of low smoothness have received much attention in the literature. Consider a bounded domain $\Omega \subset \mathbf{R}^n$. For the Dirichlet problem in dimension $n = 3$ it was shown by Solonnikov in [S] (1973) that there is a strong solution for any given intial value $u_0 \in L_3(\Omega)$ (defined for $t \in I = [0, b]$ with b depending on u_0), via results on the solution operator for the associated Stokes problem in $W_q^{(2k,k)}(\Omega \times I)$ spaces and $L_q \to L_r$ estimates. (See [S] for an account of preceding results.) Later (in 1982–85) the result was proved again and extended to general n (where the initial data are taken in L_n) by Giga and Miyakawa [Gi4,5], [G-M] for the Dirichlet problem and by Giga [Gi3] for partial Neumann problems, based on a reduction to pseudodifferential operators on the boundary and an analysis of the fractional powers of the associated Stokes operators in the solenoidal space J_0 [Gi1,2], allowing complex interpolation. ([S] in addition includes linear first order time-dependent terms in the equation.) See also Sohr and von Wahl [S-W]. Results with initial data in L_p, $p > n$, were shown by Fabes, Lewis and Rivière [F-L-R], von Wahl [W] and others.

Full Neumann problems were not treated in these works, possibly since they do not lead to simple operators on the solenoidal space J, and because the boundary condition involves the pressure besides the velocity. We shall here study the Neumann problems from the point of view developed in joint works with Solonnikov [G-S 1–4], and in [G3,4], where the basic tool is a reduction to parabolic pseudo-differential initial-boundary value problems.

We show existence of strong solutions $\{u(t), P(t)\}$ for initial data u_0 in W_p^r spaces for $r \geq \frac{n}{p} - 1$, $1 < p < \infty$, $n \geq 2$, in particular for $u_0 \in L_n$. In the construction of $u(t)$ we use a variant of a technique of [S] worked out for H_2^s estimates in [G-S3], combined with interpolation results based on [G3]. Then $P(t)$ is determined by use of explicit pseudodifferential formulas from [G-S4]. For $p > n$ also negative r can enter. (We take $r < 2 + \frac{1}{p}$, noting that the case $r > 1 - \frac{1}{p}$ has been treated more generally in [G3].) The solutions are C^∞ for $t > 0$ and satisfy suitable uniform estimates (and estimates with powers of t), see Theorems 2.4 and 2.7 for the precise statements.

The methods also work for the Dirichlet problem, giving similar results for the velocity $u(t)$, but different results for the pressure $P(t)$. In fact, $P(t)$ for the Neumann problem has a certain boundedness for $t \to 0$, that contrasts with the unboundedness for the Dirichlet problem observed for $p = 2$, $n = 2, 3$ by Heywood and Walsh [H-W], see Remark 2.9.

In preparation for the nonlinear results, we show in Section 1 that the semigroup solution operators for the associated linear problems satisfy estimates with powers of t, as operators from W_p^r spaces to W_p^s spaces with boundary conditions, for all $s \geq r > \frac{1}{p} - 1$.

1. Estimates for the Stokes problem

The Navier-Stokes problem resp. associated Stokes problem can be formulated as follows, with $\kappa = 1$ resp. $\kappa = 0$:

$$
(1.1)\qquad
\begin{array}{lrl}
\text{(i)} & \partial_t u - \Delta u + \kappa(u\cdot\nabla)u + \operatorname{grad} P = f & \text{for } (x,t)\in Q,\\
\text{(ii)} & \operatorname{div} u = 0 & \text{for } (x,t)\in Q,\\
\text{(iii)} & T\{u,P\} = \varphi & \text{for } (x,t)\in S,\\
\text{(iv)} & u|_{t=0} = u_0 & \text{for } x\in\Omega.
\end{array}
$$

Here Ω is a bounded connected open set in $\mathbf{R}^n$ with smooth boundary $\partial\Omega = \Gamma$, $\vec{n} = (n_1,...,n_n)$ is the interior unit normal vector field defined near Γ, $I =]0,b[$ for some $b \in \mathbf{R}_+$, $Q = \Omega\times I$, and $S = \Gamma\times I$; and we write $f|_\Gamma = \gamma_0 f$ and $\gamma_0\partial_\nu f = \gamma_1 f$, where $\partial_\nu f = \sum_{j=1}^n n_j\partial_j f$, $\partial_j f = \partial f/\partial x_j$. The velocity vector is denoted $u = (u_1,\dots,u_n)$, and the pressure is the scalar function P. We denote $\vec{n}\cdot u = u_\nu$, $u - u_\nu\vec{n} = u_\tau$, and $\gamma_0 u_\nu = \gamma_\nu u$. The trace operator T stands for one of the following boundary operators T_k $(k = 1,2)$,

$$
(1.2)\qquad
\begin{array}{c}
T_1\{u,P\} = \chi_1 u - \gamma_0 P\vec{n} \quad\text{or}\quad T_2\{u,P\} = \gamma_1 u - \gamma_0 P\vec{n}, \text{ where}\\
\chi_1 u = \gamma_0 S(u)\vec{n} = \gamma_0\big(\sum_j(\partial_i u_j + \partial_j u_i)n_j\big)_{i=1,\dots,n}, \quad S(u) = (\partial_i u_j + \partial_j u_i)_{i,j=1,\dots,n};
\end{array}
$$

T_2 is the simplest Neumann boundary operator for such systems, and T_1 is the appropriate version for the study of free boundary problems.

It is shown in [G-S 1–4] how the problems can be reduced to the form

$$
(1.3)\qquad
\begin{array}{rl}
\partial_t u - \Delta u + G_k u + \kappa Q u = f' & \text{in } Q_I,\\
T_k' u = \psi & \text{on } S_I,\\
u|_{t=0} = u_0 & \text{on } \Omega,
\end{array}
$$

$k = 1,2$, where

$$
(1.4)\qquad
\begin{array}{ll}
G_1 u = -2\operatorname{grad} K_D \operatorname{div}_\Gamma' \gamma_0 u, & G_2 u = -\operatorname{grad} K_D \operatorname{div}_\Gamma' \gamma_0 u,\\
T_1' u = (\chi_1 u)_\tau + \gamma_0(\operatorname{div} u)\vec{n}, & T_2' u = \gamma_1 u_\tau + \gamma_0(\operatorname{div} u)\vec{n},\\
Qu = \operatorname{pr}_J[(u\cdot\nabla)u], \quad f' = \operatorname{pr}_J f + \operatorname{grad} K_D\varphi_\nu, & \psi = \varphi_\tau.
\end{array}
$$

Here $(\,R_D \quad K_D\,) : \{g,\varrho\} \mapsto v$ is the solution operator for the Laplace Dirichlet problem $-\Delta v = g$ in Ω, $\gamma_0 v = \varrho$ on Γ; pr_J stands for the projection of $L_p(\Omega)^n$ (orthogonal in $L_2(\Omega)^n$) onto the solenoidal space $J = \{u\in L_p(\Omega)^n \mid \operatorname{div} u = 0\}$, it equals $I + \operatorname{grad} R_D \operatorname{div}$; and $\operatorname{div}_\Gamma'$ is the tangential divergence operator; cf. [G-S4] for further details. When f and φ are 0, so are f' and ψ. When u solves (1.3) with $f' = 0$, $\psi = 0$ and $\operatorname{div} u_0 = 0$, then $\{u,P\}$ with

$$
(1.5)\qquad
\begin{array}{l}
P = -2K_D \operatorname{div}_\Gamma' \gamma_0 u + \kappa R_D \operatorname{div}[(u\cdot\nabla)u], \text{ for } k = 1,\\
P = -K_D \operatorname{div}_\Gamma' \gamma_0 u + \kappa R_D \operatorname{div}[(u\cdot\nabla)u], \text{ for } k = 2,
\end{array}
$$

solves (1.1) for $f = 0$ and $\varphi = 0$.

Since the two cases $k = 1$ and 2 are quite similar (the case $k = 1$ being slightly more complicated than $k = 2$), we can in the following restrict the attention to the case $k = 1$, and we then *drop the subscript 1.* We denote

$$
(1.6)\qquad -\Delta + G = M.
$$

Let us insert a few remarks on the generalizations of L_2 Sobolev spaces to L_p, $1 < p < \infty$. On one hand, there are the so-called Bessel potential spaces H_p^s (where $(1-\Delta)^{s/2} = \mathcal{F}^{-1}(1+|\xi|^2)^{s/2}\mathcal{F}$):

$$H_p^s(\mathbf{R}^n) = \{ f \in \mathcal{S}'(\mathbf{R}^n) \mid (1-\Delta)^{s/2} f \in L_p(\mathbf{R}^n)\},\ s \in \mathbf{R}; \tag{1.7}$$

on the other hand there are the Besov spaces B_p^s, defined by

$$\begin{gathered} f \in B_p^s(\mathbf{R}^n) \iff \|f\|_{L_p}^p + \int_{\mathbf{R}^{2n}} \frac{|f(x)-f(y)|^p}{|x-y|^{n+ps}}\,dxdy < \infty, \text{ for } s \in]0,1[\,, \\ B_p^{s+t}(\mathbf{R}^n) = (1-\Delta)^{-t/2} B_p^s(\mathbf{R}^n), \text{ for } t \in \mathbf{R}; \end{gathered} \tag{1.8}$$

by restriction to Ω we get the spaces $H_p^s(\overline{\Omega})$ resp. $B_p^s(\overline{\Omega})$; moreover, $B_p^s(\Gamma)$ is defined by use of local coordinates. Here

$$\begin{gathered} B_p^s \subset H_p^s \text{ when } p \le 2, \quad H_p^s \subset B_p^s \text{ when } p \ge 2; \\ H_p^{s+\varepsilon} \subset B_p^s \subset H_p^{s-\varepsilon} \quad \text{for all } p, \text{ when } \varepsilon > 0; \end{gathered} \tag{1.9}$$

with $H_p^s = B_p^s$ if and only if $p = 2$. The classical Sobolev-Slobodetskiĭ spaces W_p^s, defined for $s \ge 0$, are related to the Bessel potential spaces and Besov spaces by:

$$W_p^s = H_p^s \quad \text{for } s \in \mathbf{N} = \{0,1,2,\dots\}, \quad W_p^s = B_p^s \quad \text{for } s \in \mathbf{R}_+ \setminus \mathbf{N}. \tag{1.10}$$

Let $\frac{1}{p'} = 1 - \frac{1}{p}$. The dual spaces are identified as follows:

$$\begin{gathered} H_p^s(\mathbf{R}^n)^* = H_{p'}^{-s}(\mathbf{R}^n), \quad B_p^s(\mathbf{R}^n)^* = B_{p'}^{-s}(\mathbf{R}^n), \text{ when } s \in \mathbf{R}, \\ H_p^s(\overline{\Omega})^* = H_{p'}^{-s}(\overline{\Omega}), \quad B_p^s(\overline{\Omega})^* = B_{p'}^{-s}(\overline{\Omega}), \text{ when } s \in]\tfrac{1}{p}-1, \tfrac{1}{p}[\,. \end{gathered} \tag{1.11}$$

(For general s, $H_p^s(\overline{\Omega})^*$ identifies with the set of distributions in $H_{p'}^{-s}(\mathbf{R}^n)$ that are supported in $\overline{\Omega}$; similar statements hold for B_p^s.) One reason that one cannot quite avoid having to deal with Besov spaces is that they are the correct boundary value spaces; in fact

$$\gamma_0 \colon H_p^s(\overline{\Omega}) \to B_p^{s-\frac{1}{p}}(\Gamma) \quad \text{and} \quad \gamma_0 \colon B_p^s(\overline{\Omega}) \to B_p^{s-\frac{1}{p}}(\Gamma),\ s > \tfrac{1}{p}, \tag{1.12}$$

are continuous and *surjective,* having continuous right inverses. Let us finally mention the interpolation properties: For $p \in]1,\infty[$, s and $t \in \mathbf{R}$ with $s \neq t$, $\theta \in]0,1[$, one has:

$$\begin{gathered} [H_p^s, H_p^t]_\theta = H_p^{(1-\theta)s+\theta t}; \\ (H_p^s, H_p^t)_{\theta,p} = (H_p^s, B_p^t)_{\theta,p} = (B_p^s, B_p^t)_{\theta,p} = B_p^{(1-\theta)s+\theta t}; \end{gathered} \tag{1.13}$$

where $[\cdot,\cdot]_\theta$ denotes complex interpolation and $(\cdot,\cdot)_{\theta,p}$ denotes a certain real interpolation. For further information and many more spaces, see e.g. Triebel [T], Bergh and Löfström [B-L], and short surveys in [G2-3]. Recall the basic feature of these types of interpolation $\{\cdot,\cdot\}_\theta$: When Λ is a linear operator from $X+Y$ to X_1+Y_1, then

$$\Lambda \colon X \xrightarrow{\sim} X_1,\ \Lambda \colon Y \xrightarrow{\sim} Y_1 \implies \Lambda \colon \{X,Y\}_\theta \xrightarrow{\sim} \{X_1,Y_1\}_\theta; \tag{1.14}$$

here $\xrightarrow{\sim}$ indicates a homeomorphism.

Assume in the rest of this section that $\kappa = 0$. Let $p \in]1, \infty[$, let A be the realization of M (1.6) in $L_p(\Omega)^n$ with domain

$$D(A) = \{\, u \in H_p^2(\overline{\Omega})^n \mid T'u = 0 \,\}, \tag{1.15}$$

and denote by A' its restriction with domain

$$D(A') = D(A) \cap J, \tag{1.16}$$

here A' maps into J (since $\operatorname{div}\operatorname{grad} K_D = 0$). In view of the resolvent estimates established in [G-K], [G3], the operator $-A$ generates a holomorphic semigroup $U(t)$ in $L_p(\Omega)^n$, and $U(t)$ preserves J (cf. [G-S4, Sect. 8]), so it defines by restriction to J a holomorphic semigroup $\widetilde{U}(t)$ in J; the latter has $-A'$ as its infinitesimal generator. The spectra of A and A' lie in $\overline{\mathbb{R}}_+$ (are in fact discrete subsets of $\overline{\mathbb{R}}_+$); this was shown for $p = 2$ in [G-S4] and extends by elliptic regularity (cf. [G2]) to general p.

It will be useful to consider also the adjoint A^* of A. Since A is elliptic and is determined by a normal boundary condition, A^* is again a realization of a normal elliptic pseudodifferential boundary problem of order and class 2. This is shown in great generality in [G1, Th. 1.6.9] where the adjoint is determined (see also [G1, Ex. 1.6.15]); since the present operator is of a relatively simple form, we shall make the calculations explicitly here rather than interpreting the general formulas there. Since the boundary condition is differential, the adjoint acts like $-\Delta$, but it has nonlocal terms in its boundary condition, and it is important for some of our arguments to find their structure.

THEOREM 1.1. 1° *The adjoint of* $K_D\colon B_p^{s-\frac1p}(\Gamma) \to H_p^s(\overline{\Omega})$ $(s \in]\frac1p - 1, \frac1p[)$ *is a trace operator of order* -1 *and class* 0, $(K_D)^*\colon H_{p'}^{-s}(\overline{\Omega}) \to B_{p'}^{-s+\frac1p}(\Gamma)$, *where the latter mapping property extends to all* $-s \in]-\frac1p, \infty[$. *Moreover,* $K_D^* \operatorname{div} = \gamma_\nu - \gamma_\nu \operatorname{pr}_J$, *where* $\gamma_\nu \operatorname{pr}_J$ *is a trace operator of class* 0.

2° *The adjoint of* A *is the (elliptic) realization of* $-\Delta$ *with domain*

$$D(A^*) = \{\, v \in L_{p'}(\Omega)^n \mid \widetilde{T}v = 0 \,\}, \quad \widetilde{T}v = \gamma_1 v + S_0\gamma_0 v + T''v, \tag{1.17}$$

where S_0 *is a differential operator on* Γ *of order* 1, *and* T'' *is a trace operator of order* 1 *and class* 0*:*

$$\begin{aligned} S_0\gamma_0 v &= [\vec{n}(\operatorname{grad}'_\Gamma)^* - s_0]\gamma_0 v_\tau + 3(\operatorname{div}'_\Gamma)^*\gamma_0 v_\nu,\\ T''v &= -2(\operatorname{div}'_\Gamma)^*\gamma_\nu \operatorname{pr}_J v. \end{aligned} \tag{1.18}$$

PROOF: Recall that $\gamma_\nu u = \gamma_0(\vec{n}\cdot u)$, and recall that an operator is of class $m \in \mathbb{N}$ precisely when it is well-defined on $H_2^m(\overline{\Omega})$ (or on $H_p^m(\overline{\Omega})$ or $B_p^m(\overline{\Omega})$ for some p, cf. [G2]). The fact that K_D^* is a trace operator of order -1 and class 0 is well-known from the general calculus; the continuity properties in L_p spaces were established in [G2].

To find the form of $K_D^* \operatorname{div}$, let $u \in C^\infty(\overline{\Omega})^n$ and $\varphi \in C^\infty(\Gamma)$, then

$$(K_D^* \operatorname{div} u, \varphi)_\Gamma = (\operatorname{div} u, K_D\varphi)_\Omega.$$

We have that $v = K_D\varphi$ satifies $\Delta v = 0$, $\gamma_0 v = \varphi$. Moreover, $w = R_D \operatorname{div} u$ satisfies

$$u = -\operatorname{grad} w + z, \text{ with } z = u + \operatorname{grad} R_D \operatorname{div} u = \operatorname{pr}_J u;$$
$$\operatorname{div} u = -\Delta w, \ \gamma_0 w = 0.$$

Then

$$\begin{aligned}(\operatorname{div} u, K_D\varphi)_\Omega &= (-\Delta w, v)_\Omega - (w, -\Delta v)_\Omega = (\gamma_1 w, \gamma_0 v)_\Gamma - (\gamma_0 w, \gamma_1 v)_\Gamma \\ &= (\gamma_1 w, \gamma_0 v)_\Gamma = (\gamma_0 \vec{n} \cdot \operatorname{grad} w, \varphi)_\Gamma = (-\gamma_\nu(u - z), \varphi)_\Gamma.\end{aligned}$$

Since φ was arbitrary, this shows that $K_D^* \operatorname{div} u = -\gamma_\nu(u-z) = -\gamma_\nu u + \gamma_\nu \operatorname{pr}_J u$; and here $\gamma_\nu \operatorname{pr}_J$ is a trace operator of class zero, since it is well-defined on all of $L_2(\Omega)^n$, cf. [G-S4, Th. 2.5 and (2.46)].

Now consider 2°. Since A^* represents an elliptic boundary problem of order and class 2 ([G1, Th. 1.6.9]), $D(A^*) \subset H^2_{p'}(\overline{\Omega})$. Recall from [G-S4, (3.20), (A.22)] that the ingredients in $T'u$ can be written in detail as follows, with $s_0(x) = \big(\partial_j n_k(x)\big)_{j,k=1,\dots,n}$,

$$(\chi_1 u)_\tau = \gamma_1 u_\tau + \operatorname{grad}'_\Gamma \gamma_0 u_\nu - s_0 \gamma_0 u_\tau,$$
$$\gamma_0 \operatorname{div} u = \operatorname{div}'_\Gamma \gamma_0 u + \gamma_1 u_\nu.$$

Let us temporarily borrow the notation $(\cdot\,,\cdot)$ from L_2 to denote also sesquilinear dualities between L_p and $L_{p'}$. For $u \in D(A)$ and $v \in H^2_{p'}(\overline{\Omega})^n$ we find since $(\chi_1 u)_\tau = 0$, $\gamma_0 \operatorname{div} u = 0$, using also the formula for $K_D^* \operatorname{div}$ established above,

$$\begin{aligned}((-\Delta + G)u, v)_\Omega - (u, -\Delta v)_\Omega &= (\gamma_1 u, \gamma_0 v)_\Gamma - (\gamma_0 u, \gamma_1 v)_\Gamma - (2 \operatorname{grad} K_D \operatorname{div}'_\Gamma \gamma_0 u, v)_\Omega \\ &= (-\operatorname{grad}'_\Gamma \gamma_0 u_\nu + s_0 \gamma_0 u_\tau, \gamma_0 v_\tau)_\Gamma + (-\operatorname{div}'_\Gamma \gamma_0 u, \gamma_0 v_\nu) \\ &\quad - (\gamma_0 u, \gamma_1 v)_\Gamma + (\gamma_0 u, 2(\operatorname{div}'_\Gamma)^* K_D^* \operatorname{div} v) \\ &= -\big(\gamma_0 u, \gamma_1 v + \vec{n}(\operatorname{grad}'_\Gamma)^* \gamma_0 v_\tau - s_0 \gamma_0 v_\tau + (\operatorname{div}'_\Gamma)^* \gamma_\nu v + 2(\operatorname{div}'_\Gamma)^*(\gamma_\nu v - \gamma_\nu \operatorname{pr}_J v)\big)_\Gamma \\ &= -(\gamma_0 u, \widetilde{T} v)_\Gamma,\end{aligned}$$

where $\widetilde{T}$ satisfies (1.17)–(1.18). Here $\gamma_0 u$ can take all values in $B_p^{2-\frac{1}{p}}(\Gamma)^n$ when $u \in D(A)$ (since $\{\gamma_0, T'\}$ is normal), so it follows that $v \in D(A^*)$ if and only if $\widetilde{T} v = 0$. ∎

We now define the following spaces:

$$\begin{aligned} H^s_{p,T',M} &= \{\, u \in H^s_p(\overline{\Omega})^n \mid T'M^l u = 0 \text{ for } l \le m-1,\ T'M^m u = 0 \text{ if } \sigma \ge 1 + \tfrac{1}{p} \,\}, \\ B^s_{p,T',M} &= \{\, u \in B^s_p(\overline{\Omega})^n \mid T'M^l u = 0 \text{ for } l \le m-1,\ T'M^m u = 0 \text{ if } \sigma \ge 1 + \tfrac{1}{p} \,\}, \\ &\quad \text{for } s = 2m + \sigma \text{ with } m \in \mathbb{N} \text{ and } \sigma \in]-1 + \tfrac{1}{p}, 3 + \tfrac{1}{p}[\end{aligned} \tag{1.19}$$

(here $l \in \mathbb{N}$); note that $H^{2m}_{p,T',M} = D(A^m)$, and that the spaces form a decreasing family. (The definitions in (1.19) are clearly consistent when s has two different representations $s = 2m + \sigma = 2(m+1) + \sigma'$.) When $\sigma = 1 + \frac{1}{p}$, the condition $T'M^m u = 0$ is understood in the sense that

$$e^+ S(M^m u)\,\vec{n} \text{ belongs to } H_p^{\frac{1}{p}}(\overline{\Sigma})^n \text{ resp. } B_p^{\frac{1}{p}}(\overline{\Sigma})^n, \tag{1.20}$$

where Σ is a tubular neighborhood of Γ and e^+ denotes extension by 0 to the part of Σ exterior to Ω; for $S(u)$ see (1.2).

$H^s_{p,T',M}$ is a closed subspace of $H^s_p(\overline{\Omega})^n$ when $s-1-\frac{1}{p}\notin 2\mathbb{N}$, otherwise it is a subspace with a stronger norm (one must add the norm of the function in (1.20)); analogous statements hold for the B^s_p scale.

When $m=0$ in (1.19), M is not used, and when moreover $\sigma\in]-1+\frac{1}{p},1+\frac{1}{p}[$, T' is not used, so we also write

$$(1.21)\qquad \begin{aligned} &H^\sigma_{p,T',M}=H^\sigma_{p,T'},\quad B^\sigma_{p,T',M}=B^\sigma_{p,T'},\quad \text{for } \sigma\in]-1+\tfrac{1}{p},3+\tfrac{1}{p}[;\text{ and}\\ &H^\sigma_{p,T'}=H^\sigma_p(\overline{\Omega})^n,\quad B^\sigma_{p,T'}=B^\sigma_p(\overline{\Omega})^n,\quad \text{for } \sigma\in]-1+\tfrac{1}{p},1+\tfrac{1}{p}[.\end{aligned}$$

In a similar way, we define spaces $H^s_{p,\widetilde{T},\Delta}$ and $B^s_{p,\widetilde{T},\Delta}$, and $H^s_{p,\gamma_1+S_0\gamma_0,\Delta}$ and $B^s_{p,\gamma_1+S_0\gamma_0,\Delta}$, where M in (1.19) is replaced by Δ, and T' is replaced by $\widetilde{T}$ or $\gamma_1+S_0\gamma_0$; the interpretation of the boundary condition in case $\sigma=1+\frac{1}{p}$ is here:

$$(1.22)\qquad \begin{aligned} &e^+(\partial_\nu\Delta^m u+[\vec{n}(\text{grad}')^*-s_0](\Delta^m u)_\tau-(\text{div}')^*((1-\text{pr}_J)\Delta^m u)_\nu)\\ &\qquad\in H^{\frac{1}{p}}_p(\overline{\Sigma})\text{ resp. } B^{\frac{1}{p}}_p(\overline{\Sigma});\text{ or}\\ &e^+(\partial_\nu\Delta^m u+[\vec{n}(\text{grad}')^*-s_0](\Delta^m u)_\tau-(\text{div}')^*(\Delta^m u)_\nu)\\ &\qquad\in H^{\frac{1}{p}}_p(\overline{\Sigma})\text{ resp. } B^{\frac{1}{p}}_p(\overline{\Sigma}).\end{aligned}$$

The resolvent estimates shown in [G3] imply in particular that for $c>0$,

$$(1.23)\qquad \begin{aligned} &(A+c)^k\colon H^s_{p,T',M}\xrightarrow{\sim}H^{s-2k}_{p,T',M}\text{ and}\\ &(A+c)^k\colon B^s_{p,T',M}\xrightarrow{\sim}B^{s-2k}_{p,T',M},\text{ for } k\in\mathbb{Z},\text{ when } s \text{ and } s-2k>\tfrac{1}{p}-1.\end{aligned}$$

Then we have in view of well-known holomorphic semigroup properties that

$$(1.24)\qquad U(t)\colon H^{2k}_{p,T',M}\to H^{2l}_{p,T',M}\text{ with norm } O(e^{ct}t^{-(k-l)})\text{ for } k\le l\in\mathbb{N}.$$

In particular, the initial value problem (1.3) with $\kappa=0$, $f'=0$, $\psi=0$, has for any $u_0\in H^{2k}_{p,T',M}$ ($k\in\mathbb{N}$) a unique solution

$$(1.25)\qquad u(t)=U(t)u_0\in C^0(\overline{\mathbb{R}}_+;H^{2k}_{p,T',M}).$$

On the other hand, the main result in [G3] on the solvability of parabolic problems implies that for $u_0\in B^{s+2-\frac{2}{p}}(\overline{\Omega})^n$ satifying the compatibility condition for $\{0,0,u_0\}$ of order s, there is a unique solution u with

$$(1.26)\qquad e^{-ct}u\in H^{(s+2,s/2+1)}_p(\overline{\Omega}\times\overline{\mathbb{R}}_+)^n\cap B^{(s+2,s/2+1)}_p(\overline{\Omega}\times\overline{\mathbb{R}}_+)^n,\text{ when } s>\tfrac{1}{p}-1.$$

For comparison, $s+2-\frac{2}{p}$ there corresponds to $2m+\sigma$ here; in fact for $s=2m+\sigma-2+\frac{2}{p}>\frac{1}{p}-1$, the compatibility condition of order s for the triple $\{0,0,u_0\}$ is precisely the condition for belonging to $B^{2m+\sigma}_{p,T',M}$. (Note that the exceptional values in (1.19), namely those where $\sigma=1+\frac{1}{p}$, correspond to the exceptional values for the compatibility conditions, where

$s+1-\frac{3}{p}\in 2\mathbb{N}$.) The semigroup solution method applies to larger initial spaces than the anisotropic result in [G3], which only goes down to $B_p^{s+2-\frac{2}{p}}(\overline{\Omega})^n$ with $s>\frac{1}{p}-1$, i.e. $s+2-\frac{2}{p}>1-\frac{1}{p}$. However, the solutions coincide when both are defined, since $B_p^{s+2,s/2+1}(\overline{Q})^n\subset C^0(\overline{I};B_p^{s+2-\frac{2}{p}}(\overline{\Omega}))\subset C^0(\overline{I};H_p^{s'+2-\frac{2}{p}}(\overline{\Omega}))$ for $s>s'>\frac{1}{p}-1>\frac{2}{p}-2$, cf. (1.9). — Initial values in H_p^σ spaces are not included in a precise way in [G3] (one has of course the consequences of (1.9)), but as we see, they enter nicely in connection with semigroup methods.

The adjoint A^* of A has just as good resolvent properties as A (its symbols have the same parameter-ellipticity properties), so $-A^*$ generates a holomorphic semigroup in $L_{p'}(\Omega)^n$, which is simply equal to $U(t)^*$. The properties (1.23)–(1.26) are likewise valid for A^* and $U(t)^*$.

Now we shall show that *in suitable intervals of* $\mathbb{R}$, the scales of spaces (1.19) ff. interpolate well:

THEOREM 1.2. *When $s>t$ and $\theta\in]0,1[$ one has the interpolation identities:*

$$\text{(1.27)}\quad\begin{aligned}
&\text{(i)} && [H^s_{p,T',M}(\overline{\Omega})^n,H^t_{p,T',M}(\overline{\Omega})^n]_\theta=H^{(1-\theta)s+\theta t}_{p,T',M}(\overline{\Omega})^n,\\
&\text{(ii)} && (H^s_{p,T',M}(\overline{\Omega})^n,H^t_{p,T',M}(\overline{\Omega})^n)_{\theta,p}=B^{(1-\theta)s+\theta t}_{p,T',M}(\overline{\Omega})^n;\\
&\text{(iii)} && [H^s_{p,\widetilde{T},\Delta}(\overline{\Omega})^n,H^t_{p,\widetilde{T},\Delta}(\overline{\Omega})^n]_\theta=H^{(1-\theta)s+\theta t}_{p,\widetilde{T},\Delta}(\overline{\Omega})^n,\\
&\text{(iv)} && (H^s_{p,\widetilde{T},\Delta}(\overline{\Omega})^n,H^t_{p,\widetilde{T},\Delta}(\overline{\Omega})^n)_{\theta,p}=B^{(1-\theta)s+\theta t}_{p,\widetilde{T},\Delta}(\overline{\Omega})^n;
\end{aligned}$$

when either s and $t\in[2m,2m+2]$, or s and $t\in]2m-1+\frac{1}{p},2m+1+\frac{1}{p}[$, for some $m\in\mathbb{N}$.

PROOF: Consider first the spaces defined from T' and M. For $m=0$, the involved trace operator is differential, so the statements with s and $t\in[0,2m]$ are known from Grisvard [Gri] (real interpolation leading to Besov spaces) and Seeley [Se] (complex interpolation of the Bessel potential spaces). The statements with s and $t\in]-1+\frac{1}{p},1+\frac{1}{p}[$ are simpler, since the spaces here are identical with the ordinary Bessel potential resp. Besov spaces, cf. (1.11), and (1.13) can be used directly. By use of (1.23) and (1.14), the statements carry over to general m; and this shows (1.27 i,ii).

Next, consider the spaces defined from $\widetilde{T}$ and Δ. Again, the statements with s and $t\in]-1+\frac{1}{p},1+\frac{1}{p}[$ are straightforward consequences of (1.13). Now consider the interval $[0,2]$. The trace operator $\widetilde{T}$ is nonlocal, but it has the very fortunate property that the nonlocalness resides only in the term $T''v$ of class zero, cf. (1.17)–(1.18). Therefore it is still possible to establish a relatively easy interpolation result, by the same device as in [G1, Th. 4.4.2 ff.]. Namely, one can use [G1, Lemma 1.6.8] to construct a homeomorphism Λ in $H_p^s(\overline{\Omega})^n$ (all $s>\frac{1}{p}-1$), belonging to the pseudodifferential boundary operator calculus, such that

$$\text{(1.28)}\qquad \gamma_1+S_0\gamma_0+T''=(\gamma_1+S_0\gamma_0)\Lambda.$$

(The statements are shown in [G1] for $p=2$, $s\geq 0$, and extend to $1<p<\infty$, $s>\frac{1}{p}-1$ by [G2].) By the definitions of $H^s_{p,\gamma_1+S_0\gamma_0,\Delta}$ and $B^s_{p,\gamma_1+S_0\gamma_0,\Delta}$, Λ defines homeomorphisms

$$\text{(1.29)}\quad \Lambda\colon H^s_{p,\widetilde{T},\Delta}\xrightarrow{\sim}H^s_{p,\gamma_1+S_0\gamma_0,\Delta},\quad \Lambda\colon B^s_{p,\widetilde{T},\Delta}\xrightarrow{\sim}B^s_{p,\gamma_1+S_0\gamma_0,\Delta},\quad \text{for } s\in]-1+\tfrac{1}{p},3+\tfrac{1}{p}[.$$

The interpolation results hold for the spaces $H^s_{p,\gamma_1+S_0\gamma_0,\Delta}$ and $B^s_{p,\gamma_1+S_0\gamma_0,\Delta}$ by the quoted results of Grisvard and Seeley, and then they carry over to the spaces $H^s_{p,\widetilde{T},\Delta}$ and $B^s_{p,\widetilde{T},\Delta}$ for the interval $[0,2]$ by use of (1.29). This shows (1.27 iii,iv) for $m=0$. Finally we get (1.27 iii,iv) for general m by using the analogue of (1.23) for A^*. (One could even extend the results to intervals of the form $]2m-1+\frac{1}{p}, 2m+3+\frac{1}{p}[$.) ∎

We expect that the interpolation results can be extended to the full scales $H^s_{p,T',M}$ and $B^s_{p,T',M}$, $s>-1+\frac{1}{p}$, but we shall not pursue the question here, since Theorem 1.2 suffices for our purposes.

Note that $H^s_{p,T',M}$ is dense in $H^r_{p,T',M}$ when $s>r$ (and similarly for the B^s_p scale); this follows from a general property of interpolation spaces, cf. e.g. [T, 1.6.2, 1.9.3]. (One can go from $H^s_{p,T',M}$ to $H^r_{p,T',M}$ in a finite number of steps of length $<\frac{1}{p}$.)

Now let us show

THEOREM 1.3. *Let $k\in\mathbb{N}$ and $s\geq 2k$. Then for $t>0$,*

$$(1.30)\quad \begin{aligned} &\text{(i)}\quad \|U(t)u_0\|_{H^s_{p,T',M}} \leq C_{s,k}e^{ct}t^{-(s-2k)/2}\|u_0\|_{H^{2k}_{p,T',M}} && \text{when } s\in\overline{\mathbb{R}}_+;\\ &\text{(ii)}\quad \|U(t)u_0\|_{B^s_{p,T',M}} \leq C_{s,k}e^{ct}t^{-(s-2k)/2}\|u_0\|_{H^{2k}_{p,T',M}} && \text{when } s\in\overline{\mathbb{R}}_+\setminus 2\mathbb{N}.\end{aligned}$$

A similar result holds for $U(t)^*$, in the spaces $H^s_{p,\widetilde{T},\Delta}$ and $B^s_{p,\widetilde{T},\Delta}$.

PROOF: The result holds for $s=2m$, $m\in\mathbb{N}$, by (1.24) with $l=m$. The basic interpolation inequality, for $0<\theta<1$,

$$(1.31)\qquad \|f\|_{\{X,Y\}_\theta} \leq C_\theta \|f\|_X^{1-\theta}\|f\|_Y^\theta,$$

holds for both real and complex interpolation. From the validity of (1.30) for $s=2m$ and $s=2m+2$, we find for $s=2(m+1-\theta)$ by complex interpolation, using (1.27 i),

$$(1.32)\quad \begin{aligned}\|U(t)u_0\|_{H^{2(m+1-\theta)}_{p,T',M}} &\leq C\|U(t)u_0\|^{1-\theta}_{H^{2(m+1)}_{p,T',M}}\|U(t)u_0\|^\theta_{H^{2m}_{p,T',M}}\\ &\leq C'e^{ct}t^{-(m+1-\theta-k))}\|u_0\|_{H^{2k}_{p,T',M}}.\end{aligned}$$

The analogous inequality for the norm in $B^{2(m+1-\theta)}_{p,T',M}$ is found by real interpolation, using (1.27 ii). For $U(t)^*$ one uses (1.27 iii,iv) in a similar way. ∎

We now get to the main result of this section, namely the extension of (1.30) to general exponents.

THEOREM 1.4. *Let $r>\frac{1}{p}-1$ and $s\in[r,\infty[$. Then for $t>0$,*

$$(1.33)\quad \begin{aligned} &\text{(i)}\quad \|U(t)u_0\|_{H^s_{p,T',M}} \leq C_{s,r}e^{ct}t^{-(s-r)/2}\|u_0\|_{H^r_{p,T',M}},\\ &\text{(ii)}\quad \|U(t)u_0\|_{B^s_{p,T',M}} \leq C_{s,r}e^{ct}t^{-(s-r)/2}\|u_0\|_{B^r_{p,T',M}}.\end{aligned}$$

A similar result holds for $U(t)^*$, in the spaces $H^s_{p,\widetilde{T},\Delta}$ and $B^s_{p,\widetilde{T},\Delta}$.

PROOF: The preceding theorem shows for the adjoint semigroup that

$$(1.34)\qquad \|U(t)^*v_0\|_{H^\sigma_{p'}(\overline{\Omega})^n} \leq C_\sigma e^{ct}t^{-\sigma/2}\|v_0\|_{L_{p'}(\Omega)^n},\ \text{for } \sigma\in[0,\tfrac{1}{p'}[;$$

here we have used that $H^{\sigma}_{p',\widetilde{T},\Delta} = H^{\sigma}(\overline{\Omega})^n$ when σ is so close to 0. By transposition, (1.34) implies, in view of (1.11),

$$\|U(t)u_0\|_{L_p(\Omega)^n} \leq C_\sigma e^{ct} t^{-\sigma/2} \|u_0\|_{H_p^{-\sigma}(\overline{\Omega})^n}, \text{ for } \sigma \in [0, \tfrac{1}{p'}[, \tag{1.35}$$

which shows that (1.33 i) holds for $s = 0$ and $r = -\sigma \in]\frac{1}{p} - 1, 0]$

Next, we show that (1.33 i) holds for $s \geq 0$ and $r \in]\frac{1}{p} - 1, 0]$. This is seen by taking (1.35) together with (1.30): Since $U(t) = U(t/2)U(t/2)$,

$$\|U(t)u_0\|_{H^s_{p,T',M}} \leq C_{s,0} e^{ct/2} t^{-s/2} \|U(t/2)u_0\|_{L_p(\Omega)^n} \leq C_{s,0} C_{0,r} e^{ct} t^{-(s-r)/2} \|u_0\|_{H^r(\overline{\Omega})^n}.$$

Here we can lift the result to the values $s + 2k$, $r + 2k$ with $s \geq 0$, $r \in]\frac{1}{p} - 1, 0]$, $k \in \mathbb{N}$, by application of $(A + c)^k$. So (1.33 i) has now been established whenever $s \geq 2k$, $r \in]2k - 1 + \frac{1}{p}, 2k]$, $k \in \mathbb{N}$. But e.g. $r \in]0, 1 + \frac{1}{p}[$ has still not been included. To get the remaining values, we proceed as follows:

Let $0 \leq \sigma < 1 - \frac{1}{p}$, then we have:

$$\|U(t)u_0\|_{H^{\sigma}_{p,T',M}} \leq C e^{ct} t^{-\sigma/2} \|u_0\|_{L_p},$$
$$\|U(t)u_0\|_{H^{2}_{p,T',M}} \leq C e^{ct} t^{-\sigma/2} \|u_0\|_{H^{2-\sigma}_{p,T',M}},$$

by two different applications of what we have just proved. Since $\sigma, 2, 0$ and $2 - \sigma$ all belong to $[0, 2]$, interpolation is allowed on both sides of the inequality sign (cf. Theorem 1.2), giving

$$\|U(t)u_0\|_{H^{\varrho}_{p,T',M}} \leq C e^{ct} t^{-\sigma/2} \|u_0\|_{H^{\varrho-\sigma}_{p,T',M}}, \text{ for } \sigma \leq \varrho \leq 2. \tag{1.36}$$

Here σ can be taken arbitrarily small in $[0, 1 - \frac{1}{p}[$.

Now let $s \geq r > 0$ be given. Consider the case where $r \in]0, 2[$, $s \in]0, 2]$. Choose a positive integer m such that $\sigma = (s - r)/m \in [0, 1 - \frac{1}{p}[$. Then we can apply (1.36) with $\varrho = s - k\sigma$ for $k = 0, 1, \ldots, m - 1$ successively by use of the fact that

$$U(mt/m) = U(t/m) \ldots U(t/m) \ (m \text{ factors}), \tag{1.37}$$

finding that

$$\begin{aligned}
\|U(mt/m)u_0\|_{H^s_{p,T',M}} &\leq C_1 e^{ct/m} t^{-\sigma/2} \|U((m-1)t/m)u_0\|_{H^{s-\sigma}_{p,T',M}} \\
&\leq \cdots \leq C_m e^{ct} t^{-m\sigma/2} \|u_0\|_{H^{s-m\sigma}_{p,T',M}} \\
&= C_m e^{ct} t^{-(s-r)/2} \|u_0\|_{H^r_{p,T',M}}.
\end{aligned}$$

If $2k < r \leq s \leq 2k + 2$, we lift the preceding inequality by use of $(A + c)^k$. Finally, if $2k - 2 < r < 2k < s$, the inequality is obtained by a combination of the cases $2k - 2 < r < s_1 = 2k$ and $r_1 = 2k < s$. All these results follow in the same way for the adjoint semigroup.

This covers (1.33 i) in all cases with $s \geq r \geq 0$, and we have earlier also treated $\frac{1}{p} - 1 < r \leq 0 \leq s$. For the last cases $\frac{1}{p} - 1 < r \leq s \leq 0$, we get the desired inequalities by duality from the results with $0 \leq r_1 = -s \leq s_1 = -r$ for the adjoint semigroup in $L_{p'}$.

The results for Besov spaces are obtained by real interpolation of the estimates for Bessel potential spaces, cf. (1.13). When $r \in]2k-2, 2k[$ and $s \in]2m-2, 2m[$, we interpolate between values $r-\varepsilon, r+\varepsilon \in]2k-2, 2k[$ and $s-\varepsilon, s+\varepsilon \in]2m-2, 2m[$, with $r - \varepsilon > \frac{1}{p} - 1$. If r or $s = 0$, we use for a small ε that $B_p^0(\overline{\Omega}) = (H_p^\varepsilon(\overline{\Omega}), H_p^{-\varepsilon}(\overline{\Omega}))_{\theta,p}$, cf. (1.13); and the result can then be lifted to higher even integers also (here one can break up in several steps, using (1.37)). ∎

Since the semigroup $\tilde{U}(t)\colon u_0 \mapsto u(t)$ for the Stokes problem (1.1) with $\kappa = 0$, $f = 0$, $\varphi = 0$, $\operatorname{div} u_0 = 0$, is simply obtained from $U(t)$ by restriction to the space J, we have as an immediate consequence of Theorems 1.3 and 1.4:

COROLLARY 1.5. *Let $r > \frac{1}{p} - 1$ and $m \in \mathbb{N}$. One has for $t > 0$, when $u_0 \in H^r_{p,T',M} \cap J$, $B^r_{p,T',M} \cap J$, resp. $H^{2m}_{p,T',M} \cap J$,*

$$
\begin{aligned}
&\text{(i)} && \|\tilde{U}(t)u_0\|_{H^s_{p,T',M}} \leq C_{s,r} e^{ct} t^{-(s-r)/2} \|u_0\|_{H^r_{p,T',M}}, && \text{when } s \geq r;\\
(1.38)\quad &\text{(ii)} && \|\tilde{U}(t)u_0\|_{B^s_{p,T',M}} \leq C_{s,r} e^{ct} t^{-(s-r)/2} \|u_0\|_{B^r_{p,T',M}}, && \text{when } s \geq r;\\
&\text{(iii)} && \|\tilde{U}(t)u_0\|_{B^s_{p,T',M}} \leq C_{s,2m} e^{ct} t^{-(s-2m)/2} \|u_0\|_{H^{2m}_{p,T',M}}, && \text{when } s > 2m.
\end{aligned}
$$

REMARK 1.6: The above deductions and results are likewise valid for *the Dirichlet problem*, where $T\{u, P\}$ in (1.1) is replaced by $\gamma_0 u$, and G_k, $\mathcal{Q}$ and T'_k in (1.3) are replaced by

$$(1.39)\qquad G_0 u = -\operatorname{grad} K_N \operatorname{div}'_\Gamma \gamma_1 u_\tau, \quad \mathcal{Q}_0 = \operatorname{pr}_{J_0}[(u \cdot \nabla)u], \quad T'_0 u = \gamma_0 u,$$

cf. [G-S4, Sect. 5]; K_N is a solution operator for the problem $\{-\Delta v, \gamma_1 v\} = \{0, \varphi\}$, and J_0 is the space of $u \in J$ with $\gamma_0 u_\nu = 0$. In this case, A^* acts like $-\Delta$ with a boundary condition of the form $\gamma_0 u + T''_0 u = 0$ for some T''_0 of class 0 (by [G1, Ex. 1.6.12]), so the adjoint can be used as above.

Note that the results are obtained without having to study fractional or imaginary powers of the realization A' (cf. [Gi2]), and that the spaces in (1.30), (1.33) are ordinary Bessel potential spaces and Besov spaces when r and s are close to zero, cf. (1.21).

2. SOLUTION OF THE NAVIER-STOKES INITIAL VALUE PROBLEMS

In this section, we assume $\kappa = 1$. Since the initial value problem has already in [G3] been solved for data in $B^r_{p,T',M}$ when $r > 1 - \frac{1}{p}$ (with $r \geq \frac{n}{p} - 1$), we restrict the attention here to small r, more precisely $r < 2 + \frac{1}{p}$ (for technical reasons). Let us denote

$$(2.1)\qquad \operatorname{pr}_J[(u \cdot \nabla)v] = \mathcal{Q}(u, v), \quad \mathcal{Q}(u, u) = \mathcal{Q}u.$$

Recall from [G3] (see the references there) the estimate

$$
\begin{aligned}
&&& \|(u \cdot \nabla)v\|_{H_p^\lambda(\overline{\Omega})^n} \leq c_{\lambda,\mu,\omega} \|u\|_{H_p^{\lambda+\mu}(\overline{\Omega})^n} \|v\|_{H_p^{\lambda+1+\omega}(\overline{\Omega})^n},\\
(2.2)\quad &&& \text{for } \mu \geq 0,\ \omega \geq 0,\ 2\lambda + \mu + \omega > \max\{0, n(\tfrac{2}{p} - 1)\} \text{ and}\\
&&& \lambda + \mu + \omega \geq \tfrac{n}{p}, \text{ where } \lambda + \mu + \omega > \tfrac{n}{p} \text{ if } \mu = 0 \text{ or } \omega = 0;
\end{aligned}
$$

it is also valid with H_p replaced by B_p. It implies:

PROPOSITION 2.1. *There are estimates*

$$\begin{aligned}\|\mathcal{Q}(u,v)\|_{H^{\lambda}_{p,T',M}} &\leq c_{\lambda,\nu}\|u\|_{H^{\nu}_{p,T',M}}\|v\|_{H^{1+\nu}_{p,T',M}},\\ \|\mathcal{Q}(u,v)\|_{B^{\lambda}_{p,T',M}} &\leq c_{\lambda,\nu}\|u\|_{B^{\nu}_{p,T',M}}\|v\|_{B^{1+\nu}_{p,T',M}},\end{aligned} \tag{2.3}$$

when λ *and* ν *are real numbers satisfying*

$$\begin{gathered}\lambda\in]\tfrac{1}{p}-1,\tfrac{1}{p}+1[,\ \nu\geq\lambda,\ \nu>\max\{0,n(\tfrac{1}{p}-\tfrac{1}{2})\},\ \nu-\tfrac{1}{p}\notin\mathbb{N}\\ \text{and } 2\nu-\lambda\geq\tfrac{n}{p},\ \text{where } 2\nu-\lambda>\tfrac{n}{p}\ \text{if } \nu=\lambda.\end{gathered} \tag{2.4}$$

PROOF: Take $\lambda+\mu=\lambda+\omega=\nu$ in (2.2), then we get the estimates in the full spaces $H^s_p(\overline{\Omega})^n$ resp. $B^s_p(\overline{\Omega})$, using that pr_J is continuous in $H^s_p(\overline{\Omega})^n$ and $B^s_p(\overline{\Omega})^n$ for $s>\frac{1}{p}-1$ (cf. [G2, Ex. 3.14]). The estimates (2.3) follow, since $H^s_{p,T',M}$ equals $H^s_p(\overline{\Omega})^n$, resp. $B^s_{p,T',M}$ equals $B^s_p(\overline{\Omega})^n$, for $s=\lambda\in]\frac{1}{p}-1,\frac{1}{p}+1[$, and is a closed subspace for $s=\nu$ and $s=\nu+1$, $\nu-\frac{1}{p}\notin\mathbb{N}$. ∎

The norm in $H^s_p(\overline{\Omega})^{n'}$, $B^s_p(\overline{\Omega})^{n'}$ ($n'\geq 1$), $H^s_{p,T',M}$ or $B^s_{p,T',M}$ will be denoted $\|\cdot\|_s$ for short, when it is clear from the context which space is considered.

We shall use the estimates (1.30) and (1.33) for the semigroup $U(t)$. Here we note that one has in addition:

LEMMA 2.2. *Let* $s>r>\frac{1}{p}-1$. *For each* $u_0\in H^r_{p,T',M}\setminus\{0\}$ *resp.* $B^r_{p,T',M}\setminus\{0\}$,

$$\begin{aligned}C_1(u_0,b,r,s)&\equiv\sup_{0<t<b}t^{\frac{s-r}{2}}e^{-ct}\|U(t)u_0\|_{H^s_{p,T',M}}\|u_0\|^{-1}_{H^r_{p,T',M}}\to 0\ \text{and}\\ C_2(u_0,b,r,s)&\equiv\sup_{0<t<b}t^{\frac{s-r}{2}}e^{-ct}\|U(t)u_0\|_{B^s_{p,T',M}}\|u_0\|^{-1}_{B^r_{p,T',M}}\to 0,\ \text{for } b\to 0;\end{aligned} \tag{2.5}$$

and if $r=2m$, $m\in\mathbb{N}$, *one also has for* $u_0\in H^r_{p,T',M}\setminus\{0\}$,

$$C_3(u_0,b,r,s)\equiv\sup_{0<t<b}t^{\frac{s-r}{2}}e^{-ct}\|U(t)u_0\|_{B^s_{p,T',M}}\|u_0\|^{-1}_{H^r_{p,T',M}}\to 0,\ \text{for } b\to 0. \tag{2.6}$$

PROOF: Let $u_0\in H^r_{p,T',M}\setminus\{0\}$ and let $0<\delta<s-r$. Since $H^{r+\delta}_{p,T',M}$ is dense in $H^r_{p,T',M}$, there is a sequence $u_k\in H^{r+\delta}_{p,T',M}$ converging to u_0 in $H^r_{p,T',M}$ for $k\to\infty$. Then by (1.33):

$$\begin{aligned}t^{\frac{s-r}{2}}e^{-ct}\|U(t)u_0\|_s&\leq t^{\frac{s-r}{2}}e^{-ct}\|U(t)(u_0-u_k)\|_s+t^{\frac{s-r}{2}}e^{-ct}\|U(t)u_k\|_s\\ &\leq t^{\frac{s-r}{2}}C_{r,s}t^{-\frac{s-r}{2}}\|u_0-u_k\|_r+t^{\frac{s-r}{2}}C_{s,r+\delta}t^{-\frac{s-r-\delta}{2}}\|u_k\|_{r+\delta}\\ &=C_{s,r}\|u_0-u_k\|_r+C_{s,r+\delta}t^{\frac{\delta}{2}}\|u_k\|_{r+\delta}.\end{aligned}$$

For any $\varepsilon>0$ we can find a b so that this is $\leq\varepsilon$ for $t\in[0,b]$, by first taking k so large that the first term is $\leq\varepsilon/2$, and then taking b so small that the second term is $\leq\varepsilon/2$ on $[0,b]$. This shows the first statement in the lemma, and the other statements follow in a similar way from (1.33) resp. (1.30). ∎

The solution operator $V(t)\colon u_0(x) \mapsto u(x,t)$ for the nonlinear problem (1.3) with $f' = 0$, $\psi = 0$, must satisfy, in appropriate spaces,

$$V(t)u_0 = U(t)u_0 - \int_0^t U(t-\tau)Q(V(\tau)u_0)\,d\tau\,. \tag{2.7}$$

as is seen by integration of the first line, with $u(t) = V(t)u_0$.

We are going to prove the solvability of this equation for $t \in [0,b]$, $u_0 \in H^r_{p,T',M}$ or $B^r_{p,T',M}$ for suitable r, with b depending on u_0. Here we shall use the arguments in [S, §10] and [G-S3] with some further precisions (and with notation close to that of [G-S3]). We begin by considering the problem with $u_0 \in H^r_{p,T',M}$, $u_0 \neq 0$. The other norms in the following are likewise taken in $H^s_{p,T',M}$ spaces (in fact in $H^s_{p,T'}$, since the exponents s will belong to $]-1+\frac{1}{p}, 3+\frac{1}{p}[$).

Let $v^{(0)} = 0$, and define successively for $k \in \mathbb{N}$,

$$v^{(k+1)}(t) = U(t)u_0 - \int_0^t U(t-\tau)Qv^{(k)}(\tau)\,dt. \tag{2.8}$$

Let λ and ν be chosen according to (2.4) and let $r \in [\lambda, \nu]$. Then we have, using (1.33) and (2.3),

$$\begin{aligned} \|v^{(k+1)}(t)\|_{1+\nu} &\le C_4 t^{-\frac{1+\nu-r}{2}} \|u_0\|_r + C_5C_6 \int_0^t (t-\tau)^{-\frac{1+\nu-\lambda}{2}} \|v^{(k)}\|_\nu \|v^{(k)}\|_{1+\nu} d\tau \\ \|v^{(k+1)}(t)\|_\nu &\le C_4 t^{-\frac{\nu-r}{2}} \|u_0\|_r + C_5C_6 \int_0^t (t-\tau)^{-\frac{\nu-\lambda}{2}} \|v^{(k)}\|_\nu \|v^{(k)}\|_{1+\nu} d\tau\,; \end{aligned} \tag{2.9}$$

where

$$C_4 = \max\{C_{\nu+1,r}, C_{\nu,r}\}e^{cb}, \quad C_5 = \max\{C_{\nu+1,\lambda}, C_{\nu,\lambda}\}e^{cb}, \quad C_6 = c_{\lambda,\nu}, \tag{2.10}$$

For brevity we now set

$$\begin{aligned} \xi_k(t) &= t^{\frac{1+\nu-r}{2}} \|v^{(k)}(t)\|_{1+\nu}, \quad \eta_k(t) = t^{\frac{\nu-r}{2}} \|v^{(k)}(t)\|_\nu\,, \\ I_1^{(k)} &= C_6 t^{\frac{1+\nu-r}{2}} \int_0^t (t-\tau)^{-\frac{1+\nu-\lambda}{2}} \tau^{-\nu+r-\frac{1}{2}} \xi_k(\tau)\eta_k(\tau)d\tau\,, \\ I_2^{(k)} &= C_6 t^{\frac{\nu-r}{2}} \int_0^t (t-\tau)^{-\frac{\nu-\lambda}{2}} \tau^{-\nu+r-\frac{1}{2}} \xi_k(\tau)\eta_k(\tau)d\tau\,; \end{aligned} \tag{2.11}$$

then (2.9) shows that

$$\begin{aligned} \xi_{k+1}(t) &\le C_4\|u_0\|_r + C_5 I_1^{(k)} \le C_4\|u_0\|_r + C_5(I_1^{(k)} + I_2^{(k)})\,, \\ \eta_{k+1}(t) &\le C_4\|u_0\|_r + C_5 I_2^{(k)} \le C_4\|u_0\|_r + C_5(I_1^{(k)} + I_2^{(k)}). \end{aligned} \tag{2.12}$$

Multiplying the inequalities for ξ_{k+1} and η_{k+1}, and setting $\omega_k(t) = \sup_{\tau\in]0,t[} \xi_k(\tau)\eta_k(\tau)$, we arrive at

$$\omega_{k+1}(t) \le (C_4\|u_0\|_r + C_5(I_1^{(k)} + I_2^{(k)}))^2 \le (C_4\|u_0\|_r + C_5\Phi(t)\omega_k(t))^2 \tag{2.13}$$

where

$$(2.14)\quad \Phi(t) = C_6 t^{\frac{1+\nu-r}{2}} \int_0^t (t-\tau)^{-\frac{1+\nu-\lambda}{2}} \tau^{-\nu+r-\frac{1}{2}} d\tau + C_6 t^{\frac{\nu-r}{2}} \int_0^t (t-\tau)^{-\frac{\nu-\lambda}{2}} \tau^{-\nu+r-\frac{1}{2}} d\tau = C_7 t^{\frac{1+\lambda+r}{2}-\nu}.$$

Now it is an elementary fact, that when a sequence $\{\omega_k\}_{k\in\mathbf{N}}$ satisfies (with $\alpha,\beta>0$)

$$\omega_0 = 0, \quad 0 \le \omega_{k+1} \le (\alpha+\beta\omega_k)^2 \quad \text{for all } k,$$

then if $4\alpha\beta < 1$,

$$\omega_k \le \omega_- = \frac{4\alpha^2}{2-4\alpha\beta+2\sqrt{1-4\alpha\beta}} \quad \text{for all } k.$$

(For, ω_- is a root in $f(\omega) = \alpha^2 + (\beta-1)\omega + \beta^2\omega^2$, so if $0 \le \omega_k \le \omega_-$, then $\omega_{k+1} \le (\alpha+\beta\omega_k)^2 \le (\alpha+\beta\omega_-)^2 = \omega_-$.) This applies to (2.13), showing that if

$$(2.15)\qquad 4C_4C_5\Phi(t)\|u_0\|_r < 1,$$

then

$$(2.16)\qquad \omega_k(t) \le \frac{4C_4^2\|u_0\|_r^2}{1+M(t)}, \quad \text{where}$$
$$M(t) = 1 - 4C_4C_5\Phi(t)\|u_0\|_r + 2\sqrt{1-4C_4C_5\Phi(t)\|u_0\|_r} > 0.$$

The condition (2.15) can be fulfilled for small t only if $\Phi(t)$ is bounded for $t\to 0$, i.e. if

$$(2.17)\qquad \frac{1+\lambda+r}{2} - \nu \ge 0.$$

Moreover, we should require for the definition of the integrals $I_i^{(k)}$ that

$$(2.18)\qquad -\frac{1+\nu-\lambda}{2} > -1 \quad \text{and} \quad -\nu+r-\frac{1}{2} > -1.$$

Let us denote

$$(2.19)\qquad r-\lambda = \varepsilon \ge 0, \quad \nu - r = \delta \ge 0,$$

then the conditions from (2.17), (2.18) and (2.4) can be written:

$$(2.20)\qquad \begin{gathered} 0 \le \delta+\varepsilon < 1, \quad 0 \le \delta < \tfrac{1}{2}, \quad 2\delta+\varepsilon \le 1, \\ r-\varepsilon \in]\tfrac{1}{p}-1, \tfrac{1}{p}+1[, \quad r+\delta > \max\{0, n(\tfrac{1}{p}-\tfrac{1}{2})\}, \quad r+\delta-\tfrac{1}{p} \notin \mathbf{N}, \\ r+2\delta+\varepsilon \ge \tfrac{n}{p}; \quad \text{with } r+2\delta+\varepsilon > \tfrac{n}{p} \text{ if } \varepsilon=\delta=0. \end{gathered}$$

Since $2\delta+\varepsilon \le 1$, r must satisfy $r \ge \frac{n}{p}-1$. Note that $r \le 0$ is not excluded, when $p \ge n$. Note moreover, that when $n \ge 2$, $n(\frac{1}{p}-\frac{1}{2}) \le \frac{n}{p}-1$, so $r+\delta > \max\{0, n(\frac{1}{p}-\frac{1}{2})\}$ reduces to $r+\delta>0$, provided δ is taken >0 when $r=\frac{n}{p}-1$. To save space, we assume $n\geq 2$ in the following (leaving the discussion of $n=1$ to the reader).

Assume that (2.15) holds. Then it follows from (2.12), (2.16) that

$$\xi_{k+1}(t)\leq C_4\|u_0\|_r+C_5\Phi(t)\frac{4C_4^2\|u_0\|_r^2}{1+M(t)}=C_4\|u_0\|_r\left(1+\frac{4C_4C_5\Phi\|u_0\|_r}{1+M}\right)$$
$$\leq C_4\|u_0\|_r\left(1+\frac{1}{1+M}\right)\leq 2C_4\|u_0\|_r\,,$$
$$\eta_{k+1}(t)\leq 2C_4\|u_0\|_r\,;$$

hence for all $k\in\mathbb{N}$,

$$\text{(2.21)}\qquad \begin{aligned}&\|v^{(k+1)}\|_{1+\nu}\leq 2C_4t^{-\frac{1+\nu-r}{2}}\|u_0\|_r=2C_4t^{-\frac{1+\delta}{2}}\|u_0\|_r\,,\\&\|v^{(k+1)}\|_\nu\leq 2C_4t^{-\frac{\delta}{2}}\|u_0\|_r\,.\end{aligned}$$

Let $r\geq\frac{n}{p}-1$. If $r>\frac{n}{p}-1$, we shall take δ and $\varepsilon\geq 0$ such that $\frac{n}{p}-r<2\delta+\varepsilon<1$; and if $r=\frac{n}{p}-1$ we shall take $\delta\in\,]0,\frac{1}{2}[$ and $\varepsilon=1-2\delta>0$ (then $2\delta+\varepsilon=1$). Note that when we take δ close to $\frac{1}{2}$, ε must be close to 0, whereas ε close to 1 requires δ close to 0.

Let us now study the convergence of the sequence $\{v^{(k)}\}$. We set $w^{(k+1)}=v^{(k+1)}-v^{(k)}$ and observe that

$$\text{(2.22)}\qquad \mathcal{Q}v^{(k)}=\mathcal{Q}v^{(k-1)}+\mathcal{Q}(v^{(k-1)},w^{(k)})+\mathcal{Q}(w^{(k)},v^{(k)}).$$

Subtracting from (2.8) the corresponding equation for $v^{(k)}$, we obtain

$$\text{(2.23)}\qquad w^{(k+1)}(t)=-\int_0^t U(t-\tau)[\mathcal{Q}(w^{(k)},v^{(k)})+\mathcal{Q}(v^{(k-1)},w^{(k)})]d\tau\,.$$

From these equations and from (1.33), (2.3), (2.10), (2.21), it follows that

$$\|w^{(k+1)}(t)\|_{1+\nu}\leq C_5C_6\int_0^t(t-\tau)^{-\frac{1+\delta+\varepsilon}{2}}(\|w^{(k)}\|_\nu\|v^{(k)}\|_{1+\nu}+\|v^{(k-1)}\|_\nu\|w^{(k)}\|_{1+\nu})d\tau$$
$$\leq 2C_4C_5C_6\|u_0\|_r\int_0^t(t-\tau)^{-\frac{1+\delta+\varepsilon}{2}}(\|w^{(k)}\|_\nu\tau^{-\frac{1+\delta}{2}}+\|w^{(k)}\|_{1+\nu}\tau^{-\frac{\delta}{2}})d\tau\,,$$
$$\|w^{(k+1)}(t)\|_\nu\leq 2C_4C_5C_6\|u_0\|_r\int_0^t(t-\tau)^{-\frac{\delta+\varepsilon}{2}}(\|w^{(k)}\|_\nu\tau^{-\frac{1+\delta}{2}}+\|w^{(k)}\|_{1+\nu}\tau^{-\frac{\delta}{2}})d\tau\,.$$

Let us set

$$\text{(2.24)}\qquad \begin{aligned}&W_k(t)=\|w^{(k)}(t)\|_{1+\nu},\quad Z_k(t)=t^{-\frac{1}{2}}\|w^{(k)}(t)\|_\nu,\\&\Sigma_N(t)=\sum_{k=1}^N(W_k(t)+Z_k(t)),\quad M_1=4C_4C_5C_6\|u_0\|_r;\end{aligned}$$

then we easily obtain

$$W_{k+1}(t)+Z_{k+1}(t)\leq M_1\int_0^t(t-\tau)^{-\frac{1+\delta+\varepsilon}{2}}\tau^{-\frac{\delta}{2}}(W_k(\tau)+Z_k(\tau))d\tau\,,\ \text{hence}$$
$$\Sigma_{N+1}(t)\leq\Sigma_1(t)+M_1\int_0^t\Sigma_N(\tau)(t-\tau)^{-\frac{1+\delta+\varepsilon}{2}}\tau^{-\frac{\delta}{2}}\,d\tau\,.$$

From this follows successively that

$$(2.25)\quad \Sigma_{N+1}(t) \le \Sigma_1(t) + M_1 \int_0^t \Sigma_1(\tau)(t-\tau)^{-\frac{1+\delta+\varepsilon}{2}} \tau^{-\frac{\delta}{2}} d\tau$$
$$+ M_1^2 \int_0^t (t-\tau_1)^{-\frac{1+\delta+\varepsilon}{2}} \tau_1^{-\frac{\delta}{2}} d\tau_1 \int_0^{\tau_1} \Sigma_1(\tau_2)(\tau_1-\tau_2)^{-\frac{1+\delta+\varepsilon}{2}} \tau_2^{-\frac{\delta}{2}} d\tau_2 + \dots$$
$$+ M_1^N \int_0^t (t-\tau_1)^{-\frac{1+\delta+\varepsilon}{2}} \tau_1^{-\frac{\delta}{2}} d\tau_1 \cdots \int_0^{\tau_{N-1}} \Sigma_1(\tau_N)(\tau_{N-1}-\tau_N)^{-\frac{1+\delta+\varepsilon}{2}} \tau_N^{-\frac{\delta}{2}} d\tau_N.$$

In view of the well-known formula

$$(2.26)\qquad \int_0^t (t-\tau)^{\alpha-1}\tau^{\beta-1} d\tau = \frac{\Gamma(\alpha)\Gamma(\beta)}{\Gamma(\alpha+\beta)} t^{\alpha+\beta-1}, \quad \alpha, \beta > 0,$$

we can write

$$\int_0^t (t-\tau)^{-\frac{1+\delta+\varepsilon}{2}} \tau^{-\frac{\delta}{2}} d\tau \le \int_0^t (t-\tau)^{\frac{1-\delta-\varepsilon}{2}-1} \tau^{\frac{1-2\delta}{2}-1} d\tau \cdot t^{\frac{1+\delta}{2}}$$
$$= \frac{\Gamma(\frac{1-\delta-\varepsilon}{2})\Gamma(\frac{1-2\delta}{2})}{\Gamma(\frac{2-3\delta-\varepsilon}{2})} t^{\frac{1-2\delta-\varepsilon}{2}} = \frac{\Gamma(\frac{1-\delta-\varepsilon}{2})\Gamma(\frac{1-2\delta}{2})}{\Gamma(\frac{1-\delta}{2}+\gamma)} t^{\gamma};$$

where $\gamma = \frac{1-2\delta-\varepsilon}{2}$. By a succession of similar considerations, we find from (2.25), using (2.21) for Σ_1,

$$\Sigma_{N+1}(t) \le 4C_4 t^{-\frac{1+\delta}{2}} \|u_0\|_r f_N(t), \text{ where}$$
$$f_N(t) = \Big\{1 + M_1 \frac{\Gamma(\frac{1-\delta-\varepsilon}{2})\Gamma(\frac{1-2\delta}{2})}{\Gamma(\frac{1-\delta}{2}+\gamma)} t^{\gamma} + \dots$$
$$(2.27)\qquad + M_1^N t^{\gamma N} \Gamma(\tfrac{1-\delta-\varepsilon}{2})^N \frac{\Gamma(\frac{1-2\delta}{2})\Gamma(\frac{1-2\delta}{2}+\gamma)\dots\Gamma(\frac{1-2\delta}{2}+(N-1)\gamma)}{\Gamma(\frac{1-\delta}{2}+\gamma)\Gamma(\frac{1-\delta}{2}+2\gamma)\dots\Gamma(\frac{1-\delta}{2}+N\gamma)}\Big\}$$
$$= \sum_{k=0}^{N} [M_1 t^{\gamma}\Gamma(\tfrac{1-\delta-\varepsilon}{2})]^k \prod_{l=1}^{k} \frac{\Gamma(\frac{1-2\delta}{2}+(l-1)\gamma)}{\Gamma(\frac{1-\delta}{2}+l\gamma)}.$$

In the case $r+1 = \frac{n}{p}$ we had $2\delta + \varepsilon = 1$, i.e. $\gamma = 0$, so $f_N(t)$ is independent of t and is the N'th partial sum in a power series, that converges to a limit f_∞ for $N \to \infty$ provided that

$$(2.28)\qquad M_2 = M_1 \frac{\Gamma(\frac{1-\delta-\varepsilon}{2})\Gamma(\frac{1-2\delta}{2})}{\Gamma(\frac{1-\delta}{2})} = 4C_4C_5C_6\|u_0\|_r \frac{\Gamma(\frac{1-\delta-\varepsilon}{2})\Gamma(\frac{1-2\delta}{2})}{\Gamma(\frac{1-\delta}{2})} < 1;$$

here $f_\infty = (1-M_2)^{-1}$. The estimate (2.27) then implies

$$(2.29)\qquad \Sigma_{N+1}(t) \le 4C_4 t^{-\frac{1+\delta}{2}} \|u_0\|_r (1-M_2)^{-1}, \text{ for all } N,\ t \in]0,b].$$

The requirements for this estimate are (2.15) and (2.28). Since in this case $\Phi(t) = C_7$ (cf. (2.14), $\frac{1+\lambda+r}{2} - \nu = \frac{1}{2}(1-2\delta-\varepsilon) = 0$), the estimate is obtained on any interval $]0,b]$ when

$$(2.30)\qquad \|u_0\|_r < \min\{(4C_4C_5C_7)^{-1}, \frac{\Gamma(\frac{1-\delta}{2})}{4C_4C_5C_6\Gamma(\frac{1-\delta-\varepsilon}{2})\Gamma(\frac{1-2\delta}{2})}\}.$$

When $\|u_0\|_r$ is larger, (2.15) and (2.28) give no further help as they stand, but then we can take recourse to Lemma 2.2: Observe from (2.9), (2.10) that for given u_0 and b, C_4 can be replaced throughout by

$$C_4' = \max\{C_1(u_0, b, \nu+1, r), C_1(u_0, b, \nu, r)\}. \tag{2.31}$$

According to Lemma 2.2, $C_4' \to 0$ for $b \to 0$, so for small enough b, (2.15) and (2.28) hold with C_4 replaced by C_4' for this particular u_0.

In the case $r+1 > \frac{n}{p}$ we had chosen δ and ε so that $2\delta + \varepsilon < 1$, $\delta < \frac{1}{2}$. We shall use

LEMMA 2.3. *Let $\varepsilon, \delta > 0$ with $2\delta + \varepsilon < 1$, $\delta < \frac{1}{2}$, and let $\gamma = \frac{1-2\delta-\varepsilon}{2}$. Then*

$$\varrho = \tfrac{1-\delta-\varepsilon}{2} > 0, \quad \gamma' = \tfrac{1+\delta}{2} > 0,$$

and for $k > l_0$, where $l_0 = [\gamma'/\gamma] + 1$,

$$\prod_{l=l_0+1}^{k} \frac{\Gamma(\frac{1-2\delta}{2} + (l-1)\gamma)}{\Gamma(\frac{1-\delta}{2} + l\gamma)} \le \prod_{l=l_0+1}^{k} \frac{1}{[(l-l_0)\gamma]^{\varrho}} = \frac{\gamma^{\varrho l_0}}{[(k-l_0)!]^{\varrho}} \gamma^{\varrho k}. \tag{2.32}$$

PROOF: We use the inequality $\Gamma(s+\theta) \le s^{\theta-1}\Gamma(s+1)$ for the Gamma function, that follows from its logarithmic convexity. The identities

$$s + \theta = \tfrac{1-2\delta}{2} + (l-1)\gamma, \quad s + 1 = \tfrac{1-\delta}{2} + l\gamma,$$

are satisfied with

$$\theta - 1 = \tfrac{-1+\delta+\varepsilon}{2} = -\varrho, \quad s = l\gamma + \tfrac{1-\delta}{2} - 1 = l\gamma - \gamma'.$$

We set $l_0 = [\gamma'/\gamma] + 1$, where $[\gamma'/\gamma]$ is the largest integer $\le \gamma'/\gamma$. Now $l\gamma - \gamma' \ge (l - l_0)\gamma$ for $l > l_0$, so then

$$\frac{\Gamma(\frac{1-2\delta}{2} + (l-1)\gamma)}{\Gamma(\frac{1-\delta}{2} + l\gamma)} = \frac{\Gamma(s+\theta)}{\Gamma(s+1)} \le s^{\theta-1} \le [(l-l_0)\gamma]^{-\varrho},$$

and (2.32) follows. ∎

Applying this estimate to the terms (from number $l_0 + 1$ on) in $f_N(t)$, we see that it converges to a positive continuous function $f_\infty(t)$ for $N \to \infty$, for $t \in \overline{\mathbb{R}}_+$; and hence

$$\Sigma_{N+1}(t) \le 4C_4 f_\infty(t) t^{-\frac{1+\delta}{2}} \|u_0\|_r, \text{ for all } N,\ t \in]0, b]. \tag{2.33}$$

The prerequisite for the estimate here was (2.15), which for any u_0 is satisfied on $[0, b]$ for a sufficiently small b depending on the norm of u_0, since $\Phi(t) \to 0$ for $t \to 0$.

The inequalities (2.29) resp. (2.33) show that the sequence $\{v^{(k)}\}$ is convergent in a weighted norm

$$\sup_{t\le b} t^{\frac{1+\nu-r}{2}} \|v(t)\|_{1+\nu} + \sup_{t \le b} t^{\frac{\nu-r}{2}} \|v(t)\|_\nu,$$

where $b > 0$ is determined as indicated above. Moreover, we can obtain an estimate of any norm $\|v^{(k)}(t)\|_{r_1}$, $r_1 \in [r, 1+\nu]$, $r_1 - \frac{1}{p} \notin \mathbb{N}$, by use of (2.8) (applying (1.33), (2.3), (2.26) again):

$$(2.34) \quad \|v^{(k+1)}\|_{r_1,\Omega} \le C_8 t^{-\frac{r_1-r}{2}} \|u_0\|_r + C_8 C_6 \int_0^t (t-\tau)^{-\frac{r_1-\lambda}{2}} \|v^{(k)}\|_\nu \|v^{(k+1)}\|_{\nu+1} d\tau$$
$$\le C_8 t^{-\frac{r_1-r}{2}} \|u_0\|_r \Big(1 + C_6 \frac{\Gamma(1-\frac{r_1-\lambda}{2})\Gamma(\frac{1-2\delta}{2})}{\Gamma(\frac{3}{2}-\frac{2\delta+r_1-\lambda}{2})} t^{\frac{1-2\delta-\varepsilon}{2}}\Big)$$
$$\le C_9(t) t^{-\frac{r_1-r}{2}} \|u_0\|_r.$$

Then by (2.23),

$$\|w^{(k+1)}\|_{r_1} \le C_{10} \int_0^t (t-\tau)^{-\frac{r_1-\lambda}{2}} (W_k(\tau) + Z_k(\tau)) \tau^{-\frac{\delta}{2}} d\tau$$

from which we can conclude that the sequence $\{v^{(k)}\}$ is convergent in the norm

$$(2.35) \qquad \sup_{t\le b} t^{\frac{r_1-r}{2}} \|v(t)\|_{r_1},$$

for any $r_1 \in [r, 1+\nu]$, $r_1 - \frac{1}{p} \notin \mathbb{N}$. Recall from (2.20) that we must have $r < 1 + \frac{1}{p} + \varepsilon$, whereas $\nu = r + \delta$, $\delta < \frac{1}{2}$, $\varepsilon + 2\delta \le 1$. Thus

$$(2.36) \qquad \begin{aligned} &\text{when } r < 1 + \tfrac{1}{p}, \text{ then } r_1 \in [r, r + \tfrac{3}{2}[, \text{ and} \\ &\text{when } 1 + \tfrac{1}{p} \le r < 2 + \tfrac{1}{p}, \text{ then } r_1 \in [r, 2 + \tfrac{r}{2} + \tfrac{1}{2p}[. \end{aligned}$$

It is seen that r_1 can reach any value $< 3 + \frac{1}{p}$ when r is close to $2 + \frac{1}{p}$. Note that for all the considered values of r and r_1, the $H_{p,T',M}$ spaces are $H_{p,T'}$ spaces, cf. (1.21).

The limiting function $u(t) = V(t)u_0$ satisfies (2.7), and the estimate (2.34) carries over to it; in particular, it is continuous in $t \in [0, b]$ with values in $H^r_{p,T'}$.

The arguments carry through *verbatim* with H^s_p norms replaced by B^s_p norms throughout. Moreover, in the very interesting case where the initial value is in $L_p(\Omega)^n$, we can replace some of the estimates above with estimates derived from (1.30 ii), allowing Besov norms in the spaces with noninteger exponent. (Note that ε and δ can always be taken $\ne 0$.)

Altogether, we have:

THEOREM 2.4. *Let $n \ge 2$ and $1 < p < \infty$, and let $r \ge \frac{n}{p} - 1$ with $r \in]-1+\frac{1}{p}, 2+\frac{1}{p}[$, $r > -\frac{1}{2}$. Then for any $u_0 \in H^r_{p,T'}$ or $B^r_{p,T'}$, the problem (1.3) with $\kappa = 1$, $f' = 0$ and $\psi = 0$, has a solution $u(t) = V(t)u_0$ on some interval $[0, b]$, satisfying*

$$(2.37) \quad u(t) \in C([0,b]; H^r_{p,T'}) \text{ resp. } C([0,b]; B^r_{p,T'}), \text{ when } u_0 \in H^r_{p,T'} \text{ resp. } u_0 \in B^r_{p,T'}.$$

Moreover, for r_1 as in (2.36), $r_1 - \frac{1}{p} \notin \mathbb{N}$,

$$(2.38) \quad \begin{aligned} &u(t) \in C(\,]0,b]; H^{r_1}_{p,T'}) \text{ resp. } C(\,]0,b]; B^{r_1}_{p,T'}), \text{ when } u_0 \in H^r_{p,T'} \text{ resp. } u_0 \in B^r_{p,T'}; \\ &u(t) \in C(\,]0,b]; H^{r_1}_{p,T'} \cap B^{r_1}_{p,T'}), \text{ when } u_0 \in L_p(\Omega)^n \text{ and } r_1 > r = 0; \\ &\text{with } \sup_{t\le b} t^{\frac{r_1-r}{2}} \|u(t)\|_{r_1} < \infty \text{ in all cases.} \end{aligned}$$

The interval length b depends on u_0 as follows, for each r: There is for each b an $R_b > 0$ such that $u(t)$ is defined on $[0, b]$ when $\|u_0\|_r < R_b$; and when $r > \frac{n}{p} - 1$, there is for each R a b_R such that $u(t)$ is defined on $[0, b_R]$ when $\|u_0\|_r < R$. When $r = \frac{n}{p} - 1$, there is for each u_0 a b_{u_0} such that $u(t)$ is defined on the interval $[0, b_{u_0}]$. (Cf. also (2.15), (2.28) ff.)

The following uniqueness holds: When $u_0 \in H^r_{p,T'}$ resp. $B^r_{p,T'}$, and ε and δ are chosen according to (2.20), there is at most one solution such that (2.37)–(2.38) hold with $r_1 = r+\delta$ and with $r_1 = r + 1 + \delta$.

PROOF: Everything has been proved except the uniqueness. Let $u_0 \in H^r_{p,T'}$, and let $u(t)$ and $u_1(t)$ be two solutions with initial value u_0 and with the stated properties. Then $z(t) = u(t) - u_1(t)$ satisfies

$$z(t) = -\int_0^t U(t-\tau)(Q(z(\tau), u(\tau)) + Q(u_1(\tau), z(\tau))\, d\tau.$$

Imitating the calculations (2.9)–(2.14) we find that (with $\nu = r + \delta$)

$$\omega(t) = \sup_{\tau \le t} \tau^{\frac{1}{2}+\delta} \|z(\tau)\|_{1+\nu} \|z(\tau)\|_\nu \quad \text{satisfies } \omega(t) \le C_{11}\omega(t)^2, \text{ for } t \le b;$$

the assumptions assure that $\omega(t)$ is bounded on $]0, b]$. Since $\omega(t) \to 0$ for $t \to 0$ in view of Lemma 2.2, it follows that $\omega(t) = 0$ in a neighborhood of 0. Replacing 0 by arbitrary points in $]0, b[$, we see that if $\omega(t) = 0$ on $]0, b_0] \subset]0, b[$, then $\omega(t) = 0$ on $]0, b_0']$ for some $b_0' > b_0$. So there is no *largest* $b_0 < b$ such that $\omega(t) = 0$ on $]0, b_0]$, and hence $\omega(t)$ must be zero on $]0, b]$. ∎

Note that $r = 0$ is allowed when $p \ge n$. (For $p > n$, even slightly negative r are allowed.) For $r_1 > r$, Lemma 2.2 can be used to show (by following the constants through the above calculations) that $\sup_{t \le b} t^{\frac{r_1 - r}{2}} \|u(t)\|_{r_1} \to 0$ for $b \to 0$, for each u_0.

We recall that [G3] gives general results for $r > 1 - \frac{1}{p}$.

Now let us show smoothness of the solution for $t > 0$.

THEOREM 2.5. *Let $u_0 \in H^r_{p,T'}$ or $B^r_{p,T'}$ with r given as in Theorem 2.4. Then the solution u defined there is C^∞ in (x, t) for $t \in]0, b[$.*

PROOF: The statement is empty unless

$$\frac{n}{p} - 1 < 2 + \frac{1}{p}, \tag{2.39}$$

so we can assume that (2.39) holds. Let $u_0 \in H^r_{p,T'}$ for an $r \in [\frac{n}{p} - 1, 2 + \frac{1}{p}[$. Let $t_0 \in]0, b[$. Using Theorem 2.4 for a succession of points $t \in]0, t_0[$, we find that $u(t_0) \in H^{r_1}_{p,T'}$ for any $r_1 < 3 + \frac{1}{p}$, and we now just have to show that u is C^∞ for $t \in]t_0, b[$. The next thing we observe is that since $3 + \frac{1}{p} - \frac{n}{p} > 0$ by (2.39), the imbedding relation (cf. e.g. [T, 2.8.1])

$$H^s_p(\overline{\Omega}) \subset H^t_q(\overline{\Omega}), \text{ when } q \ge p,\ s - \frac{n}{p} \ge t - \frac{n}{q},$$

applied with $s = r_1$ close to $3 + \frac{1}{p}$, and $t = 0$, shows that $u(t_0) \in L_q(\Omega)^n$, any $q \in [p, \infty[$. We take $q = kn$ for some integer $k > 1$, and have then reduced the problem to the case of an initial value $u(t_0) \in L_q(\Omega)^n$, $q = kn$.

Now we can again apply Theorem 2.4 to see that when $t_1 \in]t_0, b[$, $u(t_1) \in H^{3+\frac{1}{q}-\varepsilon}_{q,T'}$ for any small positive ε, and moreover, $u \in C([t_1, b]; H^{3+\frac{1}{q}-\varepsilon}_{q,T'})$. Then

$$\partial_t u = \Delta u - Gu + \mathcal{Q}u \in C([t_1, b]; H^{1+\frac{1}{q}-\varepsilon}_{q,T'}), \tag{2.40}$$

since (2.2) with $\lambda = 2$, $\mu = 1$, $\omega = 0$ (hence $\lambda + \mu + \omega = 3 > \frac{n}{kn}$) shows that $\mathcal{Q}u \in C([t_1, b]; H^2_{q,T'})$. From the statements on u and $\partial_t u$ follows in particular that u belongs to the anisotropic space $H^{(2,1)}_q(\overline{\Omega} \times [t_1, b])$. In [G3] it is shown that when $s + 3 - \frac{n+2}{p} > 0$, a solution of (1.3) in $H^{(s+2,s/2+1)}_p(\overline{\Omega} \times [0, b'])^n$ with f' and ψ in C^∞, is C^∞ for $t \in]0, b'[$. This can be used here with $s = 0$, $p = q = kn$, since $3 - \frac{n+2}{kn} > 0$. Thus $u \in C^\infty$ for $t \in]t_1, b[$, and since t_1 can be arbitrarily small, the theorem follows. ∎

REMARK 2.6: In view of Remark 1.6 and the fact that $\mathcal{Q}_0(u, v) = \mathrm{pr}_{J_0}[(u \cdot \nabla)v]$ satifies the analogue of Proposition 2.1 etc., *all the preceding deductions are likewise valid with the Neumann condition replaced by the Dirichlet condition* (for the Dirichlet problem, one can even take $c = 0$ in (1.23), so that there exist solutions with $b = \infty$ for small enough data). This gives an alternative proof of the main results on $u(t)$ with $\operatorname{div} u_0 = 0$ shown in [S] and [G-M]. More precisely, when the fractional power domains $D(A^{r/2})$ in [G-M] are identified with $H^r_{p,\gamma_0} \cap J_0$, we get a similar existence theorem as theirs but with a slightly better smoothness statement (since r_1 can be taken both as $r + \delta$ and as $r + 1 + \delta$ above, cf. (2.36)), whereas our uniqueness theorem differs by requiring estimates both for $r_1 = r + \delta$ and $r_1 = r + 1 + \delta$, but only $\|u(t)\|_{r_1} = O(t^{-\frac{r_1 - r}{2}})$, not $o(t^{-\frac{r_1 - r}{2}})$. Taking the best of each result, we altogether get an improvement. Note that our results include Besov spaces also.

For $\operatorname{div} u_0 = 0$, the function $u(t)$ determined in Theorem 2.4 is the velocity part $u(t)$ of a solution $\{u(t), P(t)\}$ to the original Navier-Stokes problem (1.1) with $f = 0$, $\varphi = 0$. Here $P(t)$ is determined by (1.5) for the Neumann problem, and we shall now see what estimates for $P(t)$ are implied from the estimates for $u(t)$.

By (2.2) with $\mu = 1 + \omega$,

$$\begin{aligned} &\|(u \cdot \nabla)u\|_\lambda \le C_{12} \|u\|^2_{\lambda+1+\omega}, \\ &\text{for } \omega \ge 0,\ 2\lambda + 1 + 2\omega > \max\{0, n(\tfrac{2}{p} - 1)\} \text{ and} \\ &\lambda + 1 + 2\omega \ge \tfrac{n}{p};\ \text{where } \lambda + 1 + 2\omega > \tfrac{n}{p} \text{ if } \omega = 0. \end{aligned} \tag{2.41}$$

The linear operators in (1.5) are continuous mappings (cf. e.g. [G2]):

$$\begin{aligned} &K_D \operatorname{div}'_\Gamma \gamma_0 \colon H^{r_1}_p(\overline{\Omega})^n + B^{r_1}_p(\overline{\Omega})^n \to H^{r_1-1}_p(\overline{\Omega}) \cap B^{r_1-1}_p(\overline{\Omega}), \text{ for } r_1 > \tfrac{1}{p}, \\ &R_D \operatorname{div} \colon H^\lambda_p(\overline{\Omega})^n \to H^{\lambda+1}_p(\overline{\Omega}), \text{ for } \lambda > \tfrac{1}{p} - 1, \\ &R_D \operatorname{div} \colon B^\lambda_p(\overline{\Omega})^n \to B^{\lambda+1}_p(\overline{\Omega}), \text{ for } \lambda > \tfrac{1}{p} - 1. \end{aligned} \tag{2.42}$$

Then we have for $\omega \ge 0$, using (2.41) with $\lambda = r_1 - \omega - 1$,

$$\begin{aligned} \|R_D \operatorname{div}[(u \cdot \nabla)u]\|_{r_1 - \omega} &\le C_{12} \|(u \cdot \nabla)u\|_{r_1 - \omega - 1}, \text{ for } r_1 - \omega > \tfrac{1}{p}, \\ &\le C_{13} \|u\|^2_{r_1}, \text{ when in addition } r_1 - \tfrac{1}{2} > \max\{0, n(\tfrac{1}{p} - \tfrac{1}{2})\} \\ &\text{and } r_1 + \omega \ge \tfrac{n}{p};\ \text{with } r_1 + \omega > \tfrac{n}{p} \text{ if } \omega = 0. \end{aligned} \tag{2.43}$$

THEOREM 2.7. *Let $r \geq \frac{n}{p} - 1$, with $r \in]-1+\frac{1}{p}, 2+\frac{1}{p}[$, $r > -\frac{1}{2}$. Then for any $u_0 \in H^r_{p,T'}$ or $B^r_{p,T'}$ with $\operatorname{div} u_0 = 0$, the Navier-Stokes Neumann problem (1.1) with $\kappa = 1$, $f = 0$ and $\varphi = 0$, has a solution $\{u(t), P(t)\}$ on some interval $[0, b]$. Here $u(t)$ is the solution described by Theorem 2.4, so the properties of $u(t)$ and the relations between u_0 and b are as listed there. $P(t)$ is derived from u by (1.5) and satifies:*

$$
\begin{aligned}
&P(t) \in C(\,]0,b]; H_p^{r_1-1}(\overline{\Omega})), \text{ when } u_0 \in H^r_{p,T'}, \text{ and}\\
&P(t) \in C(\,]0,b]; B_p^{r_1-1}(\overline{\Omega})), \text{ when } u_0 \in B^r_{p,T'} \text{ (or } u_0 \in L_p(\Omega)^n,\ r_1 > r = 0),\\
&\qquad \text{with } r_1 \text{ as in (2.36)},\ r_1 - \tfrac{1}{p} \notin \mathbb{N}, \text{ such that } r_1 - \tfrac{1}{2} > \max\{0, n(\tfrac{1}{p} - \tfrac{1}{2})\},\\
&\qquad \text{and for some } \omega \in [0,1],\ r_1 - \omega > \tfrac{1}{p},\ r_1 + \omega \geq \tfrac{n}{p}\ (> \tfrac{n}{p} \text{ if } \omega = 0);\\
&\qquad \text{here } \sup_{t \leq b} t^{r_1 - r} \|P(t)\|_{r_1-1} < \infty.
\end{aligned}
\tag{2.44}
$$

When the estimates hold with $r_1 = r$, the continuity extends to $t = 0$.

For $t > 0$, $u(t)$ and $P(t)$ are C^∞ in (x, t).

PROOF: This follows from (2.37)–(2.38) and Theorem 2.5, when we apply (2.42) to $K_D \operatorname{div}'_\Gamma \gamma_0 u(t)$ and (2.43) to $R_D \operatorname{div}[(u \cdot \nabla) u(t)]$. ∎

REMARK 2.8: For the *Dirichlet problem*, $P(t)$ is instead determined from u by

$$
P = -K_N \operatorname{div}'_\Gamma \gamma_1 u_\tau + \widetilde{G}[(u \cdot \nabla) u],
\tag{2.45}
$$

where $\widetilde{G} = R_D \operatorname{div} - K_N \gamma_\nu$ is of order -1 and class 0, cf. [G-S4, (5.39), (2.54)]. Here we have as in (2.42)–(2.43),

$$
\widetilde{G}\colon H_p^\lambda(\overline{\Omega})^n \to H_p^{\lambda+1}(\overline{\Omega}) \text{ and } \widetilde{G}\colon B_p^\lambda(\overline{\Omega})^n \to B_p^{\lambda+1}(\overline{\Omega}), \text{ for } \lambda > \tfrac{1}{p} - 1;
\tag{2.46}
$$

so that for $\omega \geq 0$,

$$
\begin{aligned}
\|\widetilde{G}[(u \cdot \nabla) u]\|_{r_1 - \omega} &\leq C_{14} \|(u \cdot \nabla) u\|_{r_1 - \omega - 1}, \text{ for } r_1 - \omega > \tfrac{1}{p},\\
&\leq C_{15} \|u\|_{r_1}^2, \text{ when in addition } r_1 - \tfrac{1}{2} > \max\{0, n(\tfrac{1}{p} - \tfrac{1}{2})\},\\
&\quad \text{and } r_1 + \omega \geq \tfrac{n}{p}; \text{ with } r_1 + \omega > \tfrac{n}{p}, \text{ if } \omega = 0.
\end{aligned}
\tag{2.47}
$$

But in contrast with (2.42),

$$
K_N \operatorname{div}'_\Gamma \gamma_1 \operatorname{pr}_\tau\colon H_p^{r_1}(\overline{\Omega})^n + B_p^{r_1}(\overline{\Omega})^n \to H_p^{r_1-1}(\overline{\Omega}) \cap B_p^{r_1-1}(\overline{\Omega}) \quad \text{only for } r_1 > \tfrac{1}{p} + 1,
\tag{2.48}
$$

since it contains γ_1. In fact, when $r_1 \leq \frac{1}{p} + 1$, one can, on the basis of [G2, Lemma 2.2], easily construct sequences $u_{0,m} \in J_0 \cap H_p^{r_1}(\overline{\Omega})^n$ such that $u_{0,m} \to 0$ there but $K_N \operatorname{div}'_\Gamma \gamma_1 \operatorname{pr}_\tau u_{0,m} \to v \neq 0$ in $H_p^{r_1-1}(\overline{\Omega})^n$; a similar statement holds with B_p. Then *for $P(t)$ in the Dirichlet case we can only conclude* (2.44) *when in addition $r_1 > \frac{1}{p} + 1$.* (Cf. [G3] for other results when $r > 1 - \frac{1}{p}$.)

REMARK 2.9: The conditions in (2.44) are a little complicated because of the different possible positions of $\frac{1}{2}$, $\frac{1}{p}$ and $\frac{n}{p}$. Let us see what it takes to have $r = r_1 = 1$ included in (2.44): The inequalities $\omega < 1 - \frac{1}{p}$ and $\omega \geq \frac{n}{p} - 1$ must be satisfied for an $\omega \in]0,1]$, and that holds precisely when $2p > n+1$. (Note that then $\frac{1}{2} > n(\frac{1}{p} - \frac{1}{2})$.) Thus for the Navier-Stokes Neumann problem,

$$(2.49) \qquad P(t) \in C([0,b]; L_p(\Omega)) \text{ holds for } u \in H_p^1(\Omega)^n \text{ when } 2p > n+1.$$

By a slight generalization, one also finds that

$$(2.50) \quad P(t) \in C([0,b]; H_p^s(\Omega)) \text{ holds for } u \in H_p^{1+s}(\Omega)^n \text{ when } 2p > \tfrac{n+1}{1+s}; \quad \text{if } s \in]0, s_0[\,,\ s_0 = \min\{\tfrac{1}{p}, \tfrac{(n-1)^2}{4n}\}.$$

(2.49) is satisfied with $p = 2$ in dimension $n = 2$, and (2.50) is satisfied with $p = 2$ in dimensions $n = 2$ and 3. In contrast with this, Heywood and Walsh have shown that *for the Dirichlet problem,* $P(t)$ need not be bounded in $L_2(\Omega)$ for $t \to 0$, when $n = 2,3$, see [H-W]. This can also be inferred from the calculations in Remark 2.8, and one can moreover show that (2.50) for $p = 2$, $n = 2,3$ is violated for the Dirichlet problem. The crucial fact is that $P(t)$ here contains $\gamma_1 u_\tau(t)$, cf. (2.45) and (2.48) ff., not just $\gamma_0 u(t)$ as in the Neumann case, cf. (1.5).

This shows an interesting difference between Dirichlet and Neumann conditions for the Navier-Stokes problem.

Let us furthermore observe that estimates of $\operatorname{grad} P$ are deduced straightforwardly from (2.44), using that $\|\operatorname{grad} P\|_{r_1-2} \leq C\|P\|_{r_1-1}$. Moreover, $\partial_t u = \Delta u - Gu - \mathcal{Q}u$, where

$$(2.51) \qquad G\colon H_p^\lambda(\overline{\Omega})^n \to H_p^{\lambda-2}(\overline{\Omega}) \text{ and } G\colon B_p^\lambda(\overline{\Omega})^n \to B_p^{\lambda-2}(\overline{\Omega}), \text{ for } \lambda > \tfrac{1}{p}.$$

Then we find *for the Neumann problems:*

$$(2.52) \quad \begin{aligned} &\operatorname{grad} P(t) \text{ and } \partial_t u(t) \in C(\,]0,b]; H_p^{r_1-2}(\overline{\Omega})), \text{ when } u_0 \in H^r_{p,T'}, \text{ and} \\ &\operatorname{grad} P(t) \text{ and } \partial_t u(t) \in C(\,]0,b]; B_p^{r_1-2}(\overline{\Omega})), \text{ when } u_0 \in B^r_{p,T'} \text{ (or } u_0 \in L_p(\Omega)^n), \\ &\qquad \text{with } \sup_{t\leq b} t^{r_1-r}\|\operatorname{grad} P(t)\|_{r_1-2} < \infty, \quad \sup_{t\leq b} t^{r_1-r}\|\partial_t u(t)\|_{r_1-2} < \infty, \\ &\qquad \text{when } r_1 \text{ is as in (2.44).} \end{aligned}$$

For the Dirichlet problem, G and $\mathcal{Q}$ are replaced by G_0 and $\mathcal{Q}_0$ as described in Remark 1.6; here $\mathcal{Q}_0$ behaves like $\mathcal{Q}$, and

$$(2.53) \qquad G_0\colon H_p^\lambda(\overline{\Omega})^n \to H_p^{\lambda-2}(\overline{\Omega}), \quad G_0\colon B_p^\lambda(\overline{\Omega})^n \to B_p^{\lambda-2}(\overline{\Omega}), \text{ for } \lambda > \tfrac{1}{p} + 1,$$

so we find, using Remark 2.8, that $\operatorname{grad} P(t)$ and $\partial_t u(t)$ behave as in (2.52) *when in addition* $r_1 > \frac{1}{p} + 1$. The case $r = r_1 = 1$ (or $= 1 + s$, small s) is discussed as in Remark 2.9.

In view of (1.9–10), the theorems obviously give solvability results in $W_p^r(\overline{\Omega})$ spaces also.

REFERENCES

[B-L]. J. Bergh and J. Löfström, "Interpolation spaces," Springer Verlag, Berlin, New York, 1976.

[F-L-R]. E. B. Fabes, J. E. Lewis and N. M. Riviere, *Boundary value problems for the Navier-Stokes equations*, Amer. J. Math. **99** (1977), 626–668.

[Gi1]. Y. Giga, *Analyticity of the semigroup generated by the Stokes operator in L_r-spaces*, Math. Zeitschr. **178** (1981), 297–329.

[Gi2]. ———, *Domains of fractional powers of the Stokes operator in L_r spaces*, Arch. Rat. Mech. Anal. **89** (1985), 251–265.

[Gi3]. ———, *The nonstationary Navier-Stokes system with some first order boundary conditions*, Proc. Jap. Acad. **58** (1982), 101–104.

[Gi4]. ———, *The Navier-Stokes initial value problem in L_p and related problems*, "Nonlinear partial differential equations in applied sciences (Tokyo 1982)," North Holland Math. Studies 81, Amsterdam-New York, 1983, pp. 37–54.

[Gi5]. ———, *Weak and strong solutions of the Navier-Stokes initial value problem*, Publ. RIMS Kyoto Univ. **19** (1983), 887–910.

[G-M]. Y. Giga and T. Miyakawa, *Solutions in L_r of the Navier-Stokes initial value problem*, Arch. Rat. Mech. Anal. **89** (1985), 267–281.

[Gri]. P. Grisvard, *Equations différentielles abstraites*, Ann. Ecole Norm. Sup. **2 (Série 4)** (1969), 311–395.

[G1]. G. Grubb, "Functional Calculus of Pseudo-Differential Boundary Problems," Progress in Math. Vol. 65, Birkhäuser, Boston, 1986.

[G2]. ———, *Pseudo-differential boundary problems in L_p spaces*, Comm. P. D. E. **15** (1990), 289–340.

[G3]. ———, *Parabolic pseudodifferential boundary problems in anisotropic L_p spaces, with applications to Navier-Stokes problems*, preprint, August 1991.

[G4]. ———, *Solution dans les espaces de Sobolev L_p anisotropes des problèmes aux limites pseudo-différentiels paraboliques et des problèmes de Stokes*, C. R. Acad. Sci. Paris **312, Série I** (1991), 89–92.

[G-K]. G. Grubb and N. J. Kokholm, *Parameter-dependent pseudodifferential boundary problems in global L_p Sobolev spaces*, preprint, August 1991.

[G-S1]. G. Grubb and V. A. Solonnikov, *Reduction of basic initial-boundary value problems for the Stokes equation to initial-boundary value problems for systems of pseudodifferential equations*, Zap. Nauchn. Sem. L.O.M.I. **163** (1987), 37–48; = J. Soviet Math. **49** (1990), 1140–1147.

[G-S2]. ———, *Solution of parabolic pseudo-differential initial-boundary value problems*, J. Diff. Equ. **87** (1990), 256–304.

[G-S3]. ———, *Reduction of basic initial-boundary value problems for the Navier-Stokes equations to nonlinear parabolic systems of pseudodifferential equations*, Zap. Nauchn. Sem. L.O.M.I. **171** (1989), 36–52; English transl. available as Copenh. Univ. Math. Dept. Report Ser. 1989 no. 5.

[G-S4]. ———, *Boundary value problems for the nonstationary Navier-Stokes equations treated by pseudo-differential methods*, Math. Scand. **69** (1991).

[H-W]. J. Heywood and O. Walsh, *A counter example concerning the pressure in the Navier-Stokes equations, as $t \to 0^+$*, preprint, August 1991 (presented at the Oberwolfach meeting on Navier-Stokes equations).

[Se]. R. T. Seeley, *Interpolation in L_p with boundary conditions*, Studia Math. **44** (1972), 47–60.

[S-W]. H. Sohr and W. von Wahl, *On the singular set and the uniqueness of weak solutions of the Navier-Stokes equations*, Manuscripta Math. **49** (1984), 27–59.

[S]. V. A. Solonnikov, *Estimates for solutions of nonstationary Navier-Stokes systems*, Zap. Nauchn. Sem. LOMI **38** (1973), 153–231; = J. Soviet Math. **8** (1977), 467–529.

[T]. H. Triebel, "Interpolation theory, function spaces, differential operators," North-Holland Publ. Co., Amsterdam, New York, 1978.

[W]. W. von Wahl, *Regularity questions for the Navier-Stokes equations*, "Approximation methods for Navier-Stokes problems," Lecture Note 771, Springer Verlag, Berlin, Heidelberg, 1980.

MATHEMATICS DEPARTMENT, UNIVERSITY OF COPENHAGEN,
UNIVERSITETSPARKEN 5, DK-2100 COPENHAGEN, DENMARK

ESTIMATES IN $C^{2l,l}$ FOR SOLTUTION OF A BOUNDARY VALUE PROBLEM FOR THE NONSTATIONARY STOKES SYSTEM WITH A SURFACE TENSION IN BOUNDARY CONDITION

IL'IA MOGILEVSKII
Department of mathematics, Tver University,
Zheliabova str.33, 170000, Tver, Russia

In [I] we studied the following free boundary problem: find a bounded domain $\Omega_t \subset R^3, t > 0$, and a vector field $v(x,t) = (v_1, v_2, v_3)$ satisfying together with a scalar function $p(x,t)$ the Navier-Stokes system of equations

$$v_t + (v \cdot \nabla)v - \nu\nabla^2 v + \nabla p = f(x,t), \quad \nabla \cdot v = 0, \quad x\epsilon\Omega_t \tag{1}$$

and the initial and boundary conditions

$$v(x,0) = v_o(x), \quad x\epsilon\Omega_t, \tag{2}$$

$$T(v,p)n - \sigma(x)Hn = 0, \quad x\epsilon\Gamma_t = \partial\Omega_t. \tag{3}$$

Here $\nabla = (\frac{\partial}{\partial x_1}, \frac{\partial}{\partial x_2}, \frac{\partial}{\partial x_3}), \nabla \cdot v = divv, \nabla p = gradp, \nabla^2 v = \Delta v, \nu = const > 0$ is a coefficient of viscosity, $\sigma(x)$ is a coefficient of surface tension which is a strictly positive smooth function: $\sigma(x) \geq \sigma_o > 0, f(x,t)$ is a given vector field of external forces,$T(v,p) = Ip + \nu S(v)$ is the stress tensor and $S(v)$ is the deformation tensor with the elements

$$S_{ij} = \frac{\partial v_i}{\partial x_j} + \frac{\partial v_j}{\partial x_i},$$

n is a unit exterior normal vector to $\Gamma_t = \partial\Omega_t$ and H is the twice mean curvature of Γ_t. The sign of H is chosen in such a way that

$$Hn = \Delta(t)\mathrm{x}, \quad x\epsilon\Gamma_t,$$

where $\Delta(t)$ is the Laplace-Beltrami operator on Γ_t and x is a radius-vector corresponding to the point x. The important part of investigation of the problem (1) - (3) is establishing of estimates for solution of the following initial boundary value problem for the nonstationary Stokes system

$$v_t - \nu\nabla^2 v + \nabla p = f(x,t), \quad \nabla \cdot v = \rho(x,t), \quad x\epsilon\Omega, \quad t\epsilon(0,T) \tag{4}$$

$$v(x,0) = v_o(x), \quad x\epsilon\Omega \tag{5}$$

$$\Pi\nu S(v)n = \Pi b, \quad x\epsilon\Gamma = \partial\Omega \tag{6a}$$

$$-p + \nu S(v)n \cdot n - \sigma(x)n \cdot \int_o^t \Delta v(x,\tau)d\tau = d + \int_0^t B(x,\tau)d\tau \tag{6b}$$

Here Ω is bounded domain in R^3 with a smooth boundary Γ, ρ is a given function. In the present paper we establish a priori estimates of the solution of a more general problem in a complete scale of Hölder spaces. Namely we consider the system

$$v_t - \nabla(\nu(x,t)S(v)) + \nabla p = f, \quad \nabla \cdot v = \rho \tag{7}$$

with initial condition (5) and boundary conditions (6). Here $\nu(x,t) \geq \nu_0 > 0$ is a given smooth function. In the case $\nu = const, \sigma = 0$ these estimates were obtained in the space $C^{2+\alpha,1+\alpha/2}$ in [2]. Estimates for the first boundary value problem for system (4) (with boundary condition $v|_{x\epsilon\Gamma} = b$) were established in [3]. Let Ω be a bounded domain in R^3 and let k be a nonnegative integer and $\alpha\epsilon(0,1)$. We denote by $C^{k+\alpha}(\Omega)$ the set of k times continuously differentiable functions defined in Ω whose $k-th$ derivatives satisfy the Hölder condition with exponent α and we supply this set with the norm

$$|u|_{C^{k+\alpha}(\Omega)} = \sum_{|\beta|\leq k} |D_x^\beta u|_\Omega + < u >_\Omega^{(k+\alpha)},$$

where

$$\beta = (\beta_1, \beta_2, \beta_3),\ |\beta| = \beta_1 + \beta_2 + \beta_3,\ D_x^\beta u = \frac{\partial^{|\beta|} u}{\partial x_1^{\beta_1}, \partial x_2^{\beta_2} \partial x_3^{\beta_3}},\quad |u|_\Omega = \sup_{x\epsilon\Omega} |v(x)|$$

and

$$< u >_\Omega^{(k+\alpha)} = \sum_{|\beta|=k} \sup_{x,y\epsilon\Omega} |x-y|^{-\alpha}\ |D_x^\beta u(x) - D_y^\beta u(y)|.$$

Let $Q_T = \Omega \times (0,T)$. By $C^{k+\alpha,\frac{k+\alpha}{2}}(Q_T)$ we mean the space of functions defined in Q_T with a finite norm

$$|u|_{Q_T}^{(k+\alpha,\frac{k+\alpha}{2})} = \sup_{t\epsilon(0,T)} |u(\cdot,t)|_{C^{k+\alpha}(\Omega)} + \sup_{x\epsilon\Omega} |u(x,\cdot)|_{C^{\frac{k+\alpha}{2}}(0,T)}$$

We shall use also seminorms

$$||u||_{Q_T}^{(1+\alpha,\gamma)} = \sup_{0<\tau<t<T} \tau^{-\frac{1+\alpha-\gamma}{2}} < u(\cdot,t) - u(\cdot,t-\tau) >_\Omega^{(\gamma)},$$

$$||u||_{Q_T}^{(2+\alpha,\gamma)} = \sup_{0<\tau<t<T} \tau^{-\frac{2+\alpha-\gamma}{2}} < u(\cdot,t) - u(\cdot,t-\tau) >_\Omega^{(\gamma)}$$

with $0 < \alpha < \gamma < 1$. It was shown in [2] that

$$||u||_{Q_T}^{(1+\alpha,\gamma)} \leq c < u >_{Q_T}^{(1+\alpha,\frac{1+\alpha}{2})},$$

$$||u||_{Q_T}^{(2+\alpha,\gamma)} \leq c < u >_{Q_T}^{(2+\alpha,1+\alpha/2)}.$$

We formulate first of all conditions for the data of the problem (7), (5), (6). Let

$$\begin{gathered} f\epsilon C^{k+\alpha,\frac{k+\alpha}{2}}(Q_T),\ \rho\epsilon C^{k+1+\alpha,\frac{k+1+\alpha}{2}}(Q_T), \\ v_o\epsilon C^{k+2+\alpha}(\Omega),\ b\epsilon C^{k+1+\alpha,\frac{k+1+\alpha}{2}}(\Gamma_T), \\ d\epsilon C^{k+1+\alpha,\frac{k+1+\alpha}{2}}(\Gamma_T),\ B\epsilon C^{k+\alpha,\frac{k+\alpha}{2}}(\Gamma_T), \\ \nu\epsilon C^{k+1+\alpha,\frac{k+1+\alpha}{2}}(Q_T),\ \sigma\epsilon C^{k+\alpha}(\Gamma). \end{gathered} \tag{8}$$

Here $\Gamma_T = \Gamma \times (0,T)$. Assume furthermore that

$$\rho_t - \nabla \cdot f = \nabla \cdot h,\ D_t^{[k/2]} h = \sum_{m=1}^3 \frac{\partial H_m}{\partial x_m} \tag{9}$$

(in the sense of distributions) with

$$h\epsilon C^{k+\alpha,\frac{k+\alpha}{2}}(Q_T) \tag{10}$$

and with

$$\sum_{m=1}^{3} ||H_m||_{Q_T}^{(\delta_k+\alpha,\gamma)} < \infty, \delta_k = \begin{cases} 1 & \text{if } k \text{ is even} \\ 2 & \text{if } k \text{ is odd} \end{cases} \tag{11}$$

Here $[l]$ is an integer part of l. Let us turn to compatibility conditions. We set

$$g^{(m)} = \frac{\partial^m}{\partial t^m} g(x,0).$$

We have the following Dirichlet problem for function $p^{(0)}(x)$

$$\Delta p^{(0)} = \nabla \cdot (\nabla[\nu^{(0)} S(v_o)] - h^{(0)}), \quad x\epsilon\Omega$$

$$p^{(0)}|_{x\epsilon\Gamma} = \nu^{(0)} S(v_0) n \cdot n|_{x\epsilon\Gamma} - d^{(0)} \tag{12}$$

It is well known (see [4] for example) that this problem has a unique solution. We obtain now from system (7)

$$v^{(1)} = f^{(0)} + \nabla[\nu^{(0)} S(v_o)] - \nabla p^{(0)} \tag{13}$$

Let us denote by $M(v)$ the matrix $[\nu(x,t)S(v)]$. $p^{(1)}(x)$ is a solution of the Dirichlet problem

$$\Delta p^{(1)} = \nabla \cdot (\nabla [M(v)]^{(1)} - h^{(1)}), x\epsilon\Omega$$

$$p^{(1)}|_{x\epsilon\Gamma} = [[M(v)]^{(1)} n \cdot n - \sigma n \cdot \Delta v_0 - d^{(1)} - B^{(0)}]_{x\epsilon\Gamma}$$

with $[M(v)]^{(1)} = \nu^{(1)} S(v_0) + \nu^{(0)} S(v^{(1)})$, therefore $p^{(1)}$ is completely defined by the data of problem (7), (5), (6). By continuing this process we find functions $p^{(m)}$ for $m = 0, 1, ..., [\frac{k+1}{2}] - 1$ and vectors $v^{(m)}$ for $m = 1, ..., [\frac{k+1}{2}]$. $p^{(m)}$ are solutions of Dirichlet problems

$$\Delta p^{(m)} = \nabla \cdot (\nabla [M(v)]^{(m)} - h^{(m)}), \quad x\epsilon\Omega$$

$$p^{(m)}|_{x\epsilon\Gamma} = [[M(v)]^{(m)} n \cdot n - \sigma n \cdot \Delta v^{(m-1)} - d^{(m)} - B^{(m-1)}]_{x\epsilon\Gamma} \tag{14}$$

for $m \geq 1$ and $p^{(0)}$ is a solution of the problem (12). Finally,

$$v^{(0)} = v_0, \quad v^{(m)} = f^{(m-1)} + \nabla [M(v)]^{(m-1)} - \nabla p^{(m-1)}, m \geq 1. \tag{15}$$

We can formulate now the compatibility conditions. Namely we assume that

$$\nabla \cdot v^{(m)} = \rho^{(m)}, \quad m = 0, ..., [\frac{k+1}{2}], \tag{16a}$$

$$\Pi([M(v)]^{(m)} n) = \Pi b^{(m)} \tag{16b}$$

where $[M(v)]^{(m)} = \sum_{i=0}^{m} C_m^i \nu^{(i)} S(v^{(m-i)}), v^{(i)}$ are defined by the formulas (15), C_m^i are binomial coefficients. We need estimates of solutions of a model boundary value problem in a half-space and of the Cauchy problem for system (4).

THEOREM 1.
Consider the initial-boundary value problem in $D_T = \{x : x \epsilon R^3, x_3 > 0\} \times (0, T)$

$$v_t - \nu_0 \nabla^2 v + \nabla p = f, \ \ \nabla \cdot v = \rho$$

$$v(x,0) = v_0, \ \ \frac{\partial v_i}{\partial x_3} + \frac{\partial v_3}{\partial x_i}|_{x_3=0} = h_i, \ \ i = 1,2$$

$$-p + 2\frac{\partial v_3}{\partial x_3} + \sigma_o \Delta \int_0^t v_3(x,\tau)d\tau = d + \int_o^t B(x,\tau)d\tau, \tag{17}$$

with positive constant σ_0 and ν_0 and with $\Delta u = \frac{\partial^2 u}{\partial x_1^2} + \frac{\partial^2 u}{\partial x_2^2}$. Suppose that conditions (8), (9), (10), (11), (16) hold with Q_T replaced by D_T, Ω replaced by $R^3_+ = \{x \epsilon R^3, x_3 > 0\}$, and Γ_T replaced by $R_T = R^2 \times (0,T)$. Then this problem has a unique solution $v \epsilon C^{2+k+\alpha, 1+\frac{k+\alpha}{2}}(D_T), \nabla p \epsilon C^{k+\alpha, \frac{k+\alpha}{2}}(D_T)$ satisfying the estimate

$$< v >_{D_T}^{(2+k+\alpha,1+\frac{k+\alpha}{2})} + < \nabla p >_{D_T}^{(k+\alpha,\frac{k+\alpha}{2})} \leq c(T)\{< f >_{D_T}^{(k+\alpha,\frac{k+\alpha}{2})} +$$

$$+ < v_0 >_{R_T^3}^{(2+k+\alpha)} + < \rho >_{D_T}^{(1+k+\alpha,\frac{1+k+\alpha}{2})} + < h >_{D_T}^{(k+\alpha,\frac{k+\alpha}{2})} +$$

$$+ < b >_{R_T}^{(1+k+\alpha\frac{1+k+\alpha}{2})} + < d >_{R_T}^{(1+k+\alpha,\frac{1+k+\alpha}{2})} + |B|_{R_T}^{(k+\alpha,\frac{k+a}{2})} + \sum_{m=1}^{3} ||H_m||_{D_T}^{(\delta_k+\alpha,\gamma)}\}, \tag{18}$$

where $C(T)$ is a nondecreasing function of T.

PROOF.
This theorem was proved in [5] for $k = 0$. A general case is investigated in the same way. By using compatibility conditions (16) we reduce problem (17) to the problem

$$u_t - \nu_0 \nabla^2 u + \nabla q = 0, \ \ \nabla \cdot u = 0$$

$$u(x,0) = 0, \ \ \frac{\partial u_i}{\partial x_3} + \frac{\partial u_3}{\partial x_i}|_{x_3=0} = a_i, i = 1,2$$

$$-q + 2\frac{\partial u_3}{\partial x_3} + \sigma_0 \Delta' \int_0^t u_3(x,\tau)d\tau = e + \int_0^t A(x,\tau)d\tau \tag{19}$$

where $a_i^{(m)} = 0, e^{(m)} = 0, A^{(m)} = 0$ for $m = 0, ..., [k/2]$. We have $(v,p) = (u,q) + (w,r)$. Vector w and function r are constructed in [5] and it is shown that

$$< w >_{D_T}^{(2+k+\alpha,1+\frac{k+\alpha}{2})} + < \nabla r >_{D_T}^{(k+\alpha,\frac{k+\alpha}{2})} \leq M(T), \tag{20}$$

where $M(T)$ is the right-hand side of (18). The solution of the problem (19) is also constructed in [5] and estimated by a theorem on Fourier multipliers in Hölder spaces. It is proved that

$$< u >_{D_T}^{(2+k+\alpha,1+\frac{k+\alpha}{2})} + < \nabla q >_{D_T}^{(k+\alpha,\frac{k+\alpha}{2})} \leq c_1(T)\{< a >_{R_T}^{(1+k+\alpha,\frac{1+k+\alpha}{2})} +$$

$$+ < e >_{R_T}^{(1+k+\alpha,\frac{1+k+\alpha}{2})} + < A >_{R_T}^{(k+\alpha,\frac{k+\alpha}{2})}\}. \tag{21}$$

Boundary data in the problem (19) have a form

$$a_i = b_i - [\frac{\partial w_i}{\partial x_3} + \frac{\partial w_3}{\partial x_i}]_{x_3=0}, \quad i = 1, 2,$$

$$e = d + [r - 2\frac{\partial w_3}{\partial x_3}]_{x_3=0}, \quad A = B - \sigma_o \Delta' w_3.$$

Inequality (18) follows from these formulas and from estimates (20), (21). The theorem is proved.

THEOREM 2.

Under conditions (8), (9), (10),(11), (16a) (with Q_T replaced by Π_T) the Cauchy problem

$$v_t - \nu_o \nabla^2 v + \nabla p = f, \quad \nabla \cdot v = \rho, \quad (x,t)\epsilon\Pi_T \equiv R^3 \times (o, T), \quad v|_{t=0} = v_0 \tag{22}$$

has a unique solution $v\epsilon C^{2+k+\alpha, 1+\frac{k+\alpha}{2}}(\Pi_T)$, $\nabla p \epsilon C^{k+\alpha, \frac{k+\alpha}{2}}(\Pi_T)$ satisfying the estimate

$$< v >_{\Pi_T}^{(2+k+\alpha, 1+\frac{k+\alpha}{2})} + < \nabla p >_{\Pi_T}^{(k+\alpha, \frac{k+\alpha}{2})} \leq c[< f >_{\Pi_T}^{(k+\alpha, \frac{k+\alpha}{2})} +$$

$$< v_o >_{R^3}^{(2+k+\alpha)} + < \rho >_{\Pi_T}^{(1+k+\alpha, \frac{1+k+\alpha}{2})} + < h >_{\Pi_T}^{(k+\alpha, \frac{k+\alpha}{2})} + \sum_{m=1}^{3} ||H_m||_{\Pi_T}^{(\delta_k+\alpha, \gamma)}] \tag{23}$$

PROOF.

The representation of a solution of the problem (22) is obtained in [2] in the form

$$v = u + \nabla \phi, \quad p = \rho - \nabla \cdot u - \phi_t,$$

where u is a solution of the Cauchy problem for the heat equation

$$u_t - \nabla^2 u = f(x,t)\epsilon\Pi_T, \quad u(x, 0) = v_o \tag{24}$$

and ϕ is a solution of the Poisson equation

$$\nabla^2 \phi = \rho - \nabla \cdot u, \quad x\epsilon R^3.$$

Using estimates for solution of the problem (24) from [6] we obtain

$$< v >_{\Pi_T}^{(2+k+\alpha, 1+\frac{k+\alpha}{2})} + < \nabla p >_{\Pi_T}^{(k+\alpha, \frac{k+\alpha}{2}, \frac{k+\alpha}{2})} \leq c[< f >_{\Pi_T}^{(k+\alpha, \frac{k+\alpha}{2})} +$$

$$< v_o >_{R^3}^{(2+k+\alpha)} + < \rho >_{\Pi_T}^{(1+k+\alpha, \frac{1+k+\alpha}{2})} + < \nabla \phi_t >_{\Pi_T}^{(k+\alpha, \frac{k+\alpha}{2}}]. \tag{25}$$

$\nabla \phi_t$ may be represented by the Newtonian potential

$$\nabla \phi_t = \nabla_x \int_{R^3} \mathrm{E}(x-y)[\rho_t - \nabla \cdot u_t] dy =$$

$$= -\nabla_x \int_{R^3} \nabla_y \mathrm{E}(x-y)[h(y,t) - \nabla_y^2 u(y,t)] dy, \quad \mathrm{E}(x) = \frac{1}{4\pi |x|}.$$

Treating this potential in the same way as in [2] we obtain

$$< \nabla \phi_t >_{\Pi_T}^{(k+\alpha, \frac{k+\alpha}{2})} \leq \mathrm{N}(T),$$

where $\mathbf{N}(T)$ is the right-hand side of (23). This inequality together with (25) leads to estimate (23). The theorem is proved.
Theorem 1 and 2 provide a basis for estimates of the solution of problem (7), (5), (6).
THEOREM 3.
Suppose that $\Gamma\epsilon C^{2+k+\alpha}$ and that conditions (8), (9), (10), (11), (16) hold. Let $\nu(x,t)\geq \nu_o>0$ and $\sigma(x)\geq\sigma_o>0$. Then any solution (v,p) of the problem (7), (5), (6) such that

$$v\epsilon C^{2+k+\alpha,1+\frac{k+\alpha}{2}}(Q_T),\nabla p\epsilon C^{k+\alpha,\frac{k+\alpha}{2}}(Q_T)$$

satisfies the inequality

$$|v|_{Q_T}^{(2+k+\alpha,1+\frac{k+\alpha}{2})}+|\nabla p|_{Q_T}^{(k+\alpha,\frac{k+\alpha}{2})}+|p|_{Q_T}^{(k+\alpha,\frac{k+\alpha}{2})}+$$

$$+|p|_{\Gamma_T}^{(1+k+\alpha,\frac{1+k+\alpha}{2})}\leq c(T)\{|f|_{Q_T}^{(k+\alpha,\frac{k+\alpha}{2})}+|\rho|_{Q_T}^{(1+k+\alpha,\frac{1+k+\alpha}{2})}$$

$$+|B|_{\Gamma_T}^{(k+\alpha,\frac{k+\alpha}{2})}+|h|_{Q_T}^{(k+\alpha,\frac{k+\alpha}{2})}+\sum_{m=1}^{3}||H_m||_{Q_T}^{(\delta_k+\alpha,\gamma)}\}\equiv c(T)F(T) \tag{26}$$

PROOF.
The proof of this theorem is based on the well-known J. Schauder's method. Let $\{\xi_j\}_{j=1}^{N_\lambda}$ be a set of points such that: $\xi_j\epsilon\Omega,\cup_{j=1}^{N_1}(\{x:|x-\xi_j|<\lambda\})\supset\Omega$, where λ is a small positive number. Let $\{\zeta_j(x)\}_{j=1}^{N_1}$ be a set of functions with the properties

$$\zeta_j\epsilon C^\infty, supp\zeta_j\subset K_{j,\lambda}\equiv\{x:|x-\xi_j|<2\lambda\},\ \ \zeta_j(x)=1\ \text{ if }\ |x-\xi_j|<\lambda,$$

$|D_x^\beta\zeta_j|\leq c_j(\beta)\lambda^{|\beta|}$. We divide the interval (0,T) by points $t_i=i\mu, i=0,...,L_\mu$, where μ is a small postive number and $\mu\cdot L_\mu\geq T$. We introduce a system of functions $\varphi_i(t)$ such that $\varphi_i\epsilon C^\infty,\varphi_o(t)=1$ if $0\leq t\leq t_1,\varphi_o(t)=0$ if $t>t_2$; for $i\geq 1$ $\varphi_i(t)=1$ if $t_i\leq t\leq t_{i+1}$ and $\varphi_i(t)=0$ if $t\epsilon[t_{i-1},t_{i+2}]$, $\frac{\partial^m}{\partial t^m}\varphi_i(t)\leq c_i(m)\mu^{-m}$.
Let $v_{ji}(x,t)=v(x,t)\zeta_j(x)\varphi_i(t), p_{ji}(x,t)=p(x,t)\zeta_j(x)\varphi_i(t)$. Then

$$|v|_{Q_T}^{(2+k+\alpha,1+\frac{k+\alpha}{2})}+|\nabla p|_{Q_T}^{(k+\alpha,\frac{k+\alpha}{2})}\leq c_1\sup_{j,i}\{|v_{ji}|_{Q_{ji}}^{(2+k+\alpha,1+\frac{k+\alpha}{2})}+|\nabla p_{ji}|_{Q_{ji}}^{(k+\alpha,\frac{k+\alpha}{2})}\}$$

where

$$Q_{ji}=(K_{j,\lambda}\cap\Omega)\times[(t_{i-1},t_{i+2})\cap(0,T)].$$

If $K_{j,\lambda}\cap\Gamma=\phi$ then we can estimate (v_{ji},p_{ji}) as a solution of the Cauchy problem (22) in $R^3\times[(t_{i-1},t_{i+2})\cap(0,T)](i=1,...,L_\mu-1)$. In the case $K_{j,\lambda}\cap\Gamma\neq\phi$ we estimate (v_{ji},p_{ji}) considering these functions as a solution of an initial-boundary value problem in a half-space. The reduction of problem (7),(5), (6) to this problem has been done in [2],[7] and [1]. By repeating arguments of these works and by using theorems 1 and 2 we obtain

$$P(T)\leq c_2[F(T)+|v|_{Q_T}], \tag{27}$$

where $P(T)$ is the left-hand side of (26) and c_2 is nondecreasing function of T. It follows from (27) that

$$|v|_{Q_T}\leq|v_o|_\Omega+\int_o^t|v_\tau(\cdot,\tau)|_\Omega d\tau\leq c_3(F(t)+\int_o^t|v|_{Q_\tau}d\tau)$$

and by the Granwall lemma

$$|v|_{Q_T} \leq c_4(t)F(t).$$

This inequality together with (27) gives estimate (26). The theorem is proved.

REMARK.

Under hypotheses of the theorem 3 problem (7), (5), (6) has a unique solution (v, p) such that

$$v \epsilon C^{2+k+\alpha, 1+\frac{k+\alpha}{2}}(Q_T), \nabla p \epsilon C^{k+\alpha, \frac{k+\alpha}{2}}(Q_T).$$

This can be established by the method described in [1].

REFERENCES

1. I.Sh. Mogilevskii, V.A. Solonnikov, On the solvatibility of a free boundary problem for the Navier-Stokes equations in the Hölder space of functions. Nonlinear Analysis, Pisa, 1991, p.257-271.
2. V A. Solonnikov, Estimates of solutions of an initial-boundary value problem for linear nonstationary system of the Navier-Stokes equations. Zap. nauchn. sem. LOMI, 1976, v. 59, p. 178-254 (in Russian) = Journ. Sov. Math. 1978, v. 10, p. 336-393.
3. V.A. Solonnikov, Estimates of solutions of a nonstationary linearized Navier-Stokes system of equations. Proc. Steklov Math. Inst. v. 70, 1964, p. 213-317. (in Russian).
4. O.A. Ladyzhenskaja, N.N. Uraltseva, Linear and quasilinear equations of elliptic type. Moscow, 1973 (in Russian).
5. I.Sh. Mogilevskij, V.A. Solonnikov, Solvability of a non-coercive inital-boundary value problem for the Stokes equation in Hölder spaces (the case of half-space). Zeitschrift für Analysis und ihre Anwendungen, 1989, Bd. 8, Hf. 4, p. 330-347 (in Russian).
6. O.A. Ladyzhenskaja, V.A. Solonnikov, N.N. Uraltseva, Linear and quasilinear equations of parabolic type. Moskow, 1967 (in Russian).
7. V.A. Solonnikov, Unsteady motion of a finite mass of fluid bounded by a free surface. Zap. nauchn.semin. LOMI, 1986, v. 152, p.137-157, (in Russian) = Journ. Sov. Math., 1988, v. 40, N. 5, p.672-686.

This paper is in final form and no similar paper has been or is being submitted elsewhere..

Decomposition of Solenoidal Fields into Poloidal Fields, Toroidal Fields and the Mean Flow. Applications to the Boussinesq-Equations

Burkhard J. Schmitt and Wolf von Wahl
Lehrstuhl für Angewandte Mathematik
Universität Bayreuth
P.O. Box 10 12 51
W-8580 Bayreuth, GERMANY

0. Introduction and Notations

Let there be given a solenoidal vector field $\underline{u}$ over an infinite layer $(x, y, z) \in \mathbb{R}^2 \times (a, b)$. $\underline{u}$ is assumed to be periodic in x, y with respect to a square or a rectangle. We prove here that $\underline{u}$ can be decomposed in a unique way into ($\underline{k} = (0,0,1)$)

$$
\begin{aligned}
\underline{u}(x,y,z) &= \operatorname{curl}\operatorname{curl}\varphi(x,y,z)\underline{k} + \operatorname{curl}\psi(x,y,z)\underline{k} + F(z) \\
&= P(x,y,z) + T(x,y,z) + F(z).
\end{aligned}
\tag{0.1}
$$

$P = \operatorname{curl}\operatorname{curl}\varphi\underline{k}$ is called the poloidal part of $\underline{u}$, $T = \operatorname{curl}\psi\underline{k}$ is the toroidal part and the field F, which depends on z only and has constant third component, is the mean flow. φ, ψ are functions which are determined uniquely if we require them to have vanishing mean value over a periodicity cell $\mathcal{P}$. $P+T$ is then nothing else but that part of $\underline{u}$ which has vanishing mean value over $\mathcal{P}$. A corresponding decomposition was already derived in the case of a spherical layer $\{\underline{x} \in \mathbb{R}^3 | r < |\underline{x}| < R\}$ in [1,2], where the assumptions imposed on $\underline{u}$ are $\nabla \cdot \underline{u} = 0$ and $\int_{|\underline{x}|=R} \underline{u} \cdot \underline{x}\, do_{\underline{x}} = 0$. In this situation it turns out that the mean flow is not needed, i.e. $F \equiv 0$. In [5, p. 236] an attempt was made to decompose $\underline{u}$ into $P + T$ in our situation, on $\underline{u} = (u_x, u_y, u_z)^{\mathrm{T}}$ the condition $\int_{\mathcal{P}} u_z(x, y, a)\, dx\, dy = 0$ was imposed. This is however not sufficient as it is exhibited by the mean flow (cf. the Remark in section 1). In fact $\int_{\mathcal{P}} u_x\, dx\, dy = \int_{\mathcal{P}} u_y\, dx\, dy = \int_{\mathcal{P}} u_z\, dx\, dy = 0$ is needed. The exact result concerning the decomposition (0.1) can be seen from Theorem 1.4 to follow. P, T, F can be understood as orthogonal projections from the L^2-space of periodic solenoidal fields (in a sense which is made precise in Theorem 1.4) onto three pairwise orthogonal subspaces. The regularity properties of these projections in terms of L^2-Sobolev spaces are studied also in Theorem 1.4.

The decomposition (0.1) is now applied to the Boussinesq-equations

$$(0.2)\quad \left\{ \begin{array}{rcl} \underline{u}' - \Delta \underline{u} + \underline{u}\cdot\nabla\underline{u} - \sqrt{R}\,\vartheta\underline{k} + \nabla\pi &=& 0, \quad \nabla\cdot\underline{u} = 0, \\ \Pr\vartheta' - \Delta\vartheta + \Pr\underline{u}\cdot\nabla\vartheta - \sqrt{R}\,u_z &=& 0 \end{array} \right.$$

over the infinite layer $\mathbb{R}^2 \times (-\frac{1}{2}, \frac{1}{2})$. $\Pr > 0$ is the Prandtl-number, $R > 0$ is the Rayleigh-number, $\underline{u}$, ϑ have the usual meaning, π is the pressure. The boundary-conditions at $z = \pm\frac{1}{2}$ are the usual ones: Stress-free boundaries or rigid boundaries. They are explained in section 2. $.'$ refers to the derivative with respect to time, and we have also to prescribe the initial values $\underline{u}_0$, ϑ_0 at time $t = 0$. $\underline{u}$, ϑ and π are required to be periodic in (x, y) with respect to a square $\mathcal{P} = (-\frac{\pi}{\alpha}, \frac{\pi}{\alpha})^2$ with wave-number α in both directions; the generalization to $\mathcal{P} = (-\frac{\pi}{\alpha}, \frac{\pi}{\alpha}) \times (-\frac{\pi}{\beta}, \frac{\pi}{\beta})$ is at hand and does not need further consideration. When applying (0.1) to $\underline{u}$ as above the boundary conditions on $\underline{u}$ go over into equivalent ones on φ, ψ and F. Moreover, $F_3 \equiv 0$. The system (0.2) itself is transformed into an equivalent one for $\Phi = (\varphi, \psi, \vartheta, F_1, F_2)^{\mathrm{T}}$. It has the form

$$(0.3)\qquad \mathcal{B}\Phi' + \mathcal{A}\Phi - \sqrt{R}\,\mathcal{C}\Phi + \mathcal{M}(\Phi) = 0$$

with matrix operators $\mathcal{B}$, $\mathcal{A}$, $\mathcal{C}$ and a nonlinear term $\mathcal{M}$. $\mathcal{A}$, $\mathcal{B}$ turn out to be diagonal and strictly positive definite selfadjoint operators in an appropriate Hilbert space H. H is simply the product $L^2_M(\Omega) \times L^2_M(\Omega) \times L^2(\Omega) \times (L^2_M((-\frac{1}{2}, \frac{1}{2})))^2$ or $L^2_M(\Omega) \times L^2_M(\Omega) \times L^2(\Omega) \times (L^2((-\frac{1}{2}, \frac{1}{2})))^2$ for stress-free boundaries or rigid boundaries with $\Omega = \mathcal{P} \times (-\frac{1}{2}, \frac{1}{2})$. The subscript $._M$ indicates that φ, ψ have vanishing mean value over $\mathcal{P}$, whereas F is required to have vanishing mean value over $(-\frac{1}{2}, \frac{1}{2})$ in the case of stress-free boundaries. The pressure is eliminated. Whereas the necessity of (0.3) is easy to show, the proof of the sufficiency throws some light on the mean flow. While it's possible to solve (0.3) in a reduced form, i.e. without mean flow (this is even easier), it's not possible to obtain back (0.2) from (0.3) in this case unless

$$(0.4)\qquad \int_{\mathcal{P}} (P+T)_x\, P_z\, dx\, dy = \int_{\mathcal{P}} (P+T)_y\, P_z\, dx\, dy = 0.$$

(0.4) holds if φ, ψ, ϑ exhibit certain symmetries. We refer for this to [3, pp. 347, 357].

Of course one can try to eliminate $\nabla\pi$ in (0.2) by using the classical tool of projecting $L^2(\Omega)$ on its divergence-free part. This has been done by Iooss in [4] for rigid boundaries. Therefore some remarks on the advantages of (0.3) are in order. While the projection Q just mentioned is a nonlocal operator and therefore yields a nonlocal nonlinearity $Q(\underline{u}\cdot\nabla\underline{u})$ when applied, the main part of $\mathcal{M}(\Phi)$ is purely local. There is only one nonlocal part in $\mathcal{M}(\Phi)$. It occurs within the subsystem for the mean flow and consists of $(\int_{\mathcal{P}}(P+T)\cdot\nabla(P+T)_x,\, dx\, dy, \int_{\mathcal{P}}(P+T)\cdot\nabla(P+T)_y,\, dx\, dy)$. In particular it's possible to study (0.3) within various subspaces which are invariant under the nonlinearity and to subject (0.3) to a numerical analysis. This was almost exclusively done by physicists. We refer for this e.g. to the paper [3, sections 2, 4]. The mathematical background for (0.3) is treated in detail in [6]. In the present paper we review in section 2 some results of [6]. (0.3) may also be used to study the regularity behaviour of Φ

near $t = 0$ by imposing suitable compatibility conditions. While this question may be of more mathematical interest, the problem of energy-stability is not. Due to the fact that the highest order derivatives of $u_z = (-\Delta_2)\varphi$, $\Delta_2 := \partial_x^2 + \partial_y^2$, are isolated in the first row of (0.3) and the pressure is eliminated at the same time, a calculus estimate already yields the precise bounds in the case of stress-free boundaries.

We introduce some notation. A vector field $\underline{u}$ or F is usually written as a column, i.e. $\underline{u} = (u_1, u_2, u_3)^{\mathrm{T}} = (u_x, u_y, u_z)^{\mathrm{T}}$, $F = (F_1, F_2, F_3)^{\mathrm{T}} = (F_x, F_y, F_z)^{\mathrm{T}}$ with the symbol $.^{\mathrm{T}}$ for transposition. $H^l(\Omega) = H^{l,2}(\Omega)$ for any open set Ω of $\mathbb{R}^n$ are the usual Sobolev spaces of integer order $l \geq 0$. In section 1 we will also introduce the Sobolev spaces $H^l_{\mathcal{P}}$ of $\mathcal{P}$-periodic functions in the plane with exponent of integration 2. If (a, b) is an open interval on the z-axis, then

$$W^k((a,b), H^l_{\mathcal{P}})$$

consists of the mappings $f: (a,b) \to H^l_{\mathcal{P}}$ with derivatives $\partial_z^p \mathrm{D}^q_{xy} f \in L^2((a,b), H^0_{\mathcal{P}}) = L^2((a,b), L^2(\mathcal{P}))$ for any integers $p, q \geq 0$ with $p \leq k$, $q \leq l$ and $p + q \leq \max\{k, l\}$. D^q_{xy} stands for any derivative of order q in the periodic variables x, y. $W^k((a,b), H^l_{\mathcal{P}})$ becomes a Hilbert space in the usual way. A selfadjoint operator A in a Hilbert space $\mathcal{H}$ is called strictly positive definite iff $(Au, u) \geq \gamma \|u\|^2$, $u \in \mathcal{D}(A)$, for some $\gamma > 0$. $C^k([a,b], \mathcal{H})$ is the usual space of k-times continuously differentiable functions on $[a, b]$ with values in the space $\mathcal{H}$.

1. A poloidal-toroidal representation for periodic solenoidal fields in an infinite layer

We want to explain how a solenoidal vector field $\underline{u}$ defined in the three-dimensional layer $L = \mathbb{R}^2 \times (a,b) \subset \mathbb{R}^3$ can be represented in terms of poloidal and toroidal fields

$$P(x,y,z) = \operatorname{curl}\operatorname{curl}\varphi(x,y,z)\underline{k}, \qquad T(x,y,z) = \operatorname{curl}\psi(x,y,z)\underline{k}$$

if $\underline{u}$ and the flux functions φ, ψ are assumed to be periodic with respect to the first two arguments.

For simplicity we restrict ourselves to the case where the lengths of the periods in x and y are equal. Moreover we will deal only with the exponent of integration 2. Thus let us set $\mathcal{P} = (-\pi, \pi)^2$ and consider the Hilbert space $L^2_{\mathcal{P}}$ consisting of all quadratically integrable (complex valued) functions on $\mathcal{P}$ which we will regard to be extended into the whole plane $\mathbb{R}^2$ periodically. Given two functions f, $f_\beta \in L^2_{\mathcal{P}}$ we will call f_β the weak β-th derivative of f *in the sense of periodic distributions in* $\mathbb{R}^2$ iff

$$(1.1) \qquad \int_{\mathcal{P}} f_\beta(x,y)\phi(x,y)\,dx\,dy = (-1)^{|\beta|}\int_{\mathcal{P}} f(x,y)\mathrm{D}^\beta\phi(x,y)\,dx\,dy \quad \forall\phi \in C^\infty_{\mathcal{P}},$$

where

$$C^\infty_{\mathcal{P}} := \left\{\phi \in C^\infty(\mathbb{R}^2, \mathbb{C}) \mid \phi \text{ periodic in } x \text{ and } y \text{ with respect to } \mathcal{P}\right\}$$

denotes the space of the $\mathcal{P}$-periodic testing functions. We will then write $\mathrm{D}^\beta f$ instead of f_β. Further we define the following Sobolev spaces of $\mathcal{P}$-periodic functions:

$$H^m_{\mathcal{P}} := \left\{f \in L^2_{\mathcal{P}} \mid \mathrm{D}^\beta f \in L^2_{\mathcal{P}} \text{ in the sense of (1.1) } \forall|\beta| \leq m\right\}$$

endowed with the norms

$$\|f\|_{\mathcal{P},m} := \Big(\sum_{|\beta|\le m} \big\|D^\beta f\big\|^2_{L^2_{\mathcal{P}}}\Big)^{1/2}, \qquad m \in \mathbb{N}_0.$$

Consider now our main device in such spaces, i.e. the Fourier expansion. Assume

$$f(x,y) = (2\pi)^{-1} \sum_{\kappa\in\mathbb{Z}^2} a_\kappa e^{i\kappa\cdot(x,y)} \quad \text{in } L^2_{\mathcal{P}}.$$

If there exists $D^\beta f \in L^2_{\mathcal{P}}$ in the sense of periodic distributions in $\mathbb{R}^2$ we will infer from (1.1) by using the testing functions $\exp(-i\kappa\cdot(x,y))$ that $a_\kappa(i\kappa)^\beta$ are the Fourier coefficients of $D^\beta f$. Therefore it is easy to see that $\|f\|_{\mathcal{P},m}$ and $\big(|a_0|^2+\sum_{\kappa\neq 0}|a_\kappa|^2|\kappa|^{2m}\big)^{1/2}$ define equivalent norms in $H^m_{\mathcal{P}}$. Especially it will suffice to show the convergence of this series in order to prove $f \in H^m_{\mathcal{P}}$ for some $f \in L^2_{\mathcal{P}}$.

Let $\Omega = \mathcal{P} \times (a,b)$ denote the three dimensional box built over the period rectangle $\mathcal{P}$. For $f \in L^2((a,b), L^2_{\mathcal{P}})$ let

$$\langle f\rangle_z := \fint_{\mathcal{P}} f(x,y,z)\,dx\,dy$$

be the normalized mean value over $\mathcal{P}$. Finally let us introduce $\Delta_2 := \partial_x^2 + \partial_y^2$, the Laplacian in two dimensions in each hyperplane $\mathbb{R}^2\times\{z\}$ of the layer L. This differential operator will arise as $-\langle \operatorname{curl}\operatorname{curl}(\,.\,\underline{k}), \underline{k}\rangle$ when multiplying a poloidal field and the vector $\underline{k}$.

Thus we start with some considerations on this operator.

Proposition 1.1:

a) *Let $f \in L^2_{\mathcal{P}}$. The problem $\Delta_2 u = f$ (in the sense of periodic distributions in $\mathbb{R}^2$) admits a solution $u \in L^2_{\mathcal{P}}$ iff $\langle f\rangle := \fint_{\mathcal{P}} f(x,y)\,dx\,dy = 0$. u is uniquely determined by its mean value $\langle u\rangle$. If $\langle u\rangle = 0$ then the estimate $\|u\|_{\mathcal{P},0} \le \|f\|_{\mathcal{P},0}$ is valid.*

b) *If $f \in H^m_{\mathcal{P}}$, then $u \in H^{m+2}_{\mathcal{P}}$, and u can be estimated via the inequality*

$$\|u\|_{\mathcal{P},m+2} \le c(m)(\|u\|_{\mathcal{P},0} + \|f\|_{\mathcal{P},m}).$$

c) *Let G be a domain in $\mathbb{R}^n$, and $\lambda \mapsto f(\cdot,\lambda) \in C^k(G, H^m_{\mathcal{P}})$ with $\langle f(\cdot,\lambda)\rangle = 0\ \forall\lambda\in G$. Then the zero mean valued solution $u(\cdot,\lambda)$ of $\Delta_2 u(\cdot,\lambda) = f(\cdot,\lambda)\ \forall\lambda$ lies in the space $C^k(G, H^{m+2}_{\mathcal{P}})$, and for all $\lambda\in G$, $|\beta|\le k$, the pointwise estimate*

$$\big\|D^\beta_\lambda u(\cdot,\lambda)\big\|_{\mathcal{P},m+2} \le c(m)\big\|D^\beta_\lambda f(\cdot,\lambda)\big\|_{\mathcal{P},m}$$

holds true.

PROOF:

a) Let u, $f \in L^2_{\mathcal{P}}$ satisfy $\Delta_2 u = f$ in the sense mentioned above. Let $(a_\kappa)_{\kappa\in\mathbb{Z}^2}$ be the Fourier coefficients of u. As explained in the preliminary examination $(-a_\kappa|\kappa|^2)_{\kappa\in\mathbb{Z}^2}$ turn out to be necessarily the Fourier coefficients of f. Thus, $\langle f\rangle = \big(-(2\pi)^{-1}a_\kappa|\kappa|^2\big)_{\kappa=0} = 0$.

Further, a_κ, $\kappa \neq 0$ are uniquely determined by f, whence u will be also unique if f and $a_0 = 2\pi\langle u\rangle$ are prescribed.

On the other hand, if f is given by Fourier coefficients $(b_\kappa)_{\kappa\in\mathbb{Z}^2}$ and $b_0 = 0$, the function

$$u(x,y) = (2\pi)^{-1}\sum_{\kappa\neq 0}\frac{-b_\kappa}{|\kappa|^2}e^{i\kappa\cdot(x,y)} + (2\pi)^{-1}a_0$$

obviously defines the required solution of $\Delta_2 u = f$ with mean value $(2\pi)^{-1}a_0$.
b) As $f \in H^m_{\mathcal{P}}$ means that $\sum_{\kappa\neq 0}|b_\kappa|^2|\kappa|^{2m} < \infty$, the fact $a_\kappa = -b_\kappa/|\kappa|^2$ for $\kappa \neq 0$ implies

$$|a_0|^2 + \sum_{\kappa\neq 0}|a_\kappa|^2|\kappa|^{2m+4} = |a_0|^2 + \sum_{\kappa\neq 0}|b_\kappa|^2|\kappa|^{2m} < \infty,$$

i.e. $u \in H^{m+2}_{\mathcal{P}}$ and the asserted estimate.
c) If $\langle u\rangle = 0$ we get $a_0 = 0$, thus b) changes to $\|u\|_{\mathcal{P},m+2} \leq c(m)\|f\|_{\mathcal{P},m}$. Apply this to the difference $u(\cdot,\lambda) - u(\cdot,\lambda')$ in order to obtain $(\lambda \mapsto u(\cdot,\lambda)) \in C(G, H^{m+2}_{\mathcal{P}})$. Now apply the same estimate to

$$\frac{1}{h}\Big(u(\cdot,\lambda_0 + he_i) - u(\cdot,\lambda_0)\Big) - u_{\lambda_0,i},$$

where $u_{\lambda_0,i}$ shall be the solution to $\mathrm{D}^{e_i}_\lambda f(\cdot,\lambda_0)$ with mean value zero. It follows that $\lambda \mapsto u(\cdot,\lambda)$ admits partial derivatives at the point λ_0. The assertion will then result inductively. □

Remark. In c) we may replace $C^k(G, H^m_{\mathcal{P}})$ by its completion $H^k(G, H^m_{\mathcal{P}})$ with respect to the norm $\|u\| = (\sum_{|\beta|\leq k}\int_G \left\|D^\beta_\lambda u(\cdot,\lambda)\right\|^2_{\mathcal{P},m} d\lambda)^{1/2}$, or by the space $W^k(G, H^m_{\mathcal{P}})$ mentioned in the introduction. In the latter case the pointwise estimate reads

$$\left\|\mathrm{D}^\beta_\lambda u(\cdot,\lambda)\right\|_{\mathcal{P},j+2} \leq c(j)\left\|\mathrm{D}^\beta_\lambda f(\cdot,\lambda)\right\|_{\mathcal{P},j}$$

for all $j \leq m$ and $|\beta| \leq k$ with $|\beta| + j + 2 \leq \max\{k, m+2\}$. In the rest of this section we will use W^k instead of the more natural H^k because W^k is needed in section 2.

If in a) the right hand side f is real valued, then the solution u will also be real valued, provided its prescribed mean value belongs to $\mathbb{R}$, too. In fact, we might as well make use of the Fourier expansion in terms of $\cos(\kappa\cdot(x,y))$ and $\sin(\kappa\cdot(x,y))$, thus we would stay properly in real function spaces. However, this would not be as handy as the present notation.

For shortness we introduce the abbreviations

$$\underline{\varepsilon} := \operatorname{curl}(\,.\,\underline{k}) = \begin{pmatrix}\partial_y\\ -\partial_x\\ 0\end{pmatrix}, \quad \underline{\delta} := \operatorname{curl}\operatorname{curl}(\,.\,\underline{k}) = \begin{pmatrix}\partial_x\partial_z\\ \partial_y\partial_z\\ -\Delta_2\end{pmatrix},$$

these operators are intended to act on functions defined in L.

Using Proposition 1.1 we obtain at once a result for the problem $\Delta_2 u = f$ in the layer L.

Corollary 1.2: *Let $f \in W^k((a,b), H^m_{\mathcal{P}})$ with $\langle f \rangle_z = 0$ $\forall' z \in (a,b)$. Then there exists exactly one $u \in L^1_{\text{loc}}((a,b), L^2_{\mathcal{P}})$ satisfying $\langle u \rangle_z = 0$ $\forall' z \in (a,b)$ and $\Delta_2 u = f$ in the sense of periodic distributions in L, i.e. when using testing functions $\zeta \in C^\infty(L)$, that are periodic in x and y with respect to $\mathcal{P}$ and vanish near the boundary $(\mathbb{R}^2 \times \{a\}) \cup (\mathbb{R}^2 \times \{b\})$. The solution u belongs to $W^k((a,b), H^{m+2}_{\mathcal{P}})$, and the following inequality holds true:*

$$\|\partial_z^\nu u\|_{\mathcal{P},j+2} \leq c(j) \|\partial_z^\nu f\|_{\mathcal{P},j} \tag{1.2}$$

$\forall' z$, $j \leq m$ and $\nu \leq k$ with $\nu + j + 2 \leq \max\{k, m+2\}$. In particular we get $\underline{\varepsilon} u \in W^k((a,b), H^{m+1}_{\mathcal{P}})$. If $k \geq 1$, then $(\underline{\delta} u)_i \in W^{k-1}((a,b), H^{m+1}_{\mathcal{P}})$ for $i = 1,2$, and $(\underline{\delta} u)_3 \in W^k((a,b), H^m_{\mathcal{P}})$. All corresponding norms can be estimated with the aid of (1.2). Moreover $\underline{\varepsilon} u$ and $\underline{\delta} u$ have zero mean values.

PROOF: As $f \in W^k((a,b), H^m_{\mathcal{P}})$ implies that the Fourier coefficients $b_\kappa(z)$ belonging to f lie in $H^k((a,b))$ and that the z-derivatives of f may be calculated termwise, the maintained existence and uniqueness as well as regularity result readily from Proposition 1.1. Expanding $\underline{\varepsilon} u$ and $\underline{\delta} u$ into their Fourier series we see that the coefficients of order zero vanish whence so do their mean values. □

The next lemma reveals the sufficiency of some necessary conditions to the flux functions φ and ψ, which will be derived from an assumed poloidal-toroidal decomposition in the main theorem named after the lemma.

Lemma 1.3: *Let $V = (V_1, V_2, V_3)^T \in L^2((a,b), L^2_{\mathcal{P}})^3$ with $\langle V \rangle_z = 0$ $\forall' z \in (a,b)$ satisfy $\langle \underline{k}, V \rangle = 0$, $\operatorname{div} V = 0$, $\underline{\varepsilon} \cdot V = 0$ in the sense of periodic distributions in L. Then $V = 0$ in $L^2((a,b), L^2_{\mathcal{P}})^3$.*

PROOF: Put $V_3 = 0$ into the presumed condition

$$\int_\Omega \langle V, \nabla \zeta \rangle \, dx\, dy\, dz = 0 = \int_\Omega \langle V, \underline{\varepsilon} \zeta \rangle \, dx\, dy\, dz,$$

$\zeta(x,y,z) = \gamma(z)\phi(x,y)$ with $\gamma \in C_0^\infty(a,b)$, $\phi \in C^\infty_{\mathcal{P}}$, in order to get the equations

$$\int_{\mathcal{P}} (\partial_x \phi(x,y) V_1(x,y,z) + \partial_y \phi(x,y) V_2(x,y,z)) \, dx\, dy = 0 \tag{1.3}$$

$$\int_{\mathcal{P}} (\partial_y \phi(x,y) V_1(x,y,z) - \partial_x \phi(x,y) V_2(x,y,z)) \, dx\, dy = 0 \tag{1.4}$$

for all $\phi \in C^\infty_{\mathcal{P}}$ and almost all $z \in (a,b)$. Choosing in particular $\phi = \partial_x \xi$ in (1.3), $\phi = \partial_y \xi$ in (1.4) and adding (1.3), (1.4) we obtain

$$\int_{\mathcal{P}} \Delta_2 \xi(x,y) V_1(x,y,z) \, dx\, dy = 0 \quad \forall \xi \in C^\infty_{\mathcal{P}}. \tag{1.5}$$

Hence, by means of Proposition 1.1, $V_1(x,y,z) = c_1(z)$ does not depend on x and y. Analogously take $\phi = \partial_y \xi$ in (1.3), $\phi = \partial_x \xi$ in (1.4) and subtract to get (1.5) with V_2 replaced by V_1 whence as above $V_2(x,y,z) = c_2(z)$. Recalling that V should have mean values zero we are set. □

Remark. The zero mean value assumption must not be removed, because otherwise it would be impossible to exclude the case

$$V(x,y,z) = \begin{pmatrix} V_1(z) \\ V_2(z) \\ 0 \end{pmatrix}.$$

In fact, every such V (being smooth enough) serves as an example for a solenoidal field with vanishing third component and $\underline{\varepsilon} \cdot V = 0$, but which is not identical to zero.

The following theorem describes the correct representation of a solenoidal field $\underline{u}$ in the layer $L = \mathbb{R}^2 \times (a,b)$.

Theorem 1.4: *Let $\underline{u} = (u_1, u_2, u_3)^{\mathrm{T}}$ be a vector field defined in the layer L, the components of the field $\underline{u}$ shall satisfy $u_1,\ u_2 \in W^k((a,b), H_{\mathcal{P}}^{m+1})$ and $u_3 \in W^{l+1}((a,b), H_{\mathcal{P}}^n)$, $k,\ l,\ m,\ n \in \mathbb{N}_0$. Assume $\operatorname{div}\underline{u} = 0$ in L. Then there exist vector fields $P,\ T,\ F \in L^2((a,b), L_{\mathcal{P}}^2)^3$ that are determined uniquely by their following properties:*

$$P = \underline{\delta}\varphi, \qquad T = \underline{\varepsilon}\psi$$

(in the sense of periodic distributions in L) with zero mean valued scalar functions φ, $\psi \in L^2((a,b), L_{\mathcal{P}}^2)$,

$$F(x,y,z) = (F_1(z), F_2(z), F_3)$$

independent of x, y, and

(1.6) $$\underline{u} = P + T + F.$$

Especially this implies $F(z) = \langle \underline{u} \rangle_z \in H^k((a,b))^3$. Moreover $\varphi \in W^{l+1}((a,b), H_{\mathcal{P}}^{n+2})$, $\psi \in W^k((a,b), H_{\mathcal{P}}^{m+2})$, and the estimates

$$\|\partial_z^\nu \varphi(\cdot,z)\|_{\mathcal{P},j+2} \le c(j) \|\partial_z^\nu u_3(\cdot,z)\|_{\mathcal{P},j}$$

for $j \le n$, $0 \le \nu \le l+1$ with $\nu + j + 2 \le \max\{l+1, n+2\}$ and

$$\|\partial_z^\nu \psi(\cdot,z)\|_{\mathcal{P},j+2} \le c(j) \|\partial_z^\nu\, \underline{\varepsilon} \cdot \underline{u}(\cdot,z)\|_{\mathcal{P},j}$$

for $j \le m$, $0 \le \nu \le k$ with $\nu + j + 2 \le \max\{k, m+2\}$ are valid. The regularity of P and T is inferred by that of φ and ψ.

Proof: Because $\underline{\delta}\varphi$ and $\underline{\varepsilon}\psi$ have zero mean values (cf. Corollary 1.2), we get $\langle \underline{u} \rangle_z = F(z)$ necessarily. Then $F_3 \equiv \text{const}$ follows from $\operatorname{div}\underline{u} = 0$. Thus let us consider $\underline{u} - F$ instead of $\underline{u}$ and assume now that $F \equiv 0$, i.e. $\underline{u}$ has mean value zero.

It is evident that for any toroidal field $T = \underline{\varepsilon}\psi$ the conditions $\operatorname{div}T = 0$, $\underline{\varepsilon} \cdot T = \Delta_2 \psi$, $\langle \underline{k}, T \rangle = 0$ must hold true. Analogously $\operatorname{div}P = 0$, $\underline{\varepsilon} \cdot P = 0$, $\langle \underline{k}, P \rangle = -\Delta_2 \varphi$ for any poloidal field $P = \underline{\delta}\varphi$. Therefore the claim $\underline{u} = P + T$ implies that $\underline{\varepsilon} \cdot T = \underline{\varepsilon} \cdot \underline{u}$ and $\langle \underline{k}, P \rangle = \langle \underline{k}, \underline{u} \rangle$. Thus by means of Lemma 1.3, P and T have to be unique. Moreover we see that necessarily

(1.7) $$-\Delta_2 \varphi = \langle \underline{k}, \underline{u} \rangle, \quad -\Delta_2 \psi = -\underline{\varepsilon} \cdot \underline{u}$$

for the flux functions φ and ψ. These equations can be solved in a unique way, as indicated by Corollary 1.2. Once that φ and ψ have been determined according to (1.7), we may apply Lemma 1.3 to $\underline{u} - P - T$. (Note that now $\underline{u}$ is assumed to be zero mean valued!) Thus we obtain $\underline{u} - P - T = 0$, i.e. the condition (1.7) is already sufficient for (1.6).

The regularity assertions stated in the theorem are an immediate consequence of Corollary 1.2. □

Remark. It should be emphasized that (1.6) actually defines an orthogonal decomposition where orthogonality is meant to be taken with respect to the inner product

$$(f,g)_z = \int_{\mathcal{P}} f(x,y,z)\overline{g(x,y,z)}\,dx\,dy,$$

for almost all $z \in (a,b)$. For, if $a_\kappa(z)$ denote the Fourier coefficients of φ and $b_\kappa(z)$ those of ψ, then the coefficient vectors of P and T are

$$c_\kappa(z) = \begin{pmatrix} \partial_z a_\kappa(z) i\kappa_1 \\ \partial_z a_\kappa(z) i\kappa_2 \\ -a_\kappa(z)|\kappa|^2 \end{pmatrix} \quad \text{and} \quad d_\kappa(z) = \begin{pmatrix} b_\kappa(z) i\kappa_2 \\ -b_\kappa(z) i\kappa_1 \\ 0 \end{pmatrix},$$

respectively. Thus,

$$\int_{\mathcal{P}} \left\langle P(x,y,z), \overline{T(x,y,z)} \right\rangle dx\,dy = \sum_{\kappa \neq 0} c_\kappa(z)^{\mathrm{T}} \overline{d_\kappa(z)} = 0.$$

2. Applications to the Boussinesq-Equations. Remarks on Energy Stability

In what follows we set $a = -\frac{1}{2}$, $b = \frac{1}{2}$, $\mathcal{P} = (-\frac{\pi}{\alpha}, \frac{\pi}{\alpha})^2$ for some $\alpha > 0$ (the wave-number). This setting turns out to be useful in what follows. It's clear that the results of the first section apply to this situation. The Boussinesq-equations (0.2) over the infinite layer $\mathbb{R}^2 \times (-\frac{1}{2}, \frac{1}{2})$ are usually connected with two types of boundary-conditions, namely:

1st *Case: Rigid boundaries:* $\underline{u} = 0$ at $z = \pm\frac{1}{2}$, $\vartheta = 0$ at $z = \pm\frac{1}{2}$.

2nd *Case: Stress-free boundaries:* $\partial_z u_x = \partial_z u_y = u_z = 0$ at $z = \pm\frac{1}{2}$, $\vartheta = 0$ at $z = \pm\frac{1}{2}$.

Now let $\underline{u} \in W^2((-\frac{1}{2}, \frac{1}{2}), H^2_{\mathcal{P}})^3$ be solenoidal. We decompose it according to Theorem 1.4 into P, T, F. If $\underline{u}$ satisfies one of the boundary-conditions above we obtain

$$F_3 \equiv 0$$

(observe that $\int_{\mathcal{P}} \Delta_2\varphi(x,y,\pm\frac{1}{2})\,dx\,dy = 0$). Exploiting that $\underline{u}$ is solenoidal we arrive in the first case at

$$\begin{aligned} \Delta_2\varphi &= 0 \quad \text{at} \quad z = \pm\tfrac{1}{2}, \\ \partial_z\Delta_2\varphi &= 0 \quad \text{at} \quad z = \pm\tfrac{1}{2}. \end{aligned}$$

This implies $\varphi = \partial_z\varphi = 0$ at $z = \pm\frac{1}{2}$. Thus $\partial_y(\partial_y\psi + F_1) = \partial_y^2\psi = 0$, $\partial_x(-\partial_x\psi + F_1) = -\partial_x^2\psi = 0$ at $z = \pm\frac{1}{2}$ and consequently $\psi = 0$, $F = 0$ at $z = \pm\frac{1}{2}$. In an analogous fashion we obtain in the second case $\varphi = \partial_z^2\varphi = 0$, $\partial_z\psi = 0$, $\partial_z F = 0$ at $z = \pm\frac{1}{2}$. Evidently our conditions on φ, ψ, ϑ and F imply the corresponding ones on $\underline{u}$, ϑ.

In the next step we express the Boussinesq-equations in terms of φ, ψ, ϑ and F. To this end we are needing certain regularity assumptions on $\underline{u}$, ϑ (which imply some regularity for $\nabla\pi$ or π). It's not necessary to state these assumptions here. We will express them in terms of φ, ψ, ϑ, F. In [6] it's shown, amongst other things, that the initial-boundary value problem in φ, ψ, ϑ, F is well posed within appropriate spaces. We will indicate here, however, how one obtains from the system for φ, ψ, ϑ and F the original system (0.2). To be sure of this step is of course an absolute necessity.

From (0.2) it follows that

$$(2.1)\quad \left\{\begin{aligned} \operatorname{curl}(\underline{u}' - \Delta\underline{u} + \underline{u}\cdot\nabla\underline{u} - \sqrt{R}\,\vartheta\underline{k}) &= 0,\\ \operatorname{div}\underline{u} &= 0,\\ \Pr\vartheta' - \Delta\vartheta + \Pr\underline{u}\cdot\nabla\vartheta - \sqrt{R}\,u_z &= 0. \end{aligned}\right.$$

By Lemma 1.3 this is equivalent to

$$(2.2)\quad \left\{\begin{aligned} \big\langle\operatorname{curl}(\underline{u}' - \Delta\underline{u} + \underline{u}\cdot\nabla\underline{u} - \sqrt{R}\,\vartheta\underline{k}), \underline{k}\big\rangle &= 0,\\ \big\langle\operatorname{curl}\operatorname{curl}(\underline{u}' - \Delta\underline{u} + \underline{u}\cdot\nabla\underline{u} - \sqrt{R}\,\vartheta\underline{k}), \underline{k}\big\rangle &= 0,\\ \int_{\mathcal{P}}\operatorname{curl}(\underline{u}' - \Delta\underline{u} + \underline{u}\cdot\nabla\underline{u} - \sqrt{R}\,\vartheta\underline{k})\,dx\,dy &= 0,\\ \operatorname{div}\underline{u} &= 0,\\ \Pr\vartheta' - \Delta\vartheta + \Pr\underline{u}\cdot\nabla\vartheta - \sqrt{R}\,u_z &= 0. \end{aligned}\right.$$

Taking the decomposition

$$\underline{u} = \operatorname{curl}\operatorname{curl}\varphi\underline{k} + \operatorname{curl}\psi\underline{k} + F = \underline{\delta}\varphi + \underline{\varepsilon}\psi + F$$

from Theorem 1.4 we infer from the first two equations in (2.2) that

$$(2.3)\quad \begin{aligned} &(-\Delta)(-\Delta_2)\varphi' + \Delta^2(-\Delta_2)\varphi - \sqrt{R}(-\Delta_2)\vartheta +\\ &+\underline{\delta}\cdot((\underline{\delta}\varphi + \underline{\varepsilon}\psi + F)\cdot\nabla(\underline{\delta}\varphi + \underline{\varepsilon}\psi + F)) = 0, \end{aligned}$$

$$(2.4)\quad \begin{aligned} &(-\Delta_2)\psi' + (-\Delta)(-\Delta_2)\psi -\\ &-\underline{\varepsilon}\cdot((\underline{\delta}\varphi + \underline{\varepsilon}\psi + F)\cdot\nabla(\underline{\delta}\varphi + \underline{\varepsilon}\psi + F)) = 0, \end{aligned}$$

whereas the last one reads

$$(2.5)\quad \Pr\vartheta' - \Delta\vartheta + \Pr(\underline{\delta}\varphi + \underline{\varepsilon}\psi + F)\cdot\nabla\vartheta - \sqrt{R}(-\Delta_2)\varphi = 0.$$

The third equation in (2.2) reads

$$(2.6)\quad \partial_z(F_1') - \partial_z^3 F_1 + (\frac{\alpha}{2\pi})^2\partial_z^2\int_{\mathcal{P}}(-\Delta_2\varphi)(\partial_x\partial_z\varphi + \partial_y\psi)\,dx\,dy = 0,$$

$$(2.7)\quad \partial_z(F_2') - \partial_z^3 F_2 + (\frac{\alpha}{2\pi})^2\partial_z^2\int_{\mathcal{P}}(-\Delta_2\varphi)(\partial_y\partial_z\varphi - \partial_x\psi)\,dx\,dy = 0.$$

Integrating the first and second row in (0.2) over $\mathcal{P}$ we obtain

$$F_1' - \partial_z^2 F_1 + (\frac{\alpha}{2\pi})^2 \int_{\mathcal{P}} \underline{u} \cdot \nabla u_x \, dx\, dy = -(\frac{\alpha}{2\pi})^2 \int_{\mathcal{P}} \partial_x \pi \, dx\, dy, \tag{2.8}$$

$$F_2' - \partial_z^2 F_2 + (\frac{\alpha}{2\pi})^2 \int_{\mathcal{P}} \underline{u} \cdot \nabla u_y \, dx\, dy = -(\frac{\alpha}{2\pi})^2 \int_{\mathcal{P}} \partial_y \pi \, dx\, dy. \tag{2.9}$$

Since

$$\begin{aligned} \int_{\mathcal{P}} \underline{u} \cdot \nabla u_x \, dx\, dy &= \partial_z \int_{\mathcal{P}} (-\Delta_2 \varphi)(\partial_x \partial_z \varphi + \partial_y \psi) \, dx\, dy \\ \int_{\mathcal{P}} \underline{u} \cdot \nabla u_y \, dx\, dy &= \partial_z \int_{\mathcal{P}} (-\Delta_2 \varphi)(\partial_y \partial_z \varphi - \partial_x \psi) \, dx\, dy \end{aligned}$$

the equations (2.6), (2.7) imply that $\int_{\mathcal{P}} \partial_x \pi \, dx\, dy$, $\int_{\mathcal{P}} \partial_y \pi \, dx\, dy$ depend on t only. Thus $\pi = \tilde{\pi} + c(t)^{\mathrm{T}} \cdot \binom{x}{y} + d(t,z)$ where the two-vector c is arbitrary and depends on t only. d is subject to the condition

$$\partial_z \Big(d + (\frac{\alpha}{2\pi})^2 \int_{\mathcal{P}} u_z^2 \, dx\, dy\Big) = \sqrt{R}(\frac{\alpha}{2\pi})^2 \int_{\mathcal{P}} \vartheta \, dx\, dy \tag{2.10}$$

which is implied by integration of the third row in (0.2) over $\mathcal{P}$. $\tilde{\pi}$ is periodic in x, y with respect to $\mathcal{P}$ and fulfills $\int_{\mathcal{P}} \tilde{\pi} \, dx\, dy = 0$.

The system we want to work with is now given by (2.3), (2.4), (2.5) and

$$F' - \partial_z^2 F + (\frac{\alpha}{2\pi})^2 \begin{pmatrix} \int_{\mathcal{P}} \underline{\tilde{u}} \cdot \nabla \tilde{u}_x \, dx\, dy \\ \int_{\mathcal{P}} \underline{\tilde{u}} \cdot \nabla \tilde{u}_y \, dx\, dy \end{pmatrix} = 0. \tag{2.11}$$

We have set $c_1 = c_2 \equiv 0$, and $\underline{\tilde{u}} = \underline{\delta}\varphi + \underline{\varepsilon}\psi$ in the decomposition $\underline{u} = \underline{\delta}\varphi + \underline{\varepsilon}\psi + F$. Observe that $\int_{\mathcal{P}} \underline{u} \cdot \nabla \underline{u} \, dx\, dy = \int_{\mathcal{P}} \underline{\tilde{u}} \cdot \nabla \underline{\tilde{u}} \, dx\, dy$ and that $\underline{\tilde{u}}$ is simply that part of $\underline{u}$ which has mean value 0 over $\mathcal{P}$. The system in question is written now in matrix-form. Set

$$\mathcal{B} = \begin{pmatrix} (-\Delta)(-\Delta_2) & 0 & 0 & 0 & 0 \\ 0 & (-\Delta_2) & 0 & 0 & 0 \\ 0 & 0 & \Pr I & 0 & 0 \\ 0 & 0 & 0 & I & 0 \\ 0 & 0 & 0 & 0 & I \end{pmatrix},$$

$$\mathcal{A} = \begin{pmatrix} \Delta^2(-\Delta_2) & 0 & 0 & 0 & 0 \\ 0 & (-\Delta)(-\Delta_2) & 0 & 0 & 0 \\ 0 & 0 & (-\Delta) & 0 & 0 \\ 0 & 0 & 0 & (-\partial_z^2) & 0 \\ 0 & 0 & 0 & 0 & (-\partial_z^2) \end{pmatrix},$$

$$\mathcal{C} = \begin{pmatrix} 0 & 0 & (-\Delta_2) & 0 & 0 \\ 0 & 0 & 0 & 0 & 0 \\ (-\Delta_2) & 0 & 0 & 0 & 0 \\ 0 & 0 & 0 & 0 & 0 \\ 0 & 0 & 0 & 0 & 0 \end{pmatrix},$$

$$\Phi = \begin{pmatrix} \varphi \\ \psi \\ \vartheta \\ F_1 \\ F_2 \end{pmatrix}.$$

The nonlinear part is denoted by $\mathcal{M}(\Phi)$. Then the system we are going to consider is simply

$$(2.12)\quad \left\{ \begin{array}{rcl} \mathcal{B}\Phi' + \mathcal{A}\Phi - \sqrt{R}\mathcal{C}\Phi + \mathcal{M}(\Phi) &=& 0 \\ \Phi(0) &=& \Phi_0 \end{array} \right.$$

under boundary conditions as indicated in the beginning of this section. To obtain (0.1) from this system we set $\underline{u} = \underline{\delta}\varphi + \underline{\varepsilon}\psi + F$. From the solution which is constructed in [6] it's easily seen that F', $\partial_z^2 F$ possess a further z-derivative for $t > 0$. Then (2.3), (2.4), (2.5), (2.6), (2.7) are at our disposal which are a reformulation of (2.2). (2.2) however yields (2.1), i.e. we obtain $\underline{u}' - \Delta\underline{u} + \underline{u}\cdot\nabla\underline{u} - \sqrt{R}\vartheta\underline{k} = -\nabla\pi$ with a periodic pressure gradient. π is decomposed as before into $\pi = \tilde{\pi} + d$, where d is subject to (2.10).

The system (2.12) is most easily treated within an appropriate Hilbert space H, where $\mathcal{A}$, $\mathcal{B}$ become strictly positive definite selfadjoint operators and $\mathcal{C}$ is hermitian. As Hilbert space H we take

$$H = \mathcal{H}_M \times \mathcal{H}_M \times \mathcal{H} \times \mathcal{H}^1 \times \mathcal{H}^1$$

with $\varphi \in \mathcal{H}_M$, $\psi \in \mathcal{H}_M$, $\vartheta \in \mathcal{H}$, $F_1 \in \mathcal{H}^1$, $F_2 \in \mathcal{H}^1$ in the case of rigid boundaries. Here

$$\begin{array}{rcl} \mathcal{H}_M &=& \left\{\tilde{\varphi} \,\middle|\, \tilde{\varphi} \in W^0((-\frac{1}{2},\frac{1}{2}), L^2_{\mathcal{P}}),\ \int_{\mathcal{P}} \tilde{\varphi}\,dx\,dy = 0\right\}, \\ \mathcal{H} &=& W^0((-\frac{1}{2},\frac{1}{2}), L^2_{\mathcal{P}}), \\ \mathcal{H}^1 &=& \left\{f \,\middle|\, f \in L^2((-\frac{1}{2},\frac{1}{2}))\right\}. \end{array}$$

$\mathcal{H}_M$, $\mathcal{H}$ are made Hilbert spaces in the usual way. For $\mathcal{H}^1$ we choose the inner product

$$(f,g) = (\frac{2\pi}{\alpha})^2 \int_{-\frac{1}{2}}^{\frac{1}{2}} f\cdot\overline{g}\,dz.$$

In the case of stress-free boundaries we take

$$H = \mathcal{H}_M \times \mathcal{H}_M \times \mathcal{H} \times \mathcal{H}^1_M \times \mathcal{H}^1_M$$

with $\mathcal{H}^1_M$ being the closed subspace of $\mathcal{H}^1$ which consists of the f having vanishing mean value over $(-\frac{1}{2},\frac{1}{2})$. Now we can define $\mathcal{A}$, $\mathcal{B}$, $\mathcal{C}$ by defining $A = \Delta^2(-\Delta_2)$, $\tilde{B} = (-\Delta)(-\Delta_2)$ for φ, $B = (-\Delta)(-\Delta_2)$ for ψ and $-\Delta$ for ϑ, $-\partial_z^2$ for F_1, F_2, $-\Delta_2$ for φ, ψ and ϑ as well. Observe that we have two different kinds of operators $(-\Delta)(-\Delta_2)$ in the case of stress-free boundaries.

Definition 2.1: *We expand φ, ψ, ϑ in series*

$$(2.13)\qquad \varphi(x,y,z) = \frac{\alpha}{2\pi}\sum_{\kappa\in\mathbb{Z}^2\setminus\{0\}} a_\kappa(z)e^{i\alpha\kappa\cdot(x,y)},$$

$$(2.14)\qquad \psi(x,y,z) = \frac{\alpha}{2\pi}\sum_{\kappa\in\mathbb{Z}^2\setminus\{0\}} b_\kappa(z)e^{i\alpha\kappa\cdot(x,y)},$$

$$(2.15)\qquad \vartheta(x,y,z) = \frac{\alpha}{2\pi}\sum_{\kappa\in\mathbb{Z}^2} c_\kappa(z)e^{i\alpha\kappa\cdot(x,y)},$$

the series being convergent in $W^0((-\frac{1}{2},\frac{1}{2}), L^2_{\mathcal{P}})$. Set

$$\begin{aligned} A_\kappa &= (\alpha^2|\kappa|^2 - \partial_z^2)^2 \\ &= \alpha^4|\kappa|^4 - 2\alpha^2|\kappa|^2\partial_z^2 + \partial_z^4, \quad \kappa \in \mathbf{Z}^2 \setminus \{0\}, \\ \mathcal{D}(A_\kappa) &= \Big\{ f \;\Big|\; f \in H^4((-\tfrac{1}{2},\tfrac{1}{2})) \textit{ with either } f = \partial_z f = 0 \\ &\qquad \textit{at } z = \pm\tfrac{1}{2} \textit{ or } f = \partial_z^2 f = 0 \textit{ at } z = \pm\tfrac{1}{2} \Big\}. \end{aligned}$$

Then A_κ is a strictly positive definite selfadjoint operator in $L^2((-\frac{1}{2},\frac{1}{2}))$. We define $A = \Delta^2(-\Delta_2)$ on

$$\begin{aligned} \mathcal{D}(A) &= \Big\{ \varphi \;\Big|\; \varphi \in \mathcal{H}_M, \quad \varphi \textit{ is expanded as in (2.13)}, \\ &\qquad a_\kappa \in \mathcal{D}(A_\kappa), \quad \kappa \in \mathbf{Z}^2 \setminus \{0\}, \\ &\qquad \sum_{\kappa \in \mathbf{Z}^2\setminus\{0\}} \int_{-\frac{1}{2}}^{\frac{1}{2}} \alpha^4|\kappa|^4 |A_\kappa a_\kappa|^2\, dz < +\infty \Big\} \end{aligned}$$

by

$$A\varphi = \frac{\alpha}{2\pi} \sum_{\kappa \in \mathbf{Z}^2\setminus\{0\}} \alpha^2|\kappa|^2 A_\kappa a_\kappa e^{i\alpha\kappa\cdot\cdot}.$$

Set

$$\begin{aligned} \tilde{B}_\kappa &= \alpha^2|\kappa|^2 - \partial_z^2, \quad \kappa \in \mathbf{Z}^2 \setminus \{0\}, \\ \mathcal{D}(\tilde{B}_\kappa) &= \Big\{ f \;\Big|\; f \in H^2((-\tfrac{1}{2},\tfrac{1}{2})) \textit{ with } f = 0 \textit{ at } z = \pm\tfrac{1}{2} \Big\}. \end{aligned}$$

Then $\tilde{B}_\kappa$ is a strictly positive definite selfadjoint operator in $L^2((-\frac{1}{2},\frac{1}{2}))$. We define $\tilde{B} = (-\Delta)(-\Delta_2)$ on

$$\begin{aligned} \mathcal{D}(\tilde{B}) &= \Big\{ \varphi \;\Big|\; \varphi \in \mathcal{H}_M, \quad \varphi \textit{ is expanded as in (2.13)}, \\ &\qquad a_\kappa \in \mathcal{D}(\tilde{B}_\kappa), \quad \kappa \in \mathbf{Z}^2 \setminus \{0\}, \\ &\qquad \sum_{\kappa \in \mathbf{Z}^2\setminus\{0\}} \int_{-\frac{1}{2}}^{\frac{1}{2}} \alpha^4|\kappa|^4 \big|\tilde{B}_\kappa a_\kappa\big|^2\, dz < +\infty \Big\} \end{aligned}$$

by

$$\tilde{B}\varphi = \frac{\alpha}{2\pi} \sum_{\kappa \in \mathbf{Z}^2\setminus\{0\}} \alpha^2|\kappa|^2 \tilde{B}_\kappa a_\kappa e^{i\alpha\kappa\cdot\cdot}.$$

Let

$$\begin{aligned} B_\kappa &= \alpha^2|\kappa|^2 - \partial_z^2, \quad \kappa \in \mathbf{Z}^2 \setminus \{0\}, \\ \mathcal{D}(B_\kappa) &= \Big\{ f \;\Big|\; f \in H^2((-\tfrac{1}{2},\tfrac{1}{2})) \textit{ with either } f = 0 \\ &\qquad \textit{at } z = \pm\tfrac{1}{2} \textit{ or } \partial_z f = 0 \textit{ at } z = \pm\tfrac{1}{2} \Big\}. \end{aligned}$$

Then B_κ is a strictly positive definite selfadjoint operator in $L^2((-\frac{1}{2},\frac{1}{2}))$ ($|\kappa| \geq 1$!). We define $B = (-\Delta)(-\Delta_2)$ on

$$\begin{aligned} \mathcal{D}(B) &= \Big\{ \psi \;\Big|\; \psi \in \mathcal{H}_M, \quad \psi \textit{ is expanded as in (2.14)}, \\ &\qquad b_\kappa \in \mathcal{D}(B_\kappa), \quad \kappa \in \mathbf{Z}^2 \setminus \{0\}, \\ &\qquad \sum_{\kappa \in \mathbf{Z}^2\setminus\{0\}} \int_{-\frac{1}{2}}^{\frac{1}{2}} \alpha^4|\kappa|^4 |B_\kappa b_\kappa|^2\, dz < +\infty \Big\} \end{aligned}$$

by

$$B\psi = \frac{\alpha}{2\pi} \sum_{\kappa \in \mathbb{Z}^2 \setminus \{0\}} \alpha^2 |\kappa|^2 B_\kappa b_\kappa e^{i\alpha\kappa \cdot} .$$

It is now obvious how $-\Delta$ is defined for ϑ, i.e. in $\mathcal{H}$, $-\Delta_2$ in $\mathcal{H}_M$ or $\mathcal{H}$, $-\partial_z^2$ in $\mathcal{H}^1$ or $\mathcal{H}_M^1$.

Next we prove

Theorem 2.2: *$\mathcal{A}$, $\mathcal{B}$ are strictly positive definite selfadjoint operators in H.*

PROOF: The assertion is proved by showing that A, B, $\tilde{B}$, $-\Delta$, $-\partial_z^2$ are strictly positive definite selfadjoint operators in the corresponding Hilbert spaces. It's sufficient to do this for A: Either the proofs are very similar to each other or the assertion is well known as in the case of $-\partial_z^2$. As for A it's clear that A is densely defined and hermitian. Now we have to show that

$$(A \pm i)\varphi = f$$

is uniquely solvable in $\mathcal{D}(A)$ for any given $f \in \mathcal{H}_M$. To this end we take the expansion

$$f = \frac{\alpha}{2\pi} \sum_{\kappa \in \mathbb{Z}^2 \setminus \{0\}} f_\kappa e^{i\alpha\kappa \cdot} .$$

and set

$$a_\kappa = \frac{1}{\alpha^2 |\kappa|^2} \Big(A_\kappa \pm \frac{i}{\alpha^2 |\kappa|^2} \Big)^{-1} f_\kappa .$$

It's clear that φ with expansion coefficients a_κ is the required solution. The strict positivity follows from Parseval's equation. □

The choice of the various Hilbert spaces of functions with vanishing mean values corresponds to the invariance properties of the nonlinear terms. For these and other invariance properties see [6, sect. IV]. The norm $\|A\,.\,\|$ is equivalent with the norm of $W^4((-\frac{1}{2},\frac{1}{2}), H^6_{\mathcal{P}})$. Corresponding equivalences hold for the other operators. See [6, sect. III]. The spaces within which we solve (2.12) are now at hand. We are looking for solutions Φ with

$$\Phi \in L^2((0,T), \mathcal{D}(\mathcal{A})), \tag{2.16}$$

$$\Phi' \in L^2((0,T), \mathcal{D}(\mathcal{B})), \tag{2.17}$$

and, as a result of interpolation,

$$\nabla \mathcal{B}\Phi \in C^0([0,T], H) \tag{2.18}$$

where ∇ refers to each component of $\mathcal{B}\Phi$. In particular Φ_0 is required to fulfill $\|\nabla\mathcal{B}\Phi_0\| < +\infty$. This construction is carried through in [6, sect. IV]. In general T is not allowed to exceed a maximal finite value, unless $\Pr = +\infty$ or the solution represents a convection roll. Imposing compatibility conditions the regularity behaviour near $t = 0$ can be studied.

When constructing the solution with properties (2.16), (2.17), (2.18) one is faced with a characteristic difficulty in the case of rigid boundaries. Δ^2 in the first row of

(2.12) is no longer the square in the operator-theoretical sense of $(-\Delta)$ in front of $(-\Delta_2)\varphi'$ as it is for stress-free boundaries. Therefore the range $\Pr \in [\frac{1}{2}, 2]$ has to be excluded. This difficulty can be removed if one considers a somewhat weaker form of solutions as is done in [6, sect. V].

Now we want to consider the energy-equality for solutions with properties (2.16), (2.17), (2.18). It reads

$$\frac{\mathrm{d}}{\mathrm{d}t}\left\|\mathcal{B}^{\frac{1}{2}}\Phi(t)\right\|^2 + 2\left\|\mathcal{A}^{\frac{1}{2}}\Phi(t)\right\|^2 - 2\sqrt{R}(\mathcal{C}\Phi(t), \Phi(t)) = 0$$

with $\left\|\mathcal{B}^{\frac{1}{2}}\Phi(t)\right\|$ as (kinetic) energy at time t. It's therefore interesting to ask for

$$\sqrt{R_{\min}(\alpha^2)} = \inf_{\Phi\in\mathcal{D}(\mathcal{A})\setminus\{0\}} \frac{(\mathcal{A}\Phi, \Phi)}{|(\mathcal{C}\Phi, \Phi)|}$$

in dependence on α^2. If $0 < R \leq R_{\min}(\alpha^2)$ then the energy can be estimated a-priori independently of R and is monotonically non increasing or even decays exponentially (if $R < R_{\min}(\alpha^2)$). This variational problem can be attacked in the way that one wants to find

$$\inf \frac{\|\nabla \underline{u}\|^2 + \|\nabla\vartheta\|^2}{2|(u_z, \vartheta)|}$$

where the infimum has to be taken over $(\underline{u}, \vartheta)$ with $\underline{u} \neq 0$, $\vartheta \neq 0$, $\nabla\cdot\underline{u} = 0$, $\underline{u}$, ϑ periodic with respect to any rectangle. The condition $\nabla \cdot \underline{u} = 0$ is covered by introducing a Lagrange-multiplier. Then the Euler-Lagrange equations are considered. After solving them one is confronted with the problem to show that the solution is a minimizer of the functional under consideration on suitable subclasses of the admissible $\underline{u}$, ϑ. Thus it seems to be easier to start with Courant's classical method of finding the eigenvalues of a compact selfadjoint operator. Here we only prove an estimate from below for $(\mathcal{A}\Phi, \Phi)$ which turns out to be sharp in the case of stress-free boundaries. We have

Proposition 2.3: *For $\Phi \in \mathcal{D}(\mathcal{A})$ the following estimate holds:*

$$(\mathcal{A}\Phi, \Phi) \geq \min_{\kappa\in\mathbb{Z}^2\setminus\{0\}} \sqrt{R_i(\alpha^2|\kappa|^2)} \cdot |(\mathcal{C}\Phi, \Phi)|, \quad i = 0, 1, \tag{2.19}$$

with

$$R_0(\alpha^2) = \frac{(\alpha^2 + \pi^2)^3}{\alpha^2}, \quad \alpha > 0,$$

in the case of stress-free boundaries, and

$$R_1(\alpha^2) = \frac{(\alpha^2 + \pi^2)}{\alpha^2}(\alpha^4 + \lambda(\alpha^2))$$

in the case of rigid boundaries. Here $\lambda(\alpha^2)$ is the smallest eigenvalue of $\partial_z^4 - 2\alpha^2\partial_z^2$ in $L^2((-\frac{1}{2}, \frac{1}{2}))$ under boundary conditions $f = \partial_z f = 0$ at $z = \pm\frac{1}{2}$. In the case of stress-free boundaries we have

$$R_{\min}(\alpha^2) = \min_{\kappa\in\mathbb{Z}^2\setminus\{0\}} R_0(\alpha^2|\kappa|^2).$$

PROOF: If we take Parseval's equation for $(\mathcal{A}\Phi, \Phi)$ and $(\mathcal{C}\Phi, \Phi)$ together with the expansions (2.13), (2.15) the estimate (2.19) is easily shown. One only has to use the extremal property of the smallest eigenvalue of $-\partial_z^2$, $(-\partial_z^2)^2$ under Dirichlet-0-conditions (which are π^2, π^4) and of $\partial_z^4 - 2\alpha^2\partial_z^2$ under boundary conditions $f = \partial_z f = 0$ at $z = \pm\frac{1}{2}$. It's easily seen that $\min_{\kappa\in\mathbb{Z}^2\setminus\{0\}} R_i(\alpha^2|\kappa|^2)$ is assumed for some κ_i, $i = 0, 1$. In the case of stress-free boundaries R_0 assumes its minimal value $R_c = 27\pi^4/4$ in $\alpha_c = \pi/\sqrt{2}$. Thus $\min_{\kappa\in\mathbb{Z}^2\setminus\{0\}} R_0(\alpha^2|\kappa|^2) = R_0(\alpha^2)$ for $\alpha \geq \alpha_c$. The functional $(\mathcal{A}\Phi, \Phi)/(\mathcal{C}\Phi, \Phi)$ assumes the value $\sqrt{R_0(\alpha^2)}$ in $\Phi = (\varphi, 0, \vartheta, 0, 0)^{\mathrm{T}}$ with $\varphi(x, y, z) = \cos\alpha x \cos\pi z$ and $\vartheta = (\alpha^2 + \pi^2)^{\frac{1}{2}}\alpha\varphi$. The situation is different for $\alpha \in (0, \alpha_c)$, since R_0 is monotonically decreasing on $(0, \alpha_c]$. If $\kappa_0 = (\kappa_{01}, \kappa_{02})$ and if $R_0(\alpha^2|\kappa_0|^2)$ is minimal, then $(\mathcal{A}\Phi, \Phi)/(\mathcal{C}\Phi, \Phi)$ assumes $\sqrt{R_0(\alpha^2|\kappa_0|^2)}$ in $\Phi = (\varphi, 0, \vartheta, 0, 0)^{\mathrm{T}}$ with $\varphi(x, y, z) = \cos\alpha\kappa_{01}x \cos\alpha\kappa_{02}y \cos\pi z$ and $\vartheta = (\alpha^2|\kappa_0|^2 + \pi^2)^{\frac{1}{2}}\alpha|\kappa_0|\varphi$. The assertion is proved. □

Anything what was said before remains true if the periodicity cell $(-\frac{\pi}{\alpha}, \frac{\pi}{\alpha})^2$ is replaced by a rectangle $(-\frac{\pi}{\alpha}, \frac{\pi}{\alpha}) \times (-\frac{\pi}{\beta}, \frac{\pi}{\beta})$, with the single modification that $\alpha^2|\kappa|^2$ has to be replaced by $\alpha^2\kappa_1^2 + \beta^2\kappa_2^2$. Thus, if one wants that the energy is monotonically non increasing for any Φ being periodic in x, y with respect to a rectangle, one needs $R \leq R_c$ in the case of stress-free boundaries.

References

[1] BACKUS, G.: A Class of Self-Sustaining Dissipative Spherical Dynamos. *Ann. Physics* **4** (1958), 372–447.

[2] BACKUS, G.: Poloidal and Toroidal Fields in Geomagnetic Field Modeling. *Rev. Geophys.* **24** (1986), 75–109.

[3] CLEVER, R.M., BUSSE, F.H.: Three-dimensional knot convection in a layer heated from below. *J. Fluid Mech.* **198** (1989), 345–363.

[4] IOOSS, G.: Théorie Non Linéaire de la Stabilité des Ecoulements Laminaires dans le Cas de "l'Echange des Stabilités". *Arch. Rational Mech. Anal.* **40** (1971), 166–208.

[5] JOSEPH, D.D.: *Stability of Fluid Motions, Vol. I.* Springer Tracts in Natural Philosophy **27**. Springer: Berlin, Heidelberg, New York (1976).

[6] WAHL, W. VON: The Boussinesq-Equations in Terms of Poloidal and Toroidal Fields and the Mean Flow. Lecture Notes. To appear in *Bayreuth. Math. Schr.*

This paper is in final form and no similar paper has been or is being submitted elsewhere.

Eddy Solutions of the Navier-Stokes Equations

Owen Walsh

We have recently noticed a class of explicit solutions to the Navier-Stokes equations, periodic in both space variables. The solutions are exponentially decaying and self-similar in the sense that the stream lines are constant in time. Figure 1 represents the stream lines in one cell of one particularly striking flow. The class of solutions is a result of an observation, Lemma 1, concerning eigenfunctions of the Stokes operator with corresponding zero pressure. Through a survey paper [1] of R. Berker, we were led to a 1923 paper [2] of G.I. Taylor in which he remarks on this class. (We would like to thank R. Finn for informing us of [1]). Taylor considered and sketched a simple example which, for historical interest, we have included as figure 2. The complex flows we noticed are a result of special eigenfunctions which are described below.

Figure 1: $\psi = (1/4)\cos(3x)\sin(4y) - (1/5)\cos(5y) - (1/5)\sin(5x)$

Consider the Navier-Stokes equations in a simply connected domain Ω. That is, for $\mathbf{x} \in \Omega$ and $t > 0$,

$$\mathbf{u}_t + \mathbf{u} \cdot \nabla \mathbf{u} = \nu \Delta \mathbf{u} - \nabla p \, , \; \nabla \cdot \mathbf{u} = 0, \tag{1}$$

with $\mathbf{u} = \mathbf{a}$ initially.

Lemma 1 *Suppose* $\mathbf{a}$ *satisfies*

$$\Delta \mathbf{a} = \lambda \mathbf{a} \, , \; \nabla \cdot \mathbf{a} = 0 \text{ in } \Omega. \tag{2}$$

Then $\mathbf{u} = e^{\nu \lambda t} \mathbf{a}$ *satisfies (1) with pressure* p *such that* $\nabla p = -\mathbf{u} \cdot \nabla \mathbf{u}$.

Proof: $\mathbf{u}$ is divergence free since $\mathbf{a}$ is. Also, $\mathbf{u}_t = \nu \lambda \mathbf{u} = \nu \Delta \mathbf{u}$ and hence we need only show that the nonlinear term is a gradient. This is equivalent to showing

$$\frac{\partial}{\partial y}\left(u_1 \frac{\partial u_1}{\partial x} + u_2 \frac{\partial u_1}{\partial y} \right) = \frac{\partial}{\partial x}\left(u_1 \frac{\partial u_2}{\partial x} + u_2 \frac{\partial u_2}{\partial y} \right),$$

which follows straightforwardly from (2).

We remark that $\mathbf{u}$ also satisfies the (time dependent) Stokes equations, the linearization of (1), with $p \equiv 0$.

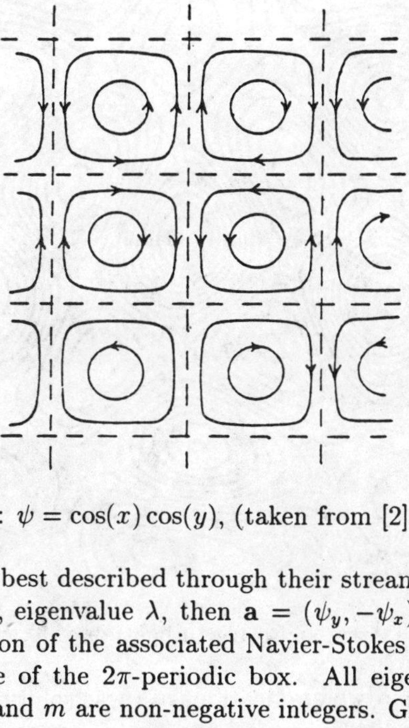

Figure 2: $\psi = \cos(x)\cos(y)$, (taken from [2], pg. 674)

The periodic flows are best described through their stream function. If ψ is an eigenfunction of the Laplacian, eigenvalue λ, then $\mathbf{a} = (\psi_y, -\psi_x)$ satisfies (2), same λ, and $e^{\nu \lambda t} \psi$ is the stream function of the associated Navier-Stokes (or Stokes) flow. Consider eigenfunctions in the case of the 2π-periodic box. All eigenvalues λ are of the form $\lambda = -(n^2 + m^2)$ where n and m are non-negative integers. Given n and m,

$$\cos(nx)\cos(my), \; \sin(nx)\sin(my), \; \cos(nx)\sin(my), \; \sin(nx)\cos(my)$$

are linearly independent eigenfunctions. Define linear combinations of these as (n, m)-eigenfunctions. Simple flows, such as Taylor's, consisting of eddies arranged in regular rectangular patterns arise when ψ is one of these (n, m)-eigenfunction. The more complex flows arise if for a given λ we have a larger basis of eigenfunctions. For instance if $n \neq m$ then we can choose ψ as a linear combination of an (n, m)-eigenfunction and an (m, n)-eigenfunction and the resulting flow does not have rectangular symmetry as in simple flows. Suppose $\lambda = -(n^2 + m^2) = -(k^2 + l^2)$ where n and m are both different from k and l. In this case we have an even larger basis of eigenfunctions and ψ can be a linear combinations of (n, m), (k, l)-eigenfunctions and their counterparts with n, m and k, l reversed. In figure 1, ψ is such an eigenfunction associated with the eigenvalue $\lambda = -25 = -(3^2 + 4^2) = -5^2$. One can continue to look for more complicated flows by finding λ with an even larger basis of eigenfunctions. This happens if one knows integers which can be written as a sum of squares in many different ways. In fact, it can be shown (see [5] for example) that integers of the form p^{2n} and p^{2n+1}, where p is prime and $p \equiv 1 \pmod 4$, can be written as a sum of squares in exactly $n + 1$ ways. For instance, $625 = 25^2 = 24^2 + 7^2 = 20^2 + 15^2$ and figure 3 is the stream lines of the flow when $\psi = \sin(25x) + \cos(25y) - \sin(24x)\cos(7y) + \cos(15x)\cos(20y) - \cos(7x)\sin(24y)$. We would like to point out that figure 3 is not one whole cell. We were forced to magnify a small region because of the intense concentration of small eddies.

Figure 3: ψ with $\lambda = -625$, (magnification of $[0, \pi/4] \times [0, \pi/4]$)

In conclusion, we mention two applications of these solutions. The first is numerical. Since periodic boundary conditions can be, and have been, implemented in many numerical schemes for both the Navier-Stokes and the Stokes equations, these provide a rich class of exact solutions to test these schemes' accuracy with. The second application is to the study of invariant manifolds. Foias and Saut, in [3] and [4], proved the existence of invariant manifolds M_k for the Navier-Stokes equations. The invariant manifold M_k

can be loosely described as the set of solutions which decay like $e^{-\lambda_k t}$ as t goes to infinity where $-\lambda_k$ is the k-th eigenvalue of the Stokes operator. Hence, for the periodic box, we have further explicit examples of solutions in M_k for each k.

I would like to thank Dr. J. Heywood for his encouragement and early interest in these examples.

References

[1] R. BERKER, Intégration des equations d'un fluide visqueux incompressible, *Handbuch der Physik* , Vol. 8/2, Springer-Verlag, Berlin, 1963, pp. 1-384.

[2] G.I. Taylor, On the decay of vortices in a viscous fluid., *Phil. Mag.*, (6) **46**, (1923), pp. 671-674

[3] C. FOIAS & J.C. SAUT, Asymptotic behavior, as $t \to +\infty$ of solutions of Navier-Stokes equations and nonlinear spectral manifolds, *Ind. Univ. Math. J.*, **33**, No. 3 (1984), 459-477.

[4] C. FOIAS & J.C. SAUT, On the smoothness of the nonlinear manifolds associated to the Navier-Stokes equations, *Ind. Univ. Math. J.*, **33**, No. 6 (1984), 911-926.

[5] L.H. HUA, *Introduction to Number Theory*, Springer-Verlag, Berlin, Heidelberg, New York, 1982.

On a Three-Norm Inequality for the Stokes Operator in Nonsmooth Domains

WENZHENG XIE

School of Mathematics, University of Minnesota
Minneapolis, MN 55455, U.S.A.

Introduction. Let Ω denote an arbitrary open set in $\mathbb{R}^3$, $C_{0*}^{\infty}(\Omega)$ denote the set of all smooth solenoidal vector fields with compact support in Ω, $\|\cdot\|$ denote the $L^2(\Omega)$ norm, $L_*^2(\Omega)$ (resp. $\hat{H}_{0*}^1(\Omega)$) denote the completion of $C_{0*}^{\infty}(\Omega)$ in norm $\|\cdot\|$ (resp. $\|\nabla\cdot\|$). The Stokes operator $\tilde{\Delta}: \hat{H}_{0*}^1(\Omega) \to L_*^2(\Omega)$ is defined by the following equation:

$$\int_{\Omega} \nabla u \cdot \nabla v \, dx = -\int_{\Omega} \tilde{\Delta} u \cdot v \, dx, \qquad \forall\, v \in C_{0*}^{\infty}(\Omega).$$

The domain of definition of $\tilde{\Delta}$ is a subspace of $\hat{H}_{0*}^1(\Omega)$ denoted by $D(\tilde{\Delta}, \Omega)$.

In this paper we try to establish the inequality

$$\sup_{\Omega} |u|^2 \le c\|\nabla u\| \|\tilde{\Delta} u\| \tag{1}$$

for any open set Ω and any $u \in D(\tilde{\Delta}, \Omega)$.

Ineq.(1) is used to estimate the trilinear form $\int_{\Omega} u \cdot \nabla u \cdot \tilde{\Delta} u \, dx$ in the theory of the Navier–Stokes equations (see e.g. [2, p.299] and [4, p.12]). Known proof of Ineq.(1) is based on the Cattabriga–Solonnikov estimate $\|u\|_{H^2(\Omega)} \le c_{\Omega} \|\tilde{\Delta} u\|$, which holds for bounded smooth domains, with constant c_{Ω} depending on the boundary curvature. It can be shown that $c_{\Omega} \to \infty$ as Ω tends to any domain with a reentrant corner.

It was suggested to the author by J. G. Heywood that Ineq.(1) should be valid in nonsmooth domains, in order to obtain a regularity theory for the Navier–Stokes equations in nonsmooth domains, using the methods of [1, 2]. The author has developed a new method and proved an analogous inequality for the Laplacian operator in arbitrary domains, and applied the inequality to the Burgers equation (which is a good model for the Navier–Stokes equations) [5, 6].

In this paper, we basically carry out the proceedure of [6] and reduce the proof of Ineq.(1) to an L^2 bound of Green's function for the spectral Stokes equations. We show that if the bound holds, then Ineq.(1) holds for any open set Ω and any $u \in D(\tilde{\Delta}, \Omega)$, with a sharp constant $c = \frac{1}{3\pi}$.

Lemma 1. *If Ineq.(1) holds with a fixed constant for every bounded smooth open set, then it also holds with the same constant for every open set.*

Proof. Let Ω be an arbitrary open set in $\mathbb{R}^3$ and let $u \in D(\tilde{\Delta}, \Omega)$. We can choose a sequence of bounded open sets Ω_n with smooth boundaries, such that $\Omega_1 \subset \Omega_2 \subset \cdots$, and $\bigcup_{n=1}^{\infty} \Omega_n = \Omega$. For each $n \geq 1$, by the Riesz Representation Theorem, there exists a unique $u_n \in \hat{H}^1_{0*}(\Omega_n)$ such that

$$\int_{\Omega_n} \nabla u_n \cdot \nabla v \, dx = \int_{\Omega_n} \nabla u \cdot \nabla v \, dx \,, \qquad \forall\, v \in \hat{H}^1_{0*}(\Omega_n) \,. \tag{2}$$

By letting $v = u_n$ and using the Schwarz inequality, we get $\|\nabla u_n\|_{L^2(\Omega_n)} \leq \|\nabla u\|_{L^2(\Omega_n)}$. By the definition of the Stokes operator, from (2) we obtain $\tilde{\Delta} u_n = (\tilde{\Delta} u)|_{\Omega_n}$. It follows from the assumption of the lemma that

$$\begin{aligned} \sup_{\Omega_n} |u_n|^2 &\leq c \|\nabla u_n\|_{L^2(\Omega_n)} \|\tilde{\Delta} u_n\|_{L^2(\Omega_n)} \\ &\leq c \|\nabla u\| \, \|\tilde{\Delta} u\| \,. \end{aligned} \tag{3}$$

Setting u_n equal to zero in $\Omega \backslash \Omega_n$, we get $u_n \in \hat{H}^1_{0*}(\Omega)$. From (2) we have

$$\lim_{n \to \infty} \int_{\Omega} \nabla u_n \cdot \nabla v \, dx = \int_{\Omega} \nabla u \cdot \nabla v \, dx \,, \qquad \forall\, v \in C^{\infty}_{0*}(\Omega) \,.$$

This and $\|\nabla u_n\| \leq \|\nabla u\|$ imply that $u_n \to u$ in $\hat{H}^1_{0*}(\Omega)$. Therefore, by a well-known Sobolev inequality (see [4]), we have

$$\int_{\Omega} |u_n - u|^6 \, dx \leq 48 \left(\int_{\Omega} |\nabla u_n - \nabla u|^2 \, dx \right)^3 \to 0 \quad \text{as} \quad n \to \infty. \tag{4}$$

Since $u \in D(\tilde{\Delta}, \Omega)$, u is continuous in Ω. Now, if (1) were not true, then there would be some $x_0 \in \Omega$ such that

$$|u(x_0)|^2 > c \|\nabla u\| \, \|\tilde{\Delta} u\| \geq \sup_{\Omega} |u_n|^2$$

holds for all n, which is obviously contradictory to (4). **Q.E.D.**

Hereafter we assume that Ω is a bounded open set with C^∞ boundary, unless explicitly stated to the contrary. Since Ω is bounded, by the Poincaré inequality, we have $\hat{H}^1_{0*}(\Omega) \subset L^2_*(\Omega)$ and hence $D(\tilde{\Delta}, \Omega) \subset L^2_*(\Omega)$. It is well known (see e.g. [4]) that the eigenfunctions of the Stokes operator can be chosen to form an orthonormal basis of $L^2_*(\Omega)$. They satisfy

$$\begin{aligned} \tilde{\Delta} \varphi_n &= -\lambda_n \varphi_n, \\ \varphi_n|_{\partial\Omega} &= 0, \end{aligned}$$

with $\lambda_n > 0$, $n = 1, 2, \ldots$.

Lemma 2. *If Ineq.(1) holds with a fixed constant for every finite linear combination of $\{\varphi_n\}$, then it also holds for every $u \in D(\widetilde{\Delta}, \Omega)$.*

Proof. Suppose that $u \in D(\widetilde{\Delta}, \Omega)$. For any positive integer m, let

$$u_m = \sum_{n=1}^{m} c_n \varphi_n, \quad \text{where } c_n = \int_\Omega u \cdot \varphi_n \, dx.$$

Integrating by parts, one obtains

$$\int_\Omega \nabla u \cdot \nabla u_m \, dx = \|\nabla u_m\|^2$$

and

$$\int_\Omega \widetilde{\Delta} u_m \cdot \widetilde{\Delta} u \, dx = \|\widetilde{\Delta} u_m\|^2.$$

Using the Schwarz inequality, we obtain $\|\nabla u_m\| \leq \|\nabla u\|$ and $\|\widetilde{\Delta} u_m\| \leq \|\widetilde{\Delta} u\|$. Hence

$$\begin{aligned} \sup_\Omega |u_m|^2 &\leq c\|\nabla u_m\| \, \|\widetilde{\Delta} u_m\| \\ &\leq c\|\nabla u\| \, \|\widetilde{\Delta} u\| \,. \end{aligned}$$

Since

$$\lim_{m \to \infty} \int_\Omega |u_m - u|^2 \, dx = 0,$$

we obtain (1) by reasoning similarly as in Lemma 1. **Q.E.D.**

Lemma 3. *For any $y \in \Omega$, any constant vector e^k and any positive integer m, there exists some $\mu > 0$ such that the inequality*

$$\frac{(e^k \cdot u(y))^2}{\|\nabla u\| \|\widetilde{\Delta} u\|} \leq 4\sqrt{\mu} \sum_{n=1}^{m} \left(\frac{e^k \cdot \varphi_n(y)}{\lambda_n + \mu} \right)^2 \tag{5}$$

holds for all functions of the form

$$u(x) = \sum_{n=1}^{m} c_n \varphi_n(x),$$

where $c_1, \cdots, c_m$ are real numbers not all equal to zero.

Remark. The use of $e^k \cdot u(y)$ instead of $|u(y)|$, is crucial here to obtaining a sharp estimate.

Proof. We have

$$\frac{(e^k \cdot u(y))^2}{\|\nabla u\| \|\widetilde{\Delta} u\|} = \frac{\left(\sum_{n=1}^{m} c_n e^k \cdot \varphi_n(y) \right)^2}{\left(\sum_{n=1}^{m} \lambda_n c_n^2 \right)^{1/2} \left(\sum_{n=1}^{m} \lambda_n^2 c_n^2 \right)^{1/2}} \,. \tag{6}$$

This quotient is a smooth and homogeneous function of $(c_1, \cdots, c_m)$ in $\mathbb{R}^m \backslash \{0\}$. Hence it attains its maximum value at some critical point $(\overline{c}_1, \cdots, \overline{c}_m)$, i.e., when the function is $\overline{u} = \sum_{n=1}^m \overline{c}_n \varphi_n$. Without loss of generality, assume $\overline{u}(y) \neq 0$. Differentiating

$$\log \frac{(e^k \cdot u(y))^2}{\|\nabla u\| \|\tilde{\Delta} u\|} \equiv \log(e^k \cdot u(y))^2 - \frac{1}{2} \log \|\nabla u\|^2 - \frac{1}{2} \log \|\tilde{\Delta} u\|^2$$

with respect to c_n at the critical point, we get

$$\frac{2e^k \cdot \varphi_n(y)}{e^k \cdot \overline{u}(y)} - \frac{\lambda_n \overline{c}_n}{\|\nabla \overline{u}\|^2} - \frac{\lambda_n^2 \overline{c}_n}{\|\tilde{\Delta} \overline{u}\|^2} = 0,$$

for $n = 1, \cdots, m$. Letting $\mu = \|\tilde{\Delta} \overline{u}\|^2 / \|\nabla \overline{u}\|^2$, we obtain

$$\frac{2e^k \cdot \varphi_n(y)}{\lambda_n + \mu} = \frac{e^k \cdot \overline{u}(y)}{\|\tilde{\Delta} \overline{u}\|^2} \lambda_n \overline{c}_n.$$

Squaring and summing over n on both sides, we obtain

$$4 \sum_{n=1}^m \left(\frac{e^k \cdot \varphi_n(y)}{\lambda_n + \mu} \right)^2 = \left(\frac{e^k \cdot \overline{u}(y)}{\|\tilde{\Delta} \overline{u}\|^2} \right)^2 \sum_{n=1}^m \lambda_n^2 \overline{c}_n^2 = \left(\frac{e^k \cdot \overline{u}(y)}{\|\tilde{\Delta} \overline{u}\|} \right)^2.$$

Therefore the maximum value of the quotient (6) is

$$\frac{(e^k \cdot \overline{u}(y))^2}{\|\nabla \overline{u}\| \|\tilde{\Delta} \overline{u}\|} = \frac{\|\tilde{\Delta} \overline{u}\|}{\|\nabla \overline{u}\|} \left(\frac{e^k \cdot \overline{u}(y)}{\|\tilde{\Delta} \overline{u}\|} \right)^2 = 4\sqrt{\mu} \sum_{n=1}^m \left(\frac{e^k \cdot \varphi_n(y)}{\lambda_n + \mu} \right)^2.$$

Thus Lemma 3 is proved. **Q.E.D.**

At this point we introduce the following spectral Stokes equations:

$$\begin{aligned} (\Delta - \mu) U^k(x, y, \mu) + \nabla P^k(x, y, \mu) &= \delta(x - y) e^k, \\ \operatorname{div} U^k(x, y, \mu) &= 0, \\ U^k(x, y, \mu)|_{x \in \partial\Omega} &= 0. \end{aligned} \tag{7}$$

Here all derivatives are with respect to x, and δ denotes the Dirac distribution. It is easy to prove the following eigenfunction expansion of the Green's function $U^k(\cdot, y, \mu)$:

$$U^k(x, y, \mu) = -\sum_{n=1}^{\infty} \frac{e^k \cdot \varphi_n(y)}{\lambda_n + \mu} \varphi_n(x).$$

Our purpose is to to bound the right hand side of (5) by using the Parseval's equality

$$\int_\Omega |U^k(x, y, \mu)|^2 \, dx = \sum_{n=1}^{\infty} \left(\frac{e^k \cdot \varphi_n(y)}{\lambda_n + \mu} \right)^2 \tag{8}$$

It is easy to verify that

$$\begin{aligned} u^k(x, y, \mu) &= -\frac{e^{-\sqrt{\mu}|x-y|}}{4\pi|x - y|} e^k + (e^k \cdot \nabla) \nabla \frac{e^{-\sqrt{\mu}|x-y|} - 1}{4\pi\mu|x - y|}, \\ p^k(x, y) &= -e^k \cdot \nabla \frac{1}{4\pi|x - y|}, \end{aligned}$$

is the Green's function in $\mathbb{R}^3$, and that

$$\int_{\mathbb{R}^3} |u^k(x,y,\mu)|^2\,dx = \frac{1}{12\pi\sqrt{\mu}}. \tag{9}$$

We propose the following

Conjecture. *For any bounded open set $\Omega \subset \mathbb{R}^3$ with C^∞ boundary, the inequality*

$$\int_\Omega |U^k(x,y,\mu)|^2\,dx \le \int_{\mathbb{R}^3} |u^k(x,y,\mu)|^2\,dx \tag{10}$$

holds for all $y \in \Omega$.

REMARK. By scaling, it is easy to see that if the Conjecture is true for one $\mu > 0$ (and every Ω), then it is also true for every $\mu > 0$. This conjecture is of independent interest. One may also consider complex values of μ. In [6] the counterpart of Ineq.(10) was proved by using a maximum principle.

Main Result. *If the above Conjecture is true, then the inequality*

$$\sup_\Omega |u|^2 \le \frac{1}{3\pi}\|\nabla u\|\,\|\tilde{\Delta} u\| \tag{11}$$

holds for every open set $\Omega \subset \mathbb{R}^3$ and every $u \in D(\tilde{\Delta},\Omega)$. The constant $\frac{1}{3\pi}$ is the best possible.

PROOF. Combining (5), (8), (9) and (10), we obtain Eq.(11) for every finite linear combination of $\{\varphi_n\}$. Then it follows from Lemma 2 and Lemma 1 that (11) also holds for every open set $\Omega \subset \mathbb{R}^3$ and every $u \in D(\tilde{\Delta},\Omega)$.

When $\Omega = \mathbb{R}^3$ and

$$u = \frac{1-e^{-|x|}}{|x|}e^k - (e^k\cdot\nabla)\nabla\left(\frac{|x|}{2} + \frac{1-e^{-|x|}}{|x|}\right),$$

the equality occurs in (11). For an arbitrary open set Ω, one can use a sequence of vector fields obtained by cutting-off and scaling u to show that the constant is the best possible. Q.E.D.

Acknowlegement. The author wishes to thank Professor J. G. Heywood for his very generous and helpful advice, and to thank Professor L. Rosen for helpful discussions.

REFERENCES

1. J. G. Heywood, *The Navier–Stokes equations: on the existence, regularity and decay of solutions*, Indiana Univ. Math. J. **29** (1980), 639–681.

2. J. G. Heywood and R. Rannacher, *Finite element approximation of the nonstationary Navier–Stokes problem. I. Regularity of solutions and second-order error estimates for spatial discretization*, SIAM J. Numer. Anal. **19** (1982), 275–311.

3. O. A. Ladyzhenskaya, *The Mathematical Theory of Viscous Incompressible Flow*, Second Edition, Gordon and Breach, New York, 1969.

4. R. Temam, *Navier–Stokes Equations and Nonlinear Functional Analysis*, SIAM, Philadelphia, 1983.

5. W. Xie, Thesis, University of British Columbia, 1991.

6. W. Xie, *A sharp pointwise bound for functions with L^2-Laplacians and zero boundary values on arbitrary three-dimensional domains*, Indiana Univ. Math. J. **40**, (1991), 1185–1192.

E-mail address: xie@s5.math.umn.edu

List of participants

Antanovski, Leonid	MARS (Microgravity Research & support) center, Via Diocleziano 328 80125 Naples, ITALY Telefon 081-725-2621 , Fax 081-725-2750
Asano, Kioshi	Inst. of Math., Yoshida College Kyoto Universität, Kyoto 606-01 JAPAN Fax 81-75-753-6767
Beale, J. Thomas	Mathematics Dept. Duke University Durham, NC 27706, USA Telefon 919-684-8124 e-mail: beale at math.duke.edu
Beirão da Veiga, Hugo	Dept. Mathematics University Pisa v. Buonarotti, 2 56100 - PISA ITALY Telefon 0039-50-500065 oder 599526 Fax: 0039-50-49344
Borchers, Wolfgang	Universität-GH Paderborn, Fachbereich Mathematik-Informatik D-4790 Paderborn, Warburger Str. 100,GERMANY Telefon 05251-602643 , Fax 05251-603836
Boudourides, M.A.	Section of Physics and Appl. Math. Democritus University of Thrace 67100 Xanthi, GREECE Telefon +30 541 20379 Fax +30 541 20275 e-mail: boudourides at xanthi.cc.duth.gr
Bühler, Karl	Institut für Strömungslehre und Strömungsmaschinen Kaiserstraße 12 D-7500 Karlsruhe 1, GERMANY Telefon 0721-608-2354, Telex 721 166 =UNIkar
Chang, Huakang	Dept. of Mathematics University of British Columbia Vancouver, B.C., CANADA Telefon 604-231-9555
Farwig, Reinhard	Rhein.Westf.Technische Hochschule Aachen Institut f. Mathematik Templergraben 55 D-5100 Aachen, GERMANY Telefon 0241-80-4923

Feistauer, Miloslav
Faculty of Mathematics and Physics
Charles University Sokolovská 83
18600 Praha 8
CZECHOSLOVAKIA
Telefon 2316034 , e-mail: FEIST at CSPGAS 11

Fischer, Thomas M.
DLR, Inst. für Theoret. Fluid Mech.
Bunsenstraße 10,
D-3400 Göttingen, GERMANY
Telefon 0551-7092294

Fursikov, Andrei
Dept. of Mechanics and Mathematics
of Moscow University,
Lenin Hills
Moscow 119899 RUSSIA
Telefon 138-66-36

Galdi, Giovanni P.
Dipartimento di Matematica
Università di Ferrara
Via Machiavelli 35
44100 Ferrara, ITALY
Telefon 0532-48859 , Fax 0532-47292

Girault, Vivette
Laboratoire d'Analyse Numérique
Tour 55 - 65, 5 ème étage
Université PARIS VI
4. Place Jussieu
75 252 - Paris Codex 05, FRANCE
Telefon 331-44 27 5114
Fax 331 44 27 4001

Gresho, Philip M.
L-262; LLNL
P.O. Box 808
Livermore, CA 94550, USA
Telefon 415-422-1812
Fax 415-422-5844

Grobbelaar, Marié
Dept. of Mathematics
University of South Africa
P.O. Box 392
0001 Pretoria, SOUTH AFRICA
Telefon 0027-12-4296266 oder 0027-11-7951750
Fax 429-3434

Grubb, Gerd
Math. Dept., University of Copenhagen
Universitetsparken 5
DK-2100 Copenhagen, DENMARK
Telefon 0045-31353133 ex 443

Hebeker, Friedrich-Karl
IBM Scientific Center
Tiergartenstraße 15
D-6900 Heidelberg , GERMANY
e-mail HEBEKER at DHDIBM1

Heywood, John G.
Dept. of Mathematics
University of British Columbia
Vancouver, B.C., CANADA
Telefon 604-231-9555

Inoue, Atsushi	Tokyo Institute of Technology 2-12-1 Okokayama, Meguro-Ku 152 JAPAN Telefon 03-3726-1111 ext 2205 E-mail inoue at math.titech.ac.jp
Kozono, Hideo	Dept. of Mathematics College of General Education Kyushu University Fukuoka, 810, JAPAN Fax 0081-92 714 0741
Kretzschmar, Horst	Fachbereich Mathematik PH Halle/Köthen Kröllwitzer Str. 44 4020 Halle, GERMANY Telefon 38608
Kröner, Dietmar	Institut für Angew. Mathematik Universität Bonn Wegelerstr. 6 D 5300 Bonn, GERMANY Telefon 0228-73 3437
Masuda, Kyûya	Dept. of. Mathematics Rikkyo University 3 Nishiikebukuro Toshimaku , Tokyo JAPAN Telefon 03-3985-2487 Fax 03-3985-2810
Mogilevskij, Ilia	Tver University Mathem. Faculty 170013, Tver yl. Zheljabova, 33 RUSSIA Telefon 08222-65320
Pileckas, Konstantin	Universität-GH Paderborn, Fachbereich Mathematik-Informatik D-4790 Paderborn, Warburger Str. 100, GERMANY Telefon 05251-602644 , Fax 05251-603836
Plotnikov , Pavel I.	Lavrentyev Institute of Hydrodynamics, Siberian Division of the USSR Academy of Sciences Novosibirsk, RUSSIA 630090 RUSSIA Tel.:35-71-62, Fax: 35-40-50
Rannacher, Rolf	Institut für Angew. Mathematik Universität Heidelberg Im Neuenheimer Feld 293 D-6900 Heidelberg, GERMANY Telefon 06221-562873 Fax 06221-565331

Rautmann, Reimund	Universität-GH Paderborn, Fachbereich Mathematik-Informatik D-4790 Paderborn, Warburger Str. 100, GERMANY Telefon 05251-602649, Fax 05251-603836
Rodenkirchen, Jürgen	Universität-GH Paderborn, Fachbereich Mathematik-Informatik D-4790 Paderborn, Warburger Str. 100, GERMANY Telefon 05251-603357, Fax 05251-603836
Roesner, Karl G.	Institut für Mechanik - Dynamik der Fluide- Hochschulstr. 1 D-6100 Darmstadt, GERMANY Telefon 6151-164328 -162992 Fax 6151-166869, e-mail: XBR1D583 at DDATHD21. BITNET
Rozhdestvensky, Boris L.	Soviet Centre of Mathematical Modelling Miusskaja SQ,4,125047 Moscow, RUSSIA Telefon 9723673, Fax 9723673
Salvi, Rodolfo	Dipartimento di Matematica Politecnico di Milano P.za L. da Vinci 32 20133 MILANO, ITALY
Schonbek, Maria Elena	Dept. of Mathematics University of California Santa Cruz CA 95064 USA Telefon 408-4594657 408-4592085 e-mail:Schonbk at UCSCC.UCSC.edu
Socolowsky, Jürgen	Fachbereich Mathematik/Informatik TH Merseburg Gensaer Straße O-4200 MERSEBURG, GERMANY, Telefon 462959
Sohr, Hermann	Universität-GH Paderborn, Fachbereich Mathematik-Informatik D-4790 Paderborn, Warburger Str. 100,GERMANY Telefon 05251-602648 , Fax 05251-603836
Solonnikov, Vsevolod	St. Petersburg Branch of V.A. Steklov Math. Inst. Acad. Sciences USSR Fontanka 27, St. Petersburg,RUSSIA Telefon 312--40-58
Süli, Endre	Oxford University Computing Laboratory 11 Keble Road Oxford OX1 3QD GREAT BRITAIN Telefon 0865-273880 oder 0865-273885 Fax 0865-273-839

Tani, Arusi

Dept. of Math., Keio University
Yokohama 223, JAPAN
Telefon 045-563-1141

Titi, Edriss S.

Dept. of Mathematics
University of California
Irvine, CA 92717 , USA
Telefon 714-856-5503 , Fax 714-856-7993
e-mail:titi at isis.ps. uci.edu

Valli, Alberto

Dipartimento di Matematica
Università di Trento
38050 POVO (Trento)
ITALY Telefon 0039-461-881580

Varnhorn, Werner

Fachbereich Mathematik
TH Darmstadt
Schloßgartenstr. 7
D-6100 Darmstadt ,GERMANY
Telefon 06151-163689, Fax 06151-164011

Velte, Waldemar

Institut für Angewandte Mathematik
und Statistik
Universität Würzburg
Am Hubland
8700 Würzburg ,GERMANY
Telefon 0931-8885043

Wahl, Wolf von

Universität Bayreuth
Mathematisches Institut
Postfach 101251
8580 Bayreuth ,GERMANY
Telefon 0921-553291

Walsh, Owen D.

Dept. of Mathematics
University of British Columbia
Vancouver, B.C.
CANADA, V6 T IY4.
Telefon 604-228-3784
e-mail wals at MTSG.UBC.CA

Wetton, Brian T.R.

Dept. of Mathematics
University of British Columbia
Vancouver, B.C.
CANADA, V6 T IY4
Telefon 604-228-3784
e-mail wetton at unixg.ubc.ca.

Wiegner, Michael

Mathematisches Institut der
Universität Bayreuth
Postfach 101251
8580 Bayreuth , GERMANY
Telefon 0921-553287
e-mail wiegner at uni.bayreuth.dbp.de.

Wolff, Michael

Fachbereich Mathematik
Institut für Angewandte Mathematik
Humboldt-Universität
O-1086 Berlin , GERMANY
Telefon 20932282 or 20932336

Xie, Wenzheng

School of Mathematics
University of Minnesota
Minneapolis, Minnesota 55455
USA
Dept. of Mathematics
University of British Columbia
Vancouver, B.C., CANADA
Telefon 604-231-9555

Printing: Druckhaus Beltz, Hemsbach
Binding: Buchbinderei Schäffer, Grünstadt

Lecture Notes in Mathematics

For information about Vols. 1–1334
please contact your bookseller or Springer-Verlag

Vol. 1380: H.P. Schlickewei, E. Wirsing (Eds.), Number Theory, Ulm 1987. Proceedings. V, 266 pages. 1989.

Vol. 1381: J.-O. Strömberg, A. Torchinsky, Weighted Hardy Spaces. V, 193 pages. 1989.

Vol. 1382: H. Reiter, Metaplectic Groups and Segal Algebras. XI, 128 pages. 1989.

Vol. 1383: D.V. Chudnovsky, G.V. Chudnovsky, H. Cohn, M.B. Nathanson (Eds.), Number Theory, New York 1985–88. Seminar. V, 256 pages. 1989.

Vol. 1384: J. Garcia-Cuerva (Ed.), Harmonic Analysis and Partial Differential Equations. Proceedings, 1987. VII, 213 pages. 1989.

Vol. 1385: A.M. Anile, Y. Choquet-Bruhat (Eds.), Relativistic Fluid Dynamics. Seminar, 1987. V, 308 pages. 1989.

Vol. 1386: A. Bellen, C.W. Gear, E. Russo (Eds.), Numerical Methods for Ordinary Differential Equations. Proceedings, 1987. VII, 136 pages. 1989.

Vol. 1387: M. Petkovi´c, Iterative Methods for Simultaneous Inclusion of Polynomial Zeros. X, 263 pages. 1989.

Vol. 1388: J. Shinoda, T.A. Slaman, T. Tugué (Eds.), Mathematical Logic and Applications. Proceedings, 1987. V, 223 pages. 1989.

Vol. 1000: Second Edition. H. Hopf, Differential Geometry in the Large. VII, 184 pages. 1989.

Vol. 1389: E. Ballico, C. Ciliberto (Eds.), Algebraic Curves and Projective Geometry. Proceedings, 1988. V, 288 pages. 1989.

Vol. 1390: G. Da Prato, L. Tubaro (Eds.), Stochastic Partial Differential Equations and Applications II. Proceedings, 1988. VI, 258 pages. 1989.

Vol. 1391: S. Cambanis, A. Weron (Eds.), Probability Theory on Vector Spaces IV. Proceedings, 1987. VIII, 424 pages. 1989.

Vol. 1392: R. Silhol, Real Algebraic Surfaces. X, 215 pages. 1989.

Vol. 1393: N. Bouleau, D. Feyel, F. Hirsch, G. Mokobodzki (Eds.), Séminaire de Théorie du Potentiel Paris, No. 9. Proceedings. VI, 265 pages. 1989.

Vol. 1394: T.L. Gill, W.W. Zachary (Eds.), Nonlinear Semigroups, Partial Differential Equations and Attractors. Proceedings, 1987. IX, 233 pages. 1989.

Vol. 1395: K. Alladi (Ed.), Number Theory, Madras 1987. Proceedings. VII, 234 pages. 1989.

Vol. 1396: L. Accardi, W. von Waldenfels (Eds.), Quantum Probability and Applications IV. Proceedings, 1987. VI, 355 pages. 1989.

Vol. 1397: P.R. Turner (Ed.), Numerical Analysis and Parallel Processing. Seminar, 1987. VI, 264 pages. 1989.

Vol. 1398: A.C. Kim, B.H. Neumann (Eds.), Groups – Korea 1988. Proceedings. V, 189 pages. 1989.

Vol. 1399: W.-P. Barth, H. Lange (Eds.), Arithmetic of Complex Manifolds. Proceedings, 1988. V, 171 pages. 1989.

Vol. 1400: U. Jannsen. Mixed Motives and Algebraic K-Theory. XIII, 246 pages. 1990.

Vol. 1401: J. Steprans, S. Watson (Eds.), Set Theory and its Applications. Proceedings, 1987. V, 227 pages. 1989.

Vol. 1402: C. Carasso, P. Charrier, B. Hanouzet, J.-L. Joly (Eds.), Nonlinear Hyperbolic Problems. Proceedings, 1988. V, 249 pages. 1989.

Vol. 1403: B. Simeone (Ed.), Combinatorial Optimization. Seminar, 1986. V, 314 pages. 1989.

Vol. 1404: M.-P. Malliavin (Ed.), Séminaire d´Algèbre Paul Dubreil et Marie-Paul Malliavin. Proceedings, 1987–1988. IV, 410 pages. 1989.

Vol. 1405: S. Dolecki (Ed.), Optimization. Proceedings, 1988. V, 223 pages. 1989. Vol. 1406: L. Jacobsen (Ed.), Analytic Theory of Continued Fractions III. Proceedings, 1988. VI, 142 pages. 1989.

Vol. 1407: W. Pohlers, Proof Theory. VI, 213 pages. 1989.

Vol. 1408: W. Lück, Transformation Groups and Algebraic K-Theory. XII, 443 pages. 1989.

Vol. 1409: E. Hairer, Ch. Lubich, M. Roche. The Numerical Solution of Differential-Algebraic Systems by Runge-Kutta Methods. VII, 139 pages. 1989.

Vol. 1410: F.J. Carreras, O. Gil-Medrano, A.M. Naveira (Eds.), Differential Geometry. Proceedings, 1988. V, 308 pages. 1989.

Vol. 1411: B. Jiang (Ed.), Topological Fixed Point Theory and Applications. Proceedings. 1988. VI, 203 pages. 1989.

Vol. 1412: V.V. Kalashnikov, V.M. Zolotarev (Eds.), Stability Problems for Stochastic Models. Proceedings, 1987. X, 380 pages. 1989.

Vol. 1413: S. Wright, Uniqueness of the Injective III_1 Factor. III, 108 pages. 1989.

Vol. 1414: E. Ramirez de Arellano (Ed.), Algebraic Geometry and Complex Analysis. Proceedings, 1987. VI, 180 pages. 1989.

Vol. 1415: M. Langevin, M. Waldschmidt (Eds.), Cinquante Ans de Polynômes. Fifty Years of Polynomials. Proceedings, 1988. IX, 235 pages.1990.

Vol. 1416: C. Albert (Ed.), Géométrie Symplectique et Mécanique. Proceedings, 1988. V, 289 pages. 1990.

Vol. 1417: A.J. Sommese, A. Biancofiore, E.L. Livorni (Eds.), Algebraic Geometry. Proceedings, 1988. V, 320 pages. 1990.

Vol. 1418: M. Mimura (Ed.), Homotopy Theory and Related Topics. Proceedings, 1988. V, 241 pages. 1990.

Vol. 1419: P.S. Bullen, P.Y. Lee, J.L. Mawhin, P. Muldowney, W.F. Pfeffer (Eds.), New Integrals. Proceedings, 1988. V, 202 pages. 1990.

Vol. 1420: M. Galbiati, A. Tognoli (Eds.), Real Analytic Geometry. Proceedings, 1988. IV, 366 pages. 1990.

Vol. 1421: H.A. Biagioni, A Nonlinear Theory of Generalized Functions, XII, 214 pages. 1990.

Vol. 1422: V. Villani (Ed.), Complex Geometry and Analysis. Proceedings, 1988. V, 109 pages. 1990.

Vol. 1423: S.O. Kochman, Stable Homotopy Groups of Spheres: A Computer-Assisted Approach. VIII, 330 pages. 1990.

Vol. 1424: F.E. Burstall, J.H. Rawnsley, Twistor Theory for Riemannian Symmetric Spaces. III, 112 pages. 1990.

Vol. 1425: R.A. Piccinini (Ed.), Groups of Self-Equivalences and Related Topics. Proceedings, 1988. V, 214 pages. 1990.

Vol. 1426: J. Azéma, P.A. Meyer, M. Yor (Eds.), Séminaire de Probabilités XXIV, 1988/89. V, 490 pages. 1990.

Vol. 1427: A. Ancona, D. Geman, N. Ikeda, École d'Eté de Probabilités de Saint Flour XVIII, 1988. Ed.: P.L. Hennequin. VII, 330 pages. 1990.

Vol. 1428: K. Erdmann, Blocks of Tame Representation Type and Related Algebras. XV. 312 pages. 1990.

Vol. 1429: S. Homer, A. Nerode, R.A. Platek, G.E. Sacks, A. Scedrov, Logic and Computer Science. Seminar, 1988. Editor: P. Odifreddi. V, 162 pages. 1990.

Vol. 1430: W. Bruns, A. Simis (Eds.), Commutative Algebra. Proceedings. 1988. V, 160 pages. 1990.

Vol. 1431: J.G. Heywood, K. Masuda, R. Rautmann, V.A. Solonnikov (Eds.), The Navier-Stokes Equations – Theory and Numerical Methods. Proceedings, 1988. VII, 238 pages. 1990.

Vol. 1432: K. Ambos-Spies, G.H. Müller, G.E. Sacks (Eds.), Recursion Theory Week. Proceedings, 1989. VI, 393 pages. 1990.

Vol. 1433: S. Lang, W. Cherry, Topics in Nevanlinna Theory. II, 174 pages.1990.

Vol. 1434: K. Nagasaka, E. Fouvry (Eds.), Analytic Number Theory. Proceedings, 1988. VI, 218 pages. 1990.

Vol. 1435: St. Ruscheweyh, E.B. Saff, L.C. Salinas, R.S. Varga (Eds.), Computational Methods and Function Theory. Proceedings, 1989. VI, 211 pages. 1990.

Vol. 1436: S. Xambó-Descamps (Ed.), Enumerative Geometry. Proceedings, 1987. V, 303 pages. 1990.

Vol. 1437: H. Inassaridze (Ed.), K-theory and Homological Algebra. Seminar, 1987–88. V, 313 pages. 1990.

Vol. 1438: P.G. Lemarié (Ed.) Les Ondelettes en 1989. Seminar. IV, 212 pages. 1990.

Vol. 1439: E. Bujalance, J.J. Etayo, J.M. Gamboa, G. Gromadzki. Automorphism Groups of Compact Bordered Klein Surfaces: A Combinatorial Approach. XIII, 201 pages. 1990.

Vol. 1440: P. Latiolais (Ed.), Topology and Combinatorial Groups Theory. Seminar, 1985–1988. VI, 207 pages. 1990.

Vol. 1441: M. Coornaert, T. Delzant, A. Papadopoulos. Géométrie et théorie des groupes. X, 165 pages. 1990.

Vol. 1442: L. Accardi, M. von Waldenfels (Eds.), Quantum Probability and Applications V. Proceedings, 1988. VI, 413 pages. 1990.

Vol. 1443: K.H. Dovermann, R. Schultz, Equivariant Surgery Theories and Their Periodicity Properties. VI, 227 pages. 1990.

Vol. 1444: H. Korezlioglu, A.S. Ustunel (Eds.), Stochastic Analysis and Related Topics Vl. Proceedings, 1988. V, 268 pages. 1990.

Vol. 1445: F. Schulz, Regularity Theory for Quasilinear Elliptic Systems and – Monge Ampère Equations in Two Dimensions. XV, 123 pages. 1990.

Vol. 1446: Methods of Nonconvex Analysis. Seminar, 1989. Editor: A. Cellina. V, 206 pages. 1990.

Vol. 1447: J.-G. Labesse, J. Schwermer (Eds), Cohomology of Arithmetic Groups and Automorphic Forms. Proceedings, 1989. V, 358 pages. 1990.

Vol. 1448: S.K. Jain, S.R. López-Permouth (Eds.), Non-Commutative Ring Theory. Proceedings, 1989. V, 166 pages. 1990.

Vol. 1449: W. Odyniec, G. Lewicki, Minimal Projections in Banach Spaces. VIII, 168 pages. 1990.

Vol. 1450: H. Fujita, T. Ikebe, S.T. Kuroda (Eds.), Functional-Analytic Methods for Partial Differential Equations. Proceedings, 1989. VII, 252 pages. 1990.

Vol. 1451: L. Alvarez-Gaumé, E. Arbarello, C. De Concini, N.J. Hitchin, Global Geometry and Mathematical Physics. Montecatini Terme 1988. Seminar. Editors: M. Francaviglia, F. Gherardelli. IX, 197 pages. 1990.

Vol. 1452: E. Hlawka, R.F. Tichy (Eds.), Number-Theoretic Analysis. Seminar, 1988–89. V, 220 pages. 1990.

Vol. 1453: Yu.G. Borisovich, Yu.E. Gliklikh (Eds.), Global Analysis – Studies and Applications IV. V, 320 pages. 1990.

Vol. 1454: F. Baldassari, S. Bosch, B. Dwork (Eds.), p-adic Analysis. Proceedings, 1989. V, 382 pages. 1990.

Vol. 1455: J.-P. Françoise, R. Roussarie (Eds.), Bifurcations of Planar Vector Fields. Proceedings, 1989. VI, 396 pages. 1990.

Vol. 1456: L.G. Kovács (Ed.), Groups – Canberra 1989. Proceedings. XII, 198 pages. 1990.

Vol. 1457: O. Axelsson, L.Yu. Kolotilina (Eds.), Preconditioned Conjugate Gradient Methods. Proceedings, 1989. V, 196 pages. 1990.

Vol. 1458: R. Schaaf, Global Solution Branches of Two Point Boundary Value Problems. XIX, 141 pages. 1990.

Vol. 1459: D. Tiba, Optimal Control of Nonsmooth Distributed Parameter Systems. VII, 159 pages. 1990.

Vol. 1460: G. Toscani, V. Boffi, S. Rionero (Eds.), Mathematical Aspects of Fluid Plasma Dynamics. Proceedings, 1988. V, 221 pages. 1991.

Vol. 1461: R. Gorenflo, S. Vessella, Abel Integral Equations. VII, 215 pages. 1991.

Vol. 1462: D. Mond, J. Montaldi (Eds.), Singularity Theory and its Applications. Warwick 1989, Part I. VIII, 405 pages. 1991.

Vol. 1463: R. Roberts, I. Stewart (Eds.), Singularity Theory and its Applications. Warwick 1989, Part II. VIII, 322 pages. 1991.

Vol. 1464: D. L. Burkholder, E. Pardoux, A. Sznitman, Ecole d'Eté de Probabilités de Saint- Flour XIX-1989. Editor: P. L. Hennequin. VI, 256 pages. 1991.

Vol. 1465: G. David, Wavelets and Singular Integrals on Curves and Surfaces. X, 107 pages. 1991.

Vol. 1466: W. Banaszczyk, Additive Subgroups of Topological Vector Spaces. VII, 178 pages. 1991.

Vol. 1467: W. M. Schmidt, Diophantine Approximations and Diophantine Equations. VIII, 217 pages. 1991.

Vol. 1468: J. Noguchi, T. Ohsawa (Eds.), Prospects in Complex Geometry. Proceedings, 1989. VII, 421 pages. 1991.

Vol. 1469: J. Lindenstrauss, V. D. Milman (Eds.), Geometric Aspects of Functional Analysis. Seminar 1989-90. XI, 191 pages. 1991.

Vol. 1470: E. Odell, H. Rosenthal (Eds.), Functional Analysis. Proceedings, 1987-89. VII, 199 pages. 1991.

Vol. 1471: A. A. Panchishkin, Non-Archimedean L-Functions of Siegel and Hilbert Modular Forms. VII, 157 pages. 1991.

Vol. 1472: T. T. Nielsen, Bose Algebras: The Complex and Real Wave Representations. V, 132 pages. 1991.

Vol. 1473: Y. Hino, S. Murakami, T. Naito, Functional Differential Equations with Infinite Delay. X, 317 pages. 1991.

Vol. 1474: S. Jackowski, B. Oliver, K. Pawałowski (Eds.), Algebraic Topology, Poznań 1989. Proceedings. VIII, 397 pages. 1991.

Vol. 1475: S. Busenberg, M. Martelli (Eds.), Delay Differential Equations and Dynamical Systems. Proceedings, 1990. VIII, 249 pages. 1991.

Vol. 1476: M. Bekkali, Topics in Set Theory. VII, 120 pages. 1991.

Vol. 1477: R. Jajte, Strong Limit Theorems in Noncommutative L_2-Spaces. X, 113 pages. 1991.

Vol. 1478: M.-P. Malliavin (Ed.), Topics in Invariant Theory. Seminar 1989-1990. VI, 272 pages. 1991.

Vol. 1479: S. Bloch, I. Dolgachev, W. Fulton (Eds.), Algebraic Geometry. Proceedings, 1989. VII, 300 pages. 1991.

Vol. 1480: F. Dumortier, R. Roussarie, J. Sotomayor, H. Żoładek, Bifurcations of Planar Vector Fields: Nilpotent Singularities and Abelian Integrals. VIII, 226 pages. 1991.

Vol. 1481: D. Ferus, U. Pinkall, U. Simon, B. Wegner (Eds.), Global Differential Geometry and Global Analysis. Proceedings, 1991. VIII, 283 pages. 1991.

Vol. 1482: J. Chabrowski, The Dirichlet Problem with L^2-Boundary Data for Elliptic Linear Equations. VI, 173 pages. 1991.

Vol. 1483: E. Reithmeier, Periodic Solutions of Nonlinear Dynamical Systems. VI, 171 pages. 1991.

Vol. 1484: H. Delfs, Homology of Locally Semialgebraic Spaces. IX, 136 pages. 1991.

Vol. 1485: J. Azéma, P. A. Meyer, M. Yor (Eds.), Séminaire de Probabilités XXV. VIII, 440 pages. 1991.

Vol. 1486: L. Arnold, H. Crauel, J.-P. Eckmann (Eds.), Lyapunov Exponents. Proceedings, 1990. VIII, 365 pages. 1991.

Vol. 1487: E. Freitag, Singular Modular Forms and Theta Relations. VI, 172 pages. 1991.

Vol. 1488: A. Carboni, M. C. Pedicchio, G. Rosolini (Eds.), Category Theory. Proceedings, 1990. VII, 494 pages. 1991.

Vol. 1489: A. Mielke, Hamiltonian and Lagrangian Flows on Center Manifolds. X, 140 pages. 1991.

Vol. 1490: K. Metsch, Linear Spaces with Few Lines. XIII, 196 pages. 1991.

Vol. 1491: E. Lluis-Puebla, J.-L. Loday, H. Gillet, C. Soulé, V. Snaith, Higher Algebraic K-Theory: an overview. IX, 164 pages. 1992.

Vol. 1492: K. R. Wicks, Fractals and Hyperspaces. VIII, 168 pages. 1991.

Vol. 1493: E. Benoît (Ed.), Dynamic Bifurcations. Proceedings, Luminy 1990. VII, 219 pages. 1991.

Vol. 1494: M.-T. Cheng, X.-W. Zhou, D.-G. Deng (Eds.), Harmonic Analysis. Proceedings, 1988. IX, 226 pages. 1991.

Vol. 1495: J. M. Bony, G. Grubb, L. Hörmander, H. Komatsu, J. Sjöstrand, Microlocal Analysis and Applications. Montecatini Terme, 1989. Editors: L. Cattabriga, L. Rodino. VII, 349 pages. 1991.

Vol. 1496: C. Foias, B. Francis, J. W. Helton, H. Kwakernaak, J. B. Pearson, H_{∞}-Control Theory. Como, 1990. Editors: E. Mosca, L. Pandolfi. VII, 336 pages. 1991.

Vol. 1497: G. T. Herman, A. K. Louis, F. Natterer (Eds.), Mathematical Methods in Tomography. Proceedings 1990. X, 268 pages. 1991.

Vol. 1498: R. Lang, Spectral Theory of Random Schrödinger Operators. X, 125 pages. 1991.

Vol. 1499: K. Taira, Boundary Value Problems and Markov Processes. IX, 132 pages. 1991.

Vol. 1500: J.-P. Serre, Lie Algebras and Lie Groups. VII, 168 pages. 1992.

Vol. 1501: A. De Masi, E. Presutti, Mathematical Methods for Hydrodynamic Limits. IX, 196 pages. 1991.

Vol. 1502: C. Simpson, Asymptotic Behavior of Monodromy. V, 139 pages. 1991.

Vol. 1503: S. Shokranian, The Selberg-Arthur Trace Formula (Lectures by J. Arthur). VII, 97 pages. 1991.

Vol. 1504: J. Cheeger, M. Gromov, C. Okonek, P. Pansu, Geometric Topology: Recent Developments. Editors: P. de Bartolomeis, F. Tricerri. VII, 197 pages. 1991.

Vol. 1505: K. Kajitani, T. Nishitani, The Hyperbolic Cauchy Problem. VII, 168 pages. 1991.

Vol. 1506: A. Buium, Differential Algebraic Groups of Finite Dimension. XV, 145 pages. 1992.

Vol. 1507: K. Hulek, T. Peternell, M. Schneider, F.-O. Schreyer (Eds.), Complex Algebraic Varieties. Proceedings, 1990. VII, 179 pages. 1992.

Vol. 1508: M. Vuorinen (Ed.), Quasiconformal Space Mappings. A Collection of Surveys 1960-1990. IX, 148 pages. 1992.

Vol. 1509: J. Aguadé, M. Castellet, F. R. Cohen (Eds.), Algebraic Topology - Homotopy and Group Cohomology. Proceedings, 1990. X, 330 pages. 1992.

Vol. 1510: P. P. Kulish (Ed.), Quantum Groups. Proceedings, 1990. XII, 398 pages. 1992.

Vol. 1511: B. S. Yadav, D. Singh (Eds.), Functional Analysis and Operator Theory. Proceedings, 1990. VIII, 223 pages. 1992.

Vol. 1512: L. M. Adleman, M.-D. A. Huang, Primality Testing and Abelian Varieties Over Finite Fields. VII, 142 pages. 1992.

Vol. 1513: L. S. Block, W. A. Coppel, Dynamics in One Dimension. VIII, 249 pages. 1992.

Vol. 1514: U. Krengel, K. Richter, V. Warstat (Eds.), Ergodic Theory and Related Topics III, Proceedings, 1990. VIII, 236 pages. 1992.

Vol. 1515: E. Ballico, F. Catanese, C. Ciliberto (Eds.), Classification of Irregular Varieties. Proceedings, 1990. VII, 149 pages. 1992.

Vol. 1516: R. A. Lorentz, Multivariate Birkhoff Interpolation. IX, 192 pages. 1992.

Vol. 1517: K. Keimel, W. Roth, Ordered Cones and Approximation. VI, 134 pages. 1992.

Vol. 1518: H. Stichtenoth, M. A. Tsfasman (Eds.), Coding Theory and Algebraic Geometry. Proceedings, 1991. VIII, 223 pages. 1992.

Vol. 1519: M. W. Short, The Primitive Soluble Permutation Groups of Degree less than 256. IX, 145 pages. 1992.

Vol. 1520: Yu. G. Borisovich, Yu. E. Gliklikh (Eds.), Global Analysis – Studies and Applications V. VII, 284 pages. 1992.

Vol. 1521: S. Busenberg, B. Forte, H. K. Kuiken, Mathematical Modelling of Industrial Process. Bari, 1990. Editors: V. Capasso, A. Fasano. VII, 162 pages. 1992.

Vol. 1522: J.-M. Delort, F. B. I. Transformation. VII, 101 pages. 1992.

Vol. 1523: W. Xue, Rings with Morita Duality. X, 168 pages. 1992.

Vol. 1524: M. Coste, L. Mahé, M.-F. Roy (Eds.), Real Algebraic Geometry. Proceedings, 1991. VIII, 418 pages. 1992.

Vol. 1525: C. Casacuberta, M. Castellet (Eds.), Mathematical Research Today and Tomorrow. VII, 112 pages. 1992.

Vol. 1526: J. Azéma, P. A. Meyer, M. Yor (Eds.), Séminaire de Probabilités XXVI. X, 633 pages. 1992.

Vol. 1527: M. I. Freidlin, J.-F. Le Gall, Ecole d'Eté de Probabilités de Saint-Flour XX – 1990. Editor: P. L. Hennequin. VIII, 244 pages. 1992.

Vol. 1528: G. Isac, Complementarity Problems. VI, 297 pages. 1992.

Vol. 1529: J. van Neerven, The Adjoint of a Semigroup of Linear Operators. X, 195 pages. 1992.

Vol. 1530: J. G. Heywood, K. Masuda, R. Rautmann, S. A. Solonnikov (Eds.), The Navier-Stokes Equations II – Theory and Numerical Methods. IX, 322 pages. 1992.